VOLUME NINETY SIX

VITAMINS AND HORMONES

Nitric Oxide

VOLUME NINETY SIX

VITAMINS AND HORMONES

Nitric Oxide

Series Editor

GERALD LITWACK, PhD

Toluca Lake, North Hollywood, California

ELSEVIER

AMSTERDAM • BOSTON • HEIDELBERG • LONDON
NEW YORK • OXFORD • PARIS • SAN DIEGO
SAN FRANCISCO • SINGAPORE • SYDNEY • TOKYO
Academic Press is an imprint of Elsevier

AP

Cover photo credit:
Girouard H., and Tabatabaei S.N.
Nitric Oxide and Cerebrovascular Regulation
Vitamins and Hormones (2014) **96**, pp. 347–386

Academic Press is an imprint of Elsevier
225 Wyman Street, Waltham, MA 02451, USA
525 B Street, Suite 1800, San Diego, CA 92101-4495, USA
32 Jamestown Road, London NW1 7BY, UK
The Boulevard, Langford Lane, Kidlington, Oxford OX5 1GB, UK

First edition 2014

Notices
Knowledge and best practice in this field are constantly changing. As new research and experience broaden our understanding, changes in research methods, professional practices, or medical treatment may become necessary.

Practitioners and researchers must always rely on their own experience and knowledge in evaluating and using any information, methods, compounds, or experiments described herein. In using such information or methods they should be mindful of their own safety and the safety of others, including parties for whom they have a professional responsibility.

To the fullest extent of the law, neither the Publisher nor the authors, contributors, or editors, assume any liability for any injury and/or damage to persons or property as a matter of products liability, negligence or otherwise, or from any use or operation of any methods, products, instructions, or ideas contained in the material herein.

ISBN: 978-0-12-800254-4
ISSN: 0083-6729

For information on all Academic Press publications
visit our website at store.elsevier.com

CONTENTS

CONTRIBUTORS

Ashok Aiyar
Stanley S. Scott Cancer Center, Louisiana State University, Health Sciences Center, New Orleans, Louisiana, USA

Alexis Bavencoffe
Center for Neuroscience and Pain Research, Department of Anesthesiology and Perioperative Medicine, The University of Texas MD Anderson Cancer Center, Houston, Texas, USA

Karin C. Calaza
Programa de Neurociências, and Departamento de Neurobiologia, Instituto de Biologia, Universidade Federal Fluminense, Niterói, RJ, Brazil

Paula Campello-Costa
Programa de Neurociências, and Departamento de Neurobiologia, Instituto de Biologia, Universidade Federal Fluminense, Niterói, RJ, Brazil

Shao-Rui Chen
Center for Neuroscience and Pain Research, Department of Anesthesiology and Perioperative Medicine, The University of Texas MD Anderson Cancer Center, Houston, Texas, USA

Angela Cheung
Division of Developmental Neurobiology, MRC National Institute for Medical Research, Mill Hill, London, United Kingdom

Marcelo Cossenza
Programa de Neurociências, Instituto de Biologia, Niterói, RJ, and Departamento de Fisiologia e Farmacologia, Instituto Biomédico, Universidade Federal Fluminense, Rio de Janeiro, Brazil

Ivan C.L. Domith
Programa de Neurociências, Instituto de Biologia, Universidade Federal Fluminense, Niterói, RJ, Brazil

Thaísa G. Encarnação
Programa de Neurociências, Instituto de Biologia, Universidade Federal Fluminense, Niterói, RJ, Brazil

David A. Geller
Department of Surgery, University of Pittsburgh, Pittsburgh, Pennsylvania, USA

Hélène Girouard
Department of Pharmacology, Faculty of Medicine, Université de Montréal, and Research Center of the Institut Universitaire de Gériatrie de Montréal, Montreal, Quebec, Canada

Luis F.H. Gladulich
Programa de Neurociências, Instituto de Biologia, Universidade Federal Fluminense, Niterói, RJ, Brazil

William P. Gray
Institute of Psychological Medicine and Clinical Neurosciences, Cardiff University, Cardiff, United Kingdom

Zhong Guo
Department of Surgery, University of Pittsburgh, Pittsburgh, Pennsylvania, USA

Bradford G. Hill
Diabetes and Obesity Center, Institute of Molecular Cardiology; Department of Physiology and Biophysics, and Department of Biochemistry and Molecular Biology, University of Louisville School of Medicine, Louisville, Kentucky, USA

Michael A. Hough
School of Biological Sciences, University of Essex, Colchester, United Kingdom

Yao Hu
Institute for Stem Cells and Neural Regeneration, School of Pharmacy, Nanjing Medical University, Nanjing, China

H.S. Jeffrey Man
Institute of Medical Science; Li Ka Shing Knowledge Institute, St. Michael's Hospital, and Divisions of Respirology and Critical Care Medicine, Department of Medicine, University of Toronto, Toronto, Ontario, Canada

Jisha Joshua
Department of Medicine, University of California, San Diego, California, USA

Hema Kalyanaraman
Department of Medicine, University of California, San Diego, California, USA

Sangwon F. Kim
Department of Psychiatry and Pharmacology, Center for Neurobiology and Behavior, The Perelman School of Medicine at University of Pennsylvania, Philadelphia, Pennsylvania, USA

Peter Kruzliak
Department of Cardiovascular Diseases, International Clinical Research Center, St. Anne's University Hospital, Brno, Czech Republic

Nisha Marathe
Department of Medicine, University of California, San Diego, California, USA

Philip A. Marsden
Institute of Medical Science; Li Ka Shing Knowledge Institute, St. Michael's Hospital, and Division of Nephrology, Department of Medicine, University of Toronto, Toronto, Ontario, Canada

Junko Maruyama
Department of Anesthesiology and Critical Care Medicine, Mie University School of Medicine, and Department of Clinical Engineering, Suzuka University of Medical Science, Mie, Japan

Kazuo Maruyama
Department of Anesthesiology and Critical Care Medicine, Mie University School of Medicine, and Department of Clinical Engineering, Suzuka University of Medical Science, Mie, Japan

Henrique R. Mendonça
Programa de Neurociências, Instituto de Biologia, Universidade Federal Fluminense, Niterói, RJ, Brazil

Roberto Paes-de-Carvalho
Programa de Neurociências, and Departamento de Neurobiologia, Instituto de Biologia, Universidade Federal Fluminense, Niterói, RJ, Brazil

Hui-Lin Pan
Center for Neuroscience and Pain Research, Department of Anesthesiology and Perioperative Medicine, The University of Texas MD Anderson Cancer Center, Houston, Texas, USA

Renate B. Pilz
Department of Medicine, University of California, San Diego, California, USA

Camila C. Portugal
Programa de Neurociências, Instituto de Biologia, Universidade Federal Fluminense, Niterói, RJ, Brazil

Brian E. Sansbury
Diabetes and Obesity Center, Institute of Molecular Cardiology, and Department of Physiology and Biophysics, Louisville, Kentucky, USA

Gary Silkstone
School of Biological Sciences, University of Essex, Colchester, United Kingdom

Renato Socodato
Programa de Neurociências, Instituto de Biologia, Universidade Federal Fluminense, Niterói, RJ, Brazil

Seyed Nasrollah Tabatabaei
Department of Computer and Software Engineering and Institute of Biomedical Engineering, École Polytechnique de Montréal

David Tate
Stanley S. Scott Cancer Center, Louisiana State University, Health Sciences Center, New Orleans, Louisiana, USA

Douglas D. Thomas
Department of Medicinal Chemistry and Pharmacognosy, University of Illinois at Chicago, Chicago, Illinois, USA

Albert K.Y. Tsui
Li Ka Shing Knowledge Institute, and Department of Anesthesia, St. Michael's Hospital, Toronto, Ontario, Canada

Laura B. Valdez
Institute of Biochemistry and Molecular Medicine (IBIMOL), Physical Chemistry Division, School of Pharmacy and Biochemistry, University of Buenos Aires, Buenos Aires, Argentina

Divya Vasudevan
Department of Medicinal Chemistry and Pharmacognosy, University of Illinois at Chicago, Chicago, Illinois, USA

Cecilia Vecoli
Institute of Clinical Physiology-CNR, Pisa, Italy

Michael T. Wilson
School of Biological Sciences, University of Essex, Colchester, United Kingdom

Jonathan Worrall
School of Biological Sciences, University of Essex, Colchester, United Kingdom

Tamara Zaobornyj
Institute of Biochemistry and Molecular Medicine (IBIMOL), Physical Chemistry Division, School of Pharmacy and Biochemistry, University of Buenos Aires, Buenos Aires, Argentina

Arnold H. Zea
Stanley S. Scott Cancer Center, Louisiana State University, Health Sciences Center, New Orleans, Louisiana, USA

Dong-Ya Zhu
Institute for Stem Cells and Neural Regeneration, and Department of Pharmacology, School of Pharmacy, Nanjing Medical University, Nanjing, China

PREFACE

Nitric oxide (nitrogen monoxide, NO), a gaseous molecule, is derived from the amino acid, L-arginine, in the human body. Its discovery was a great surprise because other hormones and regulators in the body are made up of proteins, lipid-derived compounds, and other molecules, none of which are gaseous. NO affects a number of enzyme systems and binds to heme, a cofactor in some important enzymes. In many ways, NO acts as a hormone and regulates many processes. Its activities are generally beneficial. For example, it is a powerful generator of vasodilation by suppressing vascular smooth muscle contraction. Its action is rapid as it remains in the blood for only seconds. In other activities, NO inhibits platelet aggregation and the adhesion of leukocytes to endothelia. Poorly functioning pathways involving NO are hallmarks in patients with various diseases, such as diabetes, atherosclerosis, and hypertension. Consequently, nitric oxide becomes a substance of interest in therapeutic applications. Many of the properties and actions of this new regulator, nitric oxide, are reviewed in this volume.

Reviews in this book have been ordered by first introducing the basic aspects of nitric oxide followed by chapters that involve clinical concepts. Accordingly, the first chapter is on the "Regulation of nociceptive transduction and transmission by nitric oxide" by A. Bavencoffe, S.-R. Chen, and H.-L. Pan. The next offering is by Z. Guo and D.A. Geller entitled "microRNA and human inducible nitric oxide synthase." "Heart mitochondrial nitric oxide synthase: a strategic enzyme in the regulation of cellular bioenergetics" is authored by T. Zaobornyj and L.B. Valdez. W.P. Gray and A. Cheung review "Nitric oxide regulation of adult neurogenesis." "Nitric oxide in the nervous system: biochemical, developmental, and neurobiological aspects" is the work of M. Cossenza, R. Socodato, C.C. Portugal, I.C.L. Domith, L.F.H. Gladulich, T.G. Encarnação, K.C. Kalaza, H.R. Mendonça, P. Campello-Costa, and R. Paes-de-Carvalho. Y. Hu and D.-Y. Zhu report on "Hippocampus and nitric oxide." "Nitric oxide and hypoxia signaling" is a review by H.S.J. Man, A.K.Y. Tsui, and P.A. Marsden. M.A. Hough, G. Silkstone, J. Worrall, and M.T. Wilson discuss "NO binding to the proapoptotic cytochrome *c*–cardiolipin complex." S.F. Kim writes on "The nitric oxide-mediated regulation of prostaglandin signaling in medicine."

"Nitric oxide as a mediator of estrogen effects in osteocytes" is the topic of J. Joshua, H. Kalyanaraman, N. Marathe, and R.B. Pilz.

In aspects related to disease conditions, contributions begin with "Insights into the diverse effects of nitric oxide on tumor biology" by D. Vasudevan and D.D. Thomas. A.H. Zea, A. Aiyar, and D. Tate review "Dual effect of interferon (IFNγ)-induced nitric oxide on tumorigenesis and intracellular bacteria." Next, B.E. Sansbury and B.G. Hill cover the "Anti-obesogenic role of endothelial nitric oxide synthase." H. Girouard and S.N. Tabatabaei report on "Nitric oxide and cerebrovascular regulation." Then, C. Vecoli reports on "Endothelial nitric oxide synthase gene polymorphisms in cardiovascular disease." The "Role of nitric oxide in pathophysiology and treatment of pulmonary hypertension" by P. Kruzliak, J. Maruyama, and K. Maruyama is the final chapter.

The cover illustration is Fig. 14.2 of Chapter 14.

Helene Kabes and Mary Ann Zimmerman of Elsevier, Oxford, UK were instrumental in the processing of these chapters.

Gerald Litwack
North Hollywood, California
May 8, 2014

CHAPTER ONE

Regulation of Nociceptive Transduction and Transmission by Nitric Oxide

Alexis Bavencoffe, Shao-Rui Chen, Hui-Lin Pan[1]
Center for Neuroscience and Pain Research, Department of Anesthesiology and Perioperative Medicine, The University of Texas MD Anderson Cancer Center, Houston, Texas, USA
[1]Corresponding author: e-mail address: huilinpan@mdanderson.org

Contents

Abstract

The potential involvement of nitric oxide (NO), a diffusible gaseous signaling messenger, in nociceptive transduction and transmission has been extensively investigated. However, there is no consistent and convincing evidence supporting the pronociceptive action of NO at the physiological concentration, and the discrepancies are possibly due to the nonspecificity of nitric oxide synthase inhibitors and different concentrations of NO donors used in various studies. At the spinal cord level, NO predominantly reduces synaptic transmission by inhibiting the activity of NMDA receptors and glutamate release from primary afferent terminals through S-nitrosylation of voltage-activated calcium channels. NO also promotes synaptic glycine release from inhibitory

Vitamins and Hormones, Volume 96
ISSN 0083-6729
http://dx.doi.org/10.1016/B978-0-12-800254-4.00001-5

interneurons through the cyclic guanosine monophosphate/protein kinase G signaling pathway. Thus, NO probably functions as a negative feedback regulator to reduce nociceptive transmission in the spinal dorsal horn during painful conditions.

1. INTRODUCTION

Pain receptors, also called nociceptors, are a group of sensory neurons with specialized nerve endings widely distributed in the skin, deep tissues (including the muscles and joints), and most of visceral organs. They respond to tissue injury or potentially damaging stimuli by sending nerve signals to the spinal cord and brain to begin the process of pain sensation. Nociceptors are equipped with specific molecular sensors, which detect extreme heat or cold and certain harmful chemicals. Mechanical nociceptors can also respond to tissue-damaging stimuli, such as pinching the skin or over-stretching the muscles. Activation of nociceptors generates action potentials, which are propagated along the afferent nerve axons, especially unmyelinated C-fibers and thinly myelinated Aδ-fibers. At the spinal cord level, the nociceptive nerve terminals release excitatory neurotransmitters to activate their respective postsynaptic receptors on second-order neurons. In the spinal dorsal horn, both excitatory and inhibitory interneurons can augment or attenuate nociceptive transmission (Cervero & Iggo, 1980; Zhou, Zhang, Chen, & Pan, 2007, 2008). The nociceptive signal, encoding the quality, location, and intensity of the noxious stimuli, is then conveyed via the ascending pathway to reach various brain regions to elicit pain sensation. Physiological pain responses normally protect us from tissue damage by quickly alerting us to impending injury. Unlike acute physiological pain, chronic pathological pain, including neuropathic and inflammatory pain, is often associated with increased activity and responses of spinal dorsal horn neurons, termed central sensitization (Woolf & Thompson, 1991; Xu, Dalsgaard, & Wiesenfeld-Hallin, 1992). This phenomenon is the cellular basis for hyperalgesia (increased pain response to a noxious stimulus) and allodynia (painful sensation in response to a nonnoxious stimulus).

Nitric oxide (NO) is a membrane-permeable gaseous second messenger involved in signal transduction. The physiological function of NO has been shown in a large variety of cell types and tissues, including the immune system, blood vessels, endothelial cells, and neurons. NO is produced from L-arginine by three major isoforms of nitric oxide synthase (NOS): neuronal

NOS (nNOS or NOS1), inducible NOS (iNOS or NOS2), and endothelial NOS (eNOS or NOS3) (Alderton, Cooper, & Knowles, 2001; Knowles & Moncada, 1994). Both nNOS and eNOS are constitutively expressed and activated by Ca^{2+}/calmoduline-dependent signaling, whereas iNOS is typically induced by immunostimulation, such as inflammatory cytokines and bacterial endotoxins, independent of intracellular Ca^{2+} levels. Classically, the intracellular NO effect is mediated by the NO-sensitive soluble guanylyl cyclase (sGC). When activated, sGC converts guanosine triphosphates (GTP) into cyclic guanosine monophosphates (cGMP). cGMP has different targets such as serine/threonine protein kinases G (PKG-I and PKG-II), cGMP-regulated phosphodiesterase, and cGMP-activated ion channels (Ahern, Klyachko, & Jackson, 2002; Calabrese et al., 2007). In addition, NO can promote a covalent and reversible posttranslational protein modification by interacting with the thiol side chain of cysteine residues. This chemical reaction, named S-nitrosylation, occurs without the action of any enzymes (Ahern et al., 2002; Choi et al., 2000).

The role of NO in pain signaling has been investigated in many studies using rodent models and in humans. In this chapter, we critically review the reported complex actions of NO in pain transduction and transmission. We also present recent electrophysiological evidence showing that NO inhibits nociceptive transmission at the spinal cord level and the signaling mechanisms involved.

2. ROLE OF NO IN NOCICEPTIVE TRANSDUCTION AT THE PERIPHERY

The evidence about the role of NO in pain transduction is inconsistent and conflicting. In human subjects, injection of the NO solution (>0.7 mM) into the paravascular tissues and veins (Holthusen & Arndt, 1995) or the cutaneous tissue in the forearm (Holthusen & Arndt, 1994) evokes pain in a dose-dependent fashion. However, the physiological relevance of high NO concentrations used in these studies is not clear, because the tissue level of NO is <10 μM (Wink et al., 1993). On the other hand, local infusion of the NO donor glyceryl trinitrate at the site of surgery produces a small analgesic effect in humans undergoing oral surgery (Hamza et al., 2010). Also, transdermal nitroglycerine alone does not affect pain itself but prolongs the analgesic effect of opioids in patients after knee surgery (Lauretti, de Oliveira, Reis, Mattos, & Pereira, 1999). Others have reported that inhaled NO can reduce pain intensity in patients with sickle cell crisis

(Head et al., 2010) and that local spray of isosorbide dinitrate, an NO donor, reduces pain and burning sensation, in diabetic patients (Yuen, Baker, & Rayman, 2002). Nevertheless, the potential therapeutic effect of NO donors for chronic pain treatment still needs to be confirmed in large scale clinical trials.

It has been reported that local injection of NG-methyl-L-arginine (L-NMA), an NOS inhibitor, attenuates mechanical hypersensitivity caused by injection of prostaglandin E2 into the rat skin (Chen & Levine, 1999). Also, systemic treatment with an NOS inhibitor, NG-nitro-K-arginine methyl ester (L-NAME), attenuates tactile and cold allodynia in a rat model of neuropathic pain induced by spinal nerve ligation (Yoon, Sung, & Chung, 1998). However, others have shown that the antiallodynic effect of L-NAME cannot be reversed by L-arginine (Lee, Singh, & Lodge, 2005). Furthermore, systemic treatment with other NOS inhibitors, including 7-nitroindazole, N(omega)-nitro-L-arginine (L-NNA), aminoguanidine, and LY457963, does not affect allodynia in rats subjected to spinal nerve ligation (Lee et al., 2005; Luo et al., 1999). Thus, it seems that the commonly used NOS inhibitors, such as L-NAME, may affect targets other than NOS. In support of this possibility, it has been demonstrated that the arginine analogues with alkyl or aryl esterification, including L-NAME, are in fact muscarinic receptor antagonists (Buxton et al., 1993; Chang, Chen, & Hsiue, 1997). In addition, tactile allodynia caused by nerve injury or tissue damage is attenuated in nNOS-knockout mice (Chu et al., 2005; Guan, Yaster, Raja, & Tao, 2007). sGC-knockout mice also show reduced nociceptive responses to tissue inflammation or nerve injury, but their responses to acute pain are not affected (Schmidtko et al., 2008). It should be acknowledged that in eNOS-, nNOS-, or iNOS-knockout mice, there is a developmental compensatory increase in the expression of other NOS isoforms in the spinal cord (Boettger et al., 2007; Tao et al., 2003). This compensatory upregulation of other NOS subtypes in specific NOS isoform-knockout mice complicates the interpretation of the behavioral test results.

In contrast, it has been shown that local injection of an NO donor, sodium nitroprusside, reduces mechanical hyperalgesia in the rat paw induced by prostaglandin E2 injection (Durate, Lorenzetti, & Ferreira, 1990). Interestingly, topical application of a cream containing different concentrations of NO donors, *S*-nitroso-*N*-acetylpenicillamine (SNAP) or isosorbide dinitrate, produces an opposite effect on allodynia in a rat model of postoperative pain (Prado, Schiavon, & Cunha, 2002). At a low concentration, SNAP (1–10%) or isosorbide (2.5–5%) reduces allodynia through the guanylyl cyclase. However, at a high concentration, SNAP (30%) increases allodynia induced by surgical incision, which is not blocked by a guanylyl cyclase inhibitor (Prado et al., 2002). Notably, high concentrations of NO may produce tissue-damaging effects via highly reactive oxidant species such as peroxynitrite (Koppenol, Moreno, Pryor, Ischiropoulos, & Beckman, 1992).

3. DIVERSE EFFECTS OF NO ON ION CHANNELS EXPRESSED ON PRIMARY SENSORY NEURONS

3.1. Acid-sensing ion channels

Acid-sensing ion channels (ASIC) are expressed in small-diameter sensory neurons, and their expression is increased during inflammation or by proinflammatory mediators such as bradykinin and interleukin-1 (Krishtal, 2003; Mamet, Baron, Lazdunski, & Voilley, 2002). These channels are responsible for the ion conductance activated by extracellular protons in sensory neurons (Krishtal, 2003; Krishtal & Pidoplichko, 1980). NO potentiates the activity of proton-gated ASIC channels when exposed to a modest decrease of pH to 6.3, mimicking the conditions of inflammation and anoxia, in neonatal rat dorsal root ganglion (DRG) neurons (Cadiou et al., 2007). This effect seems to be independent of cGMP/PKG pathway and requires an extracellular site of action, possibly S-nitrosylation. Topical application of an NO donor, glyceryl trinitrate, significantly increases acid-evoked pain but has no effects on heat or mechanical pain thresholds in the skin of human volunteers (Cadiou et al., 2007).

3.2. Transient receptor potential channels

3.2.1 TRPV1

Transient receptor potential (TRP) channels are cationic nonselective channels, which respond to a broad range of stimuli such as temperature, protons, osmolarity, mechanical stress, and painful stimuli. For example, transient receptor potential vanilloid type 1 (TRPV1) channels are activated by noxious heat (>43 °C), protons, and capsaicin, the active component of red chili pepper. Opening of TRPV1 channels produces a sensation of burning pain (Caterina et al., 1997, 2000). TRPV1 are expressed in small nociceptive neurons and are upregulated under inflammatory conditions. TRPV1 can be sensitized by proinflammatory mediators, including bradykinin and prostaglandins, resulting in enhanced pain sensation during inflammation.

It has been reported that TRPV1 channels can be S-nitrosylated on two cysteine residues (Cys616 and Cys621) in the N-terminal of the pore-forming region in the S5–S6 linker (Yoshida et al., 2006). Applications of SNAP or an S-nitrosylating agent (5-nitro-2-PDS) trigger Ca^{2+} influx via TRPV1 channels in HEK293 cells. Proton and heat sensitivities of TRPV1 channels are also enhanced in the presence of SNAP and 5-nitro-2-PDS. Deletion of the two cysteine residues reduces this enhancement without affecting proton and heat sensitivity and the surface expression of TRPV1 channels (Yoshida et al., 2006). However, others have shown that SNAP stimulates Ca^{2+} influx in mouse DRG neurons and in HEK293 cells expressing TRPV1 channels only at a very high concentration (>1 mM) (Miyamoto, Dubin, Petrus, & Patapoutian, 2009). In contrast, a recent study has shown that low concentrations of NO donors, such as 100 μM sodium nitroprusside (SNP) or 100 μM SNAP, inhibits TRPV1 channel activity through activation of the cGMP/PKG pathway in rat DRG neurons (Jin, Kim, & Kwak, 2012).

3.2.2 TRPA1

Originally named ANKTM1, TRPA1 channels are activated by cold temperatures (<17 °C), proinflammatory mediators, and chemical irritants such as mustard oil, wasabi, allicin, and formalin (Jordt et al., 2004). They are mainly coexpressed with TRPV1 channels in primary sensory neurons. Topical applications of TRPA1 agonists induce spontaneous pain, cold allodynia, and heat/thermal hyperalgesia (Fernandes, Fernandes, & Keeble, 2012; Nilius, Appendino, & Owsianik, 2012). It has been reported that high concentrations of SNAP (2–3 mM) can activate TRPA1 channels in DRG neurons and in HEK293 cells (Miyamoto et al., 2009).

3.2.3 TRPV3 and TRPV4

TRPV3 channels are heat-sensitive channels (>33 °C) expressed in keratinocytes and DRG neurons (Peier et al., 2002; Smith et al., 2002). TRPV3 is also activated by eugenol, camphor, and inflammatory mediators such as arachidonic acids (Hu et al., 2006). TRPV4 channels are activated by warm temperature (>27 °C), low pH, hypotonicity, and diacyl glycerol. TRPV3-knockout mice exhibit a deficit in innocuous and noxious heat sensitivity (Moqrich et al., 2005). Both SNAP and 5-nitro-2-PDS seem to stimulate TRPV3 and TRPV4 channels expressed in HEK293 cells (Yoshida et al., 2006). However, TRPV3 and TRPV4 do not have a major contribution to heat sensation (Huang, Li, Yu, Wang, & Caterina, 2011). Their physiological relevance to nociceptive transduction in sensory neurons and the skin is not clear.

3.3. K_{ATP} channels

ATP-sensitive potassium channels are regulated by the ATP/ADP ratio in a way that a drop of this ratio will activate these channels. Following their opening, the efflux of potassium will induce a hyperpolarization, decreasing the neuronal excitability. K_{ATP} channels also control neurotransmitter release by regulating neuronal excitability (Yamada & Inagaki, 2005). K_{ATP} channels are expressed in large DRG neurons and are activated by S-nitrosylation of cysteine residues in their SUR1 subunit (Kawano et al., 2009) or by the NO/cGMP/PKG pathway (Sachs, Cunha, & Ferreira, 2004). The cGMP/PKG/K_{ATP} signaling seems to be involved in the analgesic effects of noradrenaline (Romero, Guzzo, Perez, Klein, & Duarte, 2012), cannabinoid receptor 2 agonists (Negrete, Hervera, Leanez, Martin-Campos, & Pol, 2011), and morphine (Cunha et al., 2010).

4. ROLE OF NO IN REGULATING NOCICEPTIVE TRANSMISSION AT THE SPINAL CORD LEVEL

The spinal dorsal horn is a critical site for nociceptive transmission and modulation. Both nNOS and sGC are present in the superficial spinal dorsal horn (Ding & Weinberg, 2006; Schmidtko et al., 2008; Terenghi, Riveros-Moreno, Hudson, Ibrahim, & Polak, 1993). It has been reported that intrathecal injection of L-NAME decreases thermal hyperalgesia caused by carrageenan in rats, but an nNOS inhibitor, 7-nitroindazole, has no such effect (Osborne & Coderre, 1999). Also, mechanical hypersensitivity induced by spinal nerve ligation or tissue inflammation is reduced by intrathecal administration of nNOS inhibitors (L-NAME or 7-nitroindazole) (Chu et al., 2005; Guan et al., 2007). Moreover, it has been reported that intrathecal injection of an nNOS inhibitor, N(omega)-propyl-l-arginine, or an iNOS inhibitor, 2-amino-5,6-dihydro-6-methyl-4H-1,3-thiazine hydrochloride, dose-dependently reduces thermal and mechanical hypersensitivity induced by partial sciatic nerve ligation in mice (Tanabe, Nagatani, Saitoh, Takasu, & Ono, 2009). These studies suggest that spinal NO may be involved in the potentiation of central sensitization after tissue damage or nerve injury.

In contrast, other studies have shown that NO at the spinal cord level plays a role in inhibition of nociceptive transmission. For instance, intrathecal administration of the NO precursor L-arginine increases the mechanical nociceptive withdrawal threshold in rats (Zhuo, Meller, & Gebhart, 1993). Also, intrathecal administration of an NO donor, 3-morpholinosydnonimine (SIN-1; 0.1–100 μg), consistently produces an antinociceptive effects in the rat tail-flick test in a dose-dependent manner (Sousa & Prado, 2001). The analgesic effect of morphine also involves NO since the antinociceptive action of morphine is attenuated by NOS inhibitors and NO scavengers (Song, Pan, & Eisenach, 1998). Furthermore, intrathecal injection of an nNOS inhibitor (1,2-trifluoromethylphenyl imidazole, TRIM) or an NO scavenger (carboxy-PTIO), reduces the anti-allodinic effect of clonidine in rat model neuropathic pain caused by L5–L6 spinal nerve ligation (Pan, Chen, & Eisenach, 1998). We have shown that intrathecal injection of 300 μg L-arginine or 100 μg SNAP significantly increases the nociceptive threshold in response to a noxious pressure stimulus *n* rats (Jin, Chen, Cao, Li, & Pan, 2011). In addition, the anti-nociceptive effect of L-arginine and SNAP is abolished by cotreatment with TRIM and carboxy-PTIO, respectively. These findings suggest that NO inhibits nociceptive transmission at the spinal cord level.

Some studies have reported a dual effect of NO in regulating nociception at the spinal cord level. Intrathecal injection of small doses of L-arginine (10 μg) attenuates the nociceptive behaviors evoked by formalin injection in rats, whereas L-arginine at high doses (250 μg) increases nociceptive responses (Li & Qi, 2010). Similarly, intrathecal delivery of small doses (0.1–0.25 μM) of a cGMP analogue, 8-bromo-cyclic guanosine monophosphate (8-bromo-cGMP), decreases nociceptive behaviors caused by formalin injection. But intrathecal injection of a high dose of 8-bromo-cGMP (2.5 μM) increases nociceptive responses (Tegeder, Schmidtko, Niederberger, Ruth, & Geisslinger, 2002). Furthermore, in a rat model of neuropathic pain induced by sciatic nerve ligation, intrathecal administration of a low dose of SIN-1 (0.1–2 μg) reduces mechanical allodynia (Sousa & Prado, 2001). However, intrathecal injection of a high dose of SIN-1 (10–20 μg) increases allodynia (Sousa & Prado, 2001). The discrepancy regarding the complex function of NO in nociceptive processing may result from the use of different animal models of pain, the nonspecificity of NOS inhibitors, and/or different NO levels produced at the spinal cord level in various studies.

There are only a few published electrophysiological studies about the effect of NO on spinal nociceptive neurons. Spinally administered NO donors predominantly reduce the firing activity of spinal dorsal horn neurons, and treatment with L-NAME or a selective sGC inhibitor, 1H-[1,2,4]oxadiazolo[4,3-a]quinoxalin-1-one (ODQ), increases the activity of nociceptive dorsal horn neurons (Hoheisel, Unger, & Mense, 2000, 2005; Pehl & Schmid, 1997). In contrast, administration of L-NAME or 7-nitroindazole via a microdialysis fiber inserted to the spinal cord tissues blocks sensitization of spinothalamic tract neurons induced by intradermal capsaicin injection (Lin et al., 1999). Notably, NOS inhibitors have no effects on the responses of spinothalamic tract neurons to peripheral stimulation in the absence of capsaicin injection (Lin et al., 1999).

nNOS is the major isoform of NOS in the spinal dorsal horn, and its expression level is increased during peripheral inflammation and nerve injury (Dolan, Kelly, Huan, & Nolan, 2003; Ma & Eisenach, 2007). In the central nervous system, the production of NO by nNOS requires Ca^{2+} influx, which is dependent on activation of postsynaptic NMDA receptors. When NO is produced postsynaptically after *N*-methyl-D-aspartate (NMDA) receptor activation by glutamate release from nociceptive primary afferent nerve terminals, NO could diffuse back to the presynaptic terminals of primary afferent terminals in the dorsal horn to inhibit glutamate release (Jin et al., 2011). Because increased NO production is a secondary response to tissue/nerve injury-induced NMDA receptor activation, increased spinal NO release and NOS expression levels under painful conditions should not be considered supporting evidence for a pronociceptive role of NO.

5. NO REDUCES EXCITATORY, BUT POTENTIATES INHIBITORY, SYNAPTIC TRANSMISSION IN SPINAL CORDS

To further determine how NO controls excitatory and inhibitory synaptic transmission in the spinal dorsal horn, we conducted a series of patch-clamp recording experiments using a spinal cord slice preparation, which preserves the intrinsic connection between primary afferent terminals and dorsal horn neurons. High-resolution whole-cell recording of postsynaptic currents under voltage-clamp conditions provides a sensitive and accurate measure of synaptic release of glutamate, GABA, and glycine in real-time. Combined with pharmacologic and ionic manipulations, information is obtained about the dynamics of transmitter–receptor interactions, the types of postsynaptic receptors activated, the effects of drugs on transmission, and the mechanisms of synaptic plasticity. Another advantage of this technique is that experiments are performed in spinal tissue slices in the absence of any anesthetics. In the sections below, we present electrophysiological evidence about the role of NO in regulating synaptic transmission in the spinal dorsal horn.

5.1. Glutamatergic input from primary afferent nerves

Glutamate is the predominant excitatory neurotransmitter involved in nociceptive transmission. When nociceptors are activated, the primary afferent nerves release glutamate at their central terminals to superficial dorsal horn neurons of the spinal cord. Glutamate will then bind to its postsynaptic ligand-gated channels, 2-amino-3-(3-hydroxy-5-methyl-isoxazol-4-yl) propanoic acid (AMPA) and NMDA receptors, to excite second-order neurons in the spinal dorsal horn. Calcium influx after activation of NMDA receptors stimulates calcium/calmoduline-dependent signaling and nNOS activity to generate NO (Rameau, Chiu, & Ziff, 2004). Released NO can act paracrinally on neighboring neurons, glial cells or have a retrograde presynaptic effect in the spinal cord.

L-arginine significantly decreases the frequency of spontaneous excitatory postsynaptic currents (EPSCs) but has no significant effect on the frequency or amplitude of miniature EPSCs of spinal dorsal horn neurons (Jin et al., 2011). To determine the role of NO in the control of glutamate release from primary afferent terminals, the effect of L-arginine or SNAP on glutamatergic monosynaptic EPSCs evoked from the dorsal root is also examined. Both L-arginine and SNAP inhibit the amplitude of evoked monosynaptic EPSCs of dorsal horn neurons, and their effects are abolished by treatment with TRIM and carboxy-PTIO, respectively (Jin et al., 2011). These findings strongly suggest that NO inhibits glutamatergic synaptic transmission between primary afferent nerves and second-order neurons in the spinal dorsal horn.

Interestingly, sGC/cGMP is not involved in the NO effect on synaptic glutamate release to dorsal horn neurons (Jin et al., 2011). *N*-ethylmaleimide (NEM) is a specific alkylating agent of cysteine sulfhydryls, covalently modifies protein sulfhydryl groups thereby preventing subsequent S-nitrosylation of proteins (Bolotina, Najibi, Palacino, Pagano, & Cohen, 1994; Broillet & Firestein, 1996). In the presence of NEM, L-arginine or SNAP has no effect on the amplitude of evoked EPSCs of dorsal horn neurons (Jin et al., 2011). These results suggest that NO probably inhibits glutamate release from primary afferent terminals through S-nitrosylation of presynaptic proteins.

5.2. Voltage-activated calcium channels in sensory neurons

High voltage-activated calcium channels (HVACCs) are essential for neurotransmitter release and for nociceptive transmission (Smith & Cunnane, 1997). Because NO inhibits evoked glutamate release from primary afferent terminals through S-nitrosylation, NO may inhibit HVACCs in DRG neurons through S-nitrosylation. This notion is supported by the findings from using acutely dissociated DRG neurons and HEK293 cells expressing HVACCs (Jin et al., 2011). For example, L-arginine or SNAP causes a large decrease in the HVACC, but not T-type, currents in rat DRG neurons. However, this effect was not affected by ODQ, which is consistent with the report showing that NO-sensitive guanylyl cyclase is not expressed in DRG neurons (Schmidtko et al., 2008). Importantly, NEM completely blocks the inhibitory effect of L-arginine and SNAP on HVACCs in DRG neurons and in HEK293 cells (Jin et al., 2011). Therefore, NO likely inhibits the activity of HVACCs in primary sensory neurons through S-nitrosylation.

5.3. Synaptic NMDA receptors

NMDA receptors are involved in synaptic transmission and plasticity in many chronic pain conditions (Zhou, Chen, & Pan, 2011). NO can inhibit NMDA receptor currents in recombinant systems (Aizenman & Potthoff, 1999; Lei et al., 1992). In spinal dorsal horn neurons, SNAP reversibly reduces the amplitude of NMDA receptor-mediated EPSCs evoked from stimulation of spinal dorsal root and puff NMDA currents (Nicholson, Dibb, & Renton, 2004). Furthermore, it has been shown that exogenous and endogenous NO can inhibit NMDA receptor activity through S-nitrosylation of the cysteine residue 399 on the NR2A subunit (Choi et al., 2000). Because both HVACCs and NMDA receptors are critically involved in nociceptive transmission, it seems reasonable to propose that NO could act as a physiological brake to prevent over-excitation of spinal dorsal horn neurons, caused by synaptic glutamate release and NMDA receptor activation, during painful states.

5.4. Synaptic release of glycine

Both γ-aminobutryic acid (GABA) and glycine are the most predominant inhibitory neurotransmitters involved in the regulation of nociceptive transmission at the spinal cord level (Zhou et al., 2007, 2008, 2012). Activation of $GABA_A$ and glycine receptors normally causes chloride influx and hyperpolarization of mature neurons. Bath application of 100–400 μM L-arginine concentration-dependently increases the frequency of glycinergic inhibitory postsynaptic currents (IPSPs) of spinal dorsal horn neurons; and such an effect is abolished by pretreatment with TRIM (Jin et al., 2011). SNAP at 100 μM also potentiates the frequency of glycinergic spontaneous IPSCs of dorsal horn neurons, and this effect is blocked by carboxy-PTIO. L-arginine still increases the frequency of glyinergic miniature IPSCs, suggesting that NO acts at presynaptic terminals to potentiate synaptic glycine release in the spinal dorsal horn (Jin et al., 2011). Interestingly, neither L-arginine nor SNAP has any effect on the frequency and amplitude of GABAergic spontaneous IPSCs of dorsal horn neurons (Jin et al., 2011). It has also been shown that NO promotes synaptic glycine, but not GABA, release to sympathetic preganglionic neurons in the lateral spinal cord (Wu & Dun, 1996).

Further studies reveal that the stimulating effect of L-arginine on glycinergic sIPSCs is abolished by inhibition of sGC with ODQ. A membrane-permeable cGMP analogue, 8-bromo-cGMP, significantly increases the frequency, not the amplitude, of glycinergic spontaneous IPSCs. In addition, Rp-8-Br-PET-cGMPS, a specific and membrane-permeable PKG inhibitor blocks the potentiating effect of 8-bromo-cGMP or SNAP on glycinergic spontaneous IPSCs (Jin et al., 2011). Thus, the sGC/cGMP/PKG signaling cascade mediates the potentiating effect of NO on synaptic glycine release to spinal cord horn neurons.

Collectively, NO inhibits nociception transmission at the spinal cord level by inhibiting excitatory glutamatergic input from primary afferent nerves and by potentiating inhibitory glycine release. These electrophysiological data provide new insights into the underlying of cellular and signaling mechanisms of NO in the inhibition of nociceptive transmission at the spinal level.

6. CONCLUSIONS AND FUTURE DIRECTIONS

The controversial role of NO in nociceptive transduction and transmission has not been fully resolved. Despite various reports showing the involvement of NO in the nociception and nociceptor activation, there is no convincing evidence supporting the pronociceptive actions of NO at the physiological concentration. The discrepancies are possibly due to the nonspecificity of NOS inhibitors and different concentrations of NO donors used in various studies. At the spinal cord level, NO predominantly reduces pain transmission by inhibiting the activity of NMDA receptors and voltage-activated calcium channels through S-nitrosylation and by facilitating glycine release from inhibitory interneurons via the cGMP/PKG pathway (Fig. 1.1). This electrophysiological evidence indicates that NO likely

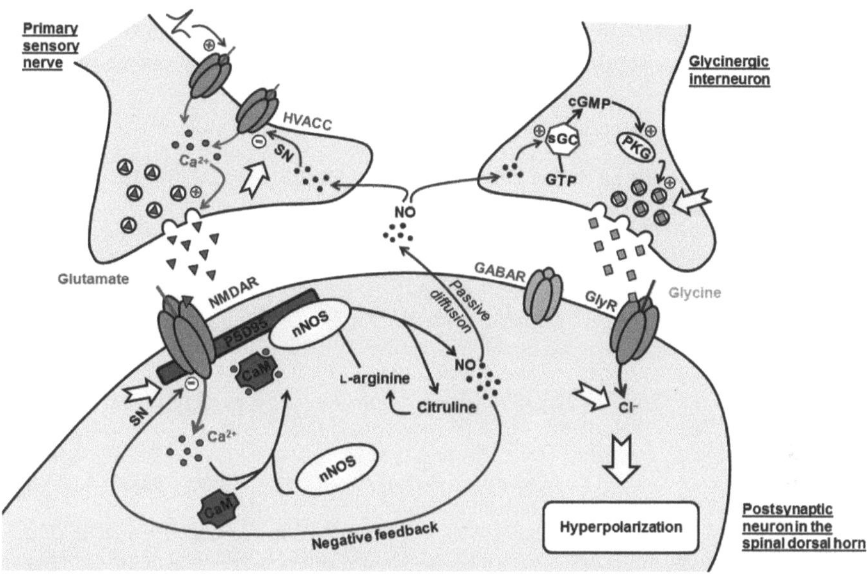

Figure 1.1 Diagram shows that NO inhibits synaptic transmission in the dorsal horn of the spinal cord. Stimulation of primary sensory nerves by painful stimuli triggers action potentials and opening of voltage-activated calcium channels, leading to calcium influx and glutamate release from presynaptic nerve terminals. Released glutamate binds to AMPA (not shown) and NMDA receptors present on postsynaptic dorsal horn neurons to increase intracellular calcium levels and to recruit nNOS to the close proximity of NMDA receptors by the scaffolding protein PSD95 and nNOS activation via calcium/calmodulin-dependent signaling. Subsequently, nNOS converts L-arginine into NO and citruline. NO then exerts a negative feedback action by inhibiting the NMDA receptor activity via S-nitrosylation. NO also diffuses passively into the presynaptic terminal and inhibits the activity of high voltage-activated calcium channels to decrease synaptic glutamate release. Moreover, NO diffuses into the neighboring glycinergic interneurons and potentiates synaptic glycine release via stimulation of an sGC/PKG-dependent pathway. Glycine activates postsynaptic glycine receptors, leading to chloride influx. Collectively, decreased glutamate release, inhibition of NMDA receptors, and increased glycine release by NO result in hyperpolarization of the postsynaptic neuron, thereby reducing nociceptive transmission at the spinal cord level. CaM, calmodulin; cGMP, cyclic guanosine monophosphate; GABAR, γ-aminobutyric acid receptor; GlyR, glycine receptor; GTP, guanosine triphosphate; HVACC, high voltage-activated calcium channel; NMDAR, N-methyl-D-aspartate receptor; nNOS, neuronal nitric oxide synthase; NO, nitric oxide; PKG, protein kinase G; PSD95, postsynaptic density protein 95; sGC, soluble guanylyl cyclase; SN, S-nitrosylation. (See the color plate.)

functions as a negative feedback regulator to reduce nociceptive transmission in the spinal dorsal horn during various painful conditions. Because of the potential confounding issues associated with pharmacological agents and conventional NOS-knockout mice, further studies are warranted to use siRNA and conditional knockout approaches to unambiguously validate the physiological role of NO in the regulation and signaling of nociception during acute and chronic pain conditions.

ACKNOWLEDGMENTS

This work was supported by the National Institutes of Health Grant NS073935 and by the N.G. and Helen T. Hawkins endowment (to H. -L. P.).

REFERENCES

Ahern, G. P., Klyachko, V. A., & Jackson, M. B. (2002). cGMP and S-nitrosylation: Two routes for modulation of neuronal excitability by NO. *Trends in Neurosciences*, *25*, 510–517.

Aizenman, E., & Potthoff, W. K. (1999). Lack of interaction between nitric oxide and the redox modulatory site of the NMDA receptor. *British Journal of Pharmacology*, *126*, 296–300.

Alderton, W. K., Cooper, C. E., & Knowles, R. G. (2001). Nitric oxide synthases: Structure, function and inhibition. *The Biochemical Journal*, *357*, 593–615.

Boettger, M. K., Uceyler, N., Zelenka, M., Schmitt, A., Reif, A., Chen, Y., et al. (2007). Differences in inflammatory pain in nNOS-, iNOS- and eNOS-deficient mice. *European Journal of Pain (London, England)*, *11*, 810–818.

Bolotina, V. M., Najibi, S., Palacino, J. J., Pagano, P. J., & Cohen, R. A. (1994). Nitric oxide directly activates calcium-dependent potassium channels in vascular smooth muscle. *Nature*, *368*, 850–853.

Broillet, M. C., & Firestein, S. (1996). Direct activation of the olfactory cyclic nucleotide-gated channel through modification of sulfhydryl groups by NO compounds. *Neuron*, *16*, 377–385.

Buxton, I. L., Cheek, D. J., Eckman, D., Westfall, D. P., Sanders, K. M., & Keef, K. D. (1993). NG-nitro L-arginine methyl ester and other alkyl esters of arginine are muscarinic receptor antagonists. *Circulation Research*, *72*, 387–395.

Cadiou, H., Studer, M., Jones, N. G., Smith, E. S., Ballard, A., McMahon, S. B., et al. (2007). Modulation of acid-sensing ion channel activity by nitric oxide. *The Journal of Neuroscience: The Official Journal of the Society for Neuroscience*, *27*, 13251–13260.

Calabrese, V., Mancuso, C., Calvani, M., Rizzarelli, E., Butterfield, D. A., & Stella, A. M. (2007). Nitric oxide in the central nervous system: Neuroprotection versus neurotoxicity. *Nature Reviews. Neuroscience*, *8*, 766–775.

Caterina, M. J., Leffler, A., Malmberg, A. B., Martin, W. J., Trafton, J., Petersen-Zeitz, K. R., et al. (2000). Impaired nociception and pain sensation in mice lacking the capsaicin receptor. *Science (New York, NY)*, *288*, 306–313.

Caterina, M. J., Schumacher, M. A., Tominaga, M., Rosen, T. A., Levine, J. D., & Julius, D. (1997). The capsaicin receptor: A heat-activated ion channel in the pain pathway. *Nature*, *389*, 816–824.

Cervero, F., & Iggo, A. (1980). The substantia gelatinosa of the spinal cord: A critical review. *Brain: A Journal of Neurology*, *103*, 717–772.

Chang, H. Y., Chen, C. W., & Hsiue, T. R. (1997). Comparative effects of L-NOARG and L-NAME on basal blood flow and ACh-induced vasodilatation in rat diaphragmatic microcirculation. *British Journal of Pharmacology, 120*, 326–332.

Chen, X., & Levine, J. D. (1999). NOS inhibitor antagonism of PGE2-induced mechanical sensitization of cutaneous C-fiber nociceptors in the rat. *Journal of Neurophysiology, 81*, 963–966.

Choi, Y. B., Tenneti, L., Le, D. A., Ortiz, J., Bai, G., Chen, H. S., et al. (2000). Molecular basis of NMDA receptor-coupled ion channel modulation by S-nitrosylation. *Nature Neuroscience, 3*, 15–21.

Chu, Y. C., Guan, Y., Skinner, J., Raja, S. N., Johns, R. A., & Tao, Y. X. (2005). Effect of genetic knockout or pharmacologic inhibition of neuronal nitric oxide synthase on complete Freund's adjuvant-induced persistent pain. *Pain, 119*, 113–123.

Cunha, T. M., Roman-Campos, D., Lotufo, C. M., Duarte, H. L., Souza, G. R., Verri, W. A., Jr., et al. (2010). Morphine peripheral analgesia depends on activation of the PI3Kgamma/AKT/nNOS/NO/KATP signaling pathway. *Proceedings of the National Academy of Sciences of the United States of America, 107*, 4442–4447.

Ding, J. D., & Weinberg, R. J. (2006). Localization of soluble guanylyl cyclase in the superficial dorsal horn. *The Journal of Comparative Neurology, 495*, 668–678.

Dolan, S., Kelly, J. G., Huan, M., & Nolan, A. M. (2003). Transient up-regulation of spinal cyclooxygenase-2 and neuronal nitric oxide synthase following surgical inflammation. *Anesthesiology, 98*, 170–180.

Durate, I. D., Lorenzetti, B. B., & Ferreira, S. H. (1990). Peripheral analgesia and activation of the nitric oxide-cyclic GMP pathway. *European Journal of Pharmacology, 186*, 289–293.

Fernandes, E. S., Fernandes, M. A., & Keeble, J. E. (2012). The functions of TRPA1 and TRPV1: Moving away from sensory nerves. *British Journal of Pharmacology, 166*, 510–521.

Guan, Y., Yaster, M., Raja, S. N., & Tao, Y. X. (2007). Genetic knockout and pharmacologic inhibition of neuronal nitric oxide synthase attenuate nerve injury-induced mechanical hypersensitivity in mice. *Molecular Pain, 3*, 29.

Hamza, M., Wang, X. M., Wu, T., Brahim, J. S., Rowan, J. S., & Dionne, R. A. (2010). Nitric oxide is negatively correlated to pain during acute inflammation. *Molecular Pain, 6*, 55.

Head, C. A., Swerdlow, P., McDade, W. A., Joshi, R. M., Ikuta, T., Cooper, M. L., et al. (2010). Beneficial effects of nitric oxide breathing in adult patients with sickle cell crisis. *American Journal of Hematology, 85*, 800–802.

Hoheisel, U., Unger, T., & Mense, S. (2000). A block of spinal nitric oxide synthesis leads to increased background activity predominantly in nociceptive dorsal horn neurones in the rat. *Pain, 88*, 249–257.

Hoheisel, U., Unger, T., & Mense, S. (2005). The possible role of the NO-cGMP pathway in nociception: Different spinal and supraspinal action of enzyme blockers on rat dorsal horn neurones. *Pain, 117*, 358–367.

Holthusen, H., & Arndt, J. O. (1994). Nitric oxide evokes pain in humans on intracutaneous injection. *Neuroscience Letters, 165*, 71–74.

Holthusen, H., & Arndt, J. O. (1995). Nitric oxide evokes pain at nociceptors of the paravascular tissue and veins in humans. *The Journal of Physiology, 487*(Pt 1), 253–258.

Hu, H. Z., Xiao, R., Wang, C., Gao, N., Colton, C. K., Wood, J. D., et al. (2006). Potentiation of TRPV3 channel function by unsaturated fatty acids. *Journal of Cellular Physiology, 208*, 201–212.

Huang, S. M., Li, X., Yu, Y., Wang, J., & Caterina, M. J. (2011). TRPV3 and TRPV4 ion channels are not major contributors to mouse heat sensation. *Molecular Pain, 7*, 37.

Jin, X. G., Chen, S. R., Cao, X. H., Li, L., & Pan, H. L. (2011). Nitric oxide inhibits nociceptive transmission by differentially regulating glutamate and glycine release to spinal dorsal horn neurons. *The Journal of Biological Chemistry*, *286*, 33190–33202.

Jin, Y., Kim, J., & Kwak, J. (2012). Activation of the cGMP/protein kinase G pathway by nitric oxide can decrease TRPV1 activity in cultured rat dorsal root ganglion neurons. *The Korean Journal of Physiology & Pharmacology*, *16*, 211–217.

Jordt, S. E., Bautista, D. M., Chuang, H. H., McKemy, D. D., Zygmunt, P. M., Hogestatt, E. D., et al. (2004). Mustard oils and cannabinoids excite sensory nerve fibres through the TRP channel ANKTM1. *Nature*, *427*, 260–265.

Kawano, T., Zoga, V., Kimura, M., Liang, M. Y., Wu, H. E., Gemes, G., et al. (2009). Nitric oxide activates ATP-sensitive potassium channels in mammalian sensory neurons: Action by direct S-nitrosylation. *Molecular Pain*, *5*, 12.

Knowles, R. G., & Moncada, S. (1994). Nitric oxide synthases in mammals. *The Biochemical Journal*, *298*(Pt 2), 249–258.

Koppenol, W. H., Moreno, J. J., Pryor, W. A., Ischiropoulos, H., & Beckman, J. S. (1992). Peroxynitrite, a cloaked oxidant formed by nitric oxide and superoxide. *Chemical Research in Toxicology*, *5*, 834–842.

Krishtal, O. (2003). The ASICs: Signaling molecules? Modulators? *Trends in Neurosciences*, *26*, 477–483.

Krishtal, O. A., & Pidoplichko, V. I. (1980). A receptor for protons in the nerve cell membrane. *Neuroscience*, *5*, 2325–2327.

Lauretti, G. R., de Oliveira, R., Reis, M. P., Mattos, A. L., & Pereira, N. L. (1999). Transdermal nitroglycerine enhances spinal sufentanil postoperative analgesia following orthopedic surgery. *Anesthesiology*, *90*, 734–739.

Lee, D. H., Singh, J. P., & Lodge, D. (2005). Experiments with nitric oxide synthase inhibitors in spinal nerve ligated rats provide no evidence of a role for nitric oxide in neuropathic mechanical allodynia. *Neuroscience Letters*, *385*, 179–183.

Lei, S. Z., Pan, Z. H., Aggarwal, S. K., Chen, H. S., Hartman, J., Sucher, N. J., et al. (1992). Effect of nitric oxide production on the redox modulatory site of the NMDA receptor-channel complex. *Neuron*, *8*, 1087–1099.

Li, K., & Qi, W. X. (2010). Effects of multiple intrathecal administration of L-arginine with different doses on formalin-induced nociceptive behavioral responses in rats. *Neuroscience Bulletin*, *26*, 211–218.

Lin, Q., Palecek, J., Paleckova, V., Peng, Y. B., Wu, J., Cui, M., et al. (1999). Nitric oxide mediates the central sensitization of primate spinothalamic tract neurons. *Journal of Neurophysiology*, *81*, 1075–1085.

Luo, Z. D., Chaplan, S. R., Scott, B. P., Cizkova, D., Calcutt, N. A., & Yaksh, T. L. (1999). Neuronal nitric oxide synthase mRNA upregulation in rat sensory neurons after spinal nerve ligation: Lack of a role in allodynia development. *The Journal of Neuroscience: The Official Journal of the Society for Neuroscience*, *19*, 9201–9208.

Ma, W., & Eisenach, J. C. (2007). Neuronal nitric oxide synthase is upregulated in a subset of primary sensory afferents after nerve injury which are necessary for analgesia from alpha2-adrenoceptor stimulation. *Brain Research*, *1127*, 52–58.

Mamet, J., Baron, A., Lazdunski, M., & Voilley, N. (2002). Proinflammatory mediators, stimulators of sensory neuron excitability via the expression of acid-sensing ion channels. *The Journal of Neuroscience: The Official Journal of the Society for Neuroscience*, *22*, 10662–10670.

Miyamoto, T., Dubin, A. E., Petrus, M. J., & Patapoutian, A. (2009). TRPV1 and TRPA1 mediate peripheral nitric oxide-induced nociception in mice. *PLoS One*, *4*, e7596.

Moqrich, A., Hwang, S. W., Earley, T. J., Petrus, M. J., Murray, A. N., Spencer, K. S., et al. (2005). Impaired thermosensation in mice lacking TRPV3, a heat and camphor sensor in the skin. *Science (New York, NY)*, *307*, 1468–1472.

Negrete, R., Hervera, A., Leanez, S., Martin-Campos, J. M., & Pol, O. (2011). The antinociceptive effects of JWH-015 in chronic inflammatory pain are produced by nitric oxide-cGMP-PKG-KATP pathway activation mediated by opioids. *PLoS One*, *6*, e26688.

Nicholson, T. E., Dibb, S., & Renton, K. W. (2004). Nitric oxide mediates an LPS-induced depression of cytochrome P450 (CYP1A) activity in astrocytes. *Brain Research*, *1029*, 148–154.

Nilius, B., Appendino, G., & Owsianik, G. (2012). The transient receptor potential channel TRPA1: From gene to pathophysiology. *Pflügers Archiv*, *464*, 425–458.

Osborne, M. G., & Coderre, T. J. (1999). Effects of intrathecal administration of nitric oxide synthase inhibitors on carrageenan-induced thermal hyperalgesia. *British Journal of Pharmacology*, *126*, 1840–1846.

Pan, H. L., Chen, S. R., & Eisenach, J. C. (1998). Role of spinal NO in antiallodynic effect of intrathecal clonidine in neuropathic rats. *Anesthesiology*, *89*, 1518–1523.

Pehl, U., & Schmid, H. A. (1997). Electrophysiological responses of neurons in the rat spinal cord to nitric oxide. *Neuroscience*, 77, 563–573.

Peier, A. M., Reeve, A. J., Andersson, D. A., Moqrich, A., Earley, T. J., Hergarden, A. C., et al. (2002). A heat-sensitive TRP channel expressed in keratinocytes. *Science (New York, NY)*, *296*, 2046–2049.

Prado, W. A., Schiavon, V. F., & Cunha, F. Q. (2002). Dual effect of local application of nitric oxide donors in a model of incision pain in rats. *European Journal of Pharmacology*, *441*, 57–65.

Rameau, G. A., Chiu, L. Y., & Ziff, E. B. (2004). Bidirectional regulation of neuronal nitric-oxide synthase phosphorylation at serine 847 by the N-methyl-D-aspartate receptor. *The Journal of Biological Chemistry*, *279*, 14307–14314.

Romero, T. R., Guzzo, L. S., Perez, A. C., Klein, A., & Duarte, I. D. (2012). Noradrenaline activates the NO/cGMP/ATP-sensitive K(+) channels pathway to induce peripheral antinociception in rats. *Nitric Oxide: Biology and Chemistry/Official Journal of The Nitric Oxide Society*, *26*, 157–161.

Sachs, D., Cunha, F. Q., & Ferreira, S. H. (2004). Peripheral analgesic blockade of hypernociception: Activation of arginine/NO/cGMP/protein kinase G/ATP-sensitive K+ channel pathway. *Proceedings of the National Academy of Sciences of the United States of America*, *101*, 3680–3685.

Schmidtko, A., Gao, W., Konig, P., Heine, S., Motterlini, R., Ruth, P., et al. (2008). cGMP produced by NO-sensitive guanylyl cyclase essentially contributes to inflammatory and neuropathic pain by using targets different from cGMP-dependent protein kinase I. *The Journal of Neuroscience: The Official Journal of the Society for Neuroscience*, *28*, 8568–8576.

Smith, A. B., & Cunnane, T. C. (1997). Multiple calcium channels control neurotransmitter release from rat postganglionic sympathetic nerve terminals. *The Journal of Physiology*, *499*(Pt 2), 341–349.

Smith, G. D., Gunthorpe, M. J., Kelsell, R. E., Hayes, P. D., Reilly, P., Facer, P., et al. (2002). TRPV3 is a temperature-sensitive vanilloid receptor-like protein. *Nature*, *418*, 186–190.

Song, H. K., Pan, H. L., & Eisenach, J. C. (1998). Spinal nitric oxide mediates antinociception from intravenous morphine. *Anesthesiology*, *89*, 215–221.

Sousa, A. M., & Prado, W. A. (2001). The dual effect of a nitric oxide donor in nociception. *Brain Research*, *897*, 9–19.

Tanabe, M., Nagatani, Y., Saitoh, K., Takasu, K., & Ono, H. (2009). Pharmacological assessments of nitric oxide synthase isoforms and downstream diversity of NO signaling in the maintenance of thermal and mechanical hypersensitivity after peripheral nerve injury in mice. *Neuropharmacology*, *56*, 702–708.

Tao, F., Tao, Y. X., Mao, P., Zhao, C., Li, D., Liaw, W. J., et al. (2003). Intact carrageenan-induced thermal hyperalgesia in mice lacking inducible nitric oxide synthase. *Neuroscience, 120*, 847–854.

Tegeder, I., Schmidtko, A., Niederberger, E., Ruth, P., & Geisslinger, G. (2002). Dual effects of spinally delivered 8-bromo-cyclic guanosine mono-phosphate (8-bromo-cGMP) in formalin-induced nociception in rats. *Neuroscience Letters, 332*, 146–150.

Terenghi, G., Riveros-Moreno, V., Hudson, L. D., Ibrahim, N. B., & Polak, J. M. (1993). Immunohistochemistry of nitric oxide synthase demonstrates immunoreactive neurons in spinal cord and dorsal root ganglia of man and rat. *Journal of the Neurological Sciences, 118*, 34–37.

Wink, D. A., Hanbauer, I., Krishna, M. C., DeGraff, W., Gamson, J., & Mitchell, J. B. (1993). Nitric oxide protects against cellular damage and cytotoxicity from reactive oxygen species. *Proceedings of the National Academy of Sciences of the United States of America, 90*, 9813–9817.

Woolf, C. J., & Thompson, S. W. (1991). The induction and maintenance of central sensitization is dependent on N-methyl-D-aspartic acid receptor activation; implications for the treatment of post-injury pain hypersensitivity states. *Pain, 44*, 293–299.

Wu, S. Y., & Dun, N. J. (1996). Potentiation of IPSCs by nitric oxide in immature rat sympathetic preganglionic neurones in vitro. *The Journal of Physiology, 495*(Pt 2), 479–490.

Xu, X. J., Dalsgaard, C. J., & Wiesenfeld-Hallin, Z. (1992). Spinal substance P and N-methyl-D-aspartate receptors are coactivated in the induction of central sensitization of the nociceptive flexor reflex. *Neuroscience, 51*, 641–648.

Yamada, K., & Inagaki, N. (2005). Neuroprotection by KATP channels. *Journal of Molecular and Cellular Cardiology, 38*, 945–949.

Yoon, Y. W., Sung, B., & Chung, J. M. (1998). Nitric oxide mediates behavioral signs of neuropathic pain in an experimental rat model. *Neuroreport, 9*, 367–372.

Yoshida, T., Inoue, R., Morii, T., Takahashi, N., Yamamoto, S., Hara, Y., et al. (2006). Nitric oxide activates TRP channels by cysteine S-nitrosylation. *Nature Chemical Biology, 2*, 596–607.

Yuen, K. C., Baker, N. R., & Rayman, G. (2002). Treatment of chronic painful diabetic neuropathy with isosorbide dinitrate spray: A double-blind placebo-controlled cross-over study. *Diabetes Care, 25*, 1699–1703.

Zhou, H. Y., Chen, S. R., Byun, H. S., Chen, H., Li, L., Han, H. D., et al. (2012). N-methyl-D-aspartate receptor- and calpain-mediated proteolytic cleavage of K+-Cl-cotransporter-2 impairs spinal chloride homeostasis in neuropathic pain. *The Journal of Biological Chemistry, 287*, 33853–33864.

Zhou, H. Y., Chen, S. R., & Pan, H. L. (2011). Targeting N-methyl-D-aspartate receptors for treatment of neuropathic pain. *Expert Review of Clinical Pharmacology, 4*, 379–388.

Zhou, H. Y., Zhang, H. M., Chen, S. R., & Pan, H. L. (2007). Increased nociceptive input rapidly modulates spinal GABAergic transmission through endogenously released glutamate. *Journal of Neurophysiology, 97*, 871–882.

Zhou, H. Y., Zhang, H. M., Chen, S. R., & Pan, H. L. (2008). Increased C-fiber nociceptive input potentiates inhibitory glycinergic transmission in the spinal dorsal horn. *The Journal of Pharmacology and Experimental Therapeutics, 324*, 1000–1010.

Zhuo, M., Meller, S. T., & Gebhart, G. F. (1993). Endogenous nitric oxide is required for tonic cholinergic inhibition of spinal mechanical transmission. *Pain, 54*, 71–78.

CHAPTER TWO

microRNA and Human Inducible Nitric Oxide Synthase

Zhong Guo, David A. Geller[1]
Department of Surgery, University of Pittsburgh, Pittsburgh, Pennsylvania, USA
[1]Corresponding author: e-mail address: gellerda@upmc.edu

Contents

Abstract

Regulation of human inducible nitric oxide synthase (iNOS) expression involves both transcriptional and posttranscriptional mechanisms. Human iNOS gene transcription is controlled in a cell type-specific manner by extracellular cytokines. Transcriptional regulation of human iNOS gene involves transcription factors NF-κB, Stat-1, AP-1, C/EBPβ, KLF6, Oct 1, and NRF. Important posttranscriptional mechanisms also regulate human iNOS mRNA stability through RNA binding proteins HuR, TTP, KSRP, and PABP. Recently, there are several miRNAs that were validated to regulate human and rodent iNOS gene expression. Among them, miR-939 and miR-26a were identified to bind with the human iNOS 3′-UTR and exert a translational blockade of human iNOS protein synthesis.

1. INTRODUCTION

The expression of inducible nitric oxide synthase (iNOS) can be induced in various cells and tissues following exposure to immunologic and inflammatory stimuli such as cytokines or lipopolysaccharide (LPS) (Kleinert, Pautz, Linker, & Schwarz, 2004). Human iNOS expression in primary human hepatocytes was originally identified by stimulating with a

Vitamins and Hormones, Volume 96
ISSN 0083-6729
http://dx.doi.org/10.1016/B978-0-12-800254-4.00002-7

cytokine mixture (CM) of TNF-α, IL-1β, IFN-γ, and LPS (Nussler et al., 1992). Subsequently, the human iNOS cDNA was cloned from LPS and cytokine-stimulated primary human hepatocytes (Geller et al., 1993). This cloned gene reveals a 4145-bp cDNA, containing a 3459-bp open reading frame that encodes 1153 amino acids with a 131-kDa molecular mass. The genomic cloning of human iNOS clarified that the completed human iNOS gene consists of 26 exons and 25 introns spanning 37 kb of genomic region and specifically mapped to chromosome 17 at position 17 cen-q11.2 (Chartrain et al., 1994).

2. REGULATION OF HUMAN iNOS GENE

2.1. Transcriptional

The expression of human iNOS is controlled in large part by transcriptional mechanisms. The functional promoter region of human iNOS gene can be extended to 16 kb (deVera et al., 1996). The validated binding sites for numerous transcription factors such as AP-1, C/EBP, CREB, GATA, HIF, IRF-1, NF-AT, NF-κB, NF-IL6, Oct-1, PARP1, p53, Sp1, KLF6, STAT-1α, and YY1 (Pautz et al., 2010). There is a classical cytokine-induced enhancer between −5 and −6 kb of human iNOS promoter (Guo, Shao, Du, Park, & Geller, 2007). The human iNOS core promoter contains a TATA box about 30 bp from the transcription start site. Near the TATA box, it also contains binding sites for the transcription factors NF-κB and C/EBP β (Pautz et al., 2010).

In human AKN-1, A549, or DLD1 cells, only human iNOS promoter fragments larger than −3.8 kb showed any significant induction with cytokines (deVera et al., 1996; Taylor et al., 1998). We have further shown that TNF-α or IL-1β signal through NF-κB binding to −5.5, −5.8, and −6.1 kb cis-acting DNA elements, while IFN-γ signals through Stat-1 by binding to motifs at −5.2 and −5.8 kb in the human iNOS promoter (Ganster, Taylor, Shao, & Geller, 2001). Another group showed that cytokine-responsiveness required 5′-flanking DNA regions extending to −8 kb and demonstrated CM-inducible activating protein (AP)-1 binding sites at −5.1 and −5.3 kb as well as a functional role for a NF-κB element located at −8.2 kb in the human iNOS promoter (Marks-Konczalik, Chu, & Moss, 1998). Interestingly, NF-κB repressing factor protein binding to a negative response element at −6.7 kb in the human iNOS promoter region was recognized to mediate constitutive silencing of human iNOS transcription (Feng et al., 2002). A far-upstream functional Oct-1 motif was also identified at

−10.2 kb in the human iNOS promoter that regulates cytokine-induced human iNOS gene transcription (Park, Guo, Shao, Du, & Geller, 2009).

The induction of hiNOS expression requires a CM including TNF-α, IL-1β, and IFN-γ (Taylor & Geller, 2000). Activation of NF-κB pathway, JAK2–STAT-1α pathway, and MAPK pathways (p42/44 MAPK, p38MAPK, JNK) are important for hiNOS expression (Ganster & Geller, 2000; Kleinert et al., 2004).

2.2. Posttranscriptional

RNA binding proteins (RNA-BP) HuR, TTP, and KSRP have been shown to be involved in the posttranscriptional mechanisms of human iNOS regulation (Pautz et al., 2010). Cytokine induction can result in *in vitro* binding of HuR to the iNOS mRNA and thus leads to iNOS mRNA stabilization (Linker et al., 2005). The KH-type splicing regulatory protein (KSRP) showed the destabilizing effect on human iNOS mRNA expression as a human iNOS 3′-UTR binding protein (Linker et al., 2005). The zinc-finger protein tristetraprolin (TTP) regulates iNOS mRNA expression indirect via an interaction with KSRP. Cytokine induction can enhance TTP–KSRP interaction and this results in reduced *in vitro* binding of KSRP to the human iNOS mRNA (Linker et al., 2005). The cytosolic poly (A)-binding protein (PABP) is recently identified to bind to the 5′-UTR and 3′-UTR of hiNOS mRNA and to posttranscriptionally enhance human iNOS mRNA stability (Casper et al., 2013).

Besides HuR, TTP, KSRP, and PABP, other RNA-BP such as hnRNP A1, PTB, and TIAR can also increase human iNOS expression by enhancement of iNOS mRNA stability. Noteworthy, the ARE/poly (U)-binding/degradation factor 1 (AUF1, also named hnRNP D), seems to regulate human iNOS expression by modulation of iNOS mRNA stability, whereas all four isoforms of the ARE/poly (U)-binding/degradation factor 1 (AUF1) negatively regulate human iNOS expression by destabilization of the mRNA (Pautz et al., 2010).

3. miRNAs REGULATION

microRNAs (miRNA) are short (~21) nucleotides that are complementary to 3′-UTR mRNA sequences and have been widely reported to regulate human gene expression (Zeng, Yi, & Cullen, 2003; John et al., 2004).

In primary human hepatocytes, cytokines induce human iNOS mRNA, protein, and NO synthesis. In contrast, primary human macrophages as well as certain human tumor cell lines demonstrate iNOS mRNA following cytokine stimulation, yet human iNOS protein is difficult to detect (Chan et al., 2005). Likewise, in primary human cardiac myocytes, human iNOS mRNA was readily seen after cytokine stimulation, but iNOS protein and NO synthesis were not identified (Luss et al., 1997). These observations are consistent with a translational blockade and suggest the possible role of miRNA in exerting negative posttranscriptional regulation.

3.1. miR-939

miR-939 was originally cloned from human cervical cancer cells (Lui, Pourmand, Patterson, & Fire, 2007). It has two mature forms: hsa-miR-939-5p and hsa-miR-939-3p. It is located at 8q24.3 genomic locus and is also imbedded in cleavage and polyadenylation specific factor 1 (CPSF1) gene. One group reported that miR-939 regulates the replication of H1N1 influenza virus in MDCK cells (Liu, Song, & Huang, 2010). We identified adjacent miR-939 binding sites in the hiNOS 3′-UTR and further proved that miR-939 binds *in vitro* and *in vivo* to exert a translational blockade of cytokine-stimulated hiNOS protein expression in primary human hepatocytes (Guo et al., 2012).

The hiNOS 3′-UTR confers posttranscriptional repression of basal and cytokine-induced hiNOS transcriptional activity in an orientation-dependent manner. miRNA-939 decreases cytokine-induced hiNOS protein expression, but does not affect hiNOS mRNA levels or hiNOS mRNA stability. The hiNOS 3′-UTR contains two functional miR-939 binding sites at +99 and +112 base pairs that are critical for miR-939 mediated translational blockade. Cytokines can induce endogenous miR-939 expression and consequently, miR-939/hiNOS mRNA complex in human hepatocytes.

Therefore, we proposed a model regarding the functional role of miR-939 in the posttranscriptional regulation of the iNOS gene in human hepatocytes (Fig. 2.1). Human primary hepatocytes can be stimulated by a combination of cytokines (TNF-α, IL-1β, and IFN-γ) to strongly express human iNOS mRNA. The cytokines activate specific transcription factors NF-κB, Stat-1, AP1, and C/EBPβ, which functionally interact with their corresponding cis-acting DNA binding sites to drive human iNOS transcription. Meanwhile, the same cytokines also increase miR-939 levels

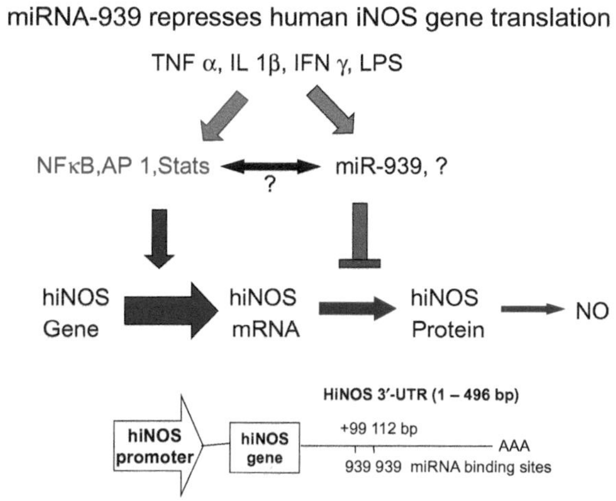

Figure 2.1 Proposed model of human iNOS gene regulation in human hepatocytes with miR-939 exerting a translational blockade for iNOS protein synthesis by binding to the iNOS 3′-UTR. (See the color plate.)

which bind to two specific miR-939 binding sites in the human iNOS 3′-UTR leading to translational inhibition of CM-induced human iNOS protein expression. Binding to both sites is likely required for maximal translational repression, because mutation of either binding site partially abrogates the inhibitory effect of exogenous miR-939 on luciferase activity. Cytokine induction of a negative miRNA regulator of human iNOS expression would theoretically serve to protect the host against untoward consequences of prolonged human iNOS overexpression in a check-and-balance system.

3.2. other miRNAs

Besides miR-939, miR-146a was also reported to regulate human iNOS gene (Table 2.1). However, unlike direct binding of miR-939 to the human iNOS 3′-UTR, miR-146a was shown to be indirectly involved in the regulation of iNOS gene expression. Exogenous supplementation of synthetic miR-146a significantly modulates inflammatory cytokines and pain-related molecules (e.g., TNFα, COX-2, iNOS, IL-6, IL8, RANTS, and ion channel, TRPV1) in human glial cells (Li et al., 2011). Recently, miR-26a was identified to suppress the expression of iNOS protein at the posttranscriptional level by interacting with 3′-UTR in cancer cells (Zhu et al., 2013).

There are at least four miRNAs that have been validated to regulate rodent iNOS genes (Table 2.1). Dai et al. reported that miR-146a, a negative regulator of Toll-like receptor (TLR) signaling, was decreased in

Table 2.1 Summary of miRNAs in regulating iNOS genes in different species

Specie	miRNA	Regulation	Mechanism	References
Human	miR-939	Down	Translational blockage by binding 3′-UTR	Guo et al. (2012)
	miR-146a	Down	Modulating inflammatory cytokines	Li et al. (2011)
	miR-26a	Down	Translational blockage by binding 3′-UTR	Zhu et al. (2013)
Mouse	miR-146a	Down	Negative regulator of Toll-like receptor (TLR) signaling	Dai et al. (2008)
	miR-155	Up	Inhibition of cytokine signal (SOCS)-1	Wang et al. (2009)
	miR-27b	Up	Targeting KSRP to increase iNOS mRNA stability	Zhou, Gong, Eischeid, and Chen (2012)
Rat	miR-34b	Down	Its inhibitor prevents IL-1β induced iNOS gene	Abouheif et al. (2010)

freshly isolated splenic lymphocytes from estrogen-treated mice. Increasing the activity of miR-146a significantly inhibited LPS-induced IFN-γ and iNOS expression in mouse splenic lymphocytes. Enhancing the activity of miR-146a also inhibited the expression of LPS-induced iNOS and NO (Dai et al., 2008). Additionally, a recent report indicates that miR-155 expression was increased in MKP-1-deficient macrophages compared with wild-type macrophages. Transfect on of miR-155 attenuated the expression of Suppressor of Cytokine Signal (SOCS)-1 and subsequently enhanced the expression of iNOS (Wang et al., 2009). miR-27b regulates the stabilization of iNOS mRNA through targeting KSRP (Zhou et al., 2012). The expression of miR-34a was significantly upregulated by IL-1β. Silencing of miR-34a significantly prevented IL-1β-induced down-regulation of Col2a1, as well as IL-1β-induced upregulation of rat iNOS gene (Abouheif et al., 2010). These studies indicate that specific miRNAs (miR-155, miR-146a, or miR-34) can indirectly up- or downregulate iNOS expression by altering upstream signal transduction pathways that subsequently effect rodent iNOS expression while miR-27b can suppress KSRP and affect iNOS mRNA stability.

4. CONCLUSION

Transcriptional and posttranscriptional regulations are two major mechanisms for hiNOS gene regulation. Pathways in human iNOS expression vary in different cells and tissues. Transcription factors such as NF-κB, AP 1, and STAT-1α functionally binding to their corresponding elements in hiNOS promoter are necessary for human iNOS transcription in most human cells. However, a quite complex network of RNA-BP (AUF1, HuR, KSRP, PTB, TIAR, TTP, and PABP) is involved in the posttranscriptional regulation of human iNOS expression. Moreover, recent data also indicates that regulation of human iNOS expression is also controlled by miRNAs such as miR-939, miR-146a, and miR-26a.

REFERENCES

Abouheif, M. M., Nakasa, T., Shibuya, H., Niimoto, T., Kongcharoensombat, W., & Ochi, M. (2010). Silencing microRNA-34a inhibits chondrocyte apoptosis in a rat osteoarthritis model in vitro. *Rheumatology*, *49*(11), 2054–2060.

Casper, I., Nowag, S., Koch, K., Hubrich, T., Bollmann, F., Henke, J., et al. (2013). Post-transcriptional regulation of the human inducible nitric oxide synthase (iNOS) expression by the cytosolic poly(A)-binding protein (PABP). *Nitric Oxide*, *33*, 6–17.

Chan, G. C., Fish, J. E., Mawji, I. A., Leung, D. D., Rachlis, A. C., & Marsden, P. A. (2005). Epigenetic basis for the transcriptional hyporesponsiveness of the human inducible nitric oxide synthase gene in vascular endothelial cells. *The Journal of Immunology*, *175*, 3846–3861.

Chartrain, N., Geller, D. A., Koty, P. P., Sitrin, N. F., Nussler, A. K., Hoffman, E. P., et al. (1994). Molecular cloning, structure, and chromosomal mapping of the human inducible nitric oxide synthase gene. *The Journal of Biological Chemistry*, *269*, 6765–6772.

Dai, R., Phillips, R. A., Zhang, Y., Khan, D., Crasta, O., & Ahmed, S. A. (2008). Suppression of LPS-induced Interferon-gamma and nitric oxide in splenic lymphocytes by select estrogen-regulated microRNAs: A novel mechanism of immune modulation. *Blood*, *112*(12), 4591–4597.

deVera, M. E., Shapiro, R. A., Nussler, A. K., Mudgett, J. S., Simmons, R. L., Morris, S. M., et al. (1996). Transcriptional regulation of human inducible nitric oxide synthase (NOS2) gene by cytokines: Initial analysis of the human NOS2 promoter. *Proceedings of the National Academy of Sciences of the United States of America*, *93*, 1054–1059.

Feng, X., Guo, Z., Nourbakhsh, M., Hauser, H., Ganster, R., Shao, L., et al. (2002). Identification of a negative response element in the human inducible nitric oxide synthase (hiNOS) promoter: The role of NF-κB repressing factor (NRF) in basal repression of the hiNOS gene. *Proceedings of the National Academy of Sciences of the United States of America*, *99*, 14212–14217.

Ganster, R. W., & Geller, D. A. (2000). *Nitric oxide*. San Diego, CA: Academic Press.

Ganster, R. W., Taylor, B. S., Shao, L., & Geller, D. A. (2001). Complex regulation of human iNOS gene transcription by Stat 1 and NF-kB. *Proceedings of the National Academy of Sciences of the United States of America*, *98*, 8638–8643.

Geller, D. A., Lowenstein, C. J., Shapiro, R. A., Nussler, A. K., Di Silvio, M., Wang, S. C., et al. (1993). Molecular cloning and expression of inducible nitric oxide synthase from human hepatocytes. *Proceedings of the National Academy of Sciences of the United States of America, 90*, 3491–3495.

Guo, Z., Shao, L., Du, Q., Park, K. S., & Geller, D. A. (2007). Identification of a classic cytokine-induced enhancer upstream in the human iNOS promoter. *FASEB Journal, 21*(2), 535–542.

Guo, Z., Shao, L., Zheng, L., Du, Q., Li, P., John, B., et al. (2012). miRNA-939 regulates human inducible nitric oxide synthase posttranscriptional gene expression in human hepatocytes. *Proceedings of the National Academy of Sciences of the United States of America, 109*(15), 5826–5831.

John, B., Enright, A. J., Aravin, A., Tuschl, T., Sander, C., & Marks, D. S. (2004). Human microRNA targets. *PLoS Biology, 2*(11), e363.

Kleinert, H., Pautz, A., Linker, K., & Schwarz, P. M. (2004). Regulation of the expression of inducible nitric oxide synthase. *European Journal of Pharmacology, 500255–266.*

Li, X., Gibson, G., Kim, J. S., Kroin, J., Xu, S., van Wijnen, A. J., et al. (2011). MicroRNA-146a is linked to pain-related pathophysiology of osteoarthritis. *Gene, 480*, 34–41.

Linker, K., Pautz, A., Fechir, M., Hubrich, T., Greeve, J., & Kleinert, H. (2005). Involvement of KSRP in the post-transcriptional regulation of human iNOS expression-complex interplay of KSRP with TTP and HuR. *Nucleic Acids Research, 33*, 4813–4827.

Liu, H., Song, L., & Huang, W. (2010). MiR26a and miR939 regulate the replication of H1N1 influenza virus in MDCK cell. *Wei Sheng Wu Xue Bao, 50*, 1399–1405.

Lui, W. O., Pourmand, N., Patterson, B. K., & Fire, A. (2007). Patterns of known and novel small RNAs in human cervical cancer. *Cancer Research, 67*, 6031–6043.

Luss, H., Li, R., Shapiro, R. A., Tzeng, E., McGowan, F. X., Yoneyama, T., et al. (1997). Dedifferentiated human ventricular cardiac myocytes express inducible nitric oxide synthase mRNA but not protein in response to IL-1, TNF, IFNg, & LPS. *Journal of Molecular and Cellular Cardiology, 29*, 1153–1165.

Marks-Konczalik, J., Chu, S. C., & Moss, J. (1998). Cytokine-mediated transcriptional induction of the human inducible nitric oxide synthase gene requires both activator protein 1 and nuclear factor kappaB-binding sites. *The Journal of Biological Chemistry, 273*, 22201–22208.

Nussler, A. K., DiSilvio, M., Billiar, T. R., Hoffman, R. A., Geller, D. A., Selby, R., et al. (1992). Stimulation of the nitric oxide pathway in human hepatocytes by cytokines and endotoxin. *The Journal of Experimental Medicine, 176*, 261–264.

Park, K. S., Guo, Z., Shao, L., Du, Q., & Geller, D. A. (2009). A far-upstream Oct-1 motif regulates cytokine-induced transcription of the human inducible nitric oxide synthase gene. *Journal of Molecular Biology, 390*, 595–603.

Pautz, A., Art, J., Hahn, S., Nowag, S., Voss, C., & Kleinert, H. (2010). Regulation of the expression of inducible nitric oxide synthase. *Nitric Oxide, 23*, 75–93.

Taylor, B. S., de Vera, M. E., Ganster, R. W., Wang, Q., Shapiro, R. A., Billiar, T. R., et al. (1998). A novel NF-κB enhancer element regulates cytokine induction of the human inducible nitric oxide synthase gene promoter. *The Journal of Biological Chemistry, 273*, 15148–15156.

Taylor, B. S., & Geller, D. A. (2000). Molecular regulation of the human inducible nitric oxide synthase (iNOS) gene (review). *Shock, 13*, 413–424.

Wang, X., Zhao, Q., Matta, R., Meng, X., Liu, X., Liu, C. G., et al. (2009). Inducible nitric-oxide synthase expression is regulated by mitogen-activated protein kinase phosphatase-1. *The Journal of Biological Chemistry, 284*, 27123–27134.

Zeng, Y., Yi, R., & Cullen, B. R. (2003). miRNAs and small interfering RNAs can inhibit mRNA expression by similar mechanisms. *Proceedings of the National Academy of Sciences of the United States of America, 100*, 9779–9784.

Zhou, R., Gong, A. Y., Eischeid, A. N., & Chen, X. M. (2012). miR-27b targets KSRP to coordinate TLR4-mediated epithelial defense against Cryptosporidium parvum infection. *PLoS Pathogens*, *8*, e1002702.

Zhu, H., Vishwamitra, D., Curry, C. V., Manshouri, R., Diao, L., Khan, A., et al. (2013). NPM-ALK up-regulates iNOS expression through a STAT3/microRNA-26a-dependent mechanism. *The Journal of Pathology*, *230*, 82–94.

CHAPTER THREE

Heart Mitochondrial Nitric Oxide Synthase: A Strategic Enzyme in the Regulation of Cellular Bioenergetics

Tamara Zaobornyj[1], Laura B. Valdez
Institute of Biochemistry and Molecular Medicine (IBIMOL), Physical Chemistry Division, School of Pharmacy and Biochemistry, University of Buenos Aires, Buenos Aires, Argentina
[1]Corresponding author: e-mail address: tamaraz@ffyb.uba.ar

Contents

Abstract

Heart mitochondria play a central role in cell energy provision and in signaling. Nitric oxide (NO) is a free radical which exerts an integral regulation of the cardiovascular system not only by adapting vascular smooth muscle tone but also by influencing ion channel function, myocyte contraction, energy metabolism, and hypertrophic myocardial remodeling. This chapter analyzes the available data about heart mitochondrial NOS (mtNOS) activity and identity. The regulation of heart mtNOS by the distinctive mitochondrial environment is described by showing the effects of Ca^{2+}, O_2, L-arginine,

Vitamins and Hormones, Volume 96
ISSN 0083-6729
http://dx.doi.org/10.1016/B978-0-12-800254-4.00003-9

NADPH, mitochondrial membrane potential (ΔΨ) and the metabolic states. Evidence about the regulation of heart mtNOS in chronic hypoxia and ischemia–reperfusion models is presented. Functional implications of heart mitochondrial NOS are delineated with emphasis on the chemical reactions through which NO interacts with mitochondrial targets and exerts some of its crucial roles.

1. INTRODUCTION

Under physiological conditions, mitochondria provide the cells with both the energy and the signals that command cell death and survival (Darley-Usmar, 2004). Indeed, heart mitochondria also play an important role in the production of reactive oxygen species, such as superoxide anion and hydrogen peroxide (O_2^- and H_2O_2, Boveris, Oshino, & Chance, 1972), and in Ca^{2+} handling (Liu & O'Rourke, 2009). A fine control of mitochondrial respiration is critical to meet energy demands of cardiac muscle. During the progression of heart disease, the loss of mitochondrial function is a key mediator of cell injury and death (Aon, Cortassa, Marbán, & O'Rourke, 2003).

Mitochondrial O_2 uptake and the coupled ATP regeneration for the homeostatic adjustment to cell metabolic needs constitute an essential and basic biological process in aerobic mammalian cells. The process of oxidative phosphorylation is wholly mitochondrial and encompasses the energy transduction of the chemiosmotic mechanism. This latter is accomplished by the specific electron and proton movements in the mitochondrial complexes I, III, and IV, in which electron transfer is coupled to proton extrusion from the matrix to the intermembrane space through the inner mitochondrial membrane. The second step of the chemiosmotic mechanism involves the exergonic reentry of the extruded protons to the mitochondrial matrix through the F_1-ATP-synthase (complex V) that produces the endergonic ADP phosphorylation.

The discovery that nitric oxide (NO) is the endothelium-derived relaxing factor (EDRF) caused a paradigm shift in the understanding of cardiovascular physiology and pathophysiology (Ignarro, Buga, Wood, Byrns, & Chaudhuri, 1987). The signaling functions of NO in the heart stem from its production by specific NO synthase (NOS) isoforms compartmentalized within different cellular microdomains and from its interactions with several biomolecules involved in distinct downstream signaling pathways (Ziolo & Bers, 2003). Nitric oxide exerts an integral regulation of the cardiovascular system not only by adapting vascular smooth muscle tone but also by influencing ion channel function, myocyte contraction, energy metabolism, and hypertrophic myocardial remodeling (Massion, Feron, Dessy, & Balligand, 2003).

Initially, endothelial NOS (eNOS) was considered to be the only isoform constitutively expressed in cardiac myocytes (Balligand et al., 1995). However, subsequent studies showed that neuronal NOS (nNOS) is targeted to the cardiac sarcoplasmic reticulum close to ion channels (Xu, Huso, Dawson, Bredt, & Becker, 1999). Additionally, iNOS has been shown to be expressed in the myocardium during inflammatory responses implicated in many pathophysiological conditions (Balligand et al., 1994). Later, NOS activity within heart mitochondria (mtNOS), nearby the energy transduction and O_2^- production machinery, was described (Kanai et al., 2001). Strategic localization of heart mtNOS in the organelles in charge of cellular energy metabolism allows for a tight control of different processes whereby mitochondria play pivotal roles.

This chapter focuses on the evidence about the NOS localization and activity in heart mitochondria, about the regulation of this enzyme by the distinctive mitochondrial environment and about the impact of mtNOS in cardiac function under physiological and pathological conditions.

2. HEART MITOCHONDRIAL NO PRODUCTION

In cardiomyocytes, eNOS is localized at the caveolae of the sarcolemma and T-tubules, where it is associated with the scaffolding protein caveolin-3 (Feron et al., 1996; Hare et al., 2000). Neuronal NOS is localized at the sarcoplasmic reticulum and it coimmunoprecipitates with ryanodine receptor (RyR). Inducible NOS is transiently expressed in the heart during ischemia–reperfusion injury (Wildhirt et al., 1999), septicemia (Ullrich et al., 2000), heart failure (Ziolo et al., 2004), as well as during aging (Yang, Larson, & Watson, 1999). During the last 15 years, different groups have used various experimental approaches and reported the presence of mtNOS in the heart (Abadir et al., 2011; Bates, Loesch, Burnstock, & Clark, 1996; Boveris, D'Amico, Lores-Arnaiz, & Costa, 2003; Dedkova & Blatter, 2009; Fellet et al., 2006; Gonzales et al., 2005; Hotta et al., 1999; Kanai et al., 2001; La Padula, Bustamante, Czerniczyniec, & Costa, 2008; Manzo-Avalos et al., 2002; Saavedra-Molina et al., 2003; Valdez et al., 2004; Zanella, Giordano, Muscari, Zini, & Guarnieri, 2004; Zaobornyj, Valdez, La Padula, Costa, & Boveris, 2005; Zenebe, Nazarewicz, Parihar, & Ghafourifar, 2007; Zorov et al., 2004). Such compartmentalization provides a unique condition whereby the enzyme produces NO nearby target effectors within the same subcellular region.

2.1. Heart mtNOS activity

The first study showing the presence of a NOS isoenzyme located in mitochondria used the silver enhanced gold immunolabelling method to stain heart, skeletal muscle, and kidney mitochondria and showed that about 85% of the heart mitochondria were positive for the eNOS label (Bates et al., 1996). This report postulated that eNOS located in mitochondria could exert a ubiquitous regulation of oxidative phosphorylation. Later, two groups almost simultaneously proved the existence of a functional mtNOS in rat liver mitochondria localized at the inner mitochondrial membrane (Ghafourifar & Richter, 1997; Giulivi, Poderoso, & Boveris, 1998). Since then, presence of NO in the mitochondria has been shown by using several methods (Table 3.1). Confocal microscopy using fluorescent probes for coupled mitochondria (MitoTracker) and NO (4,5 diaminofluorescein diacetate or DAF-2/DA) indicated the presence of NO in mitochondria (Lopez-Figueroa et al., 2000). Other studies have detected heart mitochondrial NO release using a direct spin trapping technique with electron paramagnetic resonance (EPR) spectrometry (Giulivi et al., 1998; Hotta et al., 1999). Additionally, NO steady-state concentration within mitochondria, which results from its production and consumption, has been estimated to be in the 100–360 nM NO range; after being measured as 100 nM in perfused rat heart stimulated by bradykinin (Poderoso et al., 1998) and calculated as 200 nM in state 3 and as 360 nM in state 4, considering the NO release from heart energized mitochondria (Boveris et al., 2006).

Heart mtNOS has been studied by functional experiments in which mitochondrial membrane potential-dependent NO production was measured in imaging experiments following permeabilization of cardiomycytes (Dedkova & Blatter, 2009), or directly detected from isolated cardiac mitochondria (Kanai et al., 2001; Zenebe et al., 2007). Studies of cardiac mtNOS have used various methods including immunohistochemistry (Bates et al., 1996; Hotta et al., 1999), spectrophotometry (Boveris et al., 2003; Costa et al., 2002; Fellet et al., 2006; La Padula et al., 2008; Saavedra-Molina et al., 2003; Zaobornyj et al., 2005, 2009), radiometry (Zanella et al., 2002; Zenebe et al., 2007), fluorometry (Dedkova & Blatter, 2009; Zanella et al., 2002; Zenebe et al., 2007; Zorov et al., 2004), chemiluminescence

Table 3.1 Direct and indirect detection of heart mtNOS activity

Methodology	Sensitivity [NO] (M)	References
Espectrophotometric (HbO_2 oxidation)	1×10^{-11}/ 10^{-12}	Costa et al. (2002), Boveris et al. (2003), Saavedra-Molina et al. (2003), Alvarez and Boveris (2004), Valdez et al. (2004), Gonzales et al. (2005), Zaobornyj et al. (2005), Fellet et al. (2006), Boveris, Valdez, Zaobornyj, and Bustamante (2006), Valdez, Zaobornyj, and Boveris (2006), La Padula et al. (2008), Zaobornyj et al. (2009) and Valdez et al. (2011)
EPR	1×10^{-8}	Giulivi et al. (1998)
Radiometric ([^{3}H] L-citrulline/[^{14}C] L-citrulline)	1×10^{-7}	Carreras et al. (2001), Lacza et al. (2001), Peralta et al. (2003), Zanella et al. (2004) and Zenebe et al. (2007)
Microsensor (porphyrinic and amperometric)	1×10^{-7}	Kanai et al. (2001) and Schild et al. (2003)
Fluorometric (DAF-2, DAF2-DA, DAF-FM)	1×10^{-9}	Lopez-Figueroa et al. (2000), Brodsky, Gao, Li, and Goligorsky, 2002, Zanella, Calonghi, Pagnotta, Masotti, and Guarnieri (2002), Zorov, Kobrinsky, Juhaszova, and Sollott (2004) and Dedkova and Blatter (2009)
Griess reaction (NO_2^-/NO_3^-)	1×10^{-6}	Manzo-Avalos et al. (2002)
Chemiluminescence		Zenebe et al. (2007)
Immunohistochemistry		Bates et al. (1996) and Hotta et al. (1999)

(Zenebe et al., 2007), and electrochemistry (Kanai et al., 2001; Table 3.1). Results from our laboratory showed a mtNOS activity of 0.75–1.8 nmol NO/min mg protein in heart submitochondrial membranes, as determined by using the highly sensitive and reliable oxyhemoglobin (HbO_2) assay. Heart mitochondrial NO production accounts for about 60% of total cellular NO generation, suggesting a central role of the mitochondrially produced NO in cardiomyocytes (Valdez & Boveris, 2007; Zaobornyj et al., 2009). NO production has been observed not only in mitochondrial membranes but also in coupled mitochondria as well (Boveris et al., 2006; Valdez et al., 2006).

The activity of heart mtNOS has also been assessed indirectly through the two main effects that NO exerts on mitochondrial respiratory chain function: the inhibition of O_2 consumption and the enhancement in H_2O_2 production. This mtNOS activity has been termed mtNOS functional activity (Valdez, Zaobornyj, & Boveris, 2005). Changes in mtNOS functional activity reflect variations in mitochondrial NO production and steady-state concentration.

2.2. Heart mtNOS identity

The short mitochondrial genome encodes for only few proteins, none of which resembles cytoplasmic NOS. Therefore, if mtNOS is one of the cytoplasmic NOS isoforms, one of the nuclear-encoded isoforms should be transported to the mitochondria after the protein is synthesized in the cytosol. Up to date, the immunochemical nature of heart mtNOS isoform is a subject under debate and all three NOS isoforms have been proposed to be located within heart mitochondria (Table 3.2). However, the most

Table 3.2 Immunochemical detection of heart mtNOS

Sample	Antibody	Technique	Reactivity	Reference
Rat heart mitochondria	Anti-iNOS	WB	+	Valdez et al. (2004)
Rat heart mitochondria	Anti-iNOS/-eNOS	WB	+	Zanella et al. (2004)
Rat heart mitochondria	Anti-iNOS	WB	+	Gonzales et al. (2005)
Rat heart mitochondria	Anti-iNOS	EM	+	Fellet et al. (2006)
Rat heart mitochondria	Anti-iNOS	WB	+	La Padula et al. (2008)
Rat heart mitochondria	Anti-nNOS	WB	+	Zaobornyj et al. (2009)
Guinea pig heart mitochondria	Anti-eNOS	WB	+	Hotta et al. (1999)
Cardiomyocyte mitochondria	Anti-iNOS	EM	+	Buchwalow et al. (2001)
Rabbit heart mitochondria	Anti-nNOS	WB	+	Valdez et al. (2011)

WB, Western blotting; EM, electron microscopy.

convincing data implicate nNOS as the primary candidate for the cytoplasmic NOS izosyme targeted into mitochondria (Dedkova & Blatter, 2009; Kanai et al., 2001; Zaobornyj et al., 2009). Evidence in favor of eNOS (Bates et al., 1996; Hotta et al., 1999; Zanella et al., 2004) and iNOS (Buchwalow et al., 2001; Fellet et al., 2006; Gonzales et al., 2005; La Padula et al., 2008; Valdez et al., 2004) presence in heart mitochondria has also been published.

Fellet et al. (2006) have shown the presence of gold particles in the inner mitochondrial membranes of isolated rat heart using anti-iNOS antibodies by immunogold electron microscopy. In our laboratory, we have detected an increase in heart mtNOS expression after the exposition of rats to high altitude, using anti-nNOS and anti-iNOS antibodies (Gonzales et al., 2005; Zaobornyj et al., 2009). Moreover, La Padula et al. (2008) reported a decrease in heart mtNOS expression during the regression of the cardioprotection conferred by hypoxia, using not only anti-nNOS but also anti-iNOS in Western blot experiments.

One study purified and analyzed mtNOS from rat liver by using MALDI-TOF (Elfering, Sarkela, & Giulivi, 2002). Elfering and coworkers used an anti-iNOS antibody in two-dimensional electrophoresis to identify mtNOS of liver mitochondrial proteome and found that the sequence that best matched the digested protein fragments was the one of α splice variant of the nNOS with acylation with myristic acid at the N-terminal and phosphorylation at the C-terminal region. In addition, distribution of mtNOS transcript indicated its occurrence in brain, heart, muscle, kidney, lung, testis, and spleen.

Another study showed a NOS activity in mouse cardiac mitochondria by measuring the NO production of an individual mitochondrion with a porphyrinic microsensor placed at the cytoplasmic face of the mitochondrial outer membrane. The sensor detected NO upon addition of Ca^{2+} to the reaction medium containing mitochondria isolated from control mice. The NO signal was detected for mitochondria isolated from $eNOS^{-/-}$ or $iNOS^{-/-}$ animals, while no signal was detected in mitochondria from $nNOS^{-/-}$ mice (Kanai et al., 2001). From nNOS alternative-splice variants currently described (Putzke, Seidel, Huang, & Wolf, 2000) only the nNOSα is knocked out in $nNOS^{-/-}$ mice by deleting the exon2 near the N-terminus.

Nevertheless, the exact nature of heart mtNOS isoform is still a controversial matter, a situation that extends to other detected mtNOS activities (Lacza et al., 2003). The 51–57% homology reported for nNOS, iNOS, and eNOS; and the cross-reactivity of isoform-specific anti-NOS antibodies; as well as the possible overexpression of iNOS and eNOS in experimental models could provide an explanation for the conflicting results (Dedkova et al., 2007; Kato & Giulivi, 2006).

3. REGULATION OF HEART mtNOS

A variety of physiological, pathological, and pharmacological situations can cause changes in heart mtNOS activity or expression. Thus, this enzyme seems to be critical during the adaptation of the heart mitochondria to changes in the bioenergetics of the cells. As an example, the mtNOS activity has been found upregulated in cold acclimation (Peralta et al., 2003). In addition, treatment with enalapril, an angiotensin converting enzyme inhibitor, produced an increase in the production of NO by heart mitochondrial membranes (Boveris et al., 2003). Lately, a receptor for angiotensin was found located in inner mitochondrial membrane and it was suggested that the renin–angiotensin system directly regulates mitochondrial NO production (Abadir et al., 2011). Heart mtNOS activity was also found increased in experimental endotoxemia after LPS treatment (Alvarez & Boveris, 2004).

3.1. Effect of substrates and cofactors

Interestingly, localization of NO production within mitochondria provides a distinct specific reciprocal regulation between mtNOS and intramitochondrial Ca^{2+}, pH, L-arginine, O_2, or redox state (Fig. 3.1). Heart mtNOS activity shows a relative requirement for added Ca^{2+}, an absolute requirement for NADPH and L-arginine, and an effective inhibition by the L-arginine homologues Nω-methyl-L-arginine acetate salt (L-NMMA) or Nω-nitro-L-arginine methyl ester hydrochloride (L-NAME) (Alvarez & Boveris, 2004; Valdez et al., 2004; Zaobornyj et al., 2005).

Indeed, heart mtNOS activity is dependent on Ca^{2+} and Mg^{2+} concentration in the reaction medium. Extramitochondrial free Mg^{2+}, a well-known mitochondrial Ca^{2+} uniporter blocker, inhibits NO production in rat heart mitochondria by 60%, in a dose-dependent manner (Manzo-Avalos et al., 2002; Zenebe et al., 2007). This concept is in agreement with other reports (Dedkova & Blatter, 2009; Dedkova, Ji, Lipsius, & Blatter, 2004) showing that blocking the mitochondrial Ca^{2+} uniporter with ruthenium red inhibits mitochondrial NO production. In intact energized mitochondria, elevation of extramitochondrial Ca^{2+} is sufficient to stimulate mtNOS activity and to decrease the respiration rate. Interestingly, results

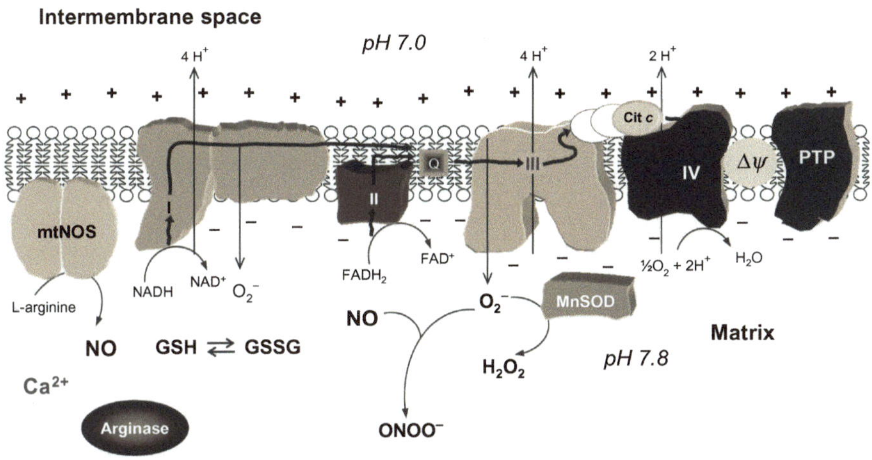

Figure 3.1 *Reciprocal regulation of mtNOS activity and mitochondrial function.* Localization of NOS within mitochondria provides a distinct specific regulation of mitochondrial NO production by intramitochondrial O_2, Ca^{2+}, pH, $\Delta\psi$, L-arginine, arginases, or redox state (GSH/GSGG balance). Nitric oxide produced by mtNOS can readily react with mitochondrial targets such as respiratory complexes I, III, or IV or mitochondrial permeability transition pore. The O_2^- is formed at the respiratory chain and undergoes a very fast reaction with NO to form $ONOO^-$ or it is catabolized by Mn-SOD to form H_2O_2.

from our laboratory showed that succinate or malate/glutamate-energized heart mitochondria produce NO without additional supplementation with Ca^{2+} (Valdez et al., 2006), indicating that the concentration of Ca^{2+} in the mitochondrial preparation may be sufficient to sustain a basal mtNOS activity. Nevertheless, the functional studies performed on cardiac mitochondria isolated from mice (Kanai et al., 2001, 2004) or rats (Manzo-Avalos et al., 2002; Saavedra-Molina et al., 2003; Zanella et al., 2004; Zaobornyj et al., 2005; Zenebe et al., 2007) demonstrated the activation of mitochondrial NO production upon addition of Ca^{2+} to the reaction medium. Concerning the mtNOS activity determined by us in heart mitochondrial membranes, NO production in the absence of Ca^{2+} in the reaction medium was 80% lower than the mitochondrial NO production obtained in the presence of Ca^{2+}. Taking into account that heart mitochondrial matrix Ca^{2+} concentrations differ from cytosolic Ca^{2+} concentrations (Dedkova & Blatter, 2008; Liu & O'Rourke, 2009), the existence of a Ca^{2+}-dependent NOS within mitochondria has significant consequences in terms of the differential regulation of this enzyme.

The mtNOS, same as cytoplasmic NOS isoforms, requires O_2, L-arginine, and certain cofactors to produce NO. The apparent O_2 K_M

for heart mtNOS is 36 μM (Alvarez, Valdez, Zaobornyj, & Boveris, 2003), suggesting that under physiological conditions whereby O_2 concentration is in the 5–20 μM range (Coburn, Ploegmakers, Gondrie, & Abboud, 1973; Gnaiger, Steinlechner-Maran, Mendez, Eberl, & Margreiter, 1995) the mtNOS activity is O_2-limited. To note, the reported K_M values for mtNOS indicate a quite low affinity of mtNOS for O_2, as compared with other NOS (K_M values of eNOS, nNOS, and iNOS are in the range 12–20 μM O_2), certainly a point that deserves further research (Boveris, Carreras, & Poderoso, 2009). In this way, the *in vivo* enzymatic activity of heart mtNOS might be 6–25% of its maximal activity (Boveris & Boveris, 2007). The study of NO production by heart mitochondrial membranes as a function of L-arginine concentration showed a hyperbolic response, with an apparent K_M value of about 35 μM, a concentration that is well within the estimated liver mitochondrial L-arginine levels of 150–300 μM (Alvarez & Boveris, 2004; Valdez et al., 2006). However, rather lower intracellular levels of L-arginine were reported in cardiomyocytes (Desrois, Sciaky, Lan, Cozzone, & Bernard, 2003). Moreover, the expression of arginase-II in mitochondria of mouse cardiac myocytes (Steppan et al., 2006) and arginase-I in feline cardiac myocytes (Jung, Kubo, Wilson, Houser, & Margulies, 2006) has been demonstrated. The activity of mtNOS may be impaired under the conditions whereby L-arginine concentration within the heart mitochondria is diminished.

Considering a total NADP(H) content in heart mitochondria of about 0.82 nmol/mg protein (Boveris & Stoppani, 1970; Chance & Williams, 1956) and assuming an intramitochondrial volume of 2 μl/mg protein, heart mitochondria NADP(H) concentration was calculated and the value obtained was 0.41 mM. Taking into account that the content of this nucleotide in the reduced state (NADPH) in heart mitochondria is about 30% of the total NADP(H), and in view of NOS K_M value for NADPH (0.1–1 μM) (Yui et al., 1991), the intramitochondrial NADPH concentration ($\cong$0.12 mM) is high enough to sustain mtNOS activity under physiological conditions.

3.2. Effect of mitochondrial metabolic state and membrane potential

Mitochondrial NO production is also influenced by mitochondrial metabolic state (Boveris et al., 2006; Valdez & Boveris, 2007; Valdez et al., 2006). During the transition from resting (state 4) to active (state 3) respiration, heart mitochondrial NO release decreases markedly (about 60%), from

2.2 to 1.2 nmol NO/min mg protein. The decrease in NO release observed for the state 4 to state 3 transition is opposite to what could be expected from the matrix acidification that follows the considered transition and the pH dependence of mtNOS activity. If proton concentration functions as a regulatory factor for mtNOS activity, NO release in state 3 (pH 7.5) should be 15% higher than in state 4 (pH 7.8) (Alvarez & Boveris, 2004), and this behavior has not been observed in intact heart mitochondria. In addition, the difference observed in NO release between states 4 and 3 cannot be explained by the variations in mitochondrial NADPH content or by the matrix L-arginine concentration, that follows the transition from state 4 to state 3.

In addition, heart mitochondrial NO release shows a strong and exponential dependence on the mitochondrial transmembrane potential ($\Delta\psi$, Boveris et al., 2006; Valdez et al., 2006). This dependence is more pronounced in the physiological range of $\Delta\psi$, that is between 150 and 180 mV, where small changes in the $\Delta\psi$ cause noticeable variations in mitochondrial NO release. In addition, a marked decrease of mtNOS activity has been observed when $\Delta\psi$ was collapsed by an uncoupler. To date, several studies have shown that abolishing $\Delta\psi$ inhibits mtNOS activity, indicating a tight regulatory interplay between mitochondrial NO production and $\Delta\psi$ (Dedkova & Blatter, 2009; Kanai et al., 2001; Zenebe et al., 2007). Considering that the permeability of mitochondrial membranes to NO is constant, the changes in NO release would reflect the effect of the membrane potential on the rate of NO production by mtNOS.

Interestingly, the rate of NO release by isolated heart mitochondria at state 4 represents about 10% of the corresponding O_2 consumption, whereas at state 3 the rate of NO release represents only about 1.5% of O_2 uptake (Boveris et al., 2006; Valdez et al., 2006). This is consistent with the hypothesis that mitochondrial respiration at physiological O_2 levels is more sensitive to exogenous NO in state 3 than in state 4 (Brookes et al., 2003).

4. EFFECTS OF NO ON HEART MITOCHONDRIAL FUNCTION

The relevance of mtNOS in mitochondrial bioenergetics and cell physiology has been the subject of many studies. While NO is highly diffusible, the distances that this free radical effectively travels are short due to the high reactivity of NO with several species abundantly available in most biological environments which include heme groups, O_2^-, and thiols. Thus cardiac mtNOS, specifically localized at mitochondria, allows the NO produced within these organelles to interact coordinately and restrictedly with the colocalized effectors. Pivotal roles of mtNOS have been postulated in chronic hypoxia, ischemia–reperfusion, cell signaling, aging, dystrophin deficiency, inflammation, and cancer.

4.1. Nitric oxide consumption reactions in mitochondria

It is well known that NO activates soluble guanylate cyclase (sGC) by binding reversibly to the heme iron of sGC and by forming a Fe–nitrosyl complex with the enzyme (Wink & Mitchell, 1998). This activation leads to the production of 3′,5′-cyclic guanosine monophosphate (cGMP) that stimulates protein kinase G (PKG) and cGMP-regulated phosphodiesterases activities (PDE, Layland, Li, & Shah, 2002).

Certain key functions of NO in signaling and regulation of cardiac function are performed through cGMP-independent pathways including those that involve mitochondria. At physiological submicromolar concentrations, most distinct effects of NO on mitochondria are exerted on the respiratory chain. First, NO competes with O_2 for the binding site at the binuclear center of cytochrome *c* oxidoreductase or complex IV. This leads to a high affinity and reversible inhibition of the enzyme (Brown & Cooper, 1994; Cleeter, Cooper, Darley-Usmar, Moncada, & Schapira, 1994). A second important effect of NO on mitochondrial respiratory chain is accomplished through the reaction of NO with the respiratory complex III, ubiquinol-cytochrome *c* oxidase. The exact mechanism underlying this effect in not fully understood. However, it has been shown that interaction of NO with mitochondrial respiratory complex III inhibits the electron transfer and enhances the production of superoxide anion (O_2^-; Poderoso et al., 1996, 1998; Valdez et al., 2005). In the matrix, the Mn-superoxide dismutase (Mn-SOD) catalyzes the reaction of two molecules of O_2^- to form hydrogen peroxide (H_2O_2), an uncharged species involved in the control of cell proliferation and death (Fridovich, 1995).

Other effects of NO in cell signaling are conveyed *via* nitrosation of proteins (Stamler, Lamas, & Fang, 2001). In nitrosation reactions, NO reacts reversibly with the nucleophilic centers in thiol residues contained within a specific consensus sequence of amino acids (Sun, Xin, Eu, Stamler, & Meissner, 2001) of a broad array of low molecular weight compounds (Scharfstein et al., 1994) or proteins (Simon et al., 1996). When mitochondria are treated with NO donors complex I is S-nitrosated resulting in significant inhibition of this respiratory complex (Borutaite, Budriunaite, & Brown, 2000; Burwell, Nadtochiy, Tompkins, Young, & Brookes, 2006).

Nitric oxide also reacts with O_2^- that is formed by the mitochondrial respiratory chain during normal or pathological oxygen metabolism, and forms peroxynitrite ($ONOO^-$) (Radi, Cassina, Hodara, Quijano, & Castro, 2002). This reaction occurs at a first order rate constant of about $2 \times 10^{10}\ M^{-1}\ s^{-1}$ and its velocity is controlled by the diffusion rate of the reactants (Kissner, Nauser, Bugnon, Lye, & Koppenol, 1997). This NO congener may hinder mitochondrial functions and cause cell death. Peroxynitrite also influences cardiac contractility and, in some cases, produces effects that are markedly different from those of NO (Katori et al., 2006; Mihm & Bauer, 2002). The switch from reversible inhibition of cellular respiration by NO to the pathological inhibition of mitochondrial function by NO-derived $ONOO^-$ has been observed in many physiopathological conditions in which intramitochondrial NO and O_2^- steady-state concentrations are enhanced.

4.2. Regulation of mitochondrial function by NO

The presence of NOS in cardiac mitochondria provides a mechanism of fine regulation of the respiratory complexes and enzymes of the citric acid or Kreb´s cycle (Fig. 3.1). The physiological NO release of heart mitochondria has been calculated as about 3.6 nmol NO/min. mg protein, considering the simultaneous oxidation of 3 mol of NADH (malate–glutamate) and 1 mol of succinate within the matrix, and taking into account that heart mitochondria oscillate between an NO-inhibited state 3 (47%) and state 4 (53%) (Boveris & Boveris, 2007). In these conditions, NO release accounts for about 3% of heart O_2 consumption.

Interestingly, the reversible inhibition of O_2 consumption by complex IV (cytochrome oxidase) shows a dependence on both O_2 and NO concentrations, competition between O_2 and NO, and different effects of NO in different mitochondrial metabolic states (Antunes, Boveris, & Cadenas, 2004, 2007). The NO concentrations that produce a half-inhibition of cytochrome oxidase activity and O_2 uptake are in the range of 80–200 nM (Boveris et al., 2000). The effect was described in mitochondria and submitochondrial particles from rat heart (Borutaite & Brown, 1996; Cassina & Radi, 1996; Poderoso et al., 1996) and in isolated beating rat heart (Poderoso et al., 1998). The mitochondrial NO steady-state concentrations in tissues, in the 50–360 nM range (Boveris et al., 2006) are in the same range that produces an effective (50%) inhibition of cytochrome oxidase activity (Cleeter et al., 1994; Boveris, Costa, Cadenas, & Poderoso, 1999). It is worth mentioning that the addition of NO to mitochondria produces, simultaneously with the effect in O_2 uptake, an inhibition of ATP synthesis and a collapse of $\Delta\psi$ (Brookes et al., 2003). The tissue P_{O_2} necessary to maintain high rates of mitochondrial respiration is a subject of permanent interest. The rate of respiration depends on P_{O_2} or O_2 concentration and on the O_2/NO ratio. In other words, the biological effect of a given NO concentration will depend on the simultaneous O_2 concentration; a lower P_{O_2} increases and extends the effects of NO (Cassina & Radi, 1996).

Mitochondria are major sources of reactive oxygen species such as O_2^- and H_2O_2 (Boveris et al., 1972). Through its interaction with components of the electron-transport chain, NO functions not only as a physiological inhibitor of cell respiration, but it also enhances the generation of reactive species (Poderoso et al., 1996) derived from the incomplete reduction of O_2 by mitochondria, and thereby triggers various mechanisms underlying the survival or death of the cells (Moncada & Erusalimsky, 2002). Nitric oxide inhibits the electron transfer chain in the cytochromes *b*–*c* segment in complex III (Poderoso et al., 1996, 1998) possibly by binding to iron–sulfur centers that are intermediates of electron transfers. This reaction increases the level of the semiquinone of ubiquinol (UQH$^\bullet$) that upon autoxidation reduces O_2 to O_2^-.

Moreover, complex I is a target for reversible inhibition by S-nitrosation of critical thiol residues, because its inhibition by NO can be reversed by light or reagents containing thiol groups (Borutaite et al., 2000). It has been suggested that a transient reversible inhibition of the mitochondrial electron transfer minimizes ischemia–reperfusion injury and that blockade of electron transport at complex I preserves respiration during reperfusion (Nadtochiy et al., 2009). Studies have shown that complex I inhibition by nitrosation protects mitochondria during hypoxia and reoxygenation and cardiomyocytes during ischemia–reperfusion (Chen, Moghaddas, Hoppel, & Lesnefsky, 2006; Shiva et al., 2007). There is evidence that S-nitrosation of complex I proteins is associated with cytoprotection from the damage caused by ischemia–reperfusion (Duranski et al., 2005; Prime et al., 2009). These findings are in accordance with the structural and functional interaction observed among complexes I, IV, and mtNOS (Franco et al., 2006; Persichini et al., 2005). A physical proximity of mtNOS with the C-terminal peptide of the Va subunit of cytochrome *c* oxidase was shown by electron microscopic immunolocalization and coimmunoprecipitation studies. Other report showed that not only complex IV but also complex I proteins immunoprecipitate with intramitochondrial and translocated nNOS, which indicates a direct molecular interaction between mtNOS and complexes I and IV. Unpublished data from our laboratory showed that heart mitochondrial inside out particles produce NO (0.62 ± 0.03 nmol/min.mg protein) supported by reversed electron flow. This mtNOS activity was still detectable in the absence of an exogenous electron donor as NADPH, suggesting that NO production could be supported by electrons derived from the respiratory chain. Moreover, rotenone inhibited mtNOS activity in inside out mitochondrial particles, but it did not inhibit the activity of isolated nNOS, indicating that its effect on NO production is due to an electron flow inhibition and not to a direct action on mtNOS structure. These results sustain the hypothesis that mtNOS could interact with complex I proteins using electrons derived from the respiratory chain for its enzymatic activity.

Nitric oxide may also act on other mitochondrial targets. It has been reported that NO increases oxidative phosphorylation efficiency (Clerc, Rigoulet, Leverve, & Fontaine, 2007), activates $mito_{KATP}$ channels (Sasaki, Sato, Ohler, O'Rourke, & Marbán, 2000), modulates mitochondrial permeability transition (MPT) pore formation (Brookes et al., 2000), triggers mitochondrial biogenesis (Nisoli et al., 2003) and regulates mitochondrial matrix pH and Ca^{2+} buffering capacity (Ghafourifar & Richter, 1999). Some reports have established important functional implications of mtNOS activity (Dedkova & Blatter, 2009; Ghafourifar & Richter, 1999; Ghafourifar, Schenk, Klein, & Richter, 1999). Those studies have shown that loading mitochondria with Ca^{2+} decreases matrix pH in a L-methyl-arginine (L-NMMA) sensitive manner (Acuna-Castroviejo, Escames, Leon, Carazo, & Khaldy, 2003). This acidification was related to a decrease in the proton extrusion from the matrix to the intermembrane space and a drop in the $\Delta\psi$, a consequence of the inhibition of the respiratory chain. Moreover, it has been suggested that Ca^{2+} uptake by respiring mitochondria stimulates mtNOS, elevates $ONOO^-$ formation, which in turn causes specific Ca^{2+} release from mitochondria followed by mtNOS inactivation. Pathological, *i.e.* excessive, levels of NO are likely to affect respiration by mechanisms qualitatively different from those observed during reversible physiological regulation. Increased $ONOO^-$ levels together with mitochondrial malfunctioning are hallmarks of hypoxia-reperfusion injury in the heart.

Additionally, exogenous and endogenously produced NO reduce mitochondrial Ca^{2+} accumulation (Dedkova & Blatter, 2005) providing a means by which mtNOS can influence mitochondrial metabolism as well as survival.

Mitochondrial biogenesis can be regulated by the sGC pathway. Then NO can modulate the response to hypoxia, depending on its concentration, through both mitochondrial-dependent and independent pathways (Mateo, Garcia-Lecea, Cadenas, Hernandez, & Moncada, 2003). Nitric oxide triggers mitochondrial biogenesis in several cell types and tissues, including those obtained from the heart, through a cGMP-dependent mechanism (Nisoli et al., 2003). Another postulated mechanism whereby O_2^- and H_2O_2 may be distinguished as prominent mediators of adaptation to hypoxia involves the stabilization of transcription factor hypoxia-inducible factor 1α (HIF-1α).

The cell injury and death observed after reoxygenation of the tissue are in accordance with several reports showing that NO can prevent or accelerate MPT (Crompton, 1999; Wang et al., 2005). This depends on whether NO concentrations are physiological or supraphysiological and whether O_2^- is generated simultaneously (Dedkova & Blatter, 2009). While supraphysiological NO concentrations sensitize mitochondrial permeability transition pore (mPTP), physiological levels of NO inhibit mPTP opening with an IC_{50} of about 10 nM (Brookes et al., 2000). The mPTP opens in response to high matrix Ca^{2+} concentrations, low ATP levels, and enhanced O_2^- production (Makazan, Saini, & Dhalla, 2007). Opening of the pore causes abrupt mitochondrial depolarization, ATP depletion, and necrotic or apoptotic cell death (Bopassa, Michel, Gateau-Roesch, Ovize, & Ferrera, 2005). In this regard, heart and liver mtNOS play important roles in mPTP regulation (Dedkova & Blatter, 2009; Leite et al., 2010; Zorov et al., 2004) suggesting relevance of mtNOS in yet another important mitochondrial function.

Figure 3.1 depicts distinct effects of NO on mitochondria, such as competitive inhibition of respiration, regulation of the rate of production of species derived from O_2 partial reduction, as well as prevention or activation of MPT, and participation in Ca^{2+} homeostasis. Such a wide spectrum of actions indicates that the interaction of NO and mitochondria plays critical roles in heart homeostasis.

4.3. Mitochondrial NOS functional activity

Many studies indicate that mitochondrial NO regulates energy metabolism and inhibits cytochrome oxidase (Antunes et al., 2004; Moncada & Erusalimsky, 2002). In studies with muscle and isolated beating heart, NO donors, and agonists such as bradykinin suppress tissue O_2 uptake and H_2O_2 production (Xie & Wolin, 1996). A recent study showed the presence of angiotensin II Type 2 receptors in inner mitochondrial membrane and postulated that activation of the mitochondrial renin–angiotensin system is coupled to mitochondrial NO production and can modulate respiration (Abadir et al., 2011).

The ability of mtNOS to regulate mitochondrial O_2 uptake and H_2O_2 production through the interaction of NO with the respiratory chain has been named mtNOS functional activity (Boveris et al., 2003; Valdez et al., 2005). This functional activity is determined by the difference in the rates of O_2 uptake or H_2O_2 production in isolated mitochondria or tissue in two conditions (Fig. 3.2). Under the condition when NO steady-state levels are the highest, such as in the presence of sufficient L-arginine and SOD, mitochondrial active O_2 consumption is impaired and mitochondrial H_2O_2 production is enhanced (Fig. 3.2A). On the contrary, when NO steady-state levels are the lowest, mitochondrial active O_2 consumption is increased and mitochondrial H_2O_2 production is decreased (Fig. 3.2B). Such situation is achieved in the absence of NOS substrates or cofactors, or in the presence of an NOS inhibitor or an NOS scavenger, *e.g.*, oxyhemoglobin. The difference between the O_2 consumption rates in these two conditions is expressed directly as a difference or referred as a fraction of state 3 O_2 consumption rate.

Supplementation of heart mitochondria in state 3 with 5 mM L-arginine decreases the respiration rate by 15–20%, while supplementation of the mitochondrial preparation with an NOS inhibitor (5–10 mM L-NAME) increases O_2 consumption by 10%. Concerning H_2O_2 production, the addition of L-arginine (1 mM) enhances H_2O_2 production by 15–20% in heart mitochondria in state 4, whereas the supplementation of the same preparation with L-NAME (5 mM) declines H_2O_2 generation by 10%.

As mentioned before, heart mtNOS functional activity is modified during physiological adaptations and after pharmacological treatments. Changes in mtNOS functional activity reveal modifications of NO production by mtNOS. For instance, treatment with enalapril produced a marked increase (1.9×) in NO production by heart mitochondria associated with an enhancement in mtNOS functional activity, measured both through a decrease in O_2 uptake (1.5×) and an increase in H_2O_2 production (1.9×) (Boveris et al., 2003). In contrast, myocardial stunning was characterized by a decreased mtNOS activity (28%) linked to a impaired mtNOS functional activity (Valdez et al., 2011). In this model, mitochondrial NOS functional activity using malate–glutamate as substrate (51%) was reduced with ischemia (34%), and with 5 min (30%) and 30 min reperfusion

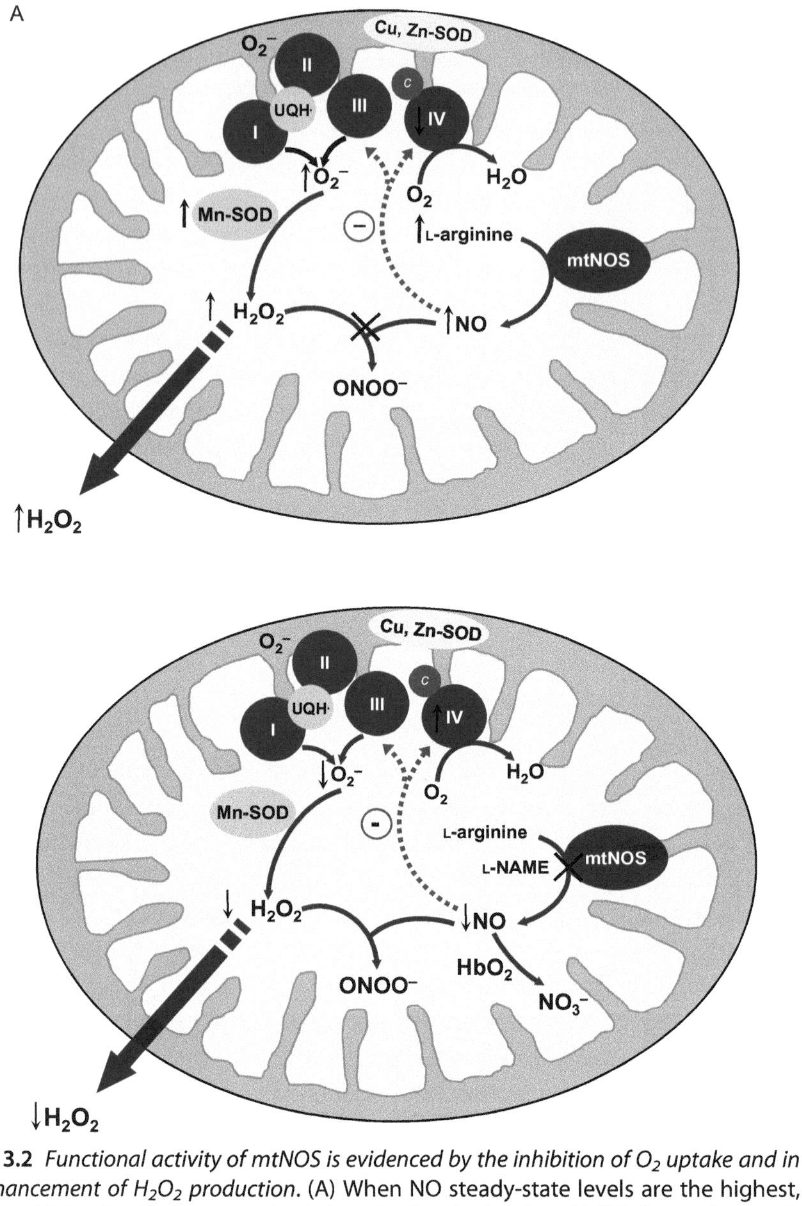

Figure 3.2 *Functional activity of mtNOS is evidenced by the inhibition of O_2 uptake and in the enhancement of H_2O_2 production.* (A) When NO steady-state levels are the highest, mitochondrial active O_2 consumption is impaired and mitochondrial H_2O_2 production is enhanced. (B) When NO steady-state levels are the lowest, mitochondrial active O_2 consumption is increased and mitochondrial H_2O_2 production is decreased.

(26%). A similar effect was observed using succinate as substrate but the lowest values of the functional activity reached were only 27% lower than the initial values, showing a milder decrease than the one observed with malate and glutamate (50%).

5. PHYSIOPATHOLOGICAL REGULATION OF HEART mtNOS

5.1. Heart chronic hypoxia

Adaptation to chronic hypoxia has been shown to confer the heart an improved tolerance to all major deleterious consequences of acute O_2 deprivation, resulting in cardioprotection, in a similar way to preconditioning (Neckar et al., 2005). Nitric oxide has been extensively proposed as one of the molecular messengers involved in this type of adaptation (Baker, Holman, Kalyanaraman, Griffith, & Pritchard, 1999). Studies performed on mitochondrial fractions indicate that chronic hypoxia upregulates mtNOS. Rats submitted to chronic hypoxia using a hypobaric chamber (simulated altitude of 5000 m, $P_{atm}=53.8$ kPa, $P_{O_2}=11.3$kPa) showed right ventricle hypertrophy, increased hematocrit values (La Padula & Costa, 2005), and about 60% enhanced heart mtNOS activity after 18 months of exposure (Zaobornyj et al., 2005). This upregulation was associated with a preservation of the papillary muscle contractile parameters upon aging and improved tolerance to postischemic contractile dysfunction. Using a similar approach, a decrease in heart mtNOS activity and expression during the regression of the cardioprotection conferred by hypoxia has been reported (La Padula et al., 2008). Other studies showed that liver mtNOS activity is upregulated after acute exposure to hypoxia (Lacza et al., 2001; Schild et al., 2003).

Concordant results were observed when rats were exposed to natural high altitude in Cerro de Pasco, Peruvian Andes (4340 m, $P_{atm}=61.3$ kPa, $P_{O_2}=12.2$kPa). Again, rat heart mtNOS activity and expression were increased after 42–84 days of exposure (Gonzales et al., 2005; Zaobornyj et al., 2009). Rats submitted to high altitude also showed right ventricle hypertrophy. The enhancement in heart mtNOS activity showed a pattern similar to the one observed for hematocrit. The effect was selective for heart mtNOS because the activities of liver mtNOS, nonmitochondrial heart NOS, and heart mitochondrial complexes were unchanged.

These observations suggested that the increase in mitochondrial NO release in chronic hypoxia, would allow O_2 to diffuse further and to reach more mitochondria (Liu, Miller, Joshi, Thomas, & Lancaster, 1998; Poderoso et al., 1998) providing a way of increasing O_2 extraction by the tissues. During chronic exposure to hypoxia, when O_2 availability is low, there is an increased requirement for energy. Additional changes in cardiomyocyte to satisfy the energy demands of the cell include an increase in the number of mitochondria, more homogenous distribution of the organelles, and a decrease in the mitochondrial size (Costa, Boveris, Koch, & Taquini, 1988; Nouette-Gaulain et al., 2005). Interestingly, the enhanced mitochondrial generation of reactive oxygen species constitutes another event implicated in the onset of cardioprotection (Bell, Emerling, & Chandel, 2005). Then, the NO produced by mtNOS may act at these stages of the sequence of events implicated in the construction of the cardioprotection established during adaptation to chronic hypoxia.

5.2. Heart acute hypoxia and ischemia-reperfusion

Acute myocardial ischemia/hypoxia can result in myocyte dysfunction and death. Therefore, restoration of the blood flow seems to be essential in order to prevent cell damage. However, the reoxygenation of the hypoxic myocytes is one of the prime mechanisms underlying cell and tissue damage in pathologic conditions including ischemic heart disease (Gottlieb, Burleson, Kloner, Babior, & Engler, 1994). A common feature of reperfusion is the increase in cellular Ca^{2+} concentration. At NO levels such as those reached during ischemia–reperfusion of cardiomyocytes, a reduction of $\Delta\psi$ is produced with the subsequent reduction of Ca^{2+} accumulation, thus protecting the cells from potentially lethal mitochondrial Ca^{2+} overload (Lyon et al., 2010; Rakhit, Mojet, Marber, & Duchen, 2001). In this situation, NO has been indicated as one of the key factors implicated in the onset of cell damage in ischemia/hypoxia and reperfusion (Borutaite & Brown, 1996). Moreover, augmented levels of oxidative species have been found under conditions associated with ischemia/hypoxia and reperfusion (Chen et al., 1998).

An *in vitro* model showed that isolated mitochondria under hypoxia or hypoxia-reoxygenation, independent of other cellular components, respond primarily by increasing the intramitochondrial Ca^{2+} levels through a shift in the balance between ionized and nonionized mitochondrial Ca^{2+} in favor of the ionized form (Zenebe et al., 2007). The mitochondrial production of NO is reduced during ischemia because NOS requires O_2 to generate NO and also because low intracellular pH generated during ischemia inhibits the enzyme. During reperfusion, elevated intramitochondrial ionized Ca^{2+} stimulates mtNOS, which produces a burst of NO that combines with O_2^- to form peroxynitrite ($ONOO^-$), a nonradical species capable of oxidizing and nitrating biomolecules and damaging irreversibly those targets (Radi et al., 2002). Peroxynitrite causes oxidative and nitrative modification of mitochondrial lipids and proteins, release of proapoptotic protein cytochrome *c* from mitochondria, and inactivation of susceptible mitochondrial enzymes (Ghafourifar et al., 1999; Nazarewicz et al., 2007; Radi et al., 2002).

Results from our laboratory using perfused rabbit hearts exposed to controlled *ex vivo* ischemia (15 min) and reperfusion (30 min) showed a decline of about 30% in mitochondrial NO production after ischemia with a recovery of mtNOS activity to control values after reperfusion, in both cases without modification of mtNOS expression (Valdez et al., 2011). Moreover, the pattern observed in mtNOS activity was in agreement with the changes detected in mtNOS functional activity, considering O_2 consumption with malate and glutamate or succinate as substrates. It is conceivable that during acute hypoxia cytochrome oxidase activity was less inhibited than in a normoxic situation. As a consequence, mtNOS activity would be immediately inhibited in the early phase of hypoxia in order to reduce the NO steady-state concentration and to maintain mitochondrial O_2 consumption. When heart reoxygenation occurs, mtNOS activity is up-triggered leading to an enhanced mitochondrial NO production. Considering that Ca^{2+} is likely one rate limiting regulator of mtNOS activity, sequential increases in the mitochondrial matrix Ca^{2+} levels and in NO steady-state concentration constitute a plausible mechanism to explain the results observed in ischemia–reperfusion. This mechanism depicted of mitochondrial damage during ischemia–reperfusion agrees with the observations of the reduction in the activity of respiratory complex I (about 20%) and with the threefold increase in nitrotyrosine levels. Furthermore, mitochondrial dysfunction developed during reperfusion has been named "complex I syndrome," that is, enzymatic complex I inactivation associated with protein nitration and oxidative damage to proteins and phospholipids (Boveris et al., 2009). It is then evident that mitochondrial NO plays a crucial role in oxidative injury and loss of cardiomyocyte viability during ischemia/hypoxia and reperfusion.

The activity of heart mtNOS was also shown to be reduced after hypovolemic shock, that is, acute heart hypoxia, when the effect of 120 min of hemorrhagic bleeding (20% of blood volume) in conscious rats

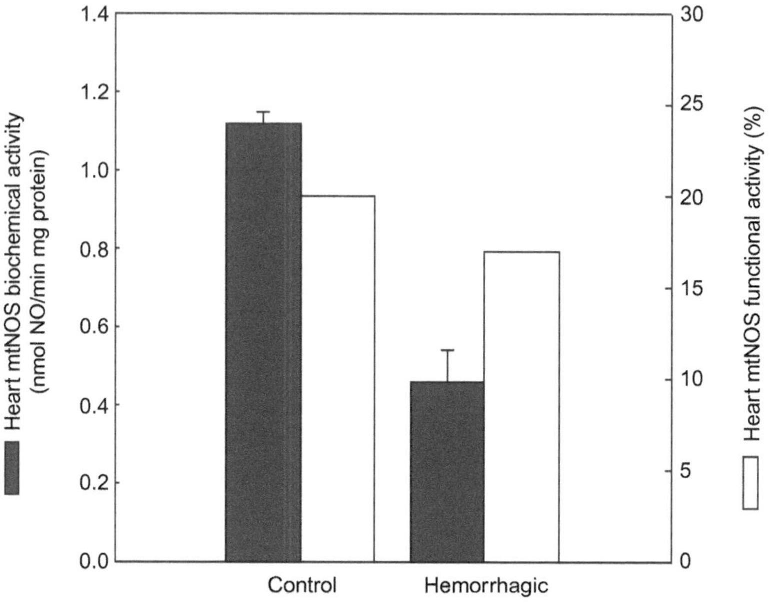

Figure 3.3 *Regulation of mtNOS by hypovolemic shock.* Heart mtNOS biochemical (gray bars) and functional (white bars) activity, measured through the regulation of mitochondrial O_2 consumption, in control and hemorrhagic rats.

was studied. Figure 3.3 shows that heart mtNOS activity declines about 60% with a reduction of mtNOS functional activity in the regulation of O_2 consumption.

6. CONCLUSIONS AND FUTURE DIRECTIONS

Strategic localization of mtNOS in organelles that are in charge of cellular energy metabolism in high energy-utilizing cells such as cardiomyocytes indicates a pivotal control of processes whereby mitochondria play essential roles. Moreover, a spatially restricted intracellular localization of NO within mitochondria permits regulation of NOS activity by the local environment within individual organelles. Certainly, further studies are required in this field for reaching a consensus about the identity of heart mtNOS. The regulation of mtNOS by $\Delta\psi$ and Ca^{2+} allows the enzyme to respond to changes in cardiomyocyte energy homeostasis. Moreover, alterations of mtNOS activity and expression during chronic and acute hypoxia and ischemia–reperfusion support the relevance of this enzyme in long-term heart adaptation to decreased O_2 or in acute ischemic insult. At the same time, NO generated within mitochondria has focused effects on mitochondrial function. Together, heart mtNOS allows NO to optimize the balance between cardiac energy production and utilization, and to regulate processes such as apoptosis, oxygen, and nitrogen free radical production and Ca^{2+} homeostasis.

ACKNOWLEDGMENTS

This work was supported by research grants from the University of Buenos Aires (UBACYT 20020100100606, 20020110100140) and Agencia Nacional de Promoción Científica y Tecnológica (PICT 2008-1138, PICT 2010-844).

REFERENCES

Abadir, P. M., Foster, D. B., Crow, M., Cooke, C. A., Rucker, J. J., Jain, A., et al. (2011). Identification and characterization of a functional mitochondrial angiotensin system. *Proceedings of the National Academy of Sciences of the United States of America*, *108*, 14849–14854.

Acuna-Castroviejo, D., Escames, G., Leon, J., Carazo, A., & Khaldy, H. (2003). Mitochondrial regulation by melatonin and its metabolites. *Advances in Experimental Medicine and Biology*, *527*, 549–557.

Alvarez, S., & Boveris, A. (2004). Mitochondrial nitric oxide metabolism in rat muscle during endotoxemia. *Free Radical Biology and Medicine*, *37*, 1472–1478.

Alvarez, S., Valdez, L. B., Zaobornyj, T., & Boveris, A. (2003). Oxygen dependence of mitochondrial nitric oxide synthase activity. *Biochemical and Biophysical Research Communications*, *305*, 771–775.

Antunes, F., Boveris, A., & Cadenas, E. (2004). On the mechanism and biology of cytochrome oxidase inhibition by nitric oxide. *Proceedings of the National Academy of Sciences of the United States of America*, *101*, 16774–16779.

Antunes, F., Boveris, A., & Cadenas, E. (2007). On the biologic role of the reaction of NO with oxidized cytochrome c oxidase. *Antioxidants & Redox Signaling*, *9*, 1569–1579.

Aon, M. A., Cortassa, S., Marbán, E., & O'Rourke, B. (2003). Synchronized whole cell oscillations in mitochondrial metabolism triggered by a local release of reactive oxygen species in cardiac myocytes. *The Journal of Biological Chemistry*, *278*, 44735–44744.

Baker, J. E., Holman, P., Kalyanaraman, B., Griffith, O. W., & Pritchard, K. A., Jr. (1999). Adaptation to chronic hypoxia confers tolerance to subsequent myocardial ischemia by increased nitric oxide production. *Annals of the New York Academy of Sciences*, *874*, 236–253.

Balligand, J. L., Kobzik, L., Han, X., Kaye, D. M., Belhassen, L., O'Hara, D. S., et al. (1995). Nitric oxide-dependent parasympathetic signaling is due to activation of constitutive endothelial (type III) nitric oxide synthase in cardiac myocytes. *The Journal of Biological Chemistry*, *270*, 14582–14586.

Balligand, J. L., Ungureanu-Longrois, D., Simmons, W. W., Pimental, D., Malinski, T. A., Kapturczak, M., et al. (1994). Cytokine-inducible nitric oxide synthase (iNOS) expression in cardiac myocytes. Characterization and regulation of iNOS expression and detection of iNOS activity in single cardiac myocytes in vitro. *The Journal of Biological Chemistry*, *269*, 27580–27588.

Bates, T. E., Loesch, A., Burnstock, G., & Clark, J. B. (1996). Mitochondrial nitric oxide synthase: A ubiquitous regulator of oxidative phosphorylation? *Biochemical and Biophysical Research Communications*, *218*, 40–44.

Bell, E. L., Emerling, B. M., & Chandel, N. S. (2005). Mitochondrial regulation of oxygen sensing. *Mitochondrion*, *5*, 322–332.

Bopassa, J. C., Michel, P., Gateau-Roesch, O., Ovize, M., & Ferrera, R. (2005). Low-pressure reperfusion alters mitochondrial permeability transition. *The American Journal of Physiology: Heart and Circulatory Physiology, 288*, H2750–H2755.

Borutaite, V., & Brown, G. C. (1996). Rapid reduction of nitric oxide by mitochondria, and reversible inhibition of mitochondrial respiration by nitric oxide. *Biochemical Journal, 315*, 295–299.

Borutaite, V., Budriunaite, A., & Brown, G. C. (2000). Reversal of nitric oxide-, peroxynitrite- and S-nitrosothiol-induced inhibition of mitochondrial respiration or complex I activity by light and thiols. *Biochimica et Biophysica Acta, 1459*, 405–412.

Boveris, A., Costa, L. E., Cadenas, E., & Poderoso, J. J. (1999). Regulation of mitochondrial respiration by adenosine diphosphate, oxygen, and nitric oxide. *Methods in Enzymology, 301*, 188–198.

Boveris, D. L., & Boveris, A. (2007). Oxygen delivery to the tissues and mitochondrial respiration. *Frontiers in Bioscience, 12*, 1014–1023.

Boveris, A., Carreras, M. C., & Poderoso, J. J. (2009). The regulation of cell energetics and mitochondrial signaling by nitric oxide. In L. J. Ignarro (Ed.), *Nitric oxide: Biology and pathobiology* (pp. 441–482). San Diego, CA: Academic Press.

Boveris, A., Costa, L. E., Poderoso, J. J., Carreras, M. C., & Cadenas, E. (2000). Regulation of mitochondrial respiration by oxygen and nitric oxide. *Annals of the New York Academy of Sciences, 899*, 121–135.

Boveris, A., D'Amico, G., Lores-Arnaiz, S., & Costa, L. E. (2003). Enalapril increases mitochondrial nitric oxide synthase activity in heart and liver. *Antioxidants & Redox Signaling, 5*, 691–697.

Boveris, A., Oshino, N., & Chance, B. (1972). The cellular production of hydrogen peroxide. *Biochemical Journal, 128*, 617–630.

Boveris, A., & Stoppani, A. O. M. (1970). *Archives of Biochemistry and Biophysics, 141*, 641.

Boveris, A., Valdez, L. B., Zaobornyj, T., & Bustamante, J. (2006). Mitochondrial metabolic states regulate nitric oxide and hydrogen peroxide diffusion to the cytosol. *Biochimica et Biophysica Acta, 1757*, 535–542.

Brodsky, S. V., Gao, S., Li, H., & Goligorsky, M. S. (2002). Hyperglycemic switch from mitochondrial nitric oxide to superoxide production in endothelial cells. *American Journal of Physiology. Heart and Circulatory Physiology, 283*, H2130–H2139.

Brookes, P. S., Kraus, D. W., Shiva, S., Doeller, J. E., Barone, M. C., Patel, R. P., et al. (2003). Control of mitochondrial respiration by NO*, effects of low oxygen and respiratory state. *The Journal of Biological Chemistry, 278*, 31603–31609.

Brookes, P. S., Salinas, E. P., Darley-Usmar, K., Eiserich, J. P., Freeman, B. A., Darley-Usmar, V. M., et al. (2000). Concentration-dependent effects of nitric oxide on mitochondrial permeability transition and cytochrome c release. *The Journal of Biological Chemistry, 275*, 20474–20479.

Brown, G. C., & Cooper, C. E. (1994). Nanomolar concentrations of nitric oxide reversibly inhibit synaptosomal respiration by competing with oxygen at cytochrome oxidase. *FEBS Letters, 356*, 295–298.

Buchwalow, I. B., Schulze, W., Karczewski, P., Kostic, M. M., Wallukat, G., Morwinski, R., et al. (2001). Inducible nitric oxide synthase in the myocard. *Molecular and Cellular Biochemistry, 217*, 73–82.

Burwell, L. S., Nadtochiy, S. M., Tompkins, A. J., Young, S., & Brookes, P. S. (2006). Direct evidence for S-nitrosation of mitochondrial complex I. *Biochemical Journal, 394*, 627–634.

Carreras, M. C., Peralta, J. G., Converso, D. P., Finocchietto, P. V., Rebagliati, I., Zaninovich, A. A., et al. (2001). Modulation of liver mitochondrial NOS is implicated in thyroid-dependent regulation of O_2 uptake. *American Journal of Physiology. Heart and Circulatory Physiology, 281*, H2282–H2288.

Cassina, A., & Radi, R. (1996). Differential inhibitory action of nitric oxide and peroxynitrite on mitochondrial electron transport. *Archives of Biochemistry and Biophysics*, *328*, 309–316.
Chance, B., & Williams, G. R. (1956). *Advances in Enzymology*, *17*, 65–134.
Chen, Q., Moghaddas, S., Hoppel, C. L., & Lesnefsky, E. J. (2006). Reversible blockade of electron transport during ischemia protects mitochondria and decreases myocardial injury following reperfusion. *Journal of Pharmacology and Experimental Therapeutics*, *319*, 1405–1412.
Chen, Z., Siu, B., Ho, Y. S., Vincent, R., Chua, C. C., Hamdy, R. C., et al. (1998). Overexpression of MnSOD protects against myocardial ischemia/reperfusion injury in transgenic mice. *Journal of Molecular and Cellular Cardiology*, *30*, 2281–2289.
Cleeter, M. W., Cooper, J. M., Darley-Usmar, V. M., Moncada, S., & Schapira, A. H. (1994). Reversible inhibition of cytochrome c oxidase, the terminal enzyme of the mitochondrial respiratory chain, by nitric oxide. Implications for neurodegenerative diseases. *FEBS Letters*, *345*, 50–54.
Clerc, P., Rigoulet, M., Leverve, X., & Fontaine, E. (2007). Nitric oxide increases oxidative phosphorylation efficiency. *Journal of Bioenergetics and Biomembranes*, *39*, 158–166.
Coburn, R. F., Ploegmakers, F., Gondrie, P., & Abboud, R. (1973). Myocardial myoglobin oxygen tension. *American Journal of Physiology*, *224*, 870–876.
Costa, L. E., Boveris, A., Koch, O. R., & Taquini, A. C. (1988). Liver and heart mitochondria in rats submitted to chronic hypobaric hypoxia. *American Journal of Physiology*, *255*, C123–C129.
Costa, L. E., La-Padula, P., Lores-Arnaiz, S., D'Amico, G., Boveris, A., Kurnjek, M. L., et al. (2002). Long-term angiotensin II inhibition increases mitochondrial nitric oxide synthase and not antioxidant enzyme activities in rat heart. *Journal of Hypertension*, *20*, 2487–2494.
Crompton, M. (1999). The mitochondrial permeability transition pore and its role in cell death. *Biochemical Journal*, *341*, 233–249.
Darley-Usmar, V. (2004). The powerhouse takes control of the cell; the role of mitochondria in signal transduction. *Free Radical Biology and Medicine*, *37*, 753–754.
Dedkova, E. N., & Blatter, L. A. (2005). Modulation of mitochondrial Ca^{2+} by nitric oxide in cultured bovine vascular endothelial cells. *American Journal of Physiology - Cell Physiology*, *289*, C836–C845.
Dedkova, E. N., & Blatter, L. A. (2008). Mitochondrial Ca^{2+} and the heart. *Cell Calcium*, *44*, 77–91.
Dedkova, E. N., & Blatter, L. A. (2009). Characteristics and function of cardiac mitochondrial nitric oxide synthase. *The Journal of Physiology*, *587*, 851–872.
Dedkova, E. N., Ji, X., Lipsius, S. L., & Blatter, L. A. (2004). Mitochondrial calcium uptake stimulates nitric oxide production in mitochondria of bovine vascular endothelial cells. *American Journal of Physiology - Cell Physiology*, *286*, C406–C415.
Dedkova, E. N., Wang, Y. G., Ji, X., Blatter, L. A., Samarel, A. M., & Lipsius, S. L. (2007). Signalling mechanisms in contraction-mediated stimulation of intracellular NO production in cat ventricular myocytes. *The Journal of Physiology*, *580*, 327–345.
Desrois, M., Sciaky, M., Lan, C., Cozzone, P. J., & Bernard, M. (2003). Preservation of amino acids during long term ischemia and subsequent reflow with supplementation of L-arginine, the nitric oxide precursor, in the rat heart. *Amino Acids*, *24*, 141–148.
Duranski, M. R., Greer, J. J., Dejam, A., Jaganmohan, S., Hogg, N., Langston, W., et al. (2005). Cytoprotective effects of nitrite during in vivo ischemia–reperfusion of the heart and liver. *The Journal of Clinical Investigation*, *115*, 1232–1240.
Elfering, S. L., Sarkela, T. M., & Giulivi, C. (2002). Biochemistry of mitochondrial nitric oxide synthase. *The Journal of Biological Chemistry*, *277*, 38079–38086.
Fellet, A. L., Balaszczuk, A. M., Arranz, C., Lopez-Costa, J. J., Boveris, A., & Bustamante, J. (2006). Autonomic regulation of pacemaker activity: Role of heart nitric oxide synthases. *The American Journal of Physiology: Heart and Circulatory Physiology*, *291*, H1246–H1254.

Feron, O., Belhassen, L., Kobzik, L., Smith, T. W., Kelly, R. A., & Michel, T. (1996). Endothelial nitric oxide synthase targeting to caveolae. Specific interactions with caveolin isoforms in cardiac myocytes and endothelial cells. *The Journal of Biological Chemistry*, *271*, 22810–22814.

Franco, M. C., Arciuch, V. G., Peralta, J. G., Galli, S., Levisman, D., Lopez, L. M., et al. (2006). Hypothyroid phenotype is contributed by mitochondrial complex I inactivation due to translocated neuronal nitric-oxide synthase. *The Journal of Biological Chemistry*, *281*, 4779–4786.

Fridovich, I. (1995). Superoxide radical and superoxide dismutases. *Annual Review of Biochemistry*, *64*, 97–112.

Ghafourifar, P., & Richter, C. (1997). Nitric oxide synthase activity in mitochondria. *FEBS Letters*, *418*, 291–296.

Ghafourifar, P., & Richter, C. (1999). Mitochondrial nitric oxide synthase regulates mitochondrial matrix pH. *Biological Chemistry*, *380*, 1025–1028.

Ghafourifar, P., Schenk, U., Klein, S. D., & Richter, C. (1999). Mitochondrial nitric-oxide synthase stimulation causes cytochrome c release from isolated mitochondria. Evidence for intramitochondrial peroxynitrite formation. *The Journal of Biological Chemistry*, *274*, 31185–31188.

Giulivi, C., Poderoso, J. J., & Boveris, A. (1998). Production of nitric oxide by mitochondria. *The Journal of Biological Chemistry*, *273*, 11038–11043.

Gnaiger, E., Steinlechner-Maran, R., Mendez, G., Eberl, T., & Margreiter, R. (1995). Control of mitochondrial and cellular respiration by oxygen. *Journal of Bioenergetics and Biomembranes*, *27*, 583–596.

Gonzales, G. F., Chung, F. A., Miranda, S., Valdez, L. B., Zaobornyj, T., Bustamante, J., et al. (2005). Heart mitochondrial nitric oxide synthase is upregulated in male rats exposed to high altitude (4,340 m). *The American Journal of Physiology: Heart and Circulatory Physiology*, *288*, H2568–H2573.

Gottlieb, R. A., Burleson, K. O., Kloner, R. A., Babior, B. M., & Engler, R. L. (1994). Reperfusion injury induces apoptosis in rabbit cardiomyocytes. *The Journal of Clinical Investigation*, *94*, 1621–1628.

Hare, J. M., Lofthouse, R. A., Juang, G. J., Colman, L., Ricker, K. M., Kim, B., et al. (2000). Contribution of caveolin protein abundance to augmented nitric oxide signaling in conscious dogs with pacing-induced heart failure. *Circulation Research*, *86*, 1085–1092.

Hotta, Y., Otsuka-Murakami, H., Fujita, M., Nakagawa, J., Yajima, M., Liu, W., et al. (1999). Protective role of nitric oxide synthase against ischemia–reperfusion injury in guinea pig myocardial mitochondria. *European Journal of Pharmacology*, *380*, 37–48.

Ignarro, L. J., Buga, G. M., Wood, K. S., Byrns, R. E., & Chaudhuri, G. (1987). Endothelium-derived relaxing factor produced and released from artery and vein is nitric oxide. *Proceedings of the National Academy of Sciences of the United States of America*, *84*, 9265–9269.

Jung, A. S., Kubo, H., Wilson, R., Houser, S. R., & Margulies, K. B. (2006). Modulation of contractility by myocyte-derived arginase in normal and hypertrophied feline myocardium. *The American Journal of Physiology: Heart and Circulatory Physiology*, *290*, H1756–H1762.

Kanai, A., Epperly, M., Pearce, L., Birder, L., Zeidel, M., Meyers, S., et al. (2004). Differing roles of mitochondrial nitric oxide synthase in cardiomyocytes and urothelial cells. *The American Journal of Physiology: Heart and Circulatory Physiology*, *286*, H13–H21.

Kanai, A. J., Pearce, L. L., Clemens, P. R., Birder, L. A., Vanbibber, M. M., Choi, S. Y., et al. (2001). Identification of a neuronal nitric oxide synthase in isolated cardiac mitochondria using electrochemical detection. *Proceedings of the National Academy of Sciences of the United States of America*, *98*, 14126–14131.

Kato, K., & Giulivi, C. (2006). Critical overview of mitochondrial nitric-oxide synthase. *Frontiers in Bioscience*, *11*, 2725–2738.

Katori, T., Donzelli, S., Tocchetti, C. G., Miranda, K. M., Cormaci, G., Thomas, D. D., et al. (2006). Peroxynitrite and myocardial contractility: In vivo versus in vitro effects. *Free Radical Biology and Medicine*, *41*, 1606–1618.

Kissner, R., Nauser, T., Bugnon, P., Lye, P. G., & Koppenol, W. H. (1997). Formation and properties of peroxynitrite as studied by laser flash photolysis, high-pressure stopped-flow technique, and pulse radiolysis. *Chemical Research in Toxicology*, *10*, 1285–1292.

Lacza, Z., Puskar, M., Figueroa, J. P., Zhang, J., Rajapakse, N., & Busija, D. W. (2001). Mitochondrial nitric oxide synthase is constitutively active and is functionally upregulated in hypoxia. *Free Radical Biology and Medicine*, *31*, 1609–1615.

Lacza, Z., Snipes, J. A., Zhang, J., Horvath, E. M., Figueroa, J. P., Szabo, C., et al. (2003). Mitochondrial nitric oxide synthase is not eNOS, nNOS or iNOS. *Free Radical Biology and Medicine*, *35*, 1217–1228.

La Padula, P., Bustamante, J., Czerniczyniec, A., & Costa, L. E. (2008). Time course of regression of the protection conferred by simulated high altitude to rat myocardium: Correlation with mtNOS. *Journal of Applied Physiology*, *105*, 951–957.

La Padula, P., & Costa, L. E. (2005). Effect of sustained hypobaric hypoxia during maturation and aging on rat myocardium. I. Mechanical activity. *Journal of Applied Physiology*, *98*, 2363–2369.

Layland, J., Li, J. M., & Shah, A. M. (2002). Role of cyclic GMP-dependent protein kinase in the contractile response to exogenous nitric oxide in rat cardiac myocytes. *The Journal of Physiology*, *540*, 457–467.

Leite, A. C., Oliveira, H. C., Utino, F. L., Garcia, R., Alberici, L. C., Fernandes, M. P., et al. (2010). Mitochondria generated nitric oxide protects against permeability transition via formation of membrane protein S-nitrosothiols. *Biochimica et Biophysica Acta*, *1797*, 1210–1216.

Liu, X., Miller, M. J., Joshi, M. S., Thomas, D. D., & Lancaster, J. R., Jr. (1998). Accelerated reaction of nitric oxide with O2 within the hydrophobic interior of biological membranes. *Proceedings of the National Academy of Sciences of the United States of America*, *95*, 2175–2179.

Liu, T., & O'Rourke, B. (2009). Regulation of mitochondrial Ca^{2+} and its effects on energetics and redox balance in normal and failing heart. *Journal of Bioenergetics and Biomembranes*, *41*, 127–132.

Lopez-Figueroa, M. O., Caamano, C., Morano, M. I., Ronn, L. C., Akil, H., & Watson, S. J. (2000). Direct evidence of nitric oxide presence within mitochondria. *Biochemical and Biophysical Research Communications*, *272*, 129–133.

Lyon, A. R., Joudrey, P. J., Jin, D., Nass, R. D., Aon, M. A., O'Rourke, B., et al. (2010). Optical imaging of mitochondrial function uncovers actively propagating waves of mitochondrial membrane potential collapse across intact heart. *Journal of Molecular and Cellular Cardiology*, *49*, 565–575.

Makazan, Z., Saini, H. K., & Dhalla, N. S. (2007). Role of oxidative stress in alterations of mitochondrial function in ischemic-reperfused hearts. *The American Journal of Physiology: Heart and Circulatory Physiology*, *292*, H1986–H1994.

Manzo-Avalos, S., Perez-Vazquez, V., Ramirez, J., Aguilera-Aguirre, L., Gonzalez-Hernandez, J. C., Clemente-Guerrero, M., et al. (2002). Regulation of the rate of synthesis of nitric oxide by Mg(2+) and hypoxia. Studies in rat heart mitochondria. *Amino Acids*, *22*, 381–389.

Massion, P. B., Feron, O., Dessy, C., & Balligand, J. L. (2003). Nitric oxide and cardiac function: Ten years after, and continuing. *Circulation Research*, *93*, 388–398.

Mateo, J., Garcia-Lecea, M., Cadenas, S., Hernandez, C., & Moncada, S. (2003). Regulation of hypoxia-inducible factor-1alpha by nitric oxide through mitochondria-dependent and -independent pathways. *Biochemical Journal*, *376*, 537–544.

Mihm, M. J., & Bauer, J. A. (2002). Peroxynitrite-induced inhibition and nitration of cardiac myofibrillar creatine kinase. *Biochimie*, *84*, 1013–1019.

Moncada, S., & Erusalimsky, J. D. (2002). Does nitric oxide modulate mitochondrial energy generation and apoptosis? *Nature Reviews. Molecular Cell Biology*, *3*, 214–220.

Nadtochiy, S. M., Burwell, L. S., Ingraham, C. A., Spencer, C. M., Friedman, A. E., Pinkert, C. A., et al. (2009). In vivo cardioprotection by S-nitroso-2-mercaptopropionyl glycine. *Journal of Molecular and Cellular Cardiology*, *46*, 960–968.

Nazarewicz, R. R., Zenebe, W. J., Parihar, A., Larson, S. K., Alidema, E., Choi, J., et al. (2007). Tamoxifen induces oxidative stress and mitochondrial apoptosis via stimulating mitochondrial nitric oxide synthase. *Cancer Research*, *67*, 1282–1290.

Neckar, J., Markova, I., Novak, F., Novakova, O., Szarszoi, O., Ost'adal, B., et al. (2005). Increased expression and altered subcellular distribution of PKC-delta in chronically hypoxic rat myocardium: Involvement in cardioprotection. *The American Journal of Physiology: Heart and Circulatory Physiology*, *288*, H1566–H1572.

Nisoli, E., Clementi, E., Paolucci, C., Cozzi, V., Tonello, C., Sciorati, C., et al. (2003). Mitochondrial biogenesis in mammals: The role of endogenous nitric oxide. *Science*, *299*, 896–899.

Nouette-Gaulain, K., Malgat, M., Rocher, C., Savineau, J. P., Marthan, R., Mazat, J. P., et al. (2005). Time course of differential mitochondrial energy metabolism adaptation to chronic hypoxia in right and left ventricles. *Cardiovascular Research*, *66*, 132–140.

Peralta, J. G., Finocchietto, P. V., Converso, D., Schöpfer, F., Carreras, M. C., & Poderoso, J. J. (2003). Modulation of mitochondrial nitric oxide synthase and energy expenditure in rats during cold acclimation. *The American Journal of Physiology: Heart and Circulatory Physiology*, *284*, H2375–H2383.

Persichini, T., Mazzone, V., Polticelli, F., Moreno, S., Venturini, G., Clementi, E., et al. (2005). Mitochondrial type I nitric oxide synthase physically interacts with cytochrome c oxidase. *Neuroscience Letters*, *384*, 254–259.

Poderoso, J. J., Carreras, M. C., Lisdero, C., Riobo, N., Schöpfer, F., & Boveris, A. (1996). Nitric oxide inhibits electron transfer and increases superoxide radical production in rat heart mitochondria and submitochondrial particles. *Archives of Biochemistry and Biophysics*, *328*, 85–92.

Poderoso, J. J., Peralta, J. G., Lisdero, C. L., Carreras, M. C., Radisic, M., Schöpfer, F., et al. (1998). Nitric oxide regulates oxygen uptake and hydrogen peroxide release by the isolated beating rat heart. *American Journal of Physiology*, *274*, C112–C119.

Prime, T. A., Blaikie, F. H., Evans, C., Nadtochiy, S. M., James, A. M., Dahm, C. C., et al. (2009). A mitochondria-targeted S-nitrosothiol modulates respiration, nitrosates thiols, and protects against ischemia–reperfusion injury. *Proceedings of the National Academy of Sciences of the United States of America*, *106*, 10764–10769.

Putzke, J., Seidel, B., Huang, P. L., & Wolf, G. (2000). Differential expression of alternatively spliced isoforms of neuronal nitric oxide synthase (nNOS) and N-methyl-D-aspartate receptors (NMDAR) in knockout mice deficient in nNOS alpha (nNOS alpha(Delta/Delta) mice). *Molecular Brain Research*, *85*, 13–23.

Radi, R., Cassina, A., Hodara, R., Quijano, C., & Castro, L. (2002). Peroxynitrite reactions and formation in mitochondria. *Free Radical Biology and Medicine*, *33*(1451–1464), 2002.

Rakhit, R. D., Mojet, M. H., Marber, M. S., & Duchen, M. R. (2001). Mitochondria as targets for nitric oxide-induced protection during simulated ischemia and reoxygenation in isolated neonatal cardiomyocytes. *Circulation*, *103*, 2617–2623.

Saavedra-Molina, A., Ramirez-Emiliano, J., Clemente-Guerrero, M., Perez-Vazquez, V., Aguilera-Aguirre, L., & Gonzalez-Hernandez, J. C. (2003). Mitochondrial nitric oxide inhibits ATP synthesis. Effect of free calcium in rat heart. *Amino Acids*, *24*, 95–102.

Sasaki, N., Sato, T., Ohler, A., O'Rourke, B., & Marbán, E. (2000). Activation of mitochondrial ATP-dependent potassium channels by nitric oxide. *Circulation*, *101*, 439–445.

Scharfstein, J. S., Keaney, J. F., Jr., Slivka, A., Welch, G. N., Vita, J. A., Stamler, J. S., et al. (1994). In vivo transfer of nitric oxide between a plasma protein-bound reservoir and low molecular weight thiols. *The Journal of Clinical Investigation, 94*, 1432–1439.

Schild, L., Reinheckel, T., Reiser, M., Horn, T. F., Wolf, G., & Augustin, W. (2003). Nitric oxide produced in rat liver mitochondria causes oxidative stress and impairment of respiration after transient hypoxia. *The FASEB Journal, 17*, 2194–2201.

Shiva, S., Sack, M. N., Greer, J. J., Duranski, M., Ringwood, L. A., Burwell, L., et al. (2007). Nitrite augments tolerance to ischemia/reperfusion injury via the modulation of mitochondrial electron transfer. *The Journal of Experimental Medicine, 204*, 2089–2102.

Simon, D. I., Mullins, M. E., Jia, L., Gaston, B., Singel, D. J., & Stamler, J. S. (1996). Polynitrosylated proteins: Characterization, bioactivity, and functional consequences. *Proceedings of the National Academy of Sciences of the United States of America, 93*, 4736–4741.

Stamler, J. S., Lamas, S., & Fang, F. C. (2001). Nitrosylation. the prototypic redox-based signaling mechanism. *Cell, 106*, 675–683.

Steppan, J., Ryoo, S., Schuleri, K. H., Gregg, C., Hasan, R. K., White, A. R., et al. (2006). Arginase modulates myocardial contractility by a nitric oxide synthase 1-dependent mechanism. *Proceedings of the National Academy of Sciences of the United States of America, 103*, 4759–4764.

Sun, J., Xin, C., Eu, J. P., Stamler, J. S., & Meissner, G. (2001). Cysteine-3635 is responsible for skeletal muscle ryanodine receptor modulation by NO. *Proceedings of the National Academy of Sciences of the United States of America, 98*, 11158–11162.

Ullrich, R., Scherrer-Crosbie, M., Bloch, K. D., Ichinose, F., Nakajima, H., Picard, M. H., et al. (2000). Congenital deficiency of nitric oxide synthase 2 protects against endotoxin-induced myocardial dysfunction in mice. *Circulation, 102*, 1440–1446.

Valdez, L. B., & Boveris, A. (2007). Mitochondrial nitric oxide synthase, a voltage-dependent enzyme, is responsible for nitric oxide diffusion to cytosol. *Frontiers in Bioscience, 12*, 1210–1219.

Valdez, L. B., Zaobornyj, T., Alvarez, S., Bustamante, J., Costa, L. E., & Boveris, A. (2004). Heart mitochondrial nitric oxide synthase. Effects of hypoxia and aging. *Molecular Aspects of Medicine, 25*, 49–59.

Valdez, L. B., Zaobornyj, T., Bombicino, S., Iglesias, D. E., Boveris, A., Donato, M., et al. (2011). Complex I syndrome in myocardial stunning and the effect of adenosine. *Free Radical Biology and Medicine, 51*, 1203–1212.

Valdez, L. B., Zaobornyj, T., & Boveris, A. (2005). Functional activity of mitochondrial nitric oxide synthase. *Methods in Enzymology, 396*, 444–455.

Valdez, L. B., Zaobornyj, T., & Boveris, A. (2006). Mitochondrial metabolic states and membrane potential modulate mtNOS activity. *Biochimica et Biophysica Acta, 1757*, 166–172.

Wang, G., Liem, D. A., Vondriska, T. M., Honda, H. M., Korge, P., Pantaleon, D. M., et al. (2005). Nitric oxide donors protect murine myocardium against infarction via modulation of mitochondrial permeability transition. *The American Journal of Physiology: Heart and Circulatory Physiology, 288*, H1290–H1295.

Wildhirt, S. M., Weismueller, S., Schulze, C., Conrad, N., Kornberg, A., & Reichart, B. (1999). Inducible nitric oxide synthase activation after ischemia/reperfusion contributes to myocardial dysfunction and extent of infarct size in rabbits: Evidence for a late phase of nitric oxide-mediated reperfusion injury. *Cardiovascular Research, 43*, 698–711.

Wink, D. A., & Mitchell, J. B. (1998). Chemical biology of nitric oxide: Insights into regulatory, cytotoxic, and cytoprotective mechanisms of nitric oxide. *Free Radical Biology and Medicine, 25*, 434–456.

Xie, Y. W., & Wolin, M. S. (1996). Role of nitric oxide and its interaction with superoxide in the suppression of cardiac muscle mitochondrial respiration. Involvement in response to hypoxia/reoxygenation. *Circulation, 94*, 2580–2586.

Xu, K. Y., Huso, D. L., Dawson, T. M., Bredt, D. S., & Becker, L. C. (1999). Nitric oxide synthase in cardiac sarcoplasmic reticulum. *Proceedings of the National Academy of Sciences of the United States of America, 96*, 657–662.

Yang, B., Larson, D. F., & Watson, R. (1999). Age-related left ventricular function in the mouse: Analysis based on in vivo pressure-volume relationships. *American Journal of Physiology, 277*, H1906–H1913.

Yui, Y., Hattori, R., Kosuga, K., Eizawa, H., Hiki, K., & Kawai, C. (1991). Purification of nitric oxide synthase from rat macrophages. *The Journal of Biological Chemistry, 266*, 12544–12547.

Zanella, B., Calonghi, N., Pagnotta, E., Masotti, L., & Guarnieri, C. (2002). Mitochondrial nitric oxide localization in H9c2 cells revealed by confocal microscopy. *Biochemical and Biophysical Research Communications, 290*, 1010–1014.

Zanella, B., Giordano, E., Muscari, C., Zini, M., & Guarnieri, C. (2004). Nitric oxide synthase activity in rat cardiac mitochondria. *Basic Research in Cardiology, 99*, 159–164.

Zaobornyj, T., Valdez, L. B., Iglesias, D. E., Gasco, M., Gonzales, G. F., & Boveris, A. (2009). Mitochondrial nitric oxide metabolism during rat heart adaptation to high altitude: Effect of sildenafil, L-NAME, and L-arginine treatments. *The American Journal of Physiology: Heart and Circulatory Physiology, 296*, H1741–H1747.

Zaobornyj, T., Valdez, L. B., La Padula, P., Costa, L. E., & Boveris, A. (2005). Effect of sustained hypobaric hypoxia during maturation and aging on rat myocardium. II. mtNOS activity. *Journal of Applied Physiology, 98*, 2370–2375.

Zenebe, W. J., Nazarewicz, R. R., Parihar, M. S., & Ghafourifar, P. (2007). Hypoxia/reoxygenation of isolated rat heart mitochondria causes cytochrome c release and oxidative stress; evidence for involvement of mitochondrial nitric oxide synthase. *Journal of Molecular and Cellular Cardiology, 43*, 411–419.

Ziolo, M. T., & Bers, D. M. (2003). The real estate of NOS signaling: Location, location, location. *Circulation Research, 92*, 1279–1281.

Ziolo, M. T., Maier, L. S., Piacentino, V., 3rd., Bossuyt, J., Houser, S. R., & Bers, D. M. (2004). Myocyte nitric oxide synthase 2 contributes to blunted beta-adrenergic response in failing human hearts by decreasing Ca^{2+} transients. *Circulation, 109*, 1886–1891.

Zorov, D. B., Kobrinsky, E., Juhaszova, M., & Sollott, S. J. (2004). Examining intracellular organelle function using fluorescent probes: From animalcules to quantum dots. *Circulation Research, 95*, 239–252.

CHAPTER FOUR

Nitric Oxide Regulation of Adult Neurogenesis

William P. Gray*[,1], Angela Cheung†[,1]
*Institute of Psychological Medicine and Clinical Neurosciences, Cardiff University, Cardiff, United Kingdom
†Division of Developmental Neurobiology, MRC National Institute for Medical Research, Mill Hill, London, United Kingdom
[1]Corresponding authors: e-mail address: graywp@cardiff.ac.uk; acheung@nimr.mrc.ac.uk

Contents

Abstract

The ubiquitous gaseous signaling molecule nitric oxide participates in the regulation of a variety of physiological and pathological processes, including adult neurogenesis. Adult neurogenesis, or the generation of new neurons in the adult brain, is a restricted event confined to areas with neurogenic capability. Although nitric oxide has been shown to mediate conflicting effects on adult neurogenesis, which may be partly explained by its unique characteristics, more studies are required in order to fully comprehend and appreciate the mechanisms involved. Neuropeptide Y, a neurotransmitter shown to be an important regulator of adult hippocampal neurogenesis, acts through intracellular nitric oxide to induce an increase in neural progenitor cell proliferation.

1. INTRODUCTION

The ubiquitous signaling molecule nitric oxide (NO) has been directly implicated in the mechanisms underlying both the early stages (neurogenesis) and advanced stages (synaptogenesis and neural map formation)

Vitamins and Hormones, Volume 96
ISSN 0083-6729
http://dx.doi.org/10.1016/B978-0-12-800254-4.00004-0

of neuronal differentiation. Indeed, NO exerts a dual role in cell proliferation by mediating both proliferative and antiproliferative effects (Villalobo, 2006). NO exerts an antiproliferative effect on cells such as vascular smooth muscle cells (Garg & Hassid, 1989) and endothelial cells (RayChaudhury, Frischer, & Malik, 1996), while mediating a proliferative effect on fibroblasts (Du et al., 1997) and myoblasts (Ulibarri, Mozdziak, Schultz, Cook, & Best, 1999). Similarly, NO mediates both neuroinhibitory and neuroproliferative effects on adult neurogenesis and observations have shown the localization of nitric oxide synthase (NOS) to sites of adult neurogenesis. This chapter first presents a short introduction to neurogenesis in the adult brain before the involvement of NO in mediating the action of a key regulator of adult neurogenesis and the seemingly dual role of NO in adult neurogenesis is discussed.

2. ADULT NEUROGENESIS

Prominent during embryonic development, the process by which new neurons are generated, termed neurogenesis, had long been considered as restricted to neural development during the prenatal period. The early theory of a fixed nervous system with "no new nerve cells after birth" and limited capacity for regeneration in the adult mammalian brain was well regarded by early neuroanatomists such as Ramón y Cajal (1928). This theory was widely accepted by the scientific community until around 50 years ago when the first evidence for neurogenesis in the adult mammalian brain was published by Altman (1962) and Altman and Das (1965). The existence of adult neurogenesis has since been confirmed by countless studies, and although initially met with scepticism, it is now a well-accepted phenomenon (Andersen, Morris, Amaral, Bliss, & O'Keefe, 2007; Ehninger & Kempermann, 2008).

Neurogenesis in adults has been confirmed in two regions of the brain, the subgranular zone (SGZ) of the dentate gyrus of the hippocampus (Altman & Das, 1965) and the subventricular zone (SVZ) of the anterior lateral ventricles (Altman, 1969). These two regions (SGZ and SVZ) are described as "neurogenic" or permissive for adult neurogenesis (Fig. 4.1).

Adult hippocampal neurogenesis differs significantly, in precursor cells, mechanism and regulation, from adult neurogenesis in the SVZ/olfactory system (Abrous, Koehl, & Le Moal, 2005). Neural precursor cells within the SGZ of the dentate gyrus of the hippocampus give rise to new granule cell neurons that populate the granule cell layer, while new neurons generated in the SVZ migrate via the rostral migratory stream to the olfactory bulb

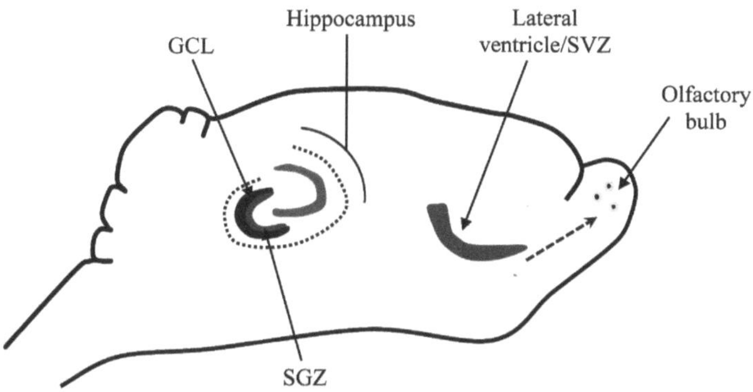

Figure 4.1 Neurogenic regions of the rat brain. The subgranular zone (SGZ; red) of the dentate gyrus of the hippocampus which populates the granule cell layer (GCL; black), and the forebrain subventricular zone (SVZ) which lines the anterior lateral ventricles (green) and provides new neurons for the olfactory bulb, are areas where adult neurogenesis have been confirmed. (See the color plate.)

where they mature into local interneurons (Altman, 1969; Altman & Das, 1965). Unlike embryonic neurogenesis, adult neurogenesis does not occur in the manner of an orchestrated homogenous population event. Instead, adult neurogenesis is highly individualistic, heterogeneous and parallel neuronal cells can be found at all developmental stages at any one time point. A highly complex, dynamic and ongoing process in the brain, adult neurogenesis is interspersed with a range of other physiological processes such as gliogenesis (Steiner et al., 2004).

Adult neurogenesis is well established in mammalian brains, including humans, and rodents in particular show significant amounts of neurogenesis (Altman & Das, 1965; Gould, Beylin, Tanapat, Reeves, & Shors, 1999). In the seminal study by Eriksson et al. (1998), adult human neurogenesis was demonstrated in postmortem brain tissue from cancer patients who had been treated, while alive, with the mitotic (S-phase) marker, bromodeoxyuridine (BrdU), to assess the proliferative activity of tumor cells. Using immunohistochemistry, the presence of BrdU$^+$ cells was verified in both the dentate gyrus and the SVZ, and by using double immunostaining for BrdU and the neuronal markers NeuN or Calbindin, the generation of new neurons from dividing progenitor cells in the dentate gyrus of the adult human hippocampus was confirmed (Eriksson et al., 1998).

The hippocampus plays important roles in memory consolidation and spatial navigation (Squire & Cave, 1991). The functional integration of newly born cells into the hippocampal circuitry during adult hippocampal neurogenesis (van Praag et al., 2002) has led to the proposition that hippocampal neurogenesis may be important in learning processes and the formation of hippocampus-dependent memories (Gould et al., 1999; Snyder, Kee, & Wojtowicz, 2001). Learning has been shown to promote adult hippocampal neurogenesis, which Gould et al. (1999) suggest to be directly involved in the transient (temporary) storage of hippocampus-dependent memories. Most notably, the ablation of adult hippocampal neurogenesis through pharmacological, genetic, or radiological procedures has been linked to the failure of certain hippocampus-dependent tasks (Ehninger & Kempermann, 2008). For example, Shors et al. (2001) ablated hippocampal precursor cells with the methylating agent methylazoxymethanol acetate (MAM) and found an impairment of a hippocampus-dependent eyeblink task but not a hippocampus-independent version of the task. Other hippocampus-dependent tasks, however, were not affected by MAM treatment, suggesting that adult hippocampal neurogenesis is functionally required for only some forms of hippocampus-dependent learning and memory (Shors et al., 2001; Shors, Townsend, Zhao, Kozorovitskiy, & Gould, 2002). More recent work has demonstrated a role for hippocampal neurogenesis in fine pattern separation (Clelland et al., 2009). Although our knowledge of the mechanisms is incomplete, adult hippocampal neurogenesis fluctuates in response to a plethora of extrinsic and intrinsic regulators including exercise (van Praag, Kempermann, & Gage, 1999); age (Kuhn, Dickinson-Anson, & Gage, 1996; Seki & Arai, 1995); stress (Gould & Tanapat, 1999); antidepressants (Malberg, Eisch, Nestler, & Duman, 2000; Santarelli et al., 2003); neurodegenerative diseases (Thompson, Boekhoorn, van Dam, & Lucassen, 2008); brain injury including seizures (Bengzon et al., 1997; Gray & Sundstrom, 1998; Parent et al., 1997), traumatic brain injury (Dash, Mach, & Moore, 2001), and stroke (Liu, Solway, Messing, & Sharp, 1998); hormones (Gould, Tanapat, Rydel, & Hastings, 2000); and growth factors (Cameron, Hazel, & McKay, 1998), to name but a few. Although increased hippocampal neurogenesis does not necessarily equate to enhanced hippocampal function, the possibility of manipulating adult neurogenesis is attractive for the development of potential therapeutic treatments, for example, by activating endogenous stem cells from within to promote repair *in vivo* (Shihabuddin, Palmer, & Gage, 1999).

3. EXPRESSION OF NOS IN NEUROGENIC REGIONS

In order for NO to play a functional role in the regulation of adult neurogenesis, NO itself must be generated by, or be generated in the vicinity of, neural precursor cells. Studies showing the localization of NOS to neurogenic areas support the role for NO in regulating adult neurogenesis (Adachi et al., 2010; Barcia et al., 2012; Islam, Kuraoka, & Kawabuchi, 2003; Moreno-López, Noval, González-Bonet, & Estrada, 2000). Through the use of NOS-specific antibodies and confocal laser scanning microscopy, for example, Islam et al. (2003) demonstrated NOS immunoreactivity intermingled, as well as colocalized, with PSA-NCAM (polysialylated neural cell adhesion molecule)-positive neuronal precursors in the granule cell layer of the dentate gyrus of adult guinea pigs. Similarly, Moreno-López et al. (2000) investigated the expression of neuronal NOS (nNOS) in proliferating (BrdU) and immature PSA-NCAM-positive precursor cells in the SVZ, olfactory and rostral migratory stream of adult mice through immunohistochemical detection. Unlike the observations in the hippocampus however, nNOS expression was only intermingled, but never colocalized, with PSA-NCAM-positive precursors in the proliferation and migration zones, suggesting that nNOS-positive neurons exerted a functional noncell autonomous influence on neuronal progenitors (Moreno-López et al., 2000). Dissimilarities in NOS expression between the SGZ and SVZ are probably not surprising, considering the mechanistic and regulatory differences between adult neurogenesis in these two areas.

4. PHARMACOLOGICAL STUDIES OF NO ON ADULT NEUROGENESIS *IN VIVO*

Many studies have investigated the role of NO in adult neurogenesis through the use of pharmacological agents *in vivo*, for example, through inhibiting the activity of NOS or by applying an exogenous source of NO. *In vivo* investigations by Packer et al. (2003) showed that the chronic inhibition of NOS with L-NAME (*N*(ω)-nitro-L-arginine methyl ester) increased proliferation in neurogenic zones such as the SGZ (+68%) and SVZ (+58%) in rats. Similarly, there was an increase in proliferating cells in response to the systemic administration of L-NAME and a different NOS inhibitor, 7-nitroindazole, in mice, although this effect was observed only in the SVZ and not the SGZ (Moreno-López et al., 2004). This disparity in region response is probably due to the presence of specific cell types or sensitivity differences (Matarredona, Murillo-Carretero, Moreno-López, & Estrada, 2005). These studies all support the idea that NO acts as a negative regulator of adult neurogenesis, although it must be noted that both L-NAME and 7-nitroindazole are relatively nonselective NOS inhibitors (Moore & Handy, 1997; Reiner & Zagvazdin, 1998).

Conversely, others have described NO as a positive regulator of adult neurogenesis. The administration of the NO donor, DETA/NONOate, increased cell proliferation in the SGZ and SVZ under both normal conditions and in response to stroke in young adult rats (Zhang et al., 2001). Likewise, the prior administration of 7-nitroindazole and the inducible NOS (iNOS) inhibitor aminoguanidine significantly reduced the number of proliferating cells in the dentate gyrus after pentylenetrazol-induced seizures in rats (Jiang, Xiao, Wang, Huang, & Zhang, 2004), supporting a role for NO in brain repair after brain injury or seizures. DETA/NONOate has also been shown to produce antidepressant effects by promoting hippocampal neurogenesis in young adult mice (Hua et al., 2008). The explanation for these paradoxical effects of NO may lie in the differential effects of the NOS isoforms (Cárdenas et al., 2005), which highlight the need for more detailed studies into the contribution by individual NOSs to adult neurogenesis.

5. NOS KNOCKOUT ANIMALS AND ADULT NEUROGENESIS

As we have discussed, pharmacological studies have shown that NO initiates both neuroinhibitory and neuroproliferative effects on adult neurogenesis (Cárdenas et al., 2005). These contradictory effects of NO may be explained by the existence of the different NOS subtypes. *In vivo* studies using NOS isoform-specific knockout mice to identify the contributions by each NOS to adult neurogenesis offer higher specificity compared to using pharmacological inhibitors, which can be relatively nonspecific, especially in the case of the endothelial NOS (eNOS) inhibitors. In general, most studies suggest an inhibitory role for NO derived from nNOS, although the cellular mechanisms underlying this are not well understood. Packer et al. (2003), for example, showed that the number of $BrdU^+$ cells was significantly increased in the neurogenic regions (olfactory subependyma and dentate gyrus) of the adult nNOS knockout mouse brain, implying that the endogenous action of nNOS negatively regulates adult neurogenesis. Similarly, Sun et al. (2005) showed that nNOS knockout mice showed reduced infarct size in response to transient focal cerebral ischemia and increased neurogenesis under both basal and ischemia-induced conditions, while Zhou et al. (2007) found that nNOS-derived NO contributes to chronic stress-induced depression by suppressing levels of hippocampal

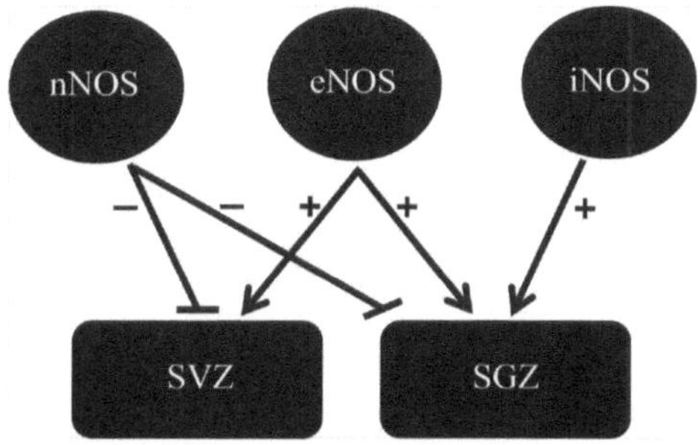

Figure 4.2 The differential effects of the NOS isoforms on adult neurogenesis *in vivo*. Studies suggest that nNOS-derived NO is inhibitory, while eNOS-derived NO is proliferative, for neurogenesis in the SGZ and SVZ under both basal and ischemia-induced conditions. iNOS-derived NO, on the other hand, promotes neurogenesis in the SGZ under ischemic conditions. (See the color plate.)

neurogenesis. On the other hand, proliferative roles have been shown by the eNOS and iNOS isoforms, which seem to be involved in promoting adult neurogenesis under basal and/or ischemic conditions. eNOS-deficient mice, for example, show a significant reduction in neuronal progenitor cell proliferation in the dentate gyrus under basal conditions (Reif et al., 2004) as well as in response to focal cerebral ischemia in the SVZ (Chen et al., 2005). iNOS knockout rodents, meanwhile, show significantly enhanced hippocampal neurogenesis in response to ischemia, but not under basal conditions (Zhu, Liu, Sun, & Lu, 2003). Overall, these studies seem to suggest that nNOS-derived NO is inhibitory, while eNOS-derived NO is proliferative, for neurogenesis in the SGZ and SVZ under both basal and ischemia-induced conditions, and iNOS-derived NO promotes neurogenesis in the SGZ under ischemic conditions (Fig. 4.2).

6. NEUROPEPTIDE Y AND NO

Neuropeptide Y (NPY) is a 36-amino acid polypeptide widely expressed within the peripheral and central nervous system. Named due to the many tyrosine (Y) residues within its structure (Tatemoto, 1982), NPY has been implicated in the regulation of a series of physiological processes including memory processes (Redrobe, Dumont, St-Pierre, & Quirion, 1999), affective disorders (Heilig, 2004) and seizure control (Erickson, Clegg, & Palmiter, 1996; Vezzani, Sperk, & Colmers, 1999). In turn, these processes have been shown to influence, or be influenced by, fluctuations in levels of adult mammalian hippocampal neurogenesis.

NPY is emerging as an important regulator of adult hippocampal neurogenesis under both normal and pathological conditions, although the mechanisms underlying this effect are unknown (Gray, 2008; Howell et al., 2007). Given the bifunctional effects of NO on neurogenesis and the localization of NOS to NPY-responsive neurogenic areas, could NO be a mediator of NPY signaling? Indeed, a connection between NO and the physiological effects of NPY was previously described by Morley, Alshaher, Farr, Flood, and Kumar (1999) and Morley and Flood (1991), who showed that a low dose of NPY was able to significantly increase NOS expression in the hypothalamus and that an NPY-induced increase in food intake was mediated via NOS. More recently, Alvaro et al. (2008) showed that NOS was involved in mediating the NPY-induced proliferation of retinal neural cells, further supporting this hypothesis. Of particular relevance is the study by Cheung, Newland, Zaben, Attard, and Gray (2012) on the neuroproliferative effect of NPY on neural precursor cells *in vitro*, which has shown that NO may indeed play a key role in mediating the effect of NPY.

NPY exerts a purely proliferative effect on nestin-positive precursor cells and β-tubulin-positive neuroblasts derived from the postnatal rat hippocampus, which is mediated via the NPY Y_1 receptor (Howell et al., 2005, 2003). Using a primary hippocampal cell culture system, Cheung et al. (2012) investigated the involvement of NO in mediating the neuroproliferative effect of NPY through the use of a range of pharmacological agonists and antagonists. While the proliferative effect of NPY on hippocampal cultures, as demonstrated through the labeling of proliferating cells using BrdU, was inhibited through the use of the (nonsubtype-selective) NOS inhibitor L-NAME, the substrate for NO synthesis, L-arginine, increased proliferation rates. Additionally, NOS inhibitors with higher selectivity for the different NOS isoforms suggested a leading involvement by the nNOS isoform. Interestingly, this proliferative effect of NPY was mediated via an intracellular release of NO as the external NO scavenger, carboxy-PTIO, exerted no effect on proliferation rates, while the supplementation of extracellular NO with the NO donor DETA/NONOate, on the other hand, exerted a negative effect on cell proliferation. Live-cell imaging studies using the NO-responsive fluorescent dye, DAF-FM DA, showed significantly enhanced DAF-FM DA fluorescence in nestin-positive precursor cells and β-tubulin-positive neuroblasts in response to an NPY pulse over time (Cheung et al., 2012).

As well as the involvement of NO, Cheung et al. (2012) also identified key players downstream of NOS/NO. NO itself activates soluble guanylate cyclase (sGC), synthesizing cGMP from cGTP, which in turn activates the cGMP-dependent protein kinase (PKG) (Cheung et al., 2012). Indeed, cGMP has been shown to be an important factor in mediating the neuroproliferative effects of NO. Neurogenesis in the SVZ and dentate gyrus of rats subjected to focal cerebral ischemia was significantly enhanced in the presence of the drug sildenafil, a phosphodiesterase type 5 inhibitor that increases the intracellular accumulation of cGMP (Zhang et al., 2003), while, as discussed previously, enhanced neurogenesis is observed in response to NO donors, which increases levels of cGMP (Zhang et al., 2001). The NO–cGMP–PKG signaling pathway proposed by Cheung et al. (2012) eventually results in the activation of the mitogen-activated protein kinases (MAPK), ERK (extracellular-regulated protein kinase) 1 and 2. Unlike the NO–cGMP–PKG pathway, ERK1/2 has long been implicated in the processes underlying the neuroproliferative effects of NPY (Alvaro et al., 2008; Hansel, Eipper, & Ronnett, 2001; Howell et al., 2005). ERK1/2 themselves mediate further pathways involved in regulating the expression of genes controlling cell proliferation and differentiation through the phosphorylation of a variety of transcription factors (Cano & Mahadevan, 1995; Lopez-Ilasaca, 1998). Indeed, NO has been shown to be essential for the proliferation of embryonic hippocampal neural stem/progenitor cells (Yoneyama, Kawada, Gotoh, Shiba, & Ogita, 2010) and the findings by Cheung et al. (2012), which identifies NPY as a selective agonist of intracellular NO signaling in postnatal hippocampal precursor cells, unite two significant modulators of adult hippocampal neurogenesis into a common signaling framework (Fig. 4.3).

7. THE DUAL ROLE OF NO IN ADULT NEUROGENESIS

What I hope we have exemplified so far is the highly complex and dual nature of NO in regulating adult neural precursors and adult neurogenesis. We have mentioned previously that while NOS knockout studies have suggested an inhibitory role for NO derived from nNOS, proliferative roles have been linked to the eNOS and iNOS isoforms. It would be simplistic to assume, however, that nNOS-derived NO can only be inhibitory in adult neurogenesis. The cell type, cell source, reactive status, timing of synthesis, and concentration are major determinants of NO's effects (Cárdenas et al., 2005). As suggested by Cárdenas et al. (2005), while low levels and early synthesis of NO, such as by eNOS, may be beneficial through a local vasodilatatory effect, high and sustained levels of NO production by nNOS, for

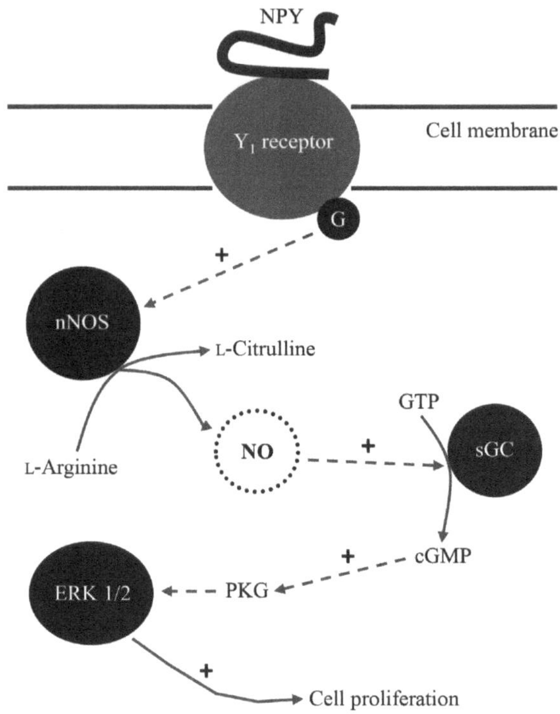

Figure 4.3 Intracellular mechanisms underlying NPY-mediated neuroproliferation. NPY released by interneurons of the dentate hilus act on NPY Y1 receptors in the SGZ to initiate a neuroproliferative effect. The Y1 GPCR activates nNOS, which synthesises NO from L-arginine. The target of NO is sGC, which converts GTP into cGMP. cGMP activates PKG, which in turn leads to the activation of a range of protein targets or kinase pathways, ultimately resulting in ERK 1/2 activation. ERK 1/2 are involved in regulating the expression of genes involved in controlling cell proliferation and differentiation (Cheung et al., 2012). (See the color plate.)

example, may be neurotoxic by contributing to oxidative stress. In fact, Park et al. (2003) demonstrated that only chronic (15 days), and not acute, NOS inhibition had a stimulatory effect on neuronal stem cell proliferation.

NO itself is a highly diffusible signaling molecule that is able to mediate both intracellular and intercellular signaling pathways through intracellular and/or extracellular release (Lancaster, 1997; Wood & Garthwaite, 1994). A previous study by Luo et al. (2010) found that the source of NO (intracellular or extracellular) was important in determining its cellular effects. nNOS-derived NO from neural stem cells acted intracellularly and promoted neural stem cell proliferation, whereas nNOS-derived NO from neurons was released extracellularly into the media, where it exerted antiproliferative effects on neural stem cell proliferation (Luo et al., 2010). Similarly, Cheung et al. (2012) also showed that NO synthesized by nNOS in response to NPY was most likely intracellular and mediated its proliferative effect, whereas extracellular NO was antiproliferative, a finding consistent with the dual role of nNOS-derived NO demonstrated by Luo et al. (2010). While high-level concentrations and sustained NO release are commonly associated with antiproliferative effects (Cárdenas et al., 2005), a low-level and localized release of NO, in response to NPY, for example, may be involved in mediating proliferative pathways. Indeed, a transient decrease in the proliferation of neural precursor cells during the early postnatal period is observed in the olfactory epithelium of nNOS knockout mice (Chen, Tu, Moon, Matarazzo, & Ronnett, 2004), which supports the involvement of nNOS generated NO in mediating neural stem/precursor cell proliferation. This dual role is a common characteristic of NO and may be enforced by the differential subcellular compartmentalization of nNOS, for example, to the nuclei of neural stem cells or to the cytoplasm of neurons (Cheung et al., 2012; Luo et al., 2010), or possibly, the existence of the nNOS splice variants such as nNOSα, nNOSβ, and nNOSγ (Alderton, Cooper, & Knowles, 2001; Corso-Diaz & Krukoff, 2010). Although the involvement of NO was previously suggested to mediate the proliferative effect of NPY on retinal neural cells by Alvaro et al. (2008), the work by Cheung et al. (2012) is the first to suggest that intracellular NO produced by nNOS signals the complete NPY Y_1 receptor-mediated proliferative effect of NPY on hippocampal nestin$^+$ precursor cells, while extracellular NO had the opposite antiproliferative effect.

On a slightly different note, the NPY Y_1 receptor has been shown to modulate NO levels during stroke in rats (Chen, Fung, & Cheung, 2002). Using a middle cerebral artery occlusion (MCAO) stroke model, Chen et al. (2002) showed that the intracerebroventricular injection of NPY or a Y_1 agonist increased the infarct volume, while the Y_1 receptor antagonist BIBP3226 reduced the infarct volume. Using electron paramagnetic resonance spectroscopy to measure NO levels, MCAO was found to increase the relative brain NO concentration to $131.94 \pm 7.99\%$, while NPY treatment increased this further to $250.94 \pm 50.48\%$. BIBP3226, however, significantly reduced the relative brain NO concentration to $69.63 \pm 8.84\%$ (Chen et al., 2002). This study suggests that Y_1-mediated NO generation during cerebral ischemia mediates ischemic damage via NO overproduction (Malinski, Bailey, Zhang, & Chopp, 1993; Matsui, Nagafuji, Kumanishi, & Asano, 1999). Surprisingly, however, ischemic injury has also been associated with increased adult neurogenesis (Lichtenwalner & Parent, 2006). The first to describe this effect was Liu et al. (1998), who observed a 12-fold increase in the number of $BrdU^{+}$ cells in the dentate SGZ after transient global ischemia in the adult gerbil. The proliferative effect of ischemia on adult hippocampal neurogenesis has also been demonstrated in rats (Jin et al., 2001) and mice (Takagi et al., 1999), occurs after a latent period of about 1–2 weeks, and is transient in its effects (Lichtenwalner & Parent, 2006; Liu et al., 1998). Indeed, contrary to the role of NO in mediating ischemic damage, the work by Zhang et al. (2001), as mentioned briefly before, has shown that the administration of the NO donor DETA/NONOate increased cell proliferation and neurogenesis in the dentate gyrus of rats under both basal and ischemic conditions. Cortical levels of cGMP were significantly increased and the administration of DETA/NONOate considerably improved the neurological outcome during recovery from ischemic damage mediated via MCAO (Zhang et al., 2001). Although the mechanisms underlying these processes are complex, this illustrates again the dual role of NO and how a single molecule of NO may be regulated to exert a range of diverse physiological effects. In this instance, the NOS subtype-producing NO and the levels of NO are probably the main determinators. nNOS probably mediates the NO injury in response to ischemia by producing high levels of NO, since, as described previously, nNOS knockout mice show reduced infarct size in response to transient focal cerebral ischemia (Sun et al., 2005).

8. CONCENTRATION-DEPENDENT EFFECTS OF NO

At a cellular level, the NO–cGMP–PKG pathway has been implicated in the regulation of apoptosis and survival in neural cells (Fiscus, 2002). Indeed, with regard to cell survival, NO can have both an apoptotic/necrotic (toxic) effect and an antiapoptotic (protective) effect depending on the NO concentration (Beckman & Koppenol, 1996; Fiscus, 2002). NO at relatively low (submicromolar) concentrations is usually correlated with mediating its protective effects and the activation of sGC leading to cGMP synthesis (Bobba, Atlante, Moro, Calissano, & Marra, 2007; Fiscus, 2002). On the other hand, excess NO production, such as in response to ischemia, is neurotoxic and can lead to cellular damage by inducing oxidative stress (Bobba et al., 2007). The oxidative stress and damage initiated by, for example, peroxynitrite, a reactive species formed as a result of NO reaction with superoxide anion, can ultimately result in cell apoptosis or necrosis (Beckman & Koppenol, 1996; Estévez, Spear, Manuel, et al., 1998). Although the cGMP–PKG pathway contributes to the proapoptotic actions of NO in certain cell types, such as vascular smooth muscle cells (Pollman, Yamada, Horiuchi, & Gibbons, 1996) and vascular endothelial cells (Suenobu, Shichiri, Iwashina, Marumo, & Hirata, 1999), cGMP has been shown to mediate the neuroprotective or antiapoptotic effects of NO in mammalian neural cells such as motor neurons (Estévez, Spear, Thompson, et al., 1998), dorsal root ganglion neurons (Thippeswamy & Morris, 1997), and cerebellar granule neurons (Bobba et al., 2007). Indeed, the addition of extracellular NO (through the application of NO donors) protected serum-deprived PC12 cells (Kim et al., 1999), as well as nerve growth factor-deprived sympathetic neurons (Farinelli, Park, & Greene, 1996), from cell death *in vitro*. Of particular significance, however, is that the neuroprotective effects of the NO donors (sodium nitroprusside or SNAP) were only observed with low concentrations (below 100 μM) of the donor (Farinelli et al., 1996; Kim et al., 1999). Low concentrations of NO activate sGC and cGMP production, which contribute to its neuroprotective effects, while higher concentrations of NO elicit toxic effects through the formation of reactive species (Fiscus, 2002). These findings support the observations by Cheung et al. (2012) regarding the dual role of NO in regulating the proliferation of neuronal precursor cells from the hippocampus. Intracellular NO probably mediates the proliferative effect of NPY, which is likely to be low level and short lived, while the long-term application of extracellular NO through the use of the NO donor DETA/NONOate was antiproliferative (Cheung et al., 2012). Similarly, inhibition of cell proliferation occurred at higher concentrations (100 μM) of DETA/NONOate and showed further inhibition with increased donor concentration (200 μM) (Cheung et al., 2012). In fact, Carreira et al. (2010) had previously reported a dual effect of DETA/NONOate, depending on the concentration, on the proliferation of neural stem cell cultures derived from the mouse SVZ. While a low concentration of DETA/NONOate (10 μM) increased cell proliferation, higher concentrations (100 μM) inhibited cell proliferation (Carreira et al., 2010). The increased cell proliferation in response to low levels of NO donor was blocked by either inhibiting the MAPK pathway with U0126 (Carreira et al., 2010) or inhibiting sGC and PKG (Carreira et al., 2013), further supporting the involvement of the sGC–PKG and ERK1/2 pathway in mediating the proliferative aspects of NO. As well as neural proliferation, low doses of DETA/NONOate (0.1 and 0.4 μM) have also been shown to promote (Chen et al., 2006), while high doses (50 μM) inhibited (Luo et al., 2010) neuronal differentiation and neurite outgrowth. Although the context may be slightly different, i.e., cell proliferation compared to cell survival, the ability to manipulate the effects of NO through regulating its properties (i.e., concentration) and the positive effects of the NO–cGMP–PKG pathway are well demonstrated in both cases.

9. CONCLUSIONS

A decline in adult hippocampal neurogenesis and cell loss has been linked with the pathogenesis of stress (Gould & Tanapat, 1999), depression (Duman, 2004; Malberg et al., 2000; Santarelli et al., 2003), and neurodegenerative diseases such as Alzheimer's disease (Haughey et al., 2002). It is only through furthering our understanding of the mechanisms controlling the proliferation of endogenous neural stem/precursor cells that new and more specific pharmacological targets can be identified for promoting adult neurogenesis, neuronal regeneration, and structural repair in the CNS in an attempt to alleviate these pathological conditions.

NO exerts a dual effect on neurogenesis that depends on many factors, including the NO concentration and NOS isoform. Further studies are required for deciphering the precise molecular mechanisms underlying this dual effect, which are not yet fully understood, although the NPY Y_1 receptor has been identified as a key target to selectively promote NO-mediated neural stem/precursor cell proliferation as a possible therapeutic intervention for promoting hippocampal neurogenesis (Cheung et al., 2012). Other potential strategies to promote neurogenesis include the use of NO-releasing drugs such as NO donors or NO-releasing nonsteroidal anti-inflammatory drugs (Carreira, Carvalho, & Araújo, 2012), but as with all processes involving the Janus-faced NO, many more studies are required to ensure that they are used at a level which will produce the optimal therapeutic effect while minimizing any side effects.

REFERENCES

Abrous, D. N., Koehl, M., & Le Moal, M. (2005). Adult neurogenesis: From precursors to network and physiology. *Physiological Reviews*, *85*, 523–569.

Adachi, M., Abe, M., Sasaki, T., Kato, H., Kasahara, J., & Araki, T. (2010). Role of inducible or neuronal nitric oxide synthase in neurogenesis of the dentate gyrus in aged mice. *Metabolic Brain Disease*, *25*, 419–424.

Alderton, W. K., Cooper, C. E., & Knowles, R. G. (2001). Nitric oxide synthases: Structure, function and inhibition. *Biochemical Journal*, *357*, 593–615.

Altman, J. (1962). Are new neurons formed in the brains of adult mammals? *Science, 135*(3509), 1127–1128.

Altman, J. (1969). Autoradiographic and histological studies of postnatal neurogenesis. IV. Cell proliferation and migration in the anterior forebrain, with special reference to persisting neurogenesis in the olfactory bulb. *Journal of Comparative Neurology, 137*(4), 433–457.

Altman, J., & Das, G. D. (1965). Autoradiographic and histological evidence of postnatal hippocampal neurogenesis in rats. *Journal of Comparative Neurology, 124*, 319–335.

Alvaro, A. R., Martins, J., Araújo, I. M., Rosmaninho-Salgado, J., Ambrósio, A. F., & Cavadas, C. (2008). Neuropeptide Y stimulates retinal neural cell proliferation—Involvement of nitric oxide. *Journal of Neurochemistry, 105*, 2501–2510.

Andersen, P., Morris, R., Amaral, D., Bliss, T., & O'Keefe, J. (2007). *The hippocampus book*. Oxford: Oxford University Press, Inc.

Barcia, J. M., Sanchez-Villarejo, M. V., Almansa, I., Lopez-Pedrajas, R., Muriach, M., & Romero, F. J. (2012). Differential hippocampal response to chronic alcohol consumption of young adult and mature adult rats. *Alcohol and Alcoholism, 47*, 219–224.

Beckman, J. S., & Koppenol, W. H. (1996). Nitric oxide, superoxide and peroxynitrite: The good, the bad, and the ugly. *American Journal of Physiology, 271*, C1424–C1437.

Bengzon, J., Kokaia, Z., Elmer, E., Nanobashvili, A., Kokaia, M., & Lindvall, O. (1997). Apoptosis and proliferation of dentate gyrus neurons after single and intermittent limbic seizures. *Proceedings of the National Academy of Sciences of the United States of America, 94*, 10432–10437.

Bobba, A., Atlante, A., Moro, L., Calissano, P., & Marra, E. (2007). Nitric oxide has dual opposite roles during early and late phases of apoptosis in cerebellar granule neurons. *Apoptosis, 12*, 1597–1610.

Cameron, H. A., Hazel, T. G., & McKay, R. D. (1998). Regulation of neurogenesis by growth factors and neurotransmitters. *Journal of Neurobiology, 36*(2), 287–306.

Cano, E., & Mahadevan, L. C. (1995). Parallel signal processing among mammalian MAPKs. *Trends in Biochemical Sciences, 20*, 117–122.

Cárdenas, A., Moro, M. A., Hurtado, O., Leza, J. C., & Lizasoain, I. (2005). Dual role of nitric oxide in adult neurogenesis. *Brain Research Reviews, 50*, 1–6.

Carreira, B. P., Carvalho, C. M., & Araújo, I. M. (2012). Regulation of injury-induced neurogenesis by nitric oxide. *Stem Cells International, 2012*, 895659.

Carreira, B. P., Morte, M. I., Inácio, A., Costa, G., Rosmaninho-Salgado, J., Agasse, F., et al. (2010). Nitric oxide stimulates the proliferation of neural stem cells bypassing the epidermal growth factor receptor. *Stem Cells, 28*, 1219–1230.

Carreira, B. P., Morte, M. I., Lourenço, A. S., Santos, A. I., Inácio, A., Ambrósio, A. F., et al. (2013). Differential contribution of the guanylyl cyclase-cyclic GMP-protein kinase G pathway to the proliferation of neural stem cells stimulated by nitric oxide. *Neurosignals, 23*, 1–13.

Chen, S. H., Fung, P. C., & Cheung, R. T. (2002). Neuropeptide Y-Y1 receptor modulates nitric oxide level during stroke in the rat. *Free Radical Biology & Medicine, 32*, 776–784.

Chen, J., Tu, Y., Moon, C., Matarazzo, V., & Ronnett, G. V. (2004). The localization of neuronal nitric oxide synthase may influence its role in neuronal precursor proliferation and synaptic maintenance. *Developmental Biology, 269*, 165–182.

Chen, J., Zacharek, A., Li, Y., Li, A., Wang, L., Katakowski, M., et al. (2006). N-cadherin mediates nitric oxide-induced neurogenesis in young and retired breeder neurospheres. *Neuroscience, 140*, 377–388.

Chen, J., Zacharek, A., Zhang, C., Jiang, H., Li, Y., Roberts, C., et al. (2005). Endothelial nitric oxide synthase regulates brain-derived neurotrophic factor expression and neurogenesis after stroke in mice. *The Journal of Neuroscience, 25*, 2366–2375.

Cheung, A., Newland, P. L., Zaben, M., Attard, G. S., & Gray, W. P. (2012). Intracellular nitric oxide mediates neuroproliferative effect of neuropeptide Y on postnatal hippocampal precursor cells. *Journal of Biological Chemistry*, *287*, 20187–20196.

Clelland, C. D., Choi, M., Romberg, C., Clemenson, G. D., Fragniere, A., Tyers, P., et al. (2009). A functional role for adult hippocampal neurogenesis in spatial pattern separation. *Science*, *325*, 210–213.

Corso-Diaz, X., & Krukoff, T. L. (2010). nNOS alpha and nNOS beta localization to aggresome-like inclusions is dependent on HSP90 activity. *Journal of Neurochemistry*, *114*, 864–872.

Dash, P. K., Mach, S. A., & Moore, A. N. (2001). Enhanced neurogenesis in the rodent hippocampus following traumatic brain injury. *Journal of Neuroscience Research*, *63*, 313–319.

Du, M., Islam, M. M., Lin, L., Ohmura, Y., Moriyama, Y., & Fujimura, S. (1997). Promotion of proliferation of murine BALB/C3T3 fibroblasts mediated by nitric oxide at lower concentrations. *International Journal of Biochemistry and Molecular Biology*, *41*, 625–631.

Duman, R. S. (2004). Depression: A case of neuronal life and death? *Biological Psychiatry*, *56*, 140–145.

Ehninger, D., & Kempermann, G. (2008). Neurogenesis in the adult hippocampus. *Cell and Tissue Research*, *331*, 243–250.

Erickson, J. C., Clegg, K. E., & Palmiter, R. D. (1996). Sensitivity to leptin and susceptibility to seizures of mice lacking neuropeptide Y. *Nature*, *381*, 415–421.

Eriksson, P. S., Perfilieva, E., Bjork-Eriksson, T., Alborn, A. M., Nordborg, C., Peterson, D. A., et al. (1998). Neurogenesis in the adult human hippocampus. *Nature Medicine*, *4*, 1313–1317.

Estévez, A. G., Spear, N., Manuel, S. M., Barbeito, L., Radi, R., & Beckman, J. S. (1998). Role of endogenous nitric oxide and peroxynitrite formation in the survival and death of motor neurons in culture. *Progress in Brain Research*, *118*, 269–280.

Estévez, A. G., Spear, N., Thompson, J. A., Cornwell, T. L., Radi, R., Barbeito, L., et al. (1998). Nitric oxide-dependent production of cGMP supports the survival of rat embryonic motor neurons cultured with brain-derived neurotrophic factor. *The Journal of Neuroscience*, *18*, 3708–3714.

Farinelli, S. E., Park, D. S., & Greene, L. A. (1996). Nitric oxide delays the death of trophic factor-deprived PC12 cells and sympathetic neurons by a cGMP-mediated mechanism. *The Journal of Neuroscience*, *16*, 2325–2334.

Fiscus, R. R. (2002). Involvement of cyclic GMP and protein kinase G in the regulation of apoptosis and survival in neural cells. *Neurosignals*, *11*, 175–190.

Garg, U. C., & Hassid, A. (1989). Nitric oxide generating vasodilators and 8-bromo-cyclic guanosine monophosphate inhibit mitogenesis and proliferation of cultured rat vascular smooth muscle cells. *Journal of Clinical Investigation*, *83*, 1774–1777.

Gould, E., Beylin, A., Tanapat, P., Reeves, A., & Shors, T. J. (1999). Learning enhances adult neurogenesis in the hippocampal formation. *Nature Neuroscience*, *2*, 260–265.

Gould, E., & Tanapat, P. (1999). Stress and hippocampal neurogenesis. *Biological Psychiatry*, *46*, 1472–1479.

Gould, E., Tanapat, P., Rydel, T., & Hastings, N. (2000). Regulation of hippocampal neurogenesis in adulthood. *Biological Psychiatry*, *48*(8), 715–720.

Gray, W. P. (2008). Neuropeptide Y signalling on hippocampal stem cells in health and disease. *Molecular and Cellular Endocrinology*, *288*, 52–62.

Gray, W. P., & Sundstrom, L. E. (1998). Kainic acid increases the proliferation of granule cell progenitors in the dentate gyrus of the adult rat. *Brain Research*, *790*, 52–59.

Hansel, D. E., Eipper, B. A., & Ronnett, G. V. (2001). Neuropeptide Y functions as a neuroproliferative factor. *Nature*, *410*, 940–944.

Haughey, N. J., Nath, A., Chan, S. L., Borchard, A. C., Rao, M. S., & Mattson, M. P. (2002). Disruption of neurogenesis by amyloid β-peptide, and perturbed neural progenitor cell homeostasis, in models of Alzheimer's disease. *Journal of Neurochemistry*, *83*, 1509–1524.

Heilig, M. (2004). The NPY system in stress, anxiety and depression. *Neuropeptides, 38*, 213–224.
Howell, O. W., Doyle, K., Goodman, J. H., Scharfman, H. E., Herzog, H., Pringle, A., et al. (2005). Neuropeptide Y stimulates neuronal precursor proliferation in the post-natal and adult dentate gyrus. *Journal of Neurochemistry, 93*, 560–570.
Howell, O. W., Scharfman, H. E., Herzog, H., Sundstrom, L. E., Beck-Sickinger, A., & Gray, W. P. (2003). Neuropeptide Y is neuroproliferative for post-natal hippocampal precursor cells. *Journal of Neurochemistry, 86*, 646–659.
Howell, O. W., Silva, S., Scharfman, H. E., Sosunov, A. A., Zaben, M., Shatya, A., et al. (2007). Neuropeptide Y is important for basal and seizure-induced precursor cell proliferation in the hippocampus. *Neurobiology of Disease, 26*, 174–188.
Hua, Y., Huang, X. Y., Zhou, L., Zhou, Q. G., Hu, Y., Luo, C. X., et al. (2008). DETA/NONOate, a nitric oxide donor, produces antidepressant effects by promoting hippocampal neurogenesis. *Psychopharmacology, 200*, 231–242.
Islam, A. T., Kuraoka, A., & Kawabuchi, M. (2003). Morphological basis of nitric oxide production and its correlation with the polysialylated precursor cells in the dentate gyrus of the adult guinea pig hippocampus. *Anatomical Science International, 78*, 98–103.
Jiang, W., Xiao, L., Wang, J. C., Huang, Y. G., & Zhang, X. (2004). Effects of nitric oxide on dentate gyrus cell proliferation after seizures induced by pentylenetrazol in the adult rat brain. *Neuroscience Letters, 367*, 344–348.
Jin, K., Minami, M., Lan, J. Q., Mao, X. O., Batteur, S., Simon, R. P., et al. (2001). Neurogenesis in dentate subgranular zone and rostral subventricular zone after focal cerebral ischemia in the rat. *Proceedings of the National Academy of Sciences of the United States of America, 98*, 4710–4715.
Kim, Y. M., Chung, H. T., Kim, S. S., Han, J. A., Yoo, Y. M., Kim, K. M., et al. (1999). Nitric oxide protects PC12 cells from serum deprivation-induced apoptosis by cGMP-dependent inhibition of caspase signaling. *The Journal of Neuroscience, 19*, 6740–6747.
Kuhn, H. G., Dickinson-Anson, H., & Gage, F. H. (1996). Neurogenesis in the dentate gyrus of the adult rat: Age-related decrease of neuronal progenitor proliferation. *The Journal of Neuroscience, 16*, 2027–2033.
Lancaster, J. R., Jr. (1997). A tutorial on the diffusibility and reactivity of free nitric oxide. *Nitric Oxide, 1*(1), 18–30.
Lichtenwalner, R. J., & Parent, J. M. (2006). Adult neurogenesis and the ischemic forebrain. *Journal of Cerebral Blood Flow & Metabolism, 26*, 1–20.
Liu, J., Solway, K., Messing, R. O., & Sharp, F. R. (1998). Increased neurogenesis in the dentate gyrus after transient global ischemia in gerbils. *The Journal of Neuroscience, 18*, 7768–7778.
Lopez-Ilasaca, M. (1998). Signaling from G-protein-coupled receptors to mitogen-activated protein (MAP)-kinase cascades. *Biochemical Pharmacology, 56*, 269–277.
Luo, C. X., Jin, X., Cao, C. C., Zhu, M. M., Wang, B., Chang, L., et al. (2010). Bidirectional regulation of neurogenesis by neuronal nitric oxide synthase derived from neurons and neural stem cells. *Stem Cells, 28*, 2041–2052.
Malberg, J. E., Eisch, A. J., Nestler, E. J., & Duman, R. S. (2000). Chronic antidepressant treatment increases neurogenesis in adult rat hippocampus. *The Journal of Neuroscience, 20*, 9104–9110.
Malinski, T., Bailey, F., Zhang, Z. G., & Chopp, M. (1993). Nitric oxide measured by a porphyrinic microsensor in rat brain after transient middle cerebral artery occlusion. *Journal of Cerebral Blood Flow & Metabolism, 13*, 355–358.
Matarredona, E. R., Murillo-Carretero, M., Moreno-López, B., & Estrada, C. (2005). Role of nitric oxide in subventricular zone neurogenesis. *Brain Research Reviews, 49*, 355–366.
Matsui, T., Nagafuji, T., Kumanishi, T., & Asano, T. (1999). Role of nitric oxide in pathogenesis underlying ischemic cerebral damage. *Cellular and Molecular Neurobiology, 19*, 177–189.

Moore, P. K., & Handy, R. L. C. (1997). Selective inhibitors of neuronal nitric oxide synthase is no NOS really good NOS for the nervous system. *Trends in Pharmacological Sciences*, *18*, 204–211.

Moreno-López, B., Noval, J. A., González-Bonet, L. G., & Estrada, C. (2000). Morphological bases for a role of nitric oxide in adult neurogenesis. *Brain Research*, *869*, 244–250.

Moreno-López, B., Romero-Grimaldi, C., Noval, J. A., Murillo-Carretero, M., Matarredona, E. R., & Estrada, C. (2004). Nitric oxide is a physiological inhibitor of neurogenesis in the adult mouse subventricular zone and olfactory bulb. *The Journal of Neuroscience*, *24*(1), 85–95.

Morley, J. E., Alshaher, M. M., Farr, S. A., Flood, J. F., & Kumar, V. B. (1999). Leptin and neuropeptide Y (NPY) modulate nitric oxide synthase: Further evidence for a role of nitric in feeding. *Peptides*, *20*, 595–600.

Morley, J. E., & Flood, J. F. (1991). Evidence that nitric oxide modulates food intake in mice. *Life Sciences*, *49*, 707.

Packer, M. A., Stasiv, Y., Benraiss, A., Chmielnicki, E., Grinberg, A., Westphal, H., et al. (2003). Nitric oxide negatively regulates mammalian adult neurogenesis. *Proceedings of the National Academy of Sciences of the United States of America*, *100*, 9566–9571.

Parent, J. M., Yu, T. W., Leibowitz, R. T., Geschwind, D. H., Sloviter, R. S., & Lowenstein, D. H. (1997). Dentate granule cell neurogenesis is increased by seizures and contributes to aberrant network reorganization in the adult rat hippocampus. *The Journal of Neuroscience*, *17*, 3727–3738.

Park, C., Sohn, Y., Shin, K. S., Kim, J., Ahn, H., & Huh, Y. (2003). The chronic inhibition of nitric oxide synthase enhances cell proliferation in the adult rat hippocampus. *Neuroscience Letters*, *339*, 9–12.

Pollman, M. J., Yamada, T., Horiuchi, M., & Gibbons, G. H. (1996). Vasoactive substances regulate vascular smooth muscle cell apoptosis. Countervailing influences of nitric oxide and angiotensin II. *Circulation Research*, *79*, 748–756.

Ramón y Cajal, S. (1928). In *Degeneration and regeneration of the nervous system Vol. 2.* (p. 750). New York: Hafner Publishing Co.

RayChaudhury, A., Frischer, H., & Malik, A. B. (1996). Inhibition of endothelial cell proliferation and bFGF induced phenotypic modulation by nitric oxide. *Journal of Cellular Biochemistry*, *63*, 125–134.

Redrobe, J. P., Dumont, Y., St-Pierre, J. A., & Quirion, R. (1999). Multiple receptors for neuropeptide Y in the hippocampus: Putative roles in seizures and cognition. *Brain Research*, *848*, 153–166.

Reif, A., Schmitt, A., Fritzen, S., Chourbaji, S., Bartsch, C., Urani, A., et al. (2004). Differential effect of endothelial nitric oxide synthase (NOS-III) on the regulation of adult neurogenesis and behaviour. *European Journal of Neuroscience*, *20*, 885–895.

Reiner, A., & Zagvazdin, Y. (1998). On the selectivity of 7-nitroindazole as an inhibitor of neuronal nitric oxide synthase. *Trends in Pharmacological Sciences*, *19*, 348–350.

Santarelli, L., Saxe, M., Gross, C., Surget, A., Battaglia, F., Dulawa, S., et al. (2003). Requirement of hippocampal neurogenesis for the behavioral effects of antidepressants. *Science*, *301*, 805–809.

Seki, T., & Arai, Y. (1995). Age-related production of new granule cells in the adult dentate gyrus. *NeuroReport*, *6*, 2479–2482.

Shihabuddin, L. S., Palmer, T. D., & Gage, F. H. (1999). The search for neural progenitor cells: Prospects for the therapy of neurodegenerative disease. *Molecular Medicine Today*, *5*, 474–480.

Shors, T. J., Miesegaes, G., Beylin, A., Zhao, M., Rydel, T., & Gould, E. (2001). Neurogenesis in the adult is involved in the formation of trace memories. *Nature*, *410*, 372–376.

Shors, T. J., Townsend, D. A., Zhao, M., Kozorovitskiy, Y., & Gould, E. (2002). Neurogenesis may relate to some but not all types of hippocampal-dependent learning. *Hippocampus*, *12*, 578–584.

Snyder, J. S., Kee, N., & Wojtowicz, J. M. (2001). Effects of adult neurogenesis on synaptic plasticity in the rat dentate gyrus. *Journal of Neurophysiology, 85*, 2423–2431.

Squire, L. R., & Cave, C. B. (1991). The hippocampus, memory, and space. *Hippocampus, 1*, 269–271.

Steiner, B., Kronenberg, G., Jessberger, S., Brandt, M. D., Reuter, K., & Kempermann, G. (2004). Differential regulation of gliogenesis in the context of adult hippocampal neurogenesis in mice. *Glia, 46*, 41–52.

Suenobu, N., Shichiri, M., Iwashina, M., Marumo, F., & Hirata, Y. (1999). Natriuretic peptides and nitric oxide induce endothelial apoptosis via a cGMP-dependent mechanism. *Arteriosclerosis, Thrombosis, and Vascular Biology, 19*, 140–146.

Sun, Y., Jin, K., Childs, J. T., Xie, L., Mao, X. O., & Greenberg, D. A. (2005). Neuronal nitric oxide synthase and ischemia-induced neurogenesis. *Journal of Cerebral Blood Flow and Metabolism, 25*, 485–492.

Takagi, Y., Nozaki, K., Takahashi, J., Yodoi, J., Ishikawa, M., & Hashimoto, N. (1999). Proliferation of neuronal precursor cells in the dentate gyrus is accelerated after transient forebrain ischemia in mice. *Brain Research, 831*, 283–287.

Tatemoto, K. (1982). Neuropeptide Y: Complete amino acid sequence of the brain peptide. *Proceedings of the National Academy of Sciences of the United States of America, 79*, 5485–5489.

Thippeswamy, T., & Morris, R. (1997). Cyclic guanosine 3',5'-monophosphate-mediated neuroprotection by nitric oxide in dissociated cultures of rat dorsal root ganglion neurons. *Brain Research, 774*, 116–122.

Thompson, A., Boekhoorn, K., van Dam, A. M., & Lucassen, P. J. (2008). Changes in adult neurogenesis in neurodegenerative diseases: Cause or consequence? *Genes, Brain and Behavior, 7*, 28–42.

Ulibarri, J. A., Mozdziak, P. E., Schultz, E., Cook, C., & Best, T. M. (1999). Nitric oxide donors, sodium nitroprusside and S-nitro-N-acetylpencillamine, stimulate myoblast proliferation in vitro. *In Vitro Cellular & Developmental Biology. Animal, 35*, 215–218.

van Praag, H., Kempermann, G., & Gage, F. H. (1999). Running increases cell proliferation and neurogenesis in the adult mouse dentate gyrus. *Nature Neuroscience, 2*, 266–270.

van Praag, H., Schinder, A. F., Christie, B. R., Toni, N., Palmer, T. D., & Gage, F. H. (2002). Functional neurogenesis in the adult hippocampus. *Nature, 415*, 1030–1034.

Vezzani, A., Sperk, G., & Colmers, W. F. (1999). Neuropeptide Y: Emerging evidence for a functional role in seizure modulation. *Trends in Neurosciences, 22*, 25–30.

Villalobo, A. (2006). Nitric oxide and cell proliferation. *FEBS Journal, 273*, 2329–2344.

Wood, J., & Garthwaite, J. (1994). Models of the diffusional spread of nitric oxide: Implications for neural nitric oxide signalling and its pharmacological properties. *Neuropharmacology, 33*, 1235–1244.

Yoneyama, M., Kawada, K., Gotoh, Y., Shiba, T., & Ogita, K. (2010). Endogenous reactive oxygen species are essential for proliferation of neural stem/progenitor cells. *Neurochemistry International, 56*, 740–746.

Zhang, R., Wang, Y., Zhang, L., Zhang, Z., Tsang, W., Lu, M., et al. (2003). Sildenafil (Viagra) induces neurogenesis and promotes functional recovery after stroke in rats. *Stroke, 33*, 2675–2680.

Zhang, R., Zhang, L., Zhang, Z., Wang, Y., Lu, M., Lapointe, M., et al. (2001). A nitric oxide donor induces neurogenesis and reduces functional deficits after stroke in rats. *Annals of Neurology, 50*, 602–611.

Zhou, Q. G., Hu, Y., Hua, Y., Hu, M., Luo, C. X., Han, X., et al. (2007). Neuronal nitric oxide synthase contributes to chronic stress-induced depression by suppressing hippocampal neurogenesis. *Journal of Neurochemistry, 103*, 1843–1854.

Zhu, D. Y., Liu, S. H., Sun, H. S., & Lu, Y. M. (2003). Expression of inducible nitric oxide synthase after focal cerebral ischemia stimulates neurogenesis in the adult rodent dentate gyrus. *Journal of Neuroscience, 23*, 223–229.

CHAPTER FIVE

Nitric Oxide in the Nervous System: Biochemical, Developmental, and Neurobiological Aspects

Marcelo Cossenza[*,†], Renato Socodato[*], Camila C. Portugal[*], Ivan C.L. Domith[*], Luis F.H. Gladulich[*], Thaísa G. Encarnação[*], Karin C. Calaza[*,‡], Henrique R. Mendonça[*], Paula Campello-Costa[*,‡], Roberto Paes-de-Carvalho[*,‡,1]

[*]Programa de Neurociências, Instituto de Biologia, Universidade Federal Fluminense, Niterói, RJ, Brazil
[†]Departamento de Fisiologia e Farmacologia, Instituto Biomédico, Universidade Federal Fluminense, Rio de Janeiro, Brazil
[‡]Departamento de Neurobiologia, Instituto de Biologia, Universidade Federal Fluminense, Niterói, RJ, Brazil
[1]Corresponding author: e-mail address: robpaes@vm.uff.br

Contents

Vitamins and Hormones, Volume 96
ISSN 0083-6729
http://dx.doi.org/10.1016/B978-0-12-800254-4.00005-2

Abstract

Nitric oxide (NO) is a very reactive molecule, and its short half-life would make it virtually invisible until its discovery. NO activates soluble guanylyl cyclase (sGC), increasing 3′,5′-cyclic guanosine monophosphate levels to activate PKGs. Although NO triggers several phosphorylation cascades due to its ability to react with Fe II in heme-containing proteins such as sGC, it also promotes a selective posttranslational modification in cysteine residues by S-nitrosylation, impacting on protein function, stability, and allocation. In the central nervous system (CNS), NO synthesis usually requires a functional coupling of nitric oxide synthase I (NOS I) and proteins such as NMDA receptors or carboxyl-terminal PDZ ligand of NOS (CAPON), which is critical for specificity and triggering of selected pathways. NO also modulates CREB (cAMP-responsive element-binding protein), ERK, AKT, and Src, with important implications for nerve cell survival and differentiation. Differences in the regulation of neuronal death or survival by NO may be explained by several mechanisms involving localization of NOS isoforms, amount of NO being produced or protein sets being modulated. A number of studies show that NO regulates neurotransmitter release and different aspects of synaptic dynamics, such as differentiation of synaptic specializations, microtubule dynamics, architecture of synaptic protein organization, and modulation of synaptic efficacy. NO has also been associated with synaptogenesis or synapse elimination, and it is required for long-term synaptic modifications taking place in axons or dendrites. In spite of tremendous advances in the knowledge of NO biological effects, a full description of its role in the CNS is far from being completely elucidated.

1. INTRODUCTION

Since the original discovery of endothelium-derived relaxing factor (EDRF) by Furchgott and Zawadzki (1980) and the demonstration by Moncada's group (Palmer, Ferrige, & Moncada, 1987) that this factor was nitric oxide (NO), a myriad of scientific reports described many functions of this molecule in biological systems and its role in various physiological phenomena, including in the nervous system. Because of this explosive interest, NO was elected "the molecule of the year" in 1992. Since this period, an increased understanding of the role played by NO in the nervous system, especially in the central nervous system (CNS), was in demand. In this chapter, we discuss some important aspects of NO biology, with especial emphasis on those involved in the biochemistry, physiology, and development of the nervous system.

1.1. Brief history and biochemistry of NOS

Pivotal works from several authors postulated EDRF as a mediator for both the tumoricidal and bactericidal actions of macrophages (Hibbs,

Vavrin, & Taintor, 1987; Ignarro, Buga, Wood, Byrns, & Chaudhuri, 1987; Palmer et al., 1987; Stuehr, Gross, Sakuma, Levi, & Nathan, 1989). Afterward, EDRF was identified as NO, a free radical that could be produced by some cells from the amino acid L-arginine (L-Arg). In fact, this finding not only explained the singularity of endothelium-dependent relaxation of smooth muscle cells, identified by Furchgott and Zawadzki in the early 1980s, but also related this chemical substance with increases of intracellular levels of 3′,5′-cyclic guanosine monophosphate (cGMP). This finding matched to the first description that in neuroblastoma cells, L-Arg could be categorized as an endogenous activator of soluble guanylyl cyclase (sGC; Deguchi & Yoshioka, 1982), an enzyme that had been previously described to respond to nitroso compounds increasing cGMP production (Arnold, Mittal, Katsuki, & Murad, 1977; Katsuki, Arnold, Mittal, & Murad, 1977; Vesely, Rovere, & Levey, 1977). Indeed, Moncada and his associates unequivocally demonstrated that sGC stimulation by L-Arg fitted with L-citrulline (L-Cit) formation in the CNS, with concomitant NADPH and calcium dependence (Knowles, Palacios, Palmer, & Moncada, 1989). Subsequently, Bredt and Snyder (1990) purified the enzyme synthesizing NO from the rat cerebellum and showed that it was a calmodulin-requiring enzyme, which was later termed nitric oxide synthase (NOS). The purification and cloning of NOS from rat cerebellum allowed the initial characterization of the neuronal isoform (nNOS or NOS I; Bredt et al., 1991). In sequence, other researchers cloned two other nonneuronal isoforms, endothelial NOS (eNOS or NOS III; Lamas, Marsden, Li, Tempst, & Michel, 1992; Sessa et al., 1992) and inducible NOS (iNOS or NOS II; Lowenstein, Glatt, Bredt, & Snyder, 1992; Lyons, Orloff, & Cunningham, 1992; Xie et al., 1992).

NO is a gaseous signaling molecule which is in most cases, but not exclusively, produced in a reaction catalyzed by NOS enzymes (Alderton, Cooper, & Knowles, 2001). Recent studies show that nitrate (NO_3^-) and nitrite (NO_2^-) are oxidized end products of NO degradation, which can be recycled *in vivo* to form NO. This represents an emerging and important alternative source for NO generation besides the classical L-Arg/NOS pathway, with particular relevance in hypoxic states (Lundberg, Weitzberg, & Gladwin, 2008) and pulmonary arterial hypertension (Sparacino-Watkins, Lai, & Gladwin, 2012). This new route is being termed *nitrate–nitrite–nitric oxide pathway*, and efforts have been employed to establish which enzymatic catalysts could be involved in this process.

Besides using L-Arg and Ca^{2+}–calmodulin, NOS activity also requires molecular oxygen, reduced NADP, flavin-derived cofactors (FMN and FAD), and tetrahydrobiopterin (BH_4), stoichiometrically producing NO and L-Cit (Wiesinger, 2001). Indeed, active NOS dimerizes through its heme group. The Ca^{2+}–calmodulin binding complex provides an additional structural stability that allows electron flux from NOS reductase domain toward its catalytic site (Alderton et al., 2001).

NOS isoforms differ significantly in tissue distribution, expression pattern, and are encoded by different genes (Alderton et al., 2001). NOS I (located at Chromosome 12) was initially isolated from the cerebellum and is predominantly expressed in neuronal tissues (Bredt et al., 1991). NOS II (located at Chromosome 17) was first cloned and isolated from macrophages (Xie et al., 1992) and then described to constitutively bind calmodulin (Cho et al., 1992), a possible mechanism explaining its low-calcium dependence. NOS II is mostly known for its regulatory roles within the immune system and has been termed inducible isoform because lipopolysaccharide (LPS) and cytokine signaling could induce its expression. Induction of NOS II usually leads to high NO production, which is generally associated with host immunity (Bogdan, 2001). NOS III (located at Chromosome 7) was first isolated from bovine aortic endothelial cells, based on cDNA cloning from the neuronal isoform, and its amino acid sequence displayed roughly 50–60% homology with the other two isoforms (Lamas et al., 1992; Nishida et al., 1992). NOS III was widely studied in the cardiovascular system, displaying robust regulatory functions in this system (Cai & Harrison, 2000).

All three NOS isoforms have several phosphorylation sites for different protein kinases, including PKA, PKC, AKT, and Ca^{2+}–calmodulin-dependent kinase (CAMK; Boehning & Snyder, 2003). NOS enzymes are extremely important for the maintenance of physiological mechanisms within an organism and genetic ablation of different NOS in mice was quite instructive in establishing functional roles of NOS-generated NO in different systems. For instance: (1) NOS $I^{-/-}$ mice display intense gastroparesis due to dysfunctional vagal innervation to stomach smooth muscle (Mashimo, Kjellin, & Goyal, 2000), decreased apoptosis induced by striatal NMDA microinjections (Ayata et al., 1997), and early impairment of hippocampal-dependent spatial memory (Kirchner et al., 2004); (2) NOS II knockout mice are more resistant to LPS-induced lung injury (Kristof, Goldberg, Laubach, & Hussain, 1998), display decreased alcohol-induced liver damage (McKim et al., 2003), diminished ischemia-induced nerve cell death in the brain (Iadecola, Zhang, Casey, Nagayama, & Ross, 1997), and delayed wound healing (Yamasaki et al., 1998); and (3) disrupting the gene coding for NOS III in mice causes spontaneous systemic and pulmonary hypertension (Huang et al., 1995), deficient vascular remodeling (Rudic et al., 1998), and inhibition of growth factor-mediated angiogenesis (Lee et al., 1999).

1.2. NO classical actions

There are numerous mechanisms by which NO has been described to act in different physiological systems and living cells. An attempt to exhaust this subject would be beyond the scope of this chapter. For such, herein we will focus in basic signaling roles of NO in nerve cells. Therefore, we decided to allocate the mechanisms of action of NO into two large groups. Such mechanisms will be separated in: (1) the reaction with iron-containing proteins and (2) selective modification of protein cysteine residues to form S-nitrosocysteine (here termed S-nitrosylation). Hence, NO mediates its downstream effects using either one or both mechanisms in neuronal or glial cells. NO, as it is currently known, includes a radical $NO^{\bullet}$ (free radical nitrogen monoxide) that undergoes interconversion to form either NO^{+} (cation nitrosonium) or NO^{-} (nitroxyl anion; Bian, Gao, Weisbrodt, & Murad, 2003). Among NO species, only the uncharged $NO^{\bullet}$ radical can activate sGC within a cell (Friebe & Koesling, 2003). sGC is the best-characterized NO target in neurons. It has an heme group containing a ferrous ion in which NO binds to, inducing a conformational change that exposes the enzyme catalytic domain (Denninger & Marletta, 1999). Upon NO binding, sGC activity is tremendously upregulated, giving rise to rapid conversion of guanosine-5′-triphosphate (GTP) into cGMP, a reaction that requires magnesium ions to occur. sGC is divided into two classes: peptide sensitive and NO sensitive, or sGC. Increased levels of cGMP lead to the opening of cyclic nucleotide-gated ion channels, activation of phosphodiesterases, which degrade cAMP and cGMP, and activation of cGMP-dependent protein kinases (PKGs). PKGs have many important biological targets that are discussed below. In that sense, increments in intracellular cGMP levels account for most of the acknowledged NO effects in cellular systems (Hanafy, Krumenacker, & Murad, 2001), with important functional roles in vascular smooth muscle physiology (Garg & Hassid, 1989), platelet aggregation (Radomski, Palmer, & Moncada, 1990), and nervous tissue function (Bredt, Hwang, & Snyder, 1990).

sGC is an enzymatic complex preferentially composed of one α1 and one β1 subunit, forming a heterodimer (Koesling, Russwurm, Mergia, Mullershausen, & Friebe, 2004), which processes catalytic activity (Krumenacker, Hanafy, & Murad, 2004). However, dimers composed of α2/β1 may have important physiological roles in the brain (Friebe & Koesling, 2003). Homodimeric conformations such as α1/α1 or β1/β1 have been found in certain organisms but are devoid of significant catalytic activity (Friebe & Koesling, 2003). The enzyme has three functional domains: an N-terminal domain, the central core, and the C-terminal catalytic domain (Friebe & Koesling, 2003). In the N-terminal resides the heme-binding domain, specifically close to the proximal histidine residue in the β1 subunit (His105). The importance of this residue for NO-mediated sGC activation is revealed by point mutation in His105, which renders the enzyme insensitive to NO stimulation (Wedel et al., 1994). An activation model predicts that NO directly binds to the ferrous heme, mediating the displacement of the ferrous atom from the proximal histidine, giving rise to a nytrosil–heme complex, which is believed to promote an initial conformational change within the β1 structure to activate the enzyme (Friebe & Koesling, 2003).

Frequently, NO/cGMP/PKG signaling cascade is classified as the canonical NO pathway. However, NO can also modulate cellular responses through a noncanonical mechanism. Given its chemical nature, NO participates in different types of reactions with nucleophilic intracellular agents. A particular type of interaction is known as S-nitrosylation, which involves the covalent attachment of an NO molecule onto thiol groups (—SH) to form S-nitrosothiol (—SNO). Generally, NO^+ is associated with this reaction because it is a target for nucleophilic attack by the sulfur atom of a thiol (Stamler, Lamas, & Fang, 2001). Just to make it clear, the correct chemical denomination to this reaction is S-nitrosation. However, the term S-nitrosylation was designed to relate it with the commonly used termination for posttranslational modifications in biological systems (Gould, Doulias, Tenopoulou, Raju, & Ischiropoulos, 2013). This reaction occurs in cysteine of proteins because this residue is unique in exhibiting a thiol. Such reaction functions as a posttranslational modification that impact on protein function, stability, and location (Gould et al., 2013). Additionally, S-nitrosylation fulfills all the major criteria to be considered an intracellular signaling mechanism, such as spatial and temporal features of NO signaling (Hess, Matsumoto, Kim, Marshall, & Stamler, 2005). In that sense, S-nitrosylation replaces the notion that free diffusion of NO within a cell could be the main mechanism for nitrergic pathway. Moreover, NO signaling by S-nitrosylation seems to be involved with several target proteins, which might be confined within subcellular compartments, and it is critical for specificity and propagation of NO signals (Stamler et al., 2001).

Interestingly, NMDA-type glutamate receptors were the first targets described to have their function controlled by S-nitrosylation. Moreover, when these receptors trigger the production of NO, their own activity may be downregulated by spatial proximity between NOS I and cysteine residues on the NMDA receptor redox site (Lipton et al., 1993). The function of S-nitrosylation has gained importance recently. Several implications have been described in exchange reactions with small thiol-containing molecules, such as S-nitrosoglutathione (GSNO), or *trans*-nitrosation reactions between proteins like S-nitrosothioredoxin. Previous findings demonstrated that S-nitrosylation might be a relevant physiological regulator in synaptic plasticity. Accordingly, it was demonstrated that NMDA-triggered NO production in neurons regulates the surface expression of GluA2 subunit of AMPA receptors by S-nitrosylation of *N*-ethylmaleimide sensitive factor (Huang et al., 1995). S-Nitrosylation of stargazine also regulates AMPA surface expression by increasing GluA1 allocation into neuronal plasma membrane (Selvakumar, Huganir, & Snyder, 2009). Furthermore, S-nitrosylation has also been associated with neuronal epigenetics. In that sense, it was demonstrated that S-nitrosylation of histone deacetylase 2 promotes chromatin remodeling in neurons increasing the activity of CREB-dependent promoters in the *bdnf* gene (Nott, Watson, Robinson, Crepaldi, & Riccio, 2008). The cessation of S-nitrosylation, which restricts the duration of this signaling process, may occur in the presence of metal ions, through an enzymatic process controlled by GSNO reductase, which accelerates the decomposition of GSNO (Gould et al., 2013) and by the thioredoxin/thioredoxin reductase system that catalyzes the denitrosylation of a number of S-nitrosoproteins (Benhar, Forrester, Hess, & Stamler, 2008). Figure 5.1 depicts a scheme of both pathways described above (NO/cGMP/PKG and S-nitrosylation signaling).

1.3. Interesting partners in the CNS: Focusing on NMDA receptors

For over 30 years, it was acknowledged the participation of receptors for amino acids as fundamental entities in excitatory synaptic transmission in the CNS. Efforts in an attempt to study the mechanisms of long-term potentiation (LTP) revealed that NMDA receptor inhibitors (specifically AP-5) could block LTP in the CA1 region of the hippocampus (Harris,

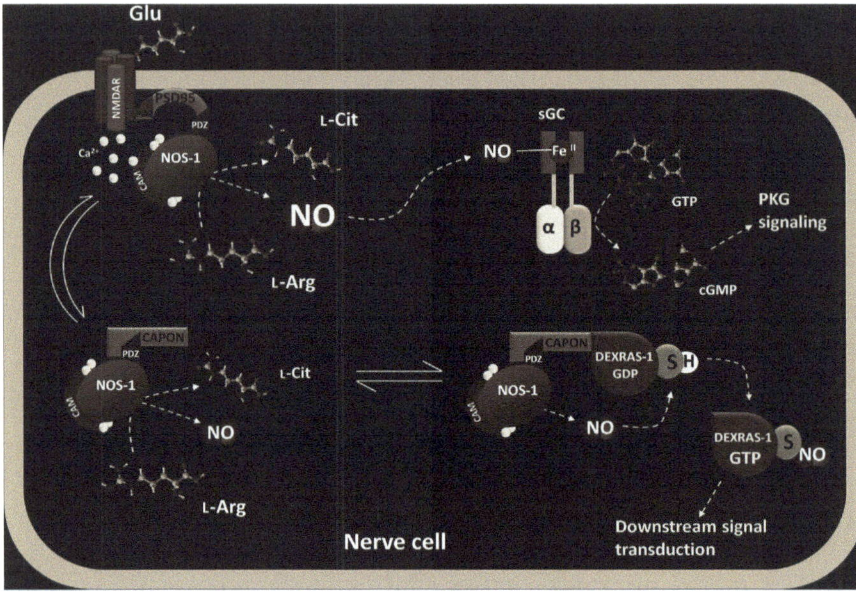

Figure 5.1 Nitric oxide synthase I-coupled platforms and NO mechanistic actions by S-nitrosylation and PKG activation in CNS. NOS I-coupled NMDA receptor through scaffold postsynaptic protein PSD-95 with PDZ domains. Opening of NMDA receptor channels promotes Ca^{2+} influx and its binding to calmodulin triggers the production of NO by L-Arg oxidation to L-citrulline. In another way, CAPON competes with PSD-95 for interaction with NOS I forcing its dissociation from the plasma membrane. DEXRAS-1 is a small GTPase protein which can interact with NOS I/CAPON complex and place them in close proximity to allow DEXRAS activation by S-nitrosylation. NO can also activate soluble guanylate cyclase (sGC) by reacting with Fe II present into its heme domain. That enzyme configuration produces rapid conversion of guanosine-5′-triphosphate (GTP) into 3′,5′-cyclic guanosine monophosphate (cGMP) to further PKG activation. This signaling mechanism is called as NO canonical pathway. (See the color plate.)

Ganong, & Cotman, 1984; Watkins & Evans, 1981). This finding was initial evidence highlighting the participation of NMDA receptors in neuronal physiology, associating synaptic plasticity with learning (Collingridge & Singer, 1990).

Up to that time, it was also known that glutamate could elicit large increases in cGMP levels in brain preparations, especially in the cerebellum, where cGMP formation had been related with glutamate-dependent activation of NMDA receptors (Ferrendelli, Chang, & Kinscherf, 1974). Garthwaite and his colleagues (Garthwaite, Charles, & Chess-Williams, 1988) were the first to demonstrate that glutamate, through NMDA receptors, could induce the release of a diffusible messenger with properties strikingly similar to that of EDRF, with Ca^{2+}–calmodulin dependence and cGMP-increasing activity. Moreover, in addition to the cerebellum, other CNS areas have also been studied in which NMDA receptor stimulation could increase cGMP levels. Similarly, NMDA stimulation could also elicit an increase in cGMP levels in hippocampal preparations (Garthwaite, 1991; Garthwaite et al., 1988) and soon became evident that NO could serve signaling purposes within nerve cells. Actually, classical stimuli for NO synthesis in the brain were believed to be Ca^{2+} dependent, probably via activity of NMDA channels.

Moreover, NO synthesis in the brain may require a further functional coupling between NOS and NMDA receptors (Brenman & Bredt, 1997). Nowadays, it is recognized that NOS I has a leading sequence in its N-terminal region, which is likely to interact with a plethora of intracellular targets. NOS I exhibits a PDZ domain, which interacts with proteins such as PSD-95 (postsynaptic density protein-95), a scaffold protein located in the postsynaptic region of neuronal cells. By facilitating the proximity of NMDA receptors to the enzyme, PSD-95 directly exposes NOS I to Ca^{2+} influx induced by NMDA receptor activation. In that sense, NMDA signaling could immediately trigger NO synthesis in PSD-95-containing neurons (Brenman & Bredt, 1997). Another scaffold protein in this platform is carboxyl-terminal PDZ ligand of NOS (CAPON), which is highly enriched in the brain (Jaffrey, Snowman, Eliasson, Cohen, & Snyder, 1998). CAPON has been described to compete with PSD-95 for interaction with NOS I, forcing the dissociation of the synthase from the plasma membrane (Esplugues, 2002). Therefore, CAPON determines the amount of NOS I docked within the postsynaptic density and, in such way, fine-tunes NO formation in neurons. This NOS I/CAPON coupling can also provide the molecular basis for additional interactions with other proteins at the postsynaptic density such as synapsins (Jaffrey, Benfenati, Snowman, Czernik, & Snyder, 2002) and the small GTPase Dexras-1 (Boehning & Snyder, 2003). Such complex places Dexras-1 in close proximity to NOS I, allowing its activation by S-nitrosylation, although the importance of this Ras-like GTPase in downstream transduction in neurons remains to be elucidated (Fig. 5.1). Also related with protein–protein interactions for modulation of NOS I activity, it has been identified a protein known as "protein inhibitor of nitric oxide synthase," or PIN. Interestingly, PIN is a member of the cellular dynein light-chain family, specifically dynein light chain 8, which is responsible for intracellular protein trafficking. PIN interacts with NOS I leading sequence, destabilizing its dimeric structure and inhibiting its enzymatic activity (Jaffrey & Snyder, 1996).

In additional experiments, performed in different neuronal tissues, NMDA receptor activation could significantly inhibit protein synthesis via activation of *e*ukaryotic *E*longation *F*actor-2 *K*inase (eEF2K), another Ca^{2+}–calmodulin-dependent enzyme (Cossenza, Cadilhe, Coutinho, & Paes-de-Carvalho, 2006; Scheetz, Nairn, & Constantine-Paton, 2000). This kinase phosphorylates a translation factor involved in both polypeptide chain elongation (eEF2 pathway) and protein synthesis. In a model of avian retinal cells in culture, the NMDA-stimulating effect in increasing NO release has been associated with direct NOS activation and L-Arg availability to further support NO synthesis (Cossenza et al., 2006).

2. NO SIGNALING PATHWAYS

Keeping in mind NO capabilities to freely travel in the vicinity of its production, it is easy to imagine that NO may display dramatic effects on cell metabolism. On the other hand, most of the known effects of NO are not directly mediated by the gas, but by intracellular signaling cascades stimulated by it. In the following sections of this chapter, we sought to discuss some of the most important signaling pathways activated by NO.

2.1. PKG modulation by NO

PKGs are kinases that demonstrate increased activity when an increase of cGMP occurs. Two isoforms are present in mammalian tissues: PKGI (subdivided into Iα and Iβ) and PKGII (Butt, Geiger, Jarchau, Lohmann, & Walter, 1993). Considering sGC as a primary NO target, PKG is often considered the primary NO effector kinase, constituting sGC/cGMP/PKGII the canonical NO signaling pathway (Wang & Robinson, 1997). When active, PKG phosphorylates many different targets, like DARPP-32, G-substrate, and inositol 1,3,4-triphosphate receptor, to name but few examples (Wang & Robinson, 1997). AKT, ERK, and Src are also downstream kinases related with cell survival, which will be more thoroughly discussed below. Interestingly, NOS I can be phosphorylated, which reduces its catalytic capabilities and works as a negative feedback mechanism for NO production (Dinerman, Steiner, Dawson, Dawson,

& Snyder, 1994). PKG activity is also constantly related with CNS development and memory formation due to its functionality on LTP, LTD (long-term depression), neurotransmitter release, neuronal survival, and transcription factor activity, all of which will be further explained in subsequent sections.

2.2. AKT modulation by NO

AKT or protein kinase B is an important serine/threonine kinase, which regulates multiple cellular functions in the CNS such as neuronal (Datta et al., 1997; Dudek et al., 1997; Mejía-García & Paes-de-Carvalho, 2007) and oligodendrocyte (Flores et al., 2000) survival, cell proliferation, cell cycle progression (Hanada, Feng, & Hemmings, 2004; Peltier, O'Neill, & Schaffer, 2007), myelin production (Flores et al., 2008), and cell differentiation (Peltier et al., 2007).

AKT belongs to the AGC subfamily of protein kinases (Manning, Whyte, Martinez, Hunter, & Sudarsanam, 2002). This subfamily includes three products of distinct genes (*akt1*, *akt2*, and *akt3*). These proteins present a conserved structure composed of three functional domains: an N-terminal pleckstrin homology domain, a central kinase domain, and a carboxyl-terminal regulatory domain containing a hydrophobic motif, which is characteristic of AGC kinases (Fayard, Tintignac, Baudry, & Hemmings, 2005; Hanada et al., 2004). There are two important phosphorylation sites on AKT, a threonine residue in the kinase domain (Thr308), whose phosphorylation is required for enzymatic activation, and a serine residue in the hydrophobic motif (Ser473), which is necessary for full catalytic function (Fayard et al., 2005; Hanada et al., 2004).

NO is also capable of regulating several cellular functions, as described above, and many of these functions might be mediated by AKT activation, as for example cell survival (Mejía-García & Paes-de-Carvalho, 2007; Mejia-Garcia, Portugal, Encarnação, Prado, & Paes-de-Carvalho, 2013). Recently, it has been demonstrated that NO mediates AKT activation in retinal cells (Fig. 5.2). It has also been demonstrated that NO is capable of inducing AKT phosphorylation at both Thr308 and Ser473, which was mediated by the NO classical pathway and PI3K activation (Mejia-Garcia et al., 2013). As previously described by Sarbassov and colleagues, the rictor–mTOR complex seems to be involved in AKT phosphorylation at Ser473 (Sarbassov, Guertin, Ali, & Sabatini, 2005). These data were confirmed when an mTOR specific inhibitor KU-0063794 was used

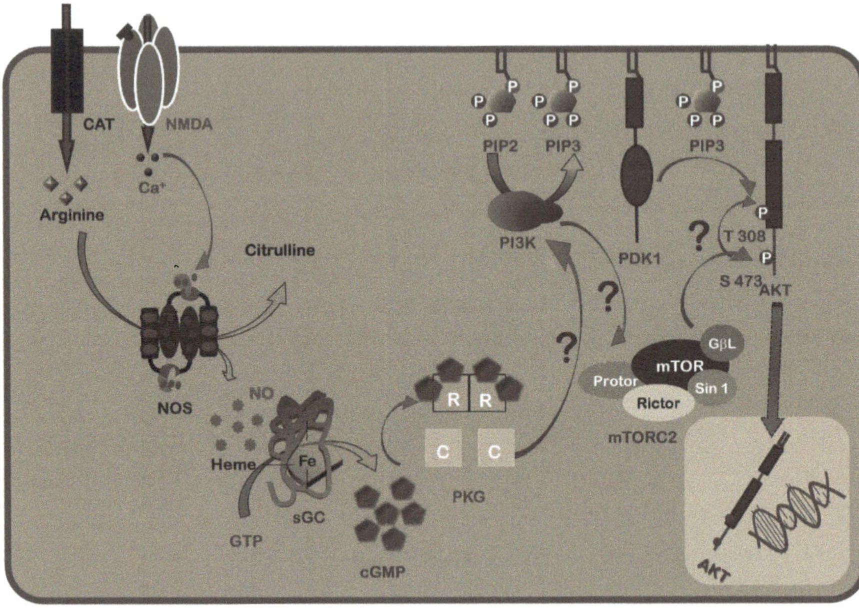

Figure 5.2 Nitric oxide increases AKT phosphorylation at residues Ser473 and Thr308. NMDA stimulation increases intracellular calcium which stimulates Nitric oxide synthase (NOS), enhancing NO production. NO interacts with soluble guanylyl cyclase (sGC) boosting cGMP production, which is capable of activating cGMP-dependent protein kinase (PKG). This kinase may enhance PI3K activity and stimulate mTORC2-dependent AKT phosphorylation at both sites, Ser473 and Thr308. Nitric oxide also enhances phospho-473 AKT localization in the nucleus. (See the color plate.)

(Garcia-Martinez et al., 2009), which completely blocked this NO-mediated AKT phosphorylation at Thr308 or Ser473. To further demonstrate NO involvement, retinal cells were incubated with L-Arg (a substrate for NOS), or glutamate, a well-known NOS stimulator, and it was observed that both treatments stimulated AKT phosphorylation (Mejia-Garcia et al., 2013). In this work, it has been observed that NO was capable of inducing AKT nuclear translocation and this effect was PI3K dependent (Fig. 5.2). In agreement with data showing the involvement of NO and AKT in neuronal survival, Ciani, Virgili, and Contestabile (2002) demonstrated that NO donors (DETA-NONOate and Glyco-SNAP-2) blocked the decrease in AKT phosphorylation mediated by L-NAME, rescuing cerebellar granule neurons from L-NAME-induced cell death. Moreover, Ha et al. (2003) reported that SNAP, an NO donor, induces Ser473 AKT phosphorylation via cGMP/PKG/PI3K pathway and that the activation of this pathway was capable of protecting PC12 cells from apoptosis induced by 6-hydroxydopamine.

2.3. ERK1/2 modulation by NO

Extracellular-regulated kinases 1 and 2 (ERKs) are a subfamily of mitogen-activated protein kinases (MAPKs), which are the classical target of MEK (1/2) in the MAP kinase cascade, a signaling pathway stimulated by activation of a broad array of receptors (Rubinfeld & Seger, 2005), which leads to cell growth and differentiation during development (Davis & Laroche, 2006; Samuels, Saitta, & Landreth, 2009).

Ca^{2+}-induced increase of NO levels induces ERK phosphorylation by rising cGMP/PKG activity (Meini et al., 2006). In cultures of developing neurons from the avian retina, both neurons and glial cells show an increase in ERK1/2 activity upon AMPA/kainate receptors stimulation. However, glial cells are dependent on NO diffused from neurons, since they do not express NOS I (Cossenza & Paes de Carvalho, 2000; Socodato, Magalhães, & Paes-de-Carvalho, 2009). It is also worth mentioning that it has been shown that ERK1/2 phosphorylation may occur independently of the NO canonical pathway, in a Ca^{2+}–calmodulin-dependent kinase II (CAMK II) and AKT-dependent manner. However, this effect has been observed in PC12 cells overexpressing NOS I (Kajiwara et al., 2013). Thus, NO regulation of ERKs is most likely to be different in other cell types, or respond differently in cells with an abnormal NOS I activity, which could explain why NO-mediated ERK1/2 phosphorylation is dependent on the sGC/PKG pathway in retinal cultures, but not in PC12 cells.

2.4. Src modulation by NO

The nonreceptor tyrosine kinase Src plays key roles in cell morphology, motility, proliferation, and survival (Roskoski, 2005). v-Src (a viral protein), encoded by the avian cancer-causing oncogene of the Rous sarcoma virus, was the first identified retroviral oncogene and Src (the cellular homologue in humans, chickens, and other animals) is encoded by a physiological gene, the first of the proto-oncogenes to be described (Martin, 2001). Two specific phosphorylation sites, Tyr416 and Tyr527, modulate Src activity within a cell. When Tyr527 is phosphorylated, by C-terminal Src kinase (Csk), or Tyr416 is dephosphorylated, by Shp-1 (Src homology 2 domain-containing tyrosine phosphatase 1) or Shp-2, Src is rendered in its inactive conformation. However, when Tyr527 dephosphorylation is induced, the enzyme autophosphorylates Tyr416, resulting in full catalytic activation (Roskoski, 2005).

In the cerebral cortex of newborn piglets, hypoxia induces an increase of Src activity by activating NOS I and increasing NO levels, which in turn inhibits Shp-2 to enhance Src phosphorylation at Tyr416 (Mishra, Ashraf, & Delivoria-Papadopoulos, 2009). In the retina, Ca^{2+}-permeable-AMPA receptors (CP-AMPARs) activation increases Src activity, leading to cell death in an NO-dependent manner. Then, activation of CP-AMPARs enhances NOS activity, increasing NO levels, which induces Src phosphorylation at Tyr416 and dephosphorylation at Tyr527, leading to Src activation through the NO canonical pathway. This Src activation is crucial to retinal cell death induced by CP-AMPAR (Socodato et al., 2012).

2.5. CREB modulation by NO

CREB is a transcription factor that is classically associated with the regulation of cell survival. In that sense, when CREB is active, it promotes the transcription of antiapoptotic genes such as Bcl-2 (Ciani, Guidi, Bartesaghi, & Contestabile, 2002). Additionally, CREB has been demonstrated to be one of ERK targets, as active ERK increases CREB-DNA binding and gene transcription (Lee, Butcher, Hoyt, Impey, & Obrietan, 2005). NO can also lead to increased CREB activity via phosphorylation by ERK1/2 through the PKG pathway in retinal neurons (Socodato et al., 2009). However, NO can also directly alter its binding to DNA via S-nitrosylation of nuclear proteins, increasing its activity through a pathway completely independent of the classical sGC/PKG cascade. This fact indicates a dual mechanism for CREB activity modulated by NO, further demonstrating that NO activity regulates transcription and promotes cell survival (Contestabile, 2008; Riccio et al., 2006).

Therefore, NO production and signaling, i.e., NOS activity in the nervous system, display a variety of effects due to its capabilities in activating different signaling cascades like AKT, ERK, Src, and CREB, which affect both NO production and surrounding cells, directly impacting cellular and tissue metabolism as represented in Fig. 5.3.

Furthermore, due to the signaling cascades most associated with NO, its effects could be related with leading to cell survival, plasticity regulation, cellular differentiation, and development in physiological conditions; however, NO can also have impact in neurotoxicity and cell death.

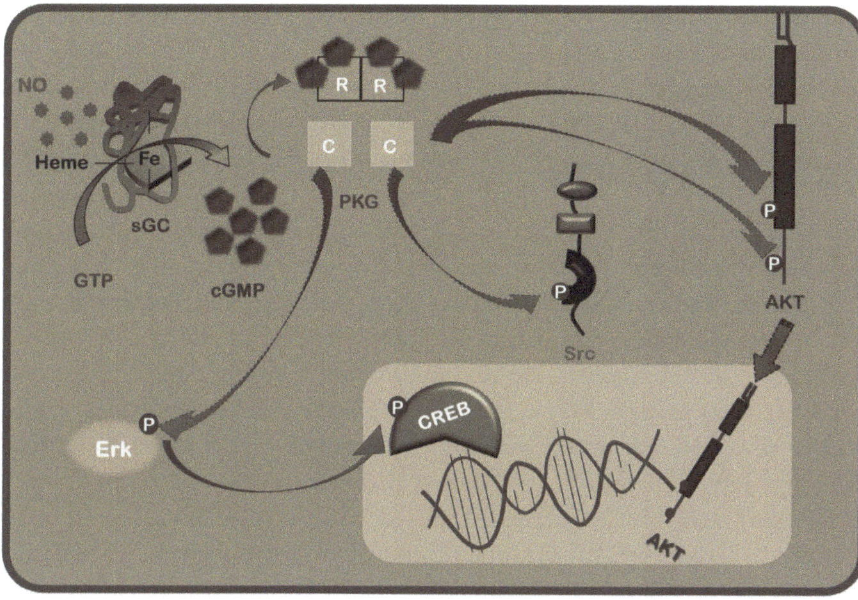

Figure 5.3 Representation of most commonly pathways activated by NO. In its canonical pathway, NO interacts with its intracellular receptor, sGC, promoting an increase of cGMP levels and PKG activation. When activated by cGMP, PKG catalytic subunits phosphorylate downstream targets, culminating in activation of Src, AKT, ERK, and CREB signaling pathways. (See the color plate.)

3. NO AND NEURONAL VIABILITY

In recent years, many efforts have been focused in an attempt to elucidate the controversial activities involving NO with neuronal cell survival or death. Some important information came from NO chemical ability to react with anion superoxide (O_2^-) giving rise to peroxynitrite ($ONOO^-$), which is a powerful oxygen specie. It is well known that aerobic metabolism constantly generates O_2^-, hydroxyl radicals ($^{\bullet}OH$), and hydrogen peroxide (H_2O_2) as end products of cellular respiration. In fact, overproduction of reactive oxygen species is related with their harmful effects on cellular components as for example in lipid peroxidation, which involves mainly oxygen free radicals. Recently, reactive nitrogen species, including $ONOO^-$, NO_2, and N_2O_3, have been highlighted as emerging deleterious agents. Two reactions might occur with such species: (1) covalent and reversible S-nitrosylation of thiol groups (discussed above) and (2) covalent and irreversible nitration of tyrosine residues in proteins.

Formation of nitrotyrosine (NO_2-Tyr) involves $ONOO^-$ and has been implicated in neuronal apoptosis (Bian et al., 2003) and various pathological conditions (Shahani & Sawa, 2012). Tyrosine nitration has been associated with changes in both disruption of protein structure and alteration in the rate of protein degradation. It has been postulated that high levels of nitrotyrosine can serve as a marker in various neurodegenerative diseases associated with aging (Shahani & Sawa, 2012). Regardless, S-nitrosylation is readily reversible and its levels depend on a balance between nitrosylation and denitrosylation. Several works are detailing the importance of S-nitrosylation in regulating processes associated with cell death. Aspects such as NO levels, their production onset within a cell, and their spatiotemporal distribution in a specific cellular compartment are decisive features for NO to fine-tune nerve cell viability. Interestingly, basal (low) NO levels support S-nitrosylation activity that may inhibit neuronal loss, while stressful stimuli producing NO are clearly related with cellular damage, although there are feedback mechanisms regulated by S-nitrosylation that might mitigate NO toxic effects. For instance, Ca^{2+} influx trough NMDARs is controlled in part by their redox site, significantly reducing glutamate excitotoxicity (Lipton et al., 1993; Shahani & Sawa, 2012).

In the context of nerve cell viability in the CNS, NO was primarily associated with neuronal damage. Studies gathered in the 1990s using inhibitors of NOS I and data from NOS I knockout mice were very instructive in this regard. Initially, it was demonstrated that selective NOS inhibitors attenuated glutamate excitotoxicity in cortical neurons (Dawson, Dawson, London, Bredt, & Snyder, 1991). Moreover, the selective NOS I inhibitor 7-nitroindazole could reduce the brain infarct area in mice subjected to acute cerebral middle artery occlusion (Yoshida, Limmroth, Irikura, & Moskowitz, 1994) and NMDA-induced striatal toxicity (Schulz et al., 1995). Works in cortical or striatal neurons from NOS I KO mice further supported the notion that NO could potentially trigger nerve cell loss in the CNS (Ayata et al., 1997; Dawson, Kizushi, Huang, Snyder, & Dawson, 1996). However, in the brain, NO displays other physiological roles, such as regulation of neuronal differentiation and synaptic plasticity (Garthwaite, 2008).

Apoptosis signal-regulating kinase 1 (ASK1) is a member of the MAPK family, capable of activating p38 MAPK, which is involved in cell death (Hattori, Naguro, Runchel, & Ichijo, 2009). ASK1 can be S-nitrosylated and activated by exogenous or endogenous NO. While endogenous NO triggers cell death by S-nitrosylated ASK1, exogenous NO is capable of promoting neuroprotection by suppressing S-nitrosylation of ASK1 during an ischemic insult (Liu et al., 2013). Furthermore, phosphatase and tensin homolog (PTEN) is an inhibitor of PI3K function, which shuts off the AKT pathway. PTEN inactivation or deletion leads to glioblastoma development, a very aggressive tumor in the brain (Nakamura et al., 2013). Although PTEN is associated with an oncogenic pathway, evidence suggests that PTEN downregulation by S-nitrosylation of Cys83, in low NO concentration, could contribute to neuronal survival by upregulating AKT, whereas in ischemic conditions, formation of S-nitrosylated AKT could lead to cell death, as seen in Fig. 5.4 (Kwak et al., 2010; Numajiri et al., 2011).

In the retina, a number of studies have also associated NO with retinal damage. NO released from different sources including from NOS II (Sennlaub, Courtois, & Goureau, 2002) or NOS I could account for NO-mediated toxicity in the retina. On the other hand, Paes-de-Carvalho

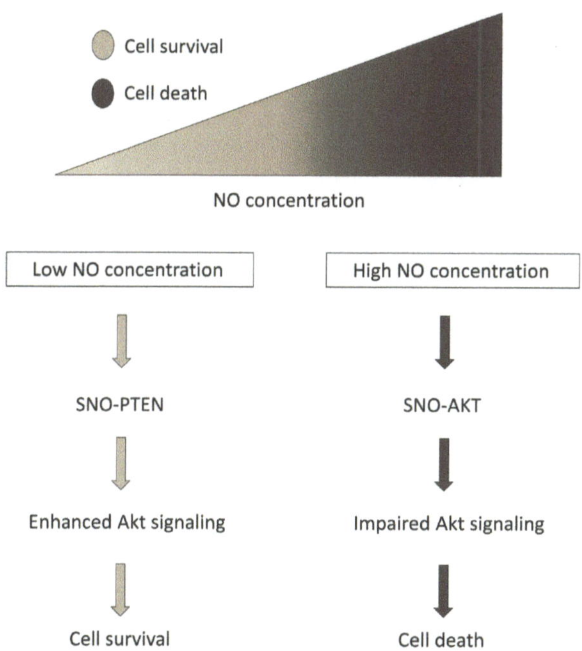

Figure 5.4 Actions of NO in neuronal signaling depend on its concentration. Low concentrations of NO in neuronal systems generally lead to cell survival by upregulating well-known protective pathways like PI3K/AKT. Higher NO concentrations are capable to promote cell death, for example, by shutting down AKT, leading cells to apoptosis. In blue, cell survival signaling; in red, cell death signaling. (See the color plate.)

and colleagues observed that in purified neuronal cultures from the developing retina, medium refeeding promotes an acute neuronal cell death, which could be prevented either via activation of A2a adenosine receptors (Paes-de-Carvalho, Maia, & Ferreira, 2003) or preincubation with SNAP, an NO donor (Mejía-García & Paes-de-Carvalho, 2007). Neuroprotection was dependent on multiple pathways like sGC, PKG, PI3K/AKT, and MEK/ERK, and inhibiting or mimicking those pathways could directly be interfered with cell death (Mejía-García & Paes-de-Carvalho, 2007). Therefore, this neuroprotective effect of NO in developing retinal neurons might actively contribute to retinal cell development. Although the protective signaling triggered by NO in the retina model, Socodato and colleagues in 2012 observed that Ca^{2+} influx by activation of CP-AMPARs could promote neuronal cell death in the retina (Socodato et al., 2012). Stimulation of CP-AMPARs has been demonstrated to activate NOS I and via sGC/PKG increased the function of Src, which in turn led to apoptotic neuronal cell death in the retina (Socodato et al., 2012).

It has been suggested that NO may be associated with neurodegenerative disorders (Sayre, Perry, & Smith, 2008). Initial studies relating NO with dopaminergic degeneration showed that 1-methyl-4-phenyl-1,2,3,6-tetrahydropyridine-mediated neuronal loss in the *substantia nigra* could be greatly reduced in either NOS II or NOS I knockout mice (Liberatore et al., 1999; Przedborski et al., 1996). Parkin is an E3 ubiquitin ligase that is responsible for targeting specific proteins for degradation and is involved in Parkinson disease onset. Upon S-nitrosylation, parkin E3 ligase activity decreases due to auto-ubiquitination (Nakamura et al., 2013). With impaired parkin activity, a robust decrease in protein degradation occurs, contributing to the formation of protein inclusions known as Lewy bodies and neuronal loss (Chung et al., 2004). Interestingly, cyclin-dependent kinase 5 (Cdk5) has no function on cell cycle in neurons, but it acts in important neuronal processes, such as survival, axonal guidance, and neuronal migration through focal adhesion kinase (Ohshima et al., 1996; Xie, Sanada, Samuels, Shih, & Tsai, 2003). S-Nitrosylation of Cys83 or Cys157 activates Cdk5, contributing to the formation of amyloid-β and neuronal loss, which has been linked with Alzheimer pathogenesis (Qu et al., 2011). Furthermore, it has been found higher levels of SNO-Cdk5 in postmortem brains of humans with Alzheimer disease (Qu et al., 2011), reinforcing the role of a Cdk5-NO linkage in neurodegeneration.

Apoptosis and necrosis are two different pathways leading to cell death. The first process is usually activated by caspases and involves ATP, cytochrome *c* release, and DNA fragmentation. The latter occurs via ATP depletion, which impairs both sodium and ATP-driven calcium pumps, culminating in phospholipase and protease activation (Brown, 2010). Experiments using purified human recombinant caspases have shown that NO is capable of inhibiting seven caspase isoforms (1, 2, 3, 4, 6, 7, and 8) by S-nitrosylating conserved cysteine residues (Li, Billiar, Talanian, & Kim, 1997). Moreover, NO decreases caspase-9 activity in cultured cortical neurons, consequently preventing the conversion of pro into active caspase-3 (Zhou, Qian, & Iadecola, 2005). These studies indicate that NO exerts an antiapoptotic function by S-nitrosylating several caspases, impairing their proteolytic activity (Shahani & Sawa, 2012).

Coculturing cerebellar granule neurons and cortical microglial cells, after microglial stimulation with LPS/IFNγ, causes necrotic neuronal cell death by NO production, which might be consistent with NO-induced decrease of ATP levels (Bal-Price & Brown, 2001). In the same model, NO production caused rapid release of glutamate, triggering cell death, which was blocked by MK-801, an NMDAR antagonist (Bal-Price & Brown, 2001). Peroxynitrite and S-nitrosothiols are capable of altering the mitochondrial membrane permeability, causing cytochrome *c* release, which triggers apoptosis by caspase activation (Brown, 2010). In primary cultures of cortical neurons and astrocytes, NO plays a critical role in the production of ATP (Almeida, Almeida, Bolanos, & Moncada, 2001). Using [(*z*)-1-[2-aminoethyl]-*N*-[2-ammonioethyl]amino]diazen-1-ium-1,2 diolate (DETA-NO), an NO donor, ATP concentration decreases approximately 25% in both cell types in 10 min of exposure but, in the next 60 min, ATP levels were higher in astrocytes compared with neurons. Since glycolysis does not take place in CNS neurons but is active in glial cells, NO-induced ATP decrease has been suggested to result in neuronal, but not glial, cell death (Almeida et al., 2001).

In cocultures of cerebellar granule neurons and astrocytes under normoxia, low NO levels cause minimum neuronal death. The same applies to 12 h of hypoxia or inflammatory activation of glia by LPS/IFN-γ (Mander, Borutaite, Moncada, & Brown, 2005). However, when hypoxia was combined either with NO or inflammatory glia activation, an extensive neuronal cell death could be observed by an increase in chromatin condensation or propidium iodide staining, and this cell death was prevented by preincubation with the NMDAR blocker MK-801 (Mander et al., 2005). NO-induced necrotic neuronal death could be further increased by deoxyglucose, a glycolysis inhibitor, which mediates energy depletion

(Mander et al., 2005). In this scenario, studies showing a synergistic effect of hypoxia and NO combination in nerve cells are of great significance for understanding better stroke, trauma, or degenerative diseases associated with metabolism dysfunction in nerve cells.

Overall, the differences in NO signaling in regulating neuronal cell death or survival may be explained by (1) neuronal or glial localization of NOS isoforms, (2) the amount of NO production within a cell upon a specific stimulus, and (3) the protein set that could be modulated by NO signaling cascade. This signaling duality relating neuronal cell death and survival is also observed in the activity of synaptic and extrasynaptic NMDA receptors in the brain (Hardingham & Bading, 2010). However, the concrete contribution of NO release, triggered by different populations of NMDA receptors in neurons, is still poorly understood and certainly deserves a much closer attention.

4. NO AND NEUROTRANSMITTERS RELEASE

As stated above, soon after being described as EDRF (Palmer et al., 1987), CNS cells were demonstrated to produce NO (Garthwaite et al., 1988; Garthwaite, Garthwaite, Palmer, & Moncada, 1989). Meanwhile, studies showed the role of NO in norepinephrine (Cohen & Weisbrod, 1988; Greenberg, Diecke, Peevy, & Tanaka, 1990; Halbrügge, Lütsch, Thyen, & Graefe, 1991b), adrenaline (Halbrügge, Lütsch, Thyen, & Graefe, 1991a), and histamine release (Masini, Salvemini, Pistelli, Mannaioni, & Vane, 1991) in the periphery. Then, Pape and Mager (1992) found that NO could control neuronal activity, whereas Ferriero, Sheldon, Black, and Chuai (1995) found the involvement of NO in neuronal cell death under pathological conditions. A number of studies have shown that NO could regulate neurotransmitter release in several CNS areas but this effect likely depended on the region studied. Actually, many of the NO effects in the CNS are due to its ability to modulate the release of different neurotransmitters, such as glutamate, GABA, and dopamine (DA) among others (Ishide, Nauli, Maher, & Ally, 2003; Kishi et al., 2001; Wang, Teschemacher, Paton, & Kasparov, 2006).

4.1. Glutamate release

Glutamate is the major excitatory neurotransmitter in the CNS, regulating many aspects of normal brain function. These functions are mediated by metabotropic and/or ionotropic glutamate receptors. Metabotropic receptors (mGluR) consist of eight subtypes, while ionotropic receptors are classified into three subtypes, which is based on their selective agonists NMDA, alpha-amino-3-hydroxy-5-methyl-4-isoazolepropionic acid (AMPA), and 2-carboxy-3-carboxymethyl-4-isopropenylpyrrolidine (kainate) receptors (Danbolt, 2001).

Glutamate release is fine-tuned by several molecules, including NO (O'dell, Hawkins, Kandel, & Arancio, 1991). It has been described that NO was capable of stimulating glutamate release in hippocampal slices (Lonart, Wang, & Johnson, 1992). Few years later, Segieth and her colleagues demonstrated that SNAP increased glutamate release in the rat hippocampus (Segieth, Getting, Biggs, & Whitton, 1995). Furthermore, it has been demonstrated in the rat dorsomedial medulla oblongata that NO could stimulate glutamate release (Lawrence & Jarrott, 1993). Still, NO-mediated glutamate release in the rat striatum is cGMP dependent and calcium independent (Guevara-Guzman, Emson, & Kendrick, 1994). In a different approach, it was demonstrated that NMDA and NOS blockers totally blocked NMDA-induced glutamate release in the hippocampus (Nei, Matsuyama, Shuntoh, & Tanaka, 1996; Segieth et al., 1995), *nucleus tractus solitarii* (Matsuo et al., 2001), cerebral cortex (Kano, Shimizu-Sasamata, Huang, Moskowitz, & Lo, 1998), and striatum (Bogdanov & Wurtman, 1997; Kendrick et al., 1996; Segovia & Mora, 1998), demonstrating that NMDA-induced glutamate release could be mediated by NO. Additionally, NOS I gene knockout impairs NMDA-induced glutamate release in both cerebral cortex and striatum (Kano et al., 1998). In the striatum and hippocampus, NMDA effect was calcium dependent (Bogdanov & Wurtman, 1997; Kendrick et al., 1996; Nei et al., 1996). Besides, kainate receptors were also capable of increasing NO-mediated glutamate release in the rat striatum in a calcium-dependent manner (Kendrick et al., 1996).

NMDA could also mediate an NO-dependent glutamate release from synaptosomal preparations (Hirsch et al., 1993; Montague, Gancayco, Winn, Marchase, & Friedlander, 1994), and NO donors were capable of stimulating glutamate release in a calcium-independent (McNaught & Brown, 1998; Sequeira, Ambrósio, Malva, Carvalho, & Carvalho, 1997) and EAAT-dependent manner (Sequeira et al., 1997). Furthermore, synaptosomes exposed to depolarizing stimuli could also enhance glutamate release, however in a calcium-dependent manner (Sequeira et al., 1997; Sistiaga, Miras-Portugal, & Sánchez-Prieto, 1997). In this experimental paradigm, NO classical pathway (sGC/cGMP/PKG) appears to inhibit this depolarization-evoked glutamate release (Sequeira et al., 1997; Sequeira,

Carvalho, & Carvalho, 1999; Sistiaga et al., 1997). In conclusion, it is likely that depolarization-evoked glutamate release is calcium dependent and inhibited by NO, while NO-mediated glutamate release is calcium independent and mediated by either EAAT reversal (Sequeira et al., 1997) or vesicular exocytosis (Meffert, Calakos, Scheller, & Schulman, 1996; Meffert, Premack, & Schulman, 1994).

Glial cells are also capable of releasing glutamate upon NO stimulation. It has been observed a rapid NO-induced glutamate release from rat astrocytes. This release was mediated by vesicular exocytosis and by both EAAT- and sGC-independent mechanisms (Bal-Price, Moneer, & Brown, 2002). Moreover, astrocytes, activated by LPS and interferon-γ, upregulate NOS II expression and augment glutamate efflux, suggesting that NO, produced by inflammation-activated astrocytes, induces glutamate release (Bal-Price et al., 2002). Microglial cells are also capable of stimulating an NO-dependent glutamate release. It was observed that rat microglial cells, activated by LPS, could enhance glutamate release via NO production, since this effect was totally abolished by an NOS inhibitor (Nakamura, Ohmaki, Murakami, & Yoneda, 2003).

4.2. GABA release

By using a number of pharmacological substances that change the availability of NO, it has been shown the involvement of NO in the regulation of GABA release in a number of different CNS areas. NO can induce an increase in GABA release, as systematically found in different brain areas. On the other hand, NO can mediate the inhibition of GABA release in the internal granule cell layer of the cerebellum and auditory cortical neurons. Some authors also investigated the role of cGMP in NO-induced GABA release since NO usually stimulates sGC. GABA release, mediated by NO, is cGMP-dependent in several hypothalamic nuclei (Yang, Chen, Li, & Pan, 2007) and in some instances PKGs may be involved as well (Li, Chen, Finnegan, & Pan, 2004). Many studies have demonstrated the involvement of NO in NMDA-stimulated GABA release (Hanania & Johnson, 1998; Ientile et al., 1997; Kano et al., 1998; Kendrick et al., 1996; Møller, Jones, & Beart, 1995) probably because NMDA can stimulate NO production by an NMDA-induced calcium influx. In the majority of these studies, NO mediates the increase in GABA release induced by NMDA. Kendrick et al. (1996) have also shown that NO decreases NMDA-stimulated GABA release in the striatum, whereas Møller et al.

(1995) suggested that NO could be linked to a negative feedback mechanism in the striatum.

It is very interesting though that in the hippocampus or retina, the role of NO in regulating GABA release may depend on its concentration. For such, Getting, Segieth, Ahmad, Biggs, and Whitton (1996) and Maggesissi et al. (2009) verified that basal or low NO concentration could decrease GABA release, whereas a high NO concentration could augment GABA release. Works associating NO-regulated GABA release in the hippocampus, however, were very inconsistent; sometimes NO mediates increase in GABA release, while in other cases it decreases the release of GABA. Data relating NO effect on GABA release in the striatum or brain stem were also very discrepant, suggesting that different NO levels could lead to different responses.

Another aspect of NO-regulated GABA release is the mechanism by which NO can alter GABA availability in the extracellular medium. Some authors described that NO stimulates GABA release in a calcium-dependent fashion, suggesting an exocytotic release (Trabace & Kendrick, 2000; Wang et al., 2006). On the other hand, some studies showed that a sodium-dependent mechanism mediates the effect of NO in GABA efflux through the reversal of GABA transporter (Hu, Zhang, Czeh, Flugge, & Zhang, 2010; Maggesissi et al., 2009). Nonetheless, some cells can use both mechanisms to release GABA in response to NO (Ohkuma, Katsura, Chen, Narihara, & Kuriyama, 1996; Yu & Eldred, 2005).

4.3. DA release

DA is a neurotransmitter present at high concentrations in the CNS, where it performs several functions (Carlsson, 2001). NO-mediated DA release was initially reported *in vitro* in the striatum where either exogenous or endogenous NO could significantly release DA (Zhu & Luo, 1992). The stimulatory effect of NO on DA release was also observed in the striatum *in vivo* (Strasser, McCarron, Ishii, Stanimirovic, & Spatz, 1994). Initial findings relating *in vitro* and *in vivo* NO-mediated DA release in the striatum were controversial since some groups observed that NO induces striatal DA release from slices in a sGC-independent manner (Büyükuysal, 1997; Lonart, Cassels, & Johnson, 1993), while cGMP analogues could stimulate DA release from the striatum *in vivo* (Guevara-Guzman et al., 1994). Furthermore, DA release induced by NO from striatal slices depends on calcium (Lonart et al., 1993) and activation of both voltage-dependent sodium and calcium channels and by facilitating DA transport reversal (Büyükuysal, 1997).

It has also been shown that DA released from the striatum *in vivo* is under the control of ionotropic glutamate receptors. Then, activation of NMDA or AMPA/kainate receptors leads to an increase in intracellular calcium, NOS stimulation, and NO production, which in turn induces DA release (Campos, Alfonso, Vidal, Faro, & Durán, 2006; Faro, Ferreira Nunes, Alfonso, Ferreira, & Durán, 2013). Besides, DA release induced by NO is also under NMDA control in other CNS regions such as the lateral olivocochlear nuclei (Halmos et al., 2008) and the medial preoptic area (Hull & Dominguez, 2006). NO also plays an important role in neuronal nicotine sensitization-dependent DA release in the *substantia nigra pars compacta* (Di Matteo et al., 2010) and nucleus accumbens (Hong, Jung, Bang, & Kim, 2006). However, in the intact retina, exposure to hydroxylamine, an NO donor, significantly decreased basal or potassium-induced DA release, while L-NA, an NOS blocker, could stimulate basal release of DA (Bugnon, Schaad, & Schorderet, 1994). Thus, DA release is under differential NO modulation in different CNS regions.

4.4. 5-Hydroxytryptamine release

Serotonin or 5-hydroxytryptamine (5-HT) is a monoamine neurotransmitter that significantly effects synaptic function, network activity, and behavior (Straub, Grant, O'Shea, & Benjamin, 2007). NO stimulates the release of 5-HT in the medial preoptic area (Lorrain & Hull, 1993), *locus coeruleus* (Prast & Philippu, 2001), and striatum (Guevara-Guzman et al., 1994). However, in the hypothalamus, NO donors modulate the release of 5-HT in a biphasic manner: while higher concentration enhanced 5-HT outflow, reperfusion with low NO donor concentration exerted the opposite effect in a sGC-dependent manner (Kaehler, Singewald, Sinner, & Philippu, 1999). 5-HT release by NO is also under the control of NMDA and AMPA/kainate receptors *in vivo* (Singewald, Kaehler, Hemeida, & Philippu, 1998) and *in vitro* in the striatum (Kendrick et al., 1996), *locus coeruleus* (Singewald et al., 1998), raphe nuclei, and prefrontal cortex (Smith & Whitton, 2000). Indeed, 5-HT release is under NO modulation in different CNS regions, and NO is likely an important transmitter coupling glutamatergic to serotoninergic neurotransmission.

5. NO AND NEUROPLASTICITY

The CNS is extremely complex and its function depends on the formation of precise connectivity patterns (Bleckert & Wong, 2011). During development, several mechanisms are important for constructing the mature connections that guide our behavior (Allen & Barres, 2009; Cohen-Cory, 2002). It is also known that the environment is capable of modeling neural circuits in a process known as neural plasticity (de Velasco et al., 2012; Tschetter et al., 2013). Synaptic reorganization is a dynamic process, which includes generation of new synapses and elimination of preexisting contacts. Besides, neural plasticity is common in response to tissue injury or neuroinflammation (Cabral-Miranda, Serfaty, & Campello-Costa, 2011; Campello-Costa, Fosse, Ribeiro, Paes-De-Carvalho, & Serfaty, 2000; Espirito-Santo et al., 2012; Mendonça et al., 2010; Oliveira-Silva et al., 2007). Moreover, it is believed that most of CNS diseases involve disorders in synaptic function or plasticity, which has been called synaptopathy (Li, Plomann, & Brundin, 2003; Won, Mah, & Kim, 2013). As a consequence, different studies have been performed to reveal the molecular mechanisms that regulate the emergence or loss of synapses, not only during normal development but also after a lesion. Both *in vitro* and *in vivo* evidence suggest that neuronal plasticity involves changes in synapse morphology and electrical activity of both pre- and postsynaptic elements, which are coordinated by extracellular signals. Of particular relevance to this chapter, NO has been suggested to be involved in different aspects of synaptic dynamics, which may be implicated in several physiological and pathological conditions (Garthwaite, 2008).

5.1. NO and structural plasticity

Neural plasticity is a highly regulated process, which requires precise coordination between pre- and postsynaptic elements. Concerning the roles played by NO, it has been shown that it includes the differentiation of synaptic specializations, microtubule dynamics, architecture of synaptic protein organization, modulation of synaptic efficacy, and regulation of gene expression (Wang et al., 2005; Fig. 5.5). Evidence for the effect of NO upon neuritogenesis or formation of functional synapses is emerging in different regions of the nervous system. Systemic blockade of NOS disrupts the elimination of transient retinotectal ipsilateral projection in the chick (Wu, Williams, & McLoon, 1994) and the development of ON/OFF

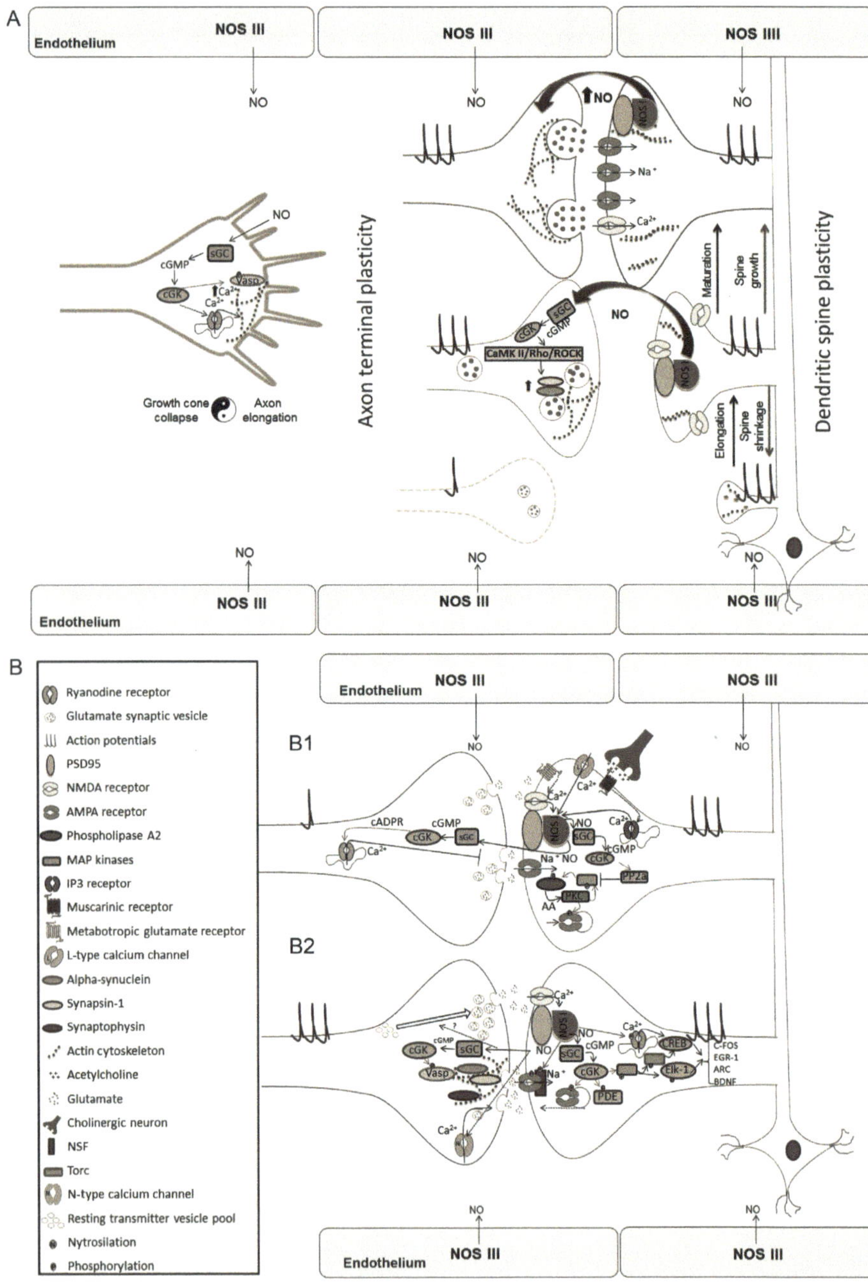

Figure 5.5 Mechanisms underlying NO-induced neuronal plasticity. (A) During early stages of development, NO modifies growth cone dynamics leading to axonal elongation or growth cone collapse (left). NO also participates in axonal and dendritic plasticity

sublaminae of ferret retinogeniculate projections (Cramer & Sur, 1999). Moreover, our group has demonstrated that NOS activity is highest during the period of topographical refinement, declining as maturation of retinotectal projections proceeds. Additionally, reduction of NOS activity within the first two postnatal weeks, but not afterward, results in anatomical reorganization of this projection with expansion of terminal fields, suggesting that NO has a stabilizing role upon primary visual targets during or shortly after the development of retinotectal maps (Campello-Costa et al., 2000). Indeed, NOS III/NOS I double-knockout mice presented a similar disorganization of uncrossed retinotectal pathways (Wu et al., 1994). A recent study has demonstrated that NOS I is expressed by specific subsets of GABAergic neocortical neurons, and the blockade of NO synthesis in the rat auditory cortex abolished presynaptic components of plastic changes at layer 2/3 pyramidal cells. These data suggest that NO-producing interneurons in the neocortex convey lateral inhibition to neighboring columns, which may shape the spatiotemporal dynamics of network activity, underlying plastic responses (Lee, Stoelzel, Chistiakova, & Volgushev, 2012). On the other hand, no change could be observed in ocular dominance column and barrel field formation after inhibition of NOS activity during the critical period of development (Finney & Shatz, 1998).

Other important data came from NO effect upon different lesion-induced plasticity models. Zhang, Granstrom, and WongRiley (1996) showed that monocular enucleation in adult rats produced a progressive downregulation of NOS at the deprived contralateral tectum, indicating that NOS level could be altered by afferent activity. We have previously shown that a reduction of NOS activity greatly enhanced the plasticity of uncrossed retinotectal pathway following a lesion to the contralateral temporal retina within the first three postnatal weeks (Campello-Costa et al., 2000). Recently, it has been demonstrated, using the snail nervous system as a

(right). Depending on spatiotemporal dynamics, both pre- and postsynaptic terminals change their morphology. Asynchronous synaptic activity leads to a retraction of axon and spine shrinkage and a synchronous activity induces NO production, neurotransmitter release, and axonal differentiation. Note that more vesicles are driven to axonal terminal. This reinforcement leads to increased NO synthesis, spine growth, and maturation, with more receptors at postsynaptic site and neurotransmitter released by presynaptic cells. (B) NO drives both LTD (B1) and LTP (B2). (B1) Noncorrelated synaptic activity leads to LTD via NO production. (B2) Correlated pre- and postsynaptic activity leads to LTP via NO production. The metabolic pathways involved are explained in the text. (See the color plate.)

model, that chronic inhibition of endogenous NO synthesis could abolish neurite outgrowth following axotomy, while chronic application of an NO donor rescued it, leading to *de novo* synapse formation and remodeling (Cooke, Mistry, Challiss, & Straub, 2013). Taken together, these data indicate that NO synthesis may act as an important mechanism involved in synaptogenesis or synapse elimination. There are now at least three different well-described synaptic targets in which NO could act as (1) presynaptic neurons in which NO could function as retrograde messenger to modulate either neurotransmitter release or presynaptic structural plasticity, including axonal differentiation; (2) postsynaptic cells where NO could modify membrane turnover of different receptors and dendritic morphology; and (3) adjacent glial cells, which have been shown to be important partners in controlling synaptic transmission and plasticity.

Concerning the effects of NO upon axons, it has been shown that it can act in different developmental stages, playing pivotal roles in neuronal growth cone morphogenesis (Nikonenko, Jourdain, & Muller, 2003), axonal guidance, and synapse formation or stabilization (Nikonenko et al., 2008). In early developmental stages, NO may directly affect growth cone motility by a transient elevation in intracellular Ca^{2+} from ryanodine-sensitive stores via sGC/PKG pathway (Fig. 5.5A). In the visual system, NOS expression is temporally linked with axonal innervation of the optic tectum, indicating that NO could play a role in pathfinding (Berman & Morris, 2011; Williams, Nordquist, & McLoon, 1994). On the other hand, NO signaling has also been associated with growth cone collapse in different neuronal cell types such as dorsal root ganglion neurons (Hess, Patterson, Smith, & Skene, 1993) and retinal ganglion cells (Ernst, Gallo, Letourneau, & McLoon, 2000).

NO affects growth processes by interfering with actin filament polymerization (Gallo, Ernst, McLoon, & Letourneau, 2002). NO produces rapid retraction of lamellipodia via PKG activity, which phosphorylates VASP, a key regulator of actin polymerization (Lindsay, Ramsey, Aitchison, Renne, & Evans, 2007). The discrepancy in the literature regarding the effects of NO on neurites during pathfinding could reflect the fact that NO effect usually depends on its concentration as well as multiple interactions between NO and other signaling molecules (Trimm & Rehder, 2004). In fact, Ernst and coworkers have shown that NO and BDNF may act together to induce growth cone stability (Ernst et al., 2000). Besides, NO may promote the differentiation of axons into varicosities, forming multi-innervated spines in hippocampal neurons (Nikonenko et al., 2008), culminating in the appearance of new synaptic contacts (Nikonenko et al., 2003). In addition to these roles on presynaptic structural plasticity, evidence shows that NO may act as a retrograde messenger, modulating transmitter release from different or even the same terminals under different conditions (Cserep et al., 2011; Garthwaite, 2008). The functional relevance of these data will be further discussed below.

Despite evidence supporting presynaptic roles, NO may also affect postsynaptic cells simultaneously. NO could S-nitrosylate different proteins, whereby it could modulate dendrite outgrowth during development (Zhang et al., 2010; Fig. 5.5A). In fact, NOS inhibition, during development, leads to disturbances in dendrite morphology and to a reduction in synapse number (Sanchez-Islas & Leon-Olea, 2004). Furthermore, NO could also affect the expression of postsynaptic receptors. NO/cGMP pathway is important for agrin/MusK signaling, which is involved in the reorganization of actin cytoskeleton with subsequent AChR aggregation at the surface of skeletal muscle cells (Godfrey & Schwarte, 2010). This means that NO coordinates presynaptic and postsynaptic function during plasticity. Actually, inhibition of NOS leads to failure of axons in responding to protrusion elimination (Nikonenko et al., 2008). In this scenario, NO could function as a postsynaptic element to drive contact formation during synaptic maturation (Yoshihara, De Roo, & Muller, 2009). It has been shown recently, using hippocampal slices, that overexpression of PSD-95 leads to NO release, which in turn acts as a retrograde messenger to induce differentiation of presynaptic buttons (Poglia, Muller, & Nikonenko, 2011).

Finally, as previously mentioned, glial cells have been disclosed as important partners that modulate synaptic transmission and plasticity (Allen & Barres, 2009). Indeed, astrocyte-secreted factors, like thrombospondins, promote normal CNS synaptogenesis (Christopherson et al., 2005). During CNS development, sensory stimulation promotes synaptic rearrangement in the optic tectum, allowing radial glia to respond to an enhancement in Ca^{2+} transients, leading to structural plasticity. Indeed, Ca^{2+} transients (Willmott, Wong, & Strong, 2000) and cytoskeletal dynamics in cultured astrocytes (Boran & Garcia, 2007) have been reported to be associated with NO/PKG pathway activation. Therefore, it is reasonable to propose that NO, in glial cells, interferes with neuron–neuron connection, which in turn may underlie plasticity. Overall, these data support the idea that NO is an important signaling molecule not only to natural plasticity, which is observed during neural circuits development, but also to lesion-induced plasticity. Alternatively, it is remarkable that NO is also associated with pathological conditions. Several neurological conditions such as autism spectrum disorders, schizophrenia, bipolar disorder associated with psychosis, and epilepsy, among others have been associated with abnormal dendritic spines (Fiala, Spacek, & Harris, 2002; Penzes, Cahill, Jones, VanLeeuwen, & Woolfrey, 2011). Hence, it is reasonable to propose that NO could be partially responsible for some of synaptic or cognitive dysfunctions in those conditions. As such, dysfunction in NO pathway would likely lead to abnormalities in axons, dendritic spines, and, ultimately, synaptogenesis.

5.2. NO and functional plasticity

Structural plasticity is accompanied by alterations in synaptic efficacy. This synaptic plasticity is the best-comprehended form of neuroplasticity, which involves pre- and postsynaptic changes. Considered in the literature as key steps in learning and memory, LTP and LTD are the most studied issues in neuroplasticity, and many forms of LTD and LTP require NMDAR activation and Ca^{2+} influx. Since the site of induction is usually postsynaptic, the processes require a messenger to dictate changes in the presynaptic site. Because of its diffusible nature and NMDAR-linked synthesis, NO seems to be good candidate for such retrograde *trans*-synaptic function. Besides, NO could also modulate postsynaptic aspects of plastic events (Garthwaite, 2008; Garthwaite et al., 1988). Here, we are going to briefly discuss the role of NO in these processes.

The hippocampus was the first locus in which those plastic phenomena were characterized, where they seem to direct spatial learning and memory. It was proposed that Ca^{2+} influx through NMDAR would act like a switch, where high Ca^{2+} levels could activate kinases and trigger LTP, while low Ca^{2+} would activate phosphatases and trigger LTD (Malenka, 1994). Recently, Bartus, Pigott, and Garthwaite (2013) studied the cellular targets of NO in the hippocampus. This study revealed that all regions expressed both sGC and cGMP, including axons and pyramidal cells, suggesting that NO can act either pre- or postsynaptically. Ratnayaka et al. (2012) showed that high-frequency stimulation (HFS) induces LTP in the postsynaptic neuron in an NMDAR- and NO-dependent fashion. This LTP, however, has a great presynaptic component, showing enhancement in the recycling pool of synaptic vesicles at the expense of the resting pool, accompanied by an increase in release probability. Moreover, glutamate application to neurons in culture increases the clusters of presynaptic proteins involved in transmitter release, such as alpha-synuclein, synapsin-1, and synaptophysin. At the postsynaptic site, glutamate induces the formation of GluR1 clusters aligned with the presynaptic cluster, thus optimizing transmission. NO regulates the formation of all these synaptic clusters via a PKG-stimulated VASP phosphorylation (Liu et al., 2004; Wang et al., 2005). Besides, Serulle et al. (2007) showed that GluR1 membrane insertion was stimulated by an NO-induced PKGII activation after LTP induction. Monfort et al. (2004) showed that CA1 LTP is dependent on NMDAR activation, followed by a sGC-induced increase in cGMP production and activation of PKG within 5 min, and then PKG-mediated phosphorylation of cGMP phosphodiesterase, which maintains cGMP below basal levels for more than 60 min. In addition, after ischemia, NO leads to LTP induction, but not its maintenance, via cGMP/PKG (Costa et al., 2011). In addition, multiple HFS trains at Schaffer collateral induced late-phase LTP. A single HFS train is sufficient to promote potentiation once exogenous NO or cGMP is applied. This late component of LTP is dependent on Ca^{2+} release from intracellular stores through ryanodine receptors (RYR), leading to CREB phosphorylation and protein synthesis at the postsynaptic site (Lu & Hawkins, 2002; Lu, Kandel, & Hawkins, 1999). However, hippocampal synapses can also undergo LTD. CA1 LTD involves the NMDA/NO/sGC/PKG pathway (Stanton et al., 2003). Subsequently, Reyes-Harde, Potter, Galione, and Stanton (1999) showed that NO/sGC-induced LTD involves ADP-cyclic ribose (cADPR)-induced Ca^{2+} influx through RYR.

Within the cerebral cortex, synaptic plasticity might govern learning, and memory storage in different cortical areas. In the auditory cortex of rats, layer V LTP requires AMPA and NMDA receptor function, NOS activity, and cGMP production (Wakatsuki et al., 1998). The presynaptic component of LTP is NOS I dependent and application of an NO donor increases the frequency, but not the amplitude of EPSPs, arguing for a role of NO in enhancing release probability (Dachtler et al., 2011). In another study, cortical cultures have been submitted to a protocol of bicuculline-induced LTP. These cultures presented a PKG-induced ERK1/2 phosphorylation, which promotes the expression of neuroplasticity-related proteins, such as c-Fos, EGR-1, Arc, BDNF, possibly via TORC-mediated CREB activation or ERK1/2-mediated Elk1 phosphorylation. The *in vivo* model of single whisker experience also promoted an increase in c-Fos, EGR-1, and BDNF expression. All these processes were dependent on the NO/sGC/PKG pathway (Gallo & Iadecola, 2011). Neocortical neurons also display LTP-dependent plasticity. Muscarinic activation of layer V pyramidal neurons from the medial prefrontal cortex induces Ca^{2+} release from intracellular stores via modulation of IP3 channels, leading to NO generation. NO diffuses to the presynaptic terminal where it reduces neurotransmitter release through sGC/PKG activation (Huang, Chan, & Hsu, 2003).

In the cerebellum, particularly in the circuit between climbing fibers (CFs), Purkinje cells (PCs), and parallel fibers (PFs), synaptic plasticity has been shown to be crucial for motor learning. Coesmans and colleagues showed that synaptic plasticity within the PF–PC circuit depended on PC Ca^{2+} levels, but differently from other brain regions, low Ca^{2+} induces LTP while high-Ca^{2+} leads to LTD (Coesmans, Weber, De Zeeuw, & Hansel, 2004). CF stimulation leads to huge Ca^{2+} transients in PC, due to the opening of voltage-gated Ca^{2+} channels. On the other hand, PF stimulation leads to discrete Ca^{2+} transients, caused by the opening of RYR in the endoplasmic reticulum (Kakizawa et al., 2012). As reviewed by Ogasawara, Doi, and Kawato (2008), PF–PC LTD can be induced by costimulation of CF and PF. NO production by the axon terminal of PF diffuses to PC dendrites inhibiting PP2a via sGC/PKG activation, which leads to a PKC-induced internalization of GluR2-containing AMPA receptors. Conversely, PF–PC LTP can be induced within the same synaptic circuit as long as only PF are stimulated (Contestabile, 2012). PF-produced NO S-nitrosylates RYR, mediating a discrete Ca^{2+} influx required for LTP (Kakizawa et al., 2012). Moreover, NSF may also be S-nitrosylated, mediating membrane insertion of GluR2-containing AMPARs (Huang et al., 2005). On the basal ganglia, Chepkova et al. (2009) showed that group 1 mGluR-induced corticostriatal LTD requires NMDA activation and NO synthesis. Together, these data support the idea that NO-dependent synaptic plasticity might be important for motor control.

Brainstem rostral ventrolateral medulla (RVLM) neurons play an important role in cardiovascular, respiratory, and nociceptive functions. Huang et al. (2003) demonstrated that application of an NO donor induced presynaptic LTP in RVLM neurons, and that co-application of the inhibitor of N-type Ca^{2+} channel, ω-conotoxin, blocked this potentiation, suggesting that the increase in transmitter release was mediated by Ca^{2+} influx through presynaptic N-type Ca^{2+} channels.

The retinocolicular system is important in driving the head sensory organs toward an object of interest, which is an important environmental adaptation in higher vertebrates including humans. LFS or HFS of the optic tract of rat pups until P14 induced LTD of the superficial layers of the superior colliculus in an L-type Ca^{2+} channel-dependent fashion (Cork, Namkung, Shin, & Mize, 2001; Lo & Mize, 2000). The authors hypothesized that activation of NMDARs could promote L-type Ca^{2+} channel activation, which provides Ca^{2+} to activate NOS, leading to synaptic depression. Confirming this hypothesis, Lo and Mize (2000) showed that retinocolicular LTD is reduced in NOS III/NOS I double-knockout mice and that this plasticity could be recovered when stimulation was combined with treatment with an NO donor. Moreover, in Xenopus tadpoles, Mu and Poo (2006) showed that retinotectal synapses undergo experience-driven spike-timing-dependent plasticity, where LTD was dependent on NO synthesis. In this same paradigm, Du, Wei, Wang, Wong, and Poo (2009) showed that LTD also depressed synapses between bipolar cells and retinal ganglion cells via PKG activation, leading to a decrease in AMPARs conductance. Since rat retinocolicular LTD has been observed until PND14 (Lo & Mize, 2000), coinciding with the period of synaptic refinement (Serfaty, Campello-Costa, & Linden, 2005; Serfaty & Linden, 1994) and high NO production (Campello-Costa et al., 2000), it should be therefore reasonable to propose that NO might drive the elimination of ectopic axons through LTD of noncorrelated synaptic partner.

REFERENCES

Alderton, W., Cooper, C., & Knowles, R. (2001). Nitric oxide synthases: Structure, function and inhibition. *The Biochemical Journal*, *357*, 593–615.

Allen, N. J., & Barres, B. A. (2009). *Intercellular communication in the nervous system* (1st ed.). Elsevier (*Glia and synapse formation: An overview*).

Almeida, A., Almeida, J., Bolanos, J. P., & Moncada, S. (2001). Different responses of astrocytes and neurons to nitric oxide: The role of glycolytically generated ATP in astrocyte protection. *Proceedings of the National Academy of Sciences of the United States of America*, *98*(26), 15294–15299.

Arnold, W. P., Mittal, C. K., Katsuki, S., & Murad, F. (1977). Nitric oxide activates guanylate cyclase and increases guanosine 3′:5′-cyclic monophosphate levels in various tissue preparations. *Proceedings of the National Academy of Sciences of the United States of America*, *74*(8), 3203–3207.

Ayata, C., Ayata, G., Hara, H., Matthews, R. T., Beal, M. F., Ferrante, R. J., et al. (1997). Mechanisms of reduced striatal NMDA excitotoxicity in type I nitric oxide synthase knock-out mice. *The Journal of Neuroscience*, *17*(18), 6908–6917.

Bal-Price, A., & Brown, G. C. (2001). Inflammatory neurodegeneration mediated by nitric oxide from activated glia-inhibiting neuronal respiration, causing glutamate release and excitotoxicity. *The Journal of Neuroscience*, *21*(17), 6480–6491.

Bal-Price, A., Moneer, Z., & Brown, G. C. (2002). Nitric oxide induces rapid, calcium-dependent release of vesicular glutamate and ATP from cultured rat astrocytes. *Glia*, *40*(3), 312–323.

Bartus, K., Pigott, B., & Garthwaite, J. (2013). Cellular targets of nitric oxide in the hippocampus. *PLoS One*, *8*(2), e57292.

Benhar, M., Forrester, M. T., Hess, D. T., & Stamler, J. S. (2008). Regulated protein denitrosylation by cytosolic and mitochondrial thioredoxins. *Science*, *320*(5879), 1050–1054.

Berman, S., & Morris, A. (2011). Nitric oxide as a putative retinal axon pathfinding and target recognition cue in *Xenopus laevis*. *Impulse (Columbia)*, *2010*, 1–12.
Bian, K., Gao, Z., Weisbrodt, N., & Murad, F. (2003). The nature of heme/iron-induced protein tyrosine nitration. *Proceedings of the National Academy of Sciences of the United States of America*, *100*(10), 5712–5717.
Bleckert, A., & Wong, R. O. (2011). Identifying roles for neurotransmission in circuit assembly: Insights gained from multiple model systems and experimental approaches. *Bioessays*, *33*(1), 61–72.
Boehning, D., & Snyder, S. H. (2003). Novel neural modulators. *Annual Review of Neuroscience*, *26*, 105–131.
Bogdan, C. (2001). Nitric oxide and the immune response. *Nature Immunology*, *2*(10), 907–916.
Bogdanov, M. B., & Wurtman, R. J. (1997). Possible involvement of nitric oxide in NMDA-induced glutamate release in the rat striatum: An in vivo microdialysis study. *Neuroscience Letters*, *221*(2), 197–201.
Boran, M. S., & Garcia, A. (2007). The cyclic GMP-protein kinase G pathway regulates cytoskeleton dynamics and motility in astrocytes. *Journal of Neurochemistry*, *102*(1), 216–230.
Bredt, D. S., Hwang, P. M., Glatt, C. E., Lowenstein, C., Reed, R. R., & Snyder, S. H. (1991). Cloned and expressed nitric oxide synthase structurally resembles cytochrome P-450 reductase. *Nature*, *351*(6329), 714–718.
Bredt, D. S., Hwang, P. M., & Snyder, S. H. (1990). Localization of nitric oxide synthase indicating a neural role for nitric oxide. *Nature*, *347*(6295), 768.
Bredt, D. S., & Snyder, S. H. (1990). Isolation of nitric oxide synthetase, a calmodulin-requiring enzyme. *Proceedings of the National Academy of Sciences of the United States of America*, *87*(2), 682–685.
Brenman, J. E., & Bredt, D. S. (1997). Synaptic signaling by nitric oxide. *Current Opinion in Neurobiology*, *7*(3), 374–378.
Brown, G. C. (2010). Nitric oxide and neuronal death. *Nitric Oxide*, *23*(3), 153–165.
Bugnon, O., Schaad, N. C., & Schorderet, M. (1994). Nitric oxide modulates endogenous dopamine release in bovine retina. *Neuroreport*, *5*(4), 401–404.
Butt, E., Geiger, J., Jarchau, T., Lohmann, S. M., & Walter, U. (1993). The cGMP-dependent protein kinase—Gene, protein, and function. *Neurochemical Research*, *18*(1), 27–42.
Büyükuysal, R. L. (1997). Effect of nitric oxide donors on endogenous dopamine release from rat striatal slices. II: The role of voltage-dependent sodium channels, calcium channel activation, reverse transport mechanism, guanylate cyclase and endogenous glutamate. *Fundamental and Clinical Pharmacology*, *11*(6), 528–536.
Cabral-Miranda, F., Serfaty, C. A., & Campello-Costa, P. (2011). A time-dependent effect of caffeine upon lesion-induced plasticity. *Neuroscience Research*, *71*(1), 99–102.
Cai, H., & Harrison, D. G. (2000). Endothelial dysfunction in cardiovascular diseases: The role of oxidant stress. *Circulation Research*, *87*(10), 840–844.
Campello-Costa, P., Fosse, A. M., Jr., Ribeiro, J. C., Paes-De-Carvalho, R., & Serfaty, C. A. (2000). Acute blockade of nitric oxide synthesis induces disorganization and amplifies lesion-induced plasticity in the rat retinotectal projection. *Journal of Neurobiology*, *44*(4), 371–381.
Campos, F., Alfonso, M., Vidal, L., Faro, L. R., & Durán, R. (2006). Mediation of glutamatergic receptors and nitric oxide on striatal dopamine release evoked by anatoxin-a. An in vivo microdialysis study. *European Journal of Pharmacology*, *548*(1–3), 90–98.
Carlsson, A. (2001). A half-century of neurotransmitter research: Impact on neurology and psychiatry. Nobel lecture. *Bioscience Reports*, *21*(6), 691–710.

Chepkova, A. N., Fleischer, W., Kazmierczak, T., Doreulee, N., Haas, H. L., & Sergeeva, O. A. (2009). Developmental alterations of DHPG-induced long-term depression of corticostriatal synaptic transmission: Switch from NMDA receptor-dependent towards CB1 receptor-dependent plasticity. *Pflügers Archiv*, *459*(1), 131–141.

Cho, H. J., Xie, Q., Calaycay, J., Mumford, R. A., Swiderek, K. M., Lee, T. D., et al. (1992). Calmodulin is a subunit of nitric oxide synthase from macrophages. *The Journal of Experimental Medicine*, *176*(2), 599–604.

Christopherson, K. S., Ullian, E. M., Stokes, C. C., Mullowney, C. E., Hell, J. W., Agah, A., et al. (2005). Thrombospondins are astrocyte-secreted proteins that promote CNS synaptogenesis. *Cell*, *120*(3), 421–433.

Chung, K. K., Thomas, B., Li, X., Pletnikova, O., Troncoso, J. C., Marsh, L., et al. (2004). S-nitrosylation of parkin regulates ubiquitination and compromises parkin's protective function. *Science*, *304*(5675), 1328–1331.

Ciani, E., Guidi, S., Bartesaghi, R., & Contestabile, A. (2002). Nitric oxide regulates cGMP-dependent cAMP-responsive element binding protein phosphorylation and Bcl-2 expression in cerebellar neurons: Implication for a survival role of nitric oxide. *Journal of Neurochemistry*, *82*(5), 1282–1289.

Ciani, E., Virgili, M., & Contestabile, A. (2002). Akt pathway mediates a cGMP-dependent survival role of nitric oxide in cerebellar granule neurones. *Journal of Neurochemistry*, *81*(2), 218–228.

Coesmans, M., Weber, J. T., De Zeeuw, C. I., & Hansel, C. (2004). Bidirectional parallel fiber plasticity in the cerebellum under climbing fiber control. *Neuron*, *44*(4), 691–700.

Cohen, R. A., & Weisbrod, R. M. (1988). Endothelium inhibits norepinephrine release from adrenergic nerves of rabbit carotid artery. *The American Journal of Physiology*, *254*(5 Pt 2), H871–H878.

Cohen-Cory, S. (2002). The developing synapse: Construction and modulation of synaptic structures and circuits. *Science*, *298*(5594), 770–776.

Collingridge, G. L., & Singer, W. (1990). Excitatory amino acid receptors and synaptic plasticity. *Trends in Pharmacological Sciences*, *11*(7), 290–296.

Contestabile, A. (2008). Regulation of transcription factors by nitric oxide in neurons and in neural-derived tumor cells. *Progress in Neurobiology*, *84*(4), 317–328.

Contestabile, A. (2012). Role of nitric oxide in cerebellar development and function: Focus on granule neurons. *Cerebellum*, *11*(1), 50–61.

Cooke, R. M., Mistry, R., Challiss, R. A., & Straub, V. A. (2013). Nitric oxide synthesis and cGMP production is important for neurite growth and synapse remodeling after axotomy. *The Journal of Neuroscience*, *33*(13), 5626–5637.

Cork, R. J., Namkung, Y., Shin, H. S., & Mize, R. R. (2001). Development of the visual pathway is disrupted in mice with a targeted disruption of the calcium channel beta (3)-subunit gene. *The Journal of Comparative Neurology*, *440*(2), 177–191.

Cossenza, M., Cadilhe, D. V., Coutinho, R. N., & Paes-de-Carvalho, R. (2006). Inhibition of protein synthesis by activation of NMDA receptors in cultured retinal cells: A new mechanism for the regulation of nitric oxide production. *Journal of Neurochemistry*, *97*(5), 1481–1493.

Cossenza, M., & Paes de Carvalho, R. (2000). L-Arginine uptake and release by cultured avian retinal cells: Differential cellular localization in relation to nitric oxide synthase. *Journal of Neurochemistry*, *74*(5), 1885–1894.

Costa, C., Tozzi, A., Siliquini, S., Galletti, F., Cardaioli, G., Tantucci, M., et al. (2011). A critical role of NO/cGMP/PKG dependent pathway in hippocampal post-ischemic LTP: Modulation by zonisamide. *Neurobiology of Disease*, *44*(2), 185–191.

Cramer, K. S., & Sur, M. (1999). The neuronal form of nitric oxide synthase is required for pattern formation by retinal afferents in the ferret lateral geniculate nucleus. *Brain Research. Developmental Brain Research*, *116*(1), 79–86.

Cserep, C., Szonyi, A., Veres, J. M., Nemeth, B., Szabadits, E., de Vente, J., et al. (2011). Nitric oxide signaling modulates synaptic transmission during early postnatal development. *Cerebral Cortex*, *21*(9), 2065–2074.

Dachtler, J., Hardingham, N. R., Glazewski, S., Wright, N. F., Blain, E. J., & Fox, K. (2011). Experience-dependent plasticity acts via GluR1 and a novel neuronal nitric oxide synthase-dependent synaptic mechanism in adult cortex. *The Journal of Neuroscience*, *31*(31), 11220–11230.

Danbolt, N. C. (2001). Glutamate uptake. *Progress in Neurobiology*, *65*(1), 1–105.

Datta, S. R., Dudek, H., Tao, X., Masters, S., Fu, H., Gotoh, Y., et al. (1997). Akt phosphorylation of BAD couples survival signals to the cell-intrinsic death machinery. *Cell*, *91*(2), 231–241.

Davis, S., & Laroche, S. (2006). Mitogen-activated protein kinase/extracellular regulated kinase signalling and memory stabilization: A review. *Genes, Brain and Behavior*, *5*(Suppl. 2), 61–72.

Dawson, V. L., Dawson, T. M., London, E. D., Bredt, D. S., & Snyder, S. H. (1991). Nitric oxide mediates glutamate neurotoxicity in primary cortical cultures. *Proceedings of the National Academy of Sciences of the United States of America*, *88*(14), 6368–6371.

Dawson, V. L., Kizushi, V. M., Huang, P. L., Snyder, S. H., & Dawson, T. M. (1996). Resistance to neurotoxicity in cortical cultures from neuronal nitric oxide synthase-deficient mice. *The Journal of Neuroscience*, *16*(8), 2479–2487.

Deguchi, T., & Yoshioka, M. (1982). L-Arginine identified as an endogenous activator for soluble guanylate cyclase from neuroblastoma cells. *Journal of Biological Chemistry*, *257*(17), 10147–10151.

Denninger, J. W., & Marletta, M. A. (1999). Guanylate cyclase and the NO/cGMP signaling pathway. *Biochimica et Biophysica Acta*, *1411*(2–3), 334–350.

de Velasco, P. C., Mendonca, H. R., Borba, J. M., Andrade da Costa, B. L., Guedes, R. C., Navarro, D. M., et al. (2012). Nutritional restriction of omega-3 fatty acids alters topographical fine tuning and leads to a delay in the critical period in the rodent visual system. *Experimental Neurology*, *234*(1), 220–229.

Di Matteo, V., Pierucci, M., Benigno, A., Esposito, E., Crescimanno, G., & Di Giovanni, G. (2010). Critical role of nitric oxide on nicotine-induced hyperactivation of dopaminergic nigrostriatal system: Electrophysiological and neurochemical evidence in rats. *CNS Neuroscience and Therapeutics*, *16*(3), 127–136.

Dinerman, J. L., Steiner, J. P., Dawson, T. M., Dawson, V., & Snyder, S. H. (1994). Cyclic nucleotide dependent phosphorylation of neuronal nitric oxide synthase inhibits catalytic activity. *Neuropharmacology*, *33*(11), 1245–1251.

Du, J. L., Wei, H. P., Wang, Z. R., Wong, S. T., & Poo, M. M. (2009). Long-range retrograde spread of LTP and LTD from optic tectum to retina. *Proceedings of the National Academy of Sciences of the United States of America*, *106*(45), 18890–18896.

Dudek, H., Datta, S. R., Franke, T. F., Birnbaum, M. J., Yao, R., Cooper, G. M., et al. (1997). Regulation of neuronal survival by the serine–threonine protein kinase Akt. *Science*, *275*(5300), 661–665.

Ernst, A. F., Gallo, G., Letourneau, P. C., & McLoon, S. C. (2000). Stabilization of growing retinal axons by the combined signaling of nitric oxide and brain-derived neurotrophic factor. *The Journal of Neuroscience*, *20*(4), 1458–1469.

Espirito-Santo, S., Mendonça, H. R., Menezes, G. D., Goulart, V. G., Gomes, A. L., Marra, C., et al. (2012). Intravitreous interleukin-2 treatment and inflammation modulates glial cells activation and uncrossed retinotectal development. *Neuroscience*, *200*, 223–236.

Esplugues, J. V. (2002). NO as a signalling molecule in the nervous system. *British Journal of Pharmacology*, *135*(5), 1079–1095.

Faro, L. R., Ferreira Nunes, B. V., Alfonso, M., Ferreira, V. M., & Durán, R. (2013). Role of glutamate receptors and nitric oxide on the effects of glufosinate ammonium, an organophosphate pesticide, on in vivo dopamine release in rat striatum. *Toxicology, 311*(3), 154–161.

Fayard, E., Tintignac, L. A., Baudry, A., & Hemmings, B. A. (2005). Protein kinase B/Akt at a glance. *Journal of Cell Science, 118*(24), 5675–5678.

Ferrendelli, J. A., Chang, M. M., & Kinscherf, D. A. (1974). Elevation of cyclic GMP levels in central nervous system by excitatory and inhibitory amino acids. *Journal of Neurochemistry, 22*(4), 535–540.

Ferriero, D. M., Sheldon, R. A., Black, S. M., & Chuai, J. (1995). Selective destruction of nitric oxide synthase neurons with quisqualate reduces damage after hypoxia–ischemia in the neonatal rat. *Pediatric Research, 38*(6), 912–918.

Fiala, J. C., Spacek, J., & Harris, K. M. (2002). Dendritic spine pathology: Cause or consequence of neurological disorders? *Brain Research. Brain Research Reviews, 39*(1), 29–54.

Finney, E. M., & Shatz, C. J. (1998). Establishment of patterned thalamocortical connections does not require nitric oxide synthase. *The Journal of Neuroscience, 18*(21), 8826–8838.

Flores, A. I., Mallon, B. S., Matsui, T., Ogawa, W., Rosenzweig, A., Okamoto, T., et al. (2000). Akt-mediated survival of oligodendrocytes induced by neuregulins. *The Journal of Neuroscience, 20*(20), 7622–7630.

Flores, A. I., Narayanan, S. P., Morse, E. N., Shick, H. E., Yin, X., Kidd, G., et al. (2008). Constitutively active Akt induces enhanced myelination in the CNS. *The Journal of Neuroscience, 28*(28), 7174–7183.

Friebe, A., & Koesling, D. (2003). Regulation of nitric oxide-sensitive guanylyl cyclase. *Circulation Research, 93*(2), 96.

Furchgott, R. F., & Zawadzki, J. V. (1980). The obligatory role of endothelial cells in the relaxation of arterial smooth muscle by acetylcholine. *Nature, 288*(5789), 373–376.

Gallo, G., Ernst, A. F., McLoon, S. C., & Letourneau, P. C. (2002). Transient PKA activity is required for initiation but not maintenance of BDNF-mediated protection from nitric oxide-induced growth-cone collapse. *The Journal of Neuroscience, 22*(12), 5016–5023.

Gallo, E. F., & Iadecola, C. (2011). Neuronal nitric oxide contributes to neuroplasticity-associated protein expression through cGMP, protein kinase G, and extracellular signal-regulated kinase. *The Journal of Neuroscience, 31*(19), 6947–6955.

Garcia-Martinez, J., Moran, J., Clarke, R., Gray, A., Cosulich, S., Chresta, C., et al. (2009). Ku-0063794 is a specific inhibitor of the mammalian target of rapamycin (mTOR). *The Biochemical Journal, 421*, 29–42.

Garg, U. C., & Hassid, A. (1989). Nitric oxide-generating vasodilators and 8-bromo-cyclic guanosine monophosphate inhibit mitogenesis and proliferation of cultured rat vascular smooth muscle cells. *The Journal of Clinical Investigation, 83*(5), 1774–1777.

Garthwaite, J. (1991). Glutamate, nitric oxide and cell–cell signalling in the nervous system. *Trends in Neurosciences, 14*(2), 60–67.

Garthwaite, J. (2008). Concepts of neural nitric oxide-mediated transmission. *The European Journal of Neuroscience, 27*(11), 2783–2802.

Garthwaite, J., Charles, S. L., & Chess-Williams, R. (1988). Endothelium-derived relaxing factor release on activation of NMDA receptors suggests role as intercellular messenger in the brain. *Nature, 336*(6197), 385–388.

Garthwaite, J., Garthwaite, G., Palmer, R. M., & Moncada, S. (1989). NMDA receptor activation induces nitric oxide synthesis from arginine in rat brain slices. *European Journal of Pharmacology, 172*(4–5), 413–416.

Getting, S. J., Segieth, J., Ahmad, S., Biggs, C. S., & Whitton, P. S. (1996). Biphasic modulation of GABA release by nitric oxide in the hippocampus of freely moving rats in vivo. *Brain Research, 717*(1–2), 196–199.

Godfrey, E. W., & Schwarte, R. C. (2010). Nitric oxide and cyclic GMP regulate early events in agrin signaling in skeletal muscle cells. *Experimental Cell Research, 316*(12), 1935–1945.

Gould, N., Doulias, P. T., Tenopoulou, M., Raju, K., & Ischiropoulos, H. (2013). Regulation of protein function and signaling by reversible cysteine S-nitrosylation. *The Journal of Biological Chemistry, 288*(37), 26473–26479.

Greenberg, S. S., Diecke, F. P., Peevy, K., & Tanaka, T. P. (1990). Release of norepinephrine from adrenergic nerve endings of blood vessels is modulated by endothelium-derived relaxing factor. *American Journal of Hypertension, 3*(3), 211–218.

Guevara-Guzman, R., Emson, P. C., & Kendrick, K. M. (1994). Modulation of in vivo striatal transmitter release by nitric oxide and cyclic GMP. *Journal of Neurochemistry, 62*(2), 807–810.

Ha, K. S., Kim, K. M., Kwon, Y. G., Bai, S. K., Nam, W. D., Yoo, Y. M., et al. (2003). Nitric oxide prevents 6-hydroxydopamine-induced apoptosis in PC12 cells through cGMP-dependent PI3 kinase/Akt activation. *The FASEB Journal, 17*(9), 1036–1047.

Halbrügge, T., Lütsch, K., Thyen, A., & Graefe, K. H. (1991a). Role of nitric oxide formation in the regulation of haemodynamics and the release of noradrenaline and adrenaline. *Naunyn-Schmiedeberg's Archives of Pharmacology, 344*(6), 720–727.

Halbrügge, T., Lütsch, K., Thyen, A., & Graefe, K. H. (1991b). Vasodilatation by endothelium-derived nitric oxide as a major determinant of noradrenaline release. *Journal of Neural Transmission, Supplement, 34*, 113–119.

Halmos, G., Horváth, T., Polony, G., Fekete, A., Kittel, A., van der Vizi, E. S., et al. (2008). The role of N-methyl-D-aspartate receptors and nitric oxide in cochlear dopamine release. *Neuroscience, 154*(2), 796–803.

Hanada, M., Feng, J., & Hemmings, B. A. (2004). Structure, regulation and function of PKB/AKT—A major therapeutic target. *Biochimica et Biophysica Acta, 1697*(1), 3–16.

Hanafy, K. A., Krumenacker, J. S., & Murad, F. (2001). NO, nitrotyrosine, and cyclic GMP in signal transduction. *Medical Science Monitor*, 7(4), 801–819.

Hanania, T., & Johnson, K. M. (1998). Regulation of neurotransmitter release by endogenous nitric oxide in striatal slices. *European Journal of Pharmacology, 359*(2–3), 111–117.

Hardingham, G. E., & Bading, H. (2010). Synaptic versus extrasynaptic NMDA receptor signalling: Implications for neurodegenerative disorders. *Nature Review Neuroscience, 11*(10), 682–696.

Harris, E. W., Ganong, A. H., & Cotman, C. W. (1984). Long-term potentiation in the hippocampus involves activation of N-methyl-D-aspartate receptors. *Brain Research, 323*(1), 132–137.

Hattori, K., Naguro, I., Runchel, C., & Ichijo, H. (2009). The roles of ASK family proteins in stress responses and diseases. *Cell Communication and Signaling*, 7, 9.

Hess, D. T., Matsumoto, A., Kim, S. O., Marshall, H. E., & Stamler, J. S. (2005). Protein S-nitrosylation: Purview and parameters. *Nature Reviews Molecular and Cell Biology, 6*(2), 150–166.

Hess, D. T., Patterson, S. I., Smith, D. S., & Skene, J. H. (1993). Neuronal growth cone collapse and inhibition of protein fatty acylation by nitric oxide. *Nature, 366*(6455), 562–565.

Hibbs, J. B., Jr., Vavrin, Z., & Taintor, R. R. (1987). L-Arginine is required for expression of the activated macrophage effector mechanism causing selective metabolic inhibition in target cells. *The Journal of Immunology, 138*(2), 550–565.

Hirsch, D. B., Steiner, J. P., Dawson, T. M., Mammen, A., Hayek, E., & Snyder, S. H. (1993). Neurotransmitter release regulated by nitric oxide in PC-12 cells and brain synaptosomes. *Current Biology, 3*(11), 749–754.

Hong, S. K., Jung, I. S., Bang, S. A., & Kim, S. E. (2006). Effect of nitric oxide synthase inhibitor and NMDA receptor antagonist on the development of nicotine sensitization of nucleus accumbens dopamine release: An in vivo microdialysis study. *Neuroscience Letters, 409*(3), 220–223.

Hu, W., Zhang, M., Czéh, B., Flügge, G., & Zhang, W. (2010). Stress impairs GABAergic network function in the hippocampus by activating nongenomic glucocorticoid receptors and affecting the integrity of the parvalbumin-expressing neuronal network. *Neuropsychopharmacology, 8*, 1693–1707.

Huang, C. C., Chan, S. H., & Hsu, K. S. (2003). cGMP/protein kinase G-dependent potentiation of glutamatergic transmission induced by nitric oxide in immature rat rostral ventrolateral medulla neurons in vitro. *Molecular Pharmacology, 64*(2), 521–532.

Huang, P. L., Huang, Z., Mashimo, H., Bloch, K. D., Moskowitz, M. A., Bevan, J. A., et al. (1995). Hypertension in mice lacking the gene for endothelial nitric oxide synthase. *Nature, 377*, 239–242.

Huang, Y., Man, H. Y., Sekine-Aizawa, Y., Han, Y., Juluri, K., Luo, H., et al. (2005). S-nitrosylation of N-ethylmaleimide sensitive factor mediates surface expression of AMPA receptors. *Neuron, 46*(4), 533–540.

Hull, E. M., & Dominguez, J. M. (2006). Getting his act together: Roles of glutamate, nitric oxide, and dopamine in the medial preoptic area. *Brain Research, 1126*(1), 66–75.

Iadecola, C., Zhang, F., Casey, R., Nagayama, M., & Ross, M. E. (1997). Delayed reduction of ischemic brain injury and neurological deficits in mice lacking the inducible nitric oxide synthase gene. *The Journal of Neuroscience, 17*(23), 9157–9164.

Ientile, R., Pedale, S., Picciurro, V., Macaione, V., Fabiano, C., & Macaione, S. (1997). Nitric oxide mediates NMDA-evoked [^{3}H]GABA release from chick retina cells. *FEBS Letters, 417*(3), 345–348.

Ignarro, L. J., Buga, G. M., Wood, K. S., Byrns, R. E., & Chaudhuri, G. (1987). Endothelium-derived relaxing factor produced and released from artery and vein is nitric oxide. *Proceedings of the National Academy of Sciences of the United States of America, 84*(24), 9265–9269.

Ishide, T., Nauli, S. M., Maher, T. J., & Ally, A. (2003). Cardiovascular responses and neurotransmitter changes following blockade of nNOS within the ventrolateral medulla during static muscle contraction. *Brain Research, 977*(1), 80–89.

Jaffrey, S. R., Benfenati, F., Snowman, A. M., Czernik, A. J., & Snyder, S. H. (2002). Neuronal nitric-oxide synthase localization mediated by a ternary complex with synapsin and CAPON. *Proceedings of the National Academy of Sciences of the United States of America, 99*(5), 3199–3204.

Jaffrey, S. R., Snowman, A. M., Eliasson, M. J., Cohen, N. A., & Snyder, S. H. (1998). CAPON: A protein associated with neuronal nitric oxide synthase that regulates its interactions with PSD95. *Neuron, 20*(1), 115–124.

Jaffrey, S. R., & Snyder, S. H. (1996). PIN: An associated protein inhibitor of neuronal nitric oxide synthase. *Science, 274*(5288), 774–777.

Kaehler, S. T., Singewald, N., Sinner, C., & Philippu, A. (1999). Nitric oxide modulates the release of serotonin in the rat hypothalamus. *Brain Research, 835*(2), 346–349.

Kajiwara, A., Tsuchiya, Y., Takata, T., Nyunoya, M., Nozaki, N., Ihara, H., et al. (2013). Nitric oxide enhances increase in cytosolic Ca^{2+} and promotes nicotine-triggered MAPK pathway in PC12 cells. *Nitric Oxide, 34*, 3–9.

Kakizawa, S., Yamazawa, T., Chen, Y., Ito, A., Murayama, T., Oyamada, H., et al. (2012). Nitric oxide-induced calcium release via ryanodine receptors regulates neuronal function. *The EMBO Journal, 31*(2), 417–428.

Kano, T., Shimizu-Sasamata, M., Huang, P., Moskowitz, M., & Lo, E. (1998). Effects of nitric oxide synthase gene knockout on neurotransmitter release in vivo. *Neuroscience*, *86*(3), 695–699.

Katsuki, S., Arnold, W., Mittal, C., & Murad, F. (1977). Stimulation of guanylate cyclase by sodium nitroprusside, nitroglycerin and nitric oxide in various tissue preparations and comparison to the effects of sodium azide and hydroxylamine. *Journal of Cyclic Nucleotide Research*, *3*(1), 23–35.

Kendrick, K. M., Guevara-Guzman, R., Riva, C., Christensen, J., Ostergaard, K., & Emson, P. C. (1996). NMDA and kainate-evoked release of nitric oxide and classical transmitters in the rat striatum: In vivo evidence that nitric oxide may play a neuroprotective role. *The European Journal of Neuroscience*, *8*(12), 2619–2634.

Kirchner, L., Weitzdoerfer, R., Hoeger, H., Url, A., Schmidt, P., Engelmann, M., et al. (2004). Impaired cognitive performance in neuronal nitric oxide synthase knockout mice is associated with hippocampal protein derangements. *Nitric Oxide*, *11*(4), 316–330.

Kishi, T., Hirooka, Y., Sakai, K., Shigematsu, H., Shimokawa, H., & Takeshita, A. (2001). Overexpression of eNOS in the RVLM causes hypotension and bradycardia via GABA release. *Hypertension*, *38*(4), 896–901.

Knowles, R. G., Palacios, M., Palmer, R. M., & Moncada, S. (1989). Formation of nitric oxide from L-arginine in the central nervous system: A transduction mechanism for stimulation of the soluble guanylate cyclase. *Proceedings of the National Academy of Sciences of the United States of America*, *86*(13), 5159–5162.

Koesling, D., Russwurm, M., Mergia, E., Mullershausen, F., & Friebe, A. (2004). Nitric oxide-sensitive guanylyl cyclase: Structure and regulation. *Neurochemistry International*, *45*(6), 813–819.

Kristof, A. S., Goldberg, P., Laubach, V., & Hussain, S. N. (1998). Role of inducible nitric oxide synthase in endotoxin-induced acute lung injury. *American Journal of Respiratory and Critical Care Medicine*, *158*(6), 1883–1889.

Krumenacker, J. S., Hanafy, K. A., & Murad, F. (2004). Regulation of nitric oxide and soluble guanylyl cyclase. *Brain Research Bulletin*, *62*(6), 505–515.

Kwak, Y. D., Ma, T., Diao, S., Zhang, X., Chen, Y., Hsu, J., et al. (2010). NO signaling and S-nitrosylation regulate PTEN inhibition in neurodegeneration. *Molecular Neurodegeneration*, *5*, 49.

Lamas, S., Marsden, P. A., Li, G. K., Tempst, P., & Michel, T. (1992). Endothelial nitric oxide synthase: Molecular cloning and characterization of a distinct constitutive enzyme isoform. *Proceedings of the National Academy of Sciences of the United States of America*, *89*(14), 6348–6352.

Lawrence, A. J., & Jarrott, B. (1993). Nitric oxide increases interstitial excitatory amino acid release in the rat dorsomedial medulla oblongata. *Neuroscience Letters*, *151*(2), 126–129.

Lee, B., Butcher, G. Q., Hoyt, K. R., Impey, S., & Obrietan, K. (2005). Activity-dependent neuroprotection and cAMP response element-binding protein (CREB): Kinase coupling, stimulus intensity, and temporal regulation of CREB phosphorylation at serine 133. *The Journal of Neuroscience*, *25*(5), 1137–1148.

Lee, P. C., Salyapongse, A. N., Bragdon, G. A., Shears, L. L., Watkins, S. C., Edington, H. D., et al. (1999). Impaired wound healing and angiogenesis in eNOS-deficient mice. *The American Journal of Physiology*, *277*(4), H1600–H1608.

Lee, C. M., Stoelzel, C., Chistiakova, M., & Volgushev, M. (2012). Heterosynaptic plasticity induced by intracellular tetanization in layer 2/3 pyramidal neurons in rat auditory cortex. *The Journal of Physiology*, *590*(10), 2253–2271.

Li, J., Billiar, T. R., Talanian, R. V., & Kim, Y. M. (1997). Nitric oxide reversibly inhibits seven members of the caspase family via S-nitrosylation. *Biochemical and Biophysical Research Communications*, *240*(2), 419–424.

Li, D. P., Chen, S. R., Finnegan, T. F., & Pan, H. L. (2004). Signalling pathway of nitric oxide in synaptic GABA release in the rat paraventricular nucleus. *The Journal of Physiology*, *554*(Pt 1), 100–110.

Li, J. Y., Plomann, M., & Brundin, P. (2003). Huntington's disease: A synaptopathy? *Trends in Molecular Medicine*, *9*(10), 414–420.

Liberatore, G. T., Jackson-Lewis, V., Vukosavic, S., Mandir, A. S., Vila, M., McAuliffe, W. G., et al. (1999). Inducible nitric oxide synthase stimulates dopaminergic neurodegeneration in the MPTP model of Parkinson disease. *Nature Medicine*, *5*(12), 1403–1409.

Lindsay, S. L., Ramsey, S., Aitchison, M., Renne, T., & Evans, T. J. (2007). Modulation of lamellipodial structure and dynamics by NO-dependent phosphorylation of VASP Ser239. *Journal of Cell Science*, *120*(17), 3011–3021.

Lipton, S. A., Choi, Y. B., Pan, Z. H., Lei, S. Z., Chen, H. S., Sucher, N. J., et al. (1993). A redox-based mechanism for the neuroprotective and neurodestructive effects of nitric oxide and related nitroso-compounds. *Nature*, *364*(6438), 626–632.

Liu, S., Ninan, I., Antonova, I., Battaglia, F., Trinchese, F., Narasanna, A., et al. (2004). Alpha-Synuclein produces a long-lasting increase in neurotransmitter release. *The EMBO Journal*, *23*(22), 4506–4516.

Liu, D. H., Yuan, F. G., Hu, S. Q., Diao, F., Wu, Y. P., Zong, Y. Y., et al. (2013). Endogenous nitric oxide induces activation of apoptosis signal-regulating kinase 1 via S-nitrosylation in rat hippocampus during cerebral ischemia–reperfusion. *Neuroscience*, *229*, 36–48.

Lo, F. S., & Mize, R. R. (2000). Synaptic regulation of L-type Ca(2+) channel activity and long-term depression during refinement of the retinocollicular pathway in developing rodent superior colliculus. *The Journal of Neuroscience*, *20*(3), RC58.

Lonart, G., Cassels, K. L., & Johnson, K. M. (1993). Nitric oxide induces calcium-dependent [^{3}H]dopamine release from striatal slices. *Journal of Neuroscience Research*, *35*(2), 192–198.

Lonart, G., Wang, J., & Johnson, K. M. (1992). Nitric oxide induces neurotransmitter release from hippocampal slices. *European Journal of Pharmacology*, *220*(2), 271–272.

Lorrain, D. S., & Hull, E. M. (1993). Nitric oxide increases dopamine and serotonin release in the medial preoptic area. *Neuroreport*, *5*(1), 87–89.

Lowenstein, C. J., Glatt, C. S., Bredt, D. S., & Snyder, S. H. (1992). Cloned and expressed macrophage nitric oxide synthase contrasts with the brain enzyme. *Proceedings of the National Academy of Sciences of the United States of America*, *89*(15), 6711–6715.

Lu, Y. F., & Hawkins, R. D. (2002). Ryanodine receptors contribute to cGMP-induced late-phase LTP and CREB phosphorylation in the hippocampus. *Journal of Neurophysiology*, *88*(3), 1270–1278.

Lu, Y. F., Kandel, E. R., & Hawkins, R. D. (1999). Nitric oxide signaling contributes to late-phase LTP and CREB phosphorylation in the hippocampus. *The Journal of Neuroscience*, *19*(23), 10250–10261.

Lundberg, J. O., Weitzberg, E., & Gladwin, M. T. (2008). The nitrate–nitrite–nitric oxide pathway in physiology and therapeutics. *Nature Reviews Drug Discovery*, 7(2), 156–167.

Lyons, C. R., Orloff, G. J., & Cunningham, J. M. (1992). Molecular cloning and functional expression of an inducible nitric oxide synthase from a murine macrophage cell line. *The Journal of Biological Chemistry*, *267*(9), 6370–6374.

Maggesissi, R. S., Gardino, P. F., Guimarães-Souza, E. M., Paes-de-Carvalho, R., Silva, R. B., & Calaza, K. C. (2009). Modulation of GABA release by nitric oxide in the chick retina: Different effects of nitric oxide depending on the cell population. *Vision Research*, *49*(20), 2494–2502.

Malenka, R. C. (1994). Synaptic plasticity in the hippocampus: LTP and LTD. *Cell*, *78*(4), 535–538.

Mander, P., Borutaite, V., Moncada, S., & Brown, G. C. (2005). Nitric oxide from inflammatory-activated glia synergizes with hypoxia to induce neuronal death. *Journal of Neuroscience Research*, *79*(1–2), 208–215.
Manning, G., Whyte, D. B., Martinez, R., Hunter, T., & Sudarsanam, S. (2002). The protein kinase complement of the human genome. *Science*, *298*(5600), 1912–1934.
Martin, A. (2001). Is tetralogy true? Lack of support for the "one-to-four rule" *Molecular Biology and Evolution*, *18*(1), 89–93.
Mashimo, H., Kjellin, A., & Goyal, R. K. (2000). Gastric stasis in neuronal nitric oxide synthase-deficient knockout mice. *Gastroenterology*, *119*(3), 766–773.
Masini, E., Salvemini, D., Pistelli, A., Mannaioni, P. F., & Vane, J. R. (1991). Rat mast cells synthesize a nitric oxide like-factor which modulates the release of histamine. *Agents and Actions*, *33*(1–2), 61–63.
Matsuo, I., Hirooka, Y., Hironaga, K., Eshima, K., Shigematsu, H., Shihara, M., et al. (2001). Glutamate release via NO production evoked by NMDA in the NTS enhances hypotension and bradycardia in vivo. *American Journal of Physiology Regulatory, Integrative and Comparative Physiology*, *280*(5), R1285–R1291.
McKim, S. E., Gäbele, E., Isayama, F., Lambert, J. C., Tucker, L. M., Wheeler, M. D., et al. (2003). Inducible nitric oxide synthase is required in alcohol-induced liver injury: Studies with knockout mice. *Gastroenterology*, *125*(6), 1834–1844.
McNaught, K. S. P., & Brown, G. C. (1998). Nitric oxide causes glutamate release from brain synaptosomes. *Journal of Neurochemistry*, *70*(4), 1541–1546.
Meffert, M., Calakos, N., Scheller, R., & Schulman, H. (1996). Nitric oxide modulates synaptic vesicle docking fusion reactions. *Neuron*, *16*(6), 1229–1236.
Meffert, M. K., Premack, B. A., & Schulman, H. (1994). Nitric oxide stimulates Ca^{2+}-independent synaptic vesicle release. *Neuron*, *12*(6), 1235–1244.
Meini, A., Garcia, J. B., Pessina, G. P., Aldinucci, C., Frosini, M., & Palmi, M. (2006). Role of intracellular Ca^{2+} and calmodulin/MAP kinase kinase/extracellular signal-regulated protein kinase signalling pathway in the mitogenic and antimitogenic effect of nitric oxide in glia- and neurone-derived cell lines. *The European Journal of Neuroscience*, *23*(7), 1690–1700.
Mejía-García, T., & Paes-de-Carvalho, R. (2007). Nitric oxide regulates cell survival in purified cultures of avian retinal neurons: Involvement of multiple transduction pathways. *Journal of Neurochemistry*, *100*(2), 382–394.
Mejia-Garcia, T. A., Portugal, C. C., Encarnação, T. G., Prado, M. A., & Paes-de-Carvalho, R. (2013). Nitric oxide regulates AKT phosphorylation and nuclear translocation in cultured retinal cells. *Cellular Signaling*, *25*(12), 2424–2439, pii: S0898-6568(13)00228-3.
Mendonça, H. R., Araujo, S. E., Gomes, A. L., Sholl-Franco, A., da Cunha Faria Melibeu, A., Serfaty, C. A., et al. (2010). Expression of GAP-43 during development and after monocular enucleation in the rat superior colliculus. *Neuroscience Letters*, *477*(1), 23–27.
Mishra, O. P., Ashraf, Q. M., & Delivoria-Papadopoulos, M. (2009). NO-mediated activation of Src kinase during hypoxia in the cerebral cortex of newborn piglets. *Neuroscience Letters*, *460*(1), 61–65.
Møller, M., Jones, N. M., & Beart, P. M. (1995). Complex involvement of nitric oxide and cGMP at N-methyl-D-aspartic acid receptors regulating gamma-[^{3}H]aminobutyric acid release from striatal slices. *Neuroscience Letters*, *190*(3), 195–198.
Monfort, P., Munoz, M. D., Kosenko, E., Llansola, M., Sanchez-Perez, A., Cauli, O., et al. (2004). Sequential activation of soluble guanylate cyclase, protein kinase G and cGMP-degrading phosphodiesterase is necessary for proper induction of long-term potentiation in CA1 of hippocampus. Alterations in hyperammonemia. *Neurochemistry International*, *45*(6), 895–901.

Montague, P. R., Gancayco, C. D., Winn, M. J., Marchase, R. B., & Friedlander, M. J. (1994). Role of NO production in NMDA receptor-mediated neurotransmitter release in cerebral cortex. *Science, 263*(5149), 973–977.

Mu, Y., & Poo, M. M. (2006). Spike timing-dependent LTP/LTD mediates visual experience-dependent plasticity in a developing retinotectal system. *Neuron, 50*(1), 115–125.

Nakamura, Y., Ohmaki, M., Murakami, K., & Yoneda, Y. (2003). Involvement of protein kinase C in glutamate release from cultured microglia. *Brain Research, 962*(1), 122–128.

Nakamura, T., Tu, S., Akhtar, M. W., Sunico, C. R., Okamoto, S., & Lipton, S. A. (2013). Aberrant protein S-nitrosylation in neurodegenerative diseases. *Neuron, 78*(4), 596–614.

Nei, K., Matsuyama, S., Shuntoh, H., & Tanaka, C. (1996). NMDA receptor activation induces glutamate release through nitric oxide synthesis in guinea pig dentate gyrus. *Brain Research, 728*(1), 105–110.

Nikonenko, I., Boda, B., Steen, S., Knott, G., Welker, E., & Muller, D. (2008). PSD-95 promotes synaptogenesis and multiinnervated spine formation through nitric oxide signaling. *The Journal of Cell Biology, 183*(6), 1115–1127.

Nikonenko, I., Jourdain, P., & Muller, D. (2003). Presynaptic remodeling contributes to activity-dependent synaptogenesis. *The Journal of Neuroscience, 23*(24), 8498–8505.

Nishida, K., Harrison, D., Navas, J., Fisher, A., Dockery, S., Uematsu, M., et al. (1992). Molecular cloning and characterization of the constitutive bovine aortic endothelial cell nitric oxide synthase. *The Journal of Clinical Investigation, 90*(5), 2092–2096.

Nott, A., Watson, P. M., Robinson, J. D., Crepaldi, L., & Riccio, A. (2008). S-nitrosylation of histone deacetylase 2 induces chromatin remodelling in neurons. *Nature, 455*(7211), 411–415.

Numajiri, N., Takasawa, K., Nishiya, T., Tanaka, H., Ohno, K., Hayakawa, W., et al. (2011). On–off system for PI3-kinase-Akt signaling through S-nitrosylation of phosphatase with sequence homology to tensin (PTEN). *Proceedings of the National Academy of Sciences of the United States of America, 108*(25), 10349–10354.

O'dell, T. J., Hawkins, R. D., Kandel, E. R., & Arancio, O. (1991). Tests of the roles of two diffusible substances in long-term potentiation: Evidence for nitric oxide as a possible early retrograde messenger. *Proceedings of the National Academy of Sciences of the United States of America, 88*(24), 11285–11289.

Ogasawara, H., Doi, T., & Kawato, M. (2008). Systems biology perspectives on cerebellar long-term depression. *Neurosignals, 16*(4), 300–317.

Ohkuma, S., Katsura, M., Chen, D. Z., Narihara, H., & Kuriyama, K. (1996). Nitric oxide-evoked [^{3}H] gamma-aminobutyric acid release is mediated by two distinct release mechanisms. *Brain Research. Molecular Brain Research, 36*(1), 137–144.

Ohshima, T., Ward, J. M., Huh, C. G., Longenecker, G., Veeranna, Pant, H. C., et al. (1996). Targeted disruption of the cyclin-dependent kinase 5 gene results in abnormal corticogenesis, neuronal pathology and perinatal death. *Proceedings of the National Academy of Sciences of the United States of America, 93*(20), 11173–11178.

Oliveira-Silva, P., Jurgilas, P. B., Trindade, P., Campello-Costa, P., Perales, J., Savino, W., et al. (2007). Matrix metalloproteinase-9 is involved in the development and plasticity of retinotectal projections in rats. *Neuroimmunomodulation, 14*(3–4), 144–149.

Paes-de-Carvalho, R., Maia, G. A., & Ferreira, J. M. (2003). Adenosine regulates the survival of avian retinal neurons and photoreceptors in culture. *Neurochemical Research, 28*(10), 1583–1590.

Palmer, R. M., Ferrige, A. G., & Moncada, S. (1987). Nitric oxide release accounts for the biological activity of endothelium-derived relaxing factor. *Nature, 327*(6122), 524–526.

Pape, H. C., & Mager, R. (1992). Nitric oxide controls oscillatory activity in thalamocortical neurons. *Neuron, 9*(3), 441–448.

Peltier, J., O'Neill, A., & Schaffer, D. V. (2007). PI3K/Akt and CREB regulate adult neural hippocampal progenitor proliferation and differentiation. *Developmental Neurobiology*, *67*(10), 1348–1361.

Penzes, P., Cahill, M. E., Jones, K. A., VanLeeuwen, J. E., & Woolfrey, K. M. (2011). Dendritic spine pathology in neuropsychiatric disorders. *Nature Neuroscience*, *14*(3), 285–293.

Poglia, L., Muller, D., & Nikonenko, I. (2011). Ultrastructural modifications of spine and synapse morphology by SAP97. *Hippocampus*, *21*(9), 990–998.

Prast, H., & Philippu, A. (2001). Nitric oxide as modulator of neuronal function. *Progress in Neurobiology*, *64*(1), 51–68.

Przedborski, S., Jackson-Lewis, V., Yokoyama, R., Shibata, T., Dawson, V. L., & Dawson, T. M. (1996). Role of neuronal nitric oxide in 1-methyl-4-phenyl-1,2,3,6-tetrahydropyridine (MPTP)-induced dopaminergic neurotoxicity. *Proceedings of the National Academy of Sciences of the United States of America*, *93*(10), 4565–4571.

Qu, J., Nakamura, T., Cao, G., Holland, E. A., McKercher, S. R., & Lipton, S. A. (2011). S-nitrosylation activates Cdk5 and contributes to synaptic spine loss induced by beta-amyloid peptide. *Proceedings of the National Academy of Sciences of the United States of America*, *108*(34), 14330–14335.

Radomski, M., Palmer, R., & Moncada, S. (1990). An L-arginine/nitric oxide pathway present in human platelets regulates aggregation. *Proceedings of the National Academy of Sciences of the United States of America*, *87*(13), 5193–5197.

Ratnayaka, A., Marra, V., Bush, D., Burden, J. J., Branco, T., & Staras, K. (2012). Recruitment of resting vesicles into recycling pools supports NMDA receptor-dependent synaptic potentiation in cultured hippocampal neurons. *The Journal of Physiology*, *590*(7), 1585–1597.

Reyes-Harde, M., Potter, B. V., Galione, A., & Stanton, P. K. (1999). Induction of hippocampal LTD requires nitric-oxide-stimulated PKG activity and Ca^{2+} release from cyclic ADP-ribose-sensitive stores. *Journal of Neurophysiology*, *82*(3), 1569–1576.

Riccio, A., Alvania, R. S., Lonze, B. E., Ramanan, N., Kim, T., Huang, Y., et al. (2006). A nitric oxide signaling pathway controls CREB-mediated gene expression in neurons. *Molecular Cell*, *21*(2), 283–294.

Roskoski, R., Jr. (2005). Src kinase regulation by phosphorylation and dephosphorylation. *Biochemical and Biophysical Research Communications*, *331*(1), 1–14.

Rubinfeld, H., & Seger, R. (2005). The ERK cascade: A prototype of MAPK signaling. *Molecular Biotechnology*, *31*(2), 151–174.

Rudic, R. D., Shesely, E. G., Maeda, N., Smithies, O., Segal, S. S., & Sessa, W. C. (1998). Direct evidence for the importance of endothelium-derived nitric oxide in vascular remodeling. *The Journal of Clinical Investigation*, *101*(4), 731–736.

Samuels, I. S., Saitta, S. C., & Landreth, G. E. (2009). MAP'ing CNS development and cognition: An ERKsome process. *Neuron*, *61*(2), 160–167.

Sanchez-Islas, E., & Leon-Olea, M. (2004). Nitric oxide synthase inhibition during synaptic maturation decreases synapsin I immunoreactivity in rat brain. *Nitric Oxide*, *10*(3), 141–149.

Sarbassov, D. D., Guertin, D. A., Ali, S. M., & Sabatini, D. M. (2005). Phosphorylation and regulation of Akt/PKB by the rictor–mTOR complex. *Science*, *307*(5712), 1098–1101.

Sayre, L. M., Perry, G., & Smith, M. A. (2008). Oxidative stress and neurotoxicity. *Chemical Research in Toxicology*, *21*(1), 172–188.

Scheetz, A. J., Nairn, A. C., & Constantine-Paton, M. (2000). NMDA receptor-mediated control of protein synthesis at developing synapses. *Nature Neuroscience*, *3*(3), 211–216.

Schulz, J. B., Matthews, R. T., Jenkins, B. G., Ferrante, R. J., Siwek, D., Henshaw, D., et al. (1995). Blockade of neuronal nitric oxide synthase protects against excitotoxicity in vivo. *The Journal of Neuroscience*, *15*(12), 8419–8429.

Segieth, J., Getting, S. J., Biggs, C. S., & Whitton, P. S. (1995). Nitric oxide regulates excitatory amino acid release in a biphasic manner in freely moving rats. *Neuroscience Letters*, *200*(2), 101–104.

Segovia, G., & Mora, F. (1998). Role of nitric oxide in modulating the release of dopamine, glutamate, and GABA in striatum of the freely moving rat. *Brain Research Bulletin*, *45*(3), 275–279.

Selvakumar, B., Huganir, R. L., & Snyder, S. H. (2009). S-nitrosylation of stargazin regulates surface expression of AMPA-glutamate neurotransmitter receptors. *Proceedings of the National Academy of Sciences of the United States of America*, *106*(38), 16440–16445.

Sennlaub, F., Courtois, Y., & Goureau, O. (2002). Inducible nitric oxide synthase mediates retinal apoptosis in ischemic proliferative retinopathy. *The Journal of Neuroscience*, *22*(10), 3987–3993.

Sequeira, S. M., Ambrósio, A. F., Malva, J. O., Carvalho, A. P., & Carvalho, C. M. (1997). Modulation of glutamate release from rat hippocampal synaptosomes by nitric oxide. *Nitric Oxide*, *1*(4), 315–329.

Sequeira, S. M., Carvalho, A. P., & Carvalho, C. M. (1999). Both protein kinase G dependent and independent mechanisms are involved in the modulation of glutamate release by nitric oxide in rat hippocampal nerve terminals. *Neuroscience Letters*, *261*(1), 29–32.

Serfaty, C. A., Campello-Costa, P., & Linden, R. (2005). Rapid and long-term plasticity in the neonatal and adult retinotectal pathways following a retinal lesion. *Brain Research. Bulletin*, *66*(2), 128–134.

Serfaty, C. A., & Linden, R. (1994). Development of abnormal lamination and binocular segregation in the retinotectal pathways of the rat. *Brain Research. Developmental Brain Research*, *82*(1–2), 35–44.

Serulle, Y., Zhang, S., Ninan, I., Puzzo, D., McCarthy, M., Khatri, L., et al. (2007). A GluR1–cGKII interaction regulates AMPA receptor trafficking. *Neuron*, *56*(4), 670–688.

Sessa, W. C., Harrison, J. K., Barber, C. M., Zeng, D., Durieux, M. E., D'Angelo, D. D., et al. (1992). Molecular cloning and expression of a cDNA encoding endothelial cell nitric oxide synthase. *The Journal of Biological Chemistry*, *267*(22), 15274–15276.

Shahani, N., & Sawa, A. (2012). Protein S-nitrosylation: Role for nitric oxide signaling in neuronal death. *Biochimica et Biophysica Acta*, *1820*(6), 736–742.

Singewald, N., Kaehler, S. T., Hemeida, R., & Philippu, A. (1998). Influence of excitatory amino acids on basal and sensory stimuli-induced release of 5-HT in the locus coeruleus. *British Journal of Pharmacology*, *123*(4), 746–752.

Sistiaga, A., Miras-Portugal, T., & Sánchez-Prieto, J. (1997). Modulation of glutamate release by a nitric oxide/cyclic GMP-dependent pathway. *European Journal of Pharmacology*, *321*(2), 247–257.

Smith, J. C., & Whitton, P. S. (2000). Nitric oxide modulates N-methyl-D-aspartate-evoked serotonin release in the raphe nuclei and frontal cortex of the freely moving rat. *Neuroscience Letters*, *291*(1), 5–8.

Socodato, R. E. S., Magalhães, C. R., & Paes-de-Carvalho, R. (2009). Glutamate and nitric oxide modulate ERK and CREB phosphorylation in the avian retina: Evidence for direct signaling from neurons to Müller glial cells. *Journal of Neurochemistry*, *108*(2), 417–429.

Socodato, R., Santiago, F. N., Portugal, C. C., Domingues, A. F., Santiago, A. R., Relvas, J. B., et al. (2012). Calcium-permeable alpha-amino-3-hydroxy-5-methyl-4-isoxazolepropionic acid receptors trigger neuronal nitric-oxide synthase activation to promote nerve cell death in an Src kinase-dependent fashion. *The Journal of Biological Chemistry*, *287*(46), 38680–38694.

Sparacino-Watkins, C. E., Lai, Y. C., & Gladwin, M. T. (2012). Nitrate–nitrite–nitric oxide pathway in pulmonary arterial hypertension therapeutics. *Circulation*, *125*(23), 2824–2826.

Stamler, J. S., Lamas, S., & Fang, F. C. (2001). Nitrosylation. The prototypic redox-based signaling mechanism. *Cell, 106*(6), 675–683.
Stanton, P. K., Winterer, J., Bailey, C. P., Kyrozis, A., Raginov, I., Laube, G., et al. (2003). Long-term depression of presynaptic release from the readily releasable vesicle pool induced by NMDA receptor-dependent retrograde nitric oxide. *The Journal of Neuroscience, 23*(13), 5936–5944.
Strasser, A., McCarron, R. M., Ishii, H., Stanimirovic, D., & Spatz, M. (1994). L-Arginine induces dopamine release from the striatum in vivo. *Neuroreport, 5*(17), 2298–2300.
Straub, V. A., Grant, J., O'Shea, M., & Benjamin, P. R. (2007). Modulation of serotonergic neurotransmission by nitric oxide. *Journal of Neurophysiology, 97*(2), 1088–1099.
Stuehr, D. J., Gross, S. S., Sakuma, I., Levi, R., & Nathan, C. F. (1989). Activated murine macrophages secrete a metabolite of arginine with the bioactivity of endothelium-derived relaxing factor and the chemical reactivity of nitric oxide. *The Journal of Experimental Medicine, 169*(3), 1011–1020.
Trabace, L., & Kendrick, K. M. (2000). Nitric oxide can differentially modulate striatal neurotransmitter concentrations via soluble guanylate cyclase and peroxynitrite formation. *Journal of Neurochemistry, 75*(4), 1664–1674.
Trimm, K. R., & Rehder, V. (2004). Nitric oxide acts as a slow-down and search signal in developing neurites. *The European Journal of Neuroscience, 19*(4), 809–818.
Tschetter, W. W., Alam, N. M., Yee, C. W., Gorz, M., Douglas, R. M., Sagdullaev, B., et al. (2013). Experience-enabled enhancement of adult visual cortex function. *The Journal of Neuroscience, 33*(12), 5362–5366.
Vesely, D. L., Rovere, L. E., & Levey, G. S. (1977). Activation of guanylate cyclase by streptozotocin and 1-methyl-1-nitrosourea. *Cancer Research, 37*(1), 28–31.
Wakatsuki, H., Gomi, H., Kudoh, M., Kimura, S., Takahashi, K., Takeda, M., et al. (1998). Layer-specific NO dependence of long-term potentiation and biased NO release in layer V in the rat auditory cortex. *The Journal of Physiology, 513*(1), 71–81.
Wang, H. G., Lu, F. M., Jin, I., Udo, H., Kandel, E. R., de Vente, J., et al. (2005). Presynaptic and postsynaptic roles of NO, cGK, and RhoA in long-lasting potentiation and aggregation of synaptic proteins. *Neuron, 45*(3), 389–403.
Wang, X., & Robinson, P. J. (1997). Cyclic GMP-dependent protein kinase and cellular signaling in the nervous system. *Journal of Neurochemistry, 68*(2), 443–456.
Wang, S., Teschemacher, A. G., Paton, J. F., & Kasparov, S. (2006). Mechanism of nitric oxide action on inhibitory GABAergic signaling within the nucleus tractus solitarii. *The FASEB Journal, 20*(9), 1537–1539.
Watkins, J. C., & Evans, R. H. (1981). Excitatory amino acid transmitters. *Annual Review of Pharmacology and Toxicology, 21*, 165–204.
Wedel, B., Humbert, P., Harteneck, C., Foerster, J., Malkewitz, J., Böhme, E., et al. (1994). Mutation of His-105 in the beta 1 subunit yields a nitric oxide-insensitive form of soluble guanylyl cyclase. *Proceedings of the National Academy of Sciences of the United States of America, 91*(7), 2592–2596.
Wiesinger, H. (2001). Arginine metabolism and the synthesis of nitric oxide in the nervous system. *Progress in Neurobiology, 64*(4), 365–391.
Williams, C. V., Nordquist, D., & McLoon, S. C. (1994). Correlation of nitric oxide synthase expression with changing patterns of axonal projections in the developing visual system. *The Journal of Neuroscience, 14*(3–2), 1746–1755.
Willmott, N. J., Wong, K., & Strong, A. J. (2000). A fundamental role for the nitric oxide-G-kinase signaling pathway in mediating intercellular Ca(2+) waves in glia. *The Journal of Neuroscience, 20*(5), 1767–1779.
Won, H., Mah, W., & Kim, E. (2013). Autism spectrum disorder causes, mechanisms, and treatments: Focus on neuronal synapses. *Frontiers in Molecular Neuroscience, 6*, 1–19.

Wu, H. H., Williams, C. V., & McLoon, S. C. (1994). Involvement of nitric oxide in the elimination of a transient retinotectal projection in development. *Science*, *265*(5178), 1593–1596.

Xie, Q.-W., Cho, H. J., Calaycay, J., Mumford, R. A., Swiderek, K. M., Lee, T. D., et al. (1992). Cloning and characterization of inducible nitric oxide synthase from mouse macrophages. *Science*, *256*(5054), 225–228.

Xie, Z., Sanada, K., Samuels, B. A., Shih, H., & Tsai, L. H. (2003). Serine 732 phosphorylation of FAK by Cdk5 is important for microtubule organization, nuclear movement, and neuronal migration. *Cell*, *114*(4), 469–482.

Yamasaki, K., Edington, H., McClosky, C., Tzeng, E., Lizonova, A., Kovesdi, I., et al. (1998). Reversal of impaired wound repair in iNOS-deficient mice by topical adenoviral-mediated iNOS gene transfer. *The Journal of Clinical Investigation*, *101*(5), 967–971.

Yang, Q., Chen, S. R., Li, D. P., & Pan, H. L. (2007). Kv1.1/1.2 channels are downstream effectors of nitric oxide on synaptic GABA release to preautonomic neurons in the paraventricular nucleus. *Neuroscience*, *149*(2), 315–327.

Yoshida, T., Limmroth, V., Irikura, K., & Moskowitz, M. A. (1994). The NOS inhibitor, 7-nitroindazole, decreases focal infarct volume but not the response to topical acetylcholine in pial vessels. *Journal of Cerebral Blood Flow and Metabolism*, *14*(6), 924–929.

Yoshihara, Y., De Roo, M., & Muller, D. (2009). Dendritic spine formation and stabilization. *Current Opinion in Neurobiology*, *19*(2), 146–153.

Yu, D., & Eldred, W. D. (2005). Nitric oxide stimulates gamma-aminobutyric acid release and inhibits glycine release in retina. *The Journal of Comparative Neurology*, *483*(3), 278–291.

Zhang, C., Granstrom, L., & Wong-Riley, M. T. (1996). Deafferentation leads to a down-regulation of nitric oxide synthase in the rat visual system. *Neuroscience Letters*, *211*(1), 61–64.

Zhang, P., Yu, P. C., Tsang, A. H., Chen, Y., Fu, A. K., Fu, W. Y., et al. (2010). S-nitrosylation of cyclin-dependent kinase 5 (cdk5) regulates its kinase activity and dendrite growth during neuronal development. *The Journal of Neuroscience*, *30*(43), 14366–14370.

Zhou, P., Qian, L., & Iadecola, C. (2005). Nitric oxide inhibits caspase activation and apoptotic morphology but does not rescue neuronal death. *Journal of Cerebral Blood Flow and Metabolism*, *25*(3), 348–357.

Zhu, X. Z., & Luo, L. G. (1992). Effect of nitroprusside (nitric oxide) on endogenous dopamine release from rat striatal slices. *Journal of Neurochemistry*, *59*(3), 932–935.

FURTHER READING

Huang, C. C., & Hsu, K. S. (2010). Activation of muscarinic acetylcholine receptors induces a nitric oxide-dependent long-term depression in rat medial prefrontal cortex. *Cerebral Cortex*, *20*(4), 982–996.

Mize, R. R., & Lo, F. (2000). Nitric oxide, impulse activity, and neurotrophins in visual system development. *Brain Research*, *886*(1–2), 15–32.

CHAPTER SIX

Hippocampus and Nitric Oxide

Yao Hu*, Dong-Ya Zhu*,†,1

*Institute for Stem Cells and Neural Regeneration, School of Pharmacy, Nanjing Medical University, Nanjing, China
†Department of Pharmacology, School of Pharmacy, Nanjing Medical University, Nanjing, China
[1]Corresponding author: e-mail address: dyzhu@njmu.edu.cn

Contents

Abstract

Since it was first identified to play an important role in relaxation of blood vessels, nitric oxide has been demonstrated to regulate many biological processes, especially in the central nervous system. Of the three types of enzymes that produce nitric oxide in humans and rodents, neuronal type is found almost exclusively in the nervous system. This gaseous molecule is a nonclassical neurotransmitter, which maintains the activities of neural cells and regulates the normal functions of brain. It appears to play a role in promoting the transfer of nerve signals from one neuron to another, maintaining the synaptic strength. Meanwhile, nitric oxide is a unique regulator on neurogenesis and synaptogenesis, producing the positive or negative effects upon different signal pathways or cellular origins and locations. Based on its significant roles in neural plasticity, nitric oxide is involved in a number of central nervous diseases, such as ischemia, depression, anxiety, and Alzheimer's disease. Clarifying the profiles of nitric oxide in the brain tissues and its participation in pathophysiological processes opens a new

Vitamins and Hormones, Volume 96
ISSN 0083-6729
http://dx.doi.org/10.1016/B978-0-12-800254-4.00006-4

avenue for development of new therapeutic strategies. Thus, this chapter specifies the effects of nitric oxide in the hippocampus, a key structure implicated in the modulation of mood and memories, exhibiting the trend of future research on nitric oxide.

1. INTRODUCTION

Nitric oxide (NO) is a free radical gaseous molecule involved in a wide range of physiological and pathological processes. In the central nervous system (CNS), it is an important regulator which has a series of functions, such as the regulation of cytogenesis, synaptic plasticity, neurotransmitter release, and hormone secretion. On the other side, it is involved in a number of diseases, especially neurodegenerative disorders, brain insults, and mood disorders.

The earliest research on NO in modern times began in 1916. Mitchell and colleagues noted that the high concentration of nitrites and nitrates in human urine could not be accounted for by dietary intake alone (Mitchell, Shonle, & Grindley, 1916). Since then, scientists started researches in the field of nitrite and nitrate biosynthesis. Although we have known that nitrite and nitrate are the metabolic products of NO now, the mechanism underlying NO production had not been understood at that time. In 1977, it was reported that NO causes an increase in cyclic GMP in certain tissues, as a breakdown product of nitroglycerine (Katsuki, Arnold, & Murad, 1977). In 1979, Ignarro and colleagues demonstrated that NO induces vascular smooth muscle relaxation (Gruetter et al., 1979). Next year, Furchgott and Zawadzki reported an endothelium-derived relaxing factor (EDRF) produced by the vascular endothelium up stimulation (Furchgott & Zawadzki, 1980). Following studies show that EDRF shares many similarities with NO. In 1987, Ignarro and Palmer independently demonstrated that EDRF is indeed NO (Ignarro, Buga, Wood, Byrns, & Chaudhuri, 1987; Palmer, Ferrige, & Moncada, 1987). And then, Marletta and colleagues reported that L-arginine is oxidized to nitrite with the intermediated formation of NO (Marletta, Yoon, Iyengar, Leaf, & Wishnok, 1988). In the 1980s, NO was demonstrated as a nonadrenergic and noncholinergic neurotransmitter (Bowman, Gillespie, & Pollock, 1982). Activation of glutamatergic *N*-methy-D-aspartate (NMDA) receptors induces NO production, and an enzyme in the brain can synthesize nitrite from L-arginine in the presence of NAPDH (Garthwaite, Charles, & Chess-Williams, 1988; Schmidt, Wilke, Evers, & Böhme, 1989). In the 1990s,

Bredt and Snyder isolated nitric oxide synthase (NOS) in the rat brain. Subsequently, NOS was isolated from endothelial cells and macrophages. For their brilliant works in the field of NO, Murad, Ignarro, and Furchgott shared the Nobel Prize in Physiology or Medicine in 1998.

These landmark works contribute to the progressive research on NO in the CNS. Over the past 20 years, a number of pioneering works were performed. Now, the profile of cellular and molecular prosperities of NOS/NO has been sketched, and the implications of NO in the physiological functions and pathological actions in the CNS have been revealed.

The hippocampus is an area of primitive cortex, or allocortex, hidden on the underside of the temporal lobe. It is connected to a set of immediately adjacent cortical areas, including the perirhinal, entorhinal, and parahippocampal cortices (Zola-Morgan & Squire, 1993). The hippocampus has three subdivisions: CA3, CA2, and CA1 (CA is the abbreviations of cornu ammonis). The other regions of the hippocampal formation include the dentate gyrus (DG), subiculum, presubiculum, parasubiculum, and entorhinal cortex. These regions of hippocampal formation are linked, next to next, by unique and largely unidirectional neuronal pathways. The classical trisynaptic circuit described the ensemble of pathways like the following sketch: EC → DG (synapse 1), DG → CA3 (synapse 2), and CA3 → CA1 (synapse 3). Moreover, there are robust projections from CA1 to the subiculum and entorhinal cortex, and the major projections from the entorhinal cortex to the neocortex.

Since the groundbreaking case of patient H.M., who lost much of his memory when his medial temporal lobe was removed for the treatment of his intractable epilepsy, a huge amount of data has linked the hippocampus to memory in humans and various animal models (Squire, 1992). The hippocampus plays a critical role in the formation of new memories about experienced events (Eichenbaum & Cohen, 1993). Actually, severe damage to the hippocampi in both hemispheres results in profound difficulties in forming new memories and often also affects memories formed before the damage occurred. Although there are hippocampus-dependent or -independent memories, there can be no doubt that several forms of memories suffer after removal of the hippocampus, such as episodic memory, spatial memory, and contextual fear memory. The NMDA-mediated plasticity in the hippocampus is important for the cognitive function (Fanselow & Dong, 2010). Moreover, Klüver and Bucy found that removal of the medial temporal lobe caused profound emotional disturbances in monkeys, suggesting that the hippocampus is involved in states of emotion (Klüver & Bucy, 1937). The hippocampus exerts strong regulatory control of the hypothalamic–pituitary–adrenal (HPA) axis, and elevation of stress hormones leads to hippocampal dysfunction in humans and rodents (Herman, Ostrander, Mueller, & Figueiredo, 2005; McEwen et al., 1997). Generally, the dorsal hippocampus is considered to perform primarily cognitive functions, while the ventral contributes to emotion regulations. Strikingly, gene expression in the dorsal hippocampus correlates with cortical regions involved in information processing, while genes expressed in the ventral hippocampus correlate with regions involved in amygdala and hypothalamus (Fanselow & Dong, 2010). Additionally, the subgranular zone (SGZ) of DG in the hippocampus is one of the main sites where the neurogenesis occurs (Ehninger & Kempermann, 2008).

Given the importance of the hippocampus in regulating the functions of cognition and emotion, and affecting the incidence and therapy of CNS disorders, we lay out our current understanding in the regulatory of NO on functional and structural hippocampal plasticity, and the implications of hippocampal NO in CNS disorders.

2. PROFILES OF NO IN THE CNS

2.1. NO formation

NO has been accepted as a nonclassical neurotransmitter since it was identified. Be different with the classical neurotransmitter, NO is not packaged and released from synaptic vesicles. Instead, it is generated by a complex family of NOS. The NOS enzymatic activity is dependent on substrate and cofactor availability, as well as the rate of electron transfer. In the presence of oxygen, NADPH and cofactors such as flavin mononeucleotide, flavin adenine dinucleotide, heme, and tetrahydrobiopterin, NOS catalyzes the oxidation of the terminal guanidinyl nitrogen of the amino acid L-arginine to form L-citrulline and NO (Bredt, 1999; Knowles, 1997). When NO formed, it diffuses within the cell or across the cells, involved in the autocrine and paracrine activities.

The NOS family consists of three isoforms: neuronal NOS (nNOS, I), endothelial NOS (eNOS, II), and inducible NOS (iNOS, III). Both nNOS and eNOS are constitutively expressed and their activities are Ca^{2+} dependent, whereas the activity of iNOS is fully activated at basal intracellular calcium concentration, so its activity is Ca^{2+} independent. Therefore, the generation and the local concentration of NO produced by these three isoforms of NOS are different. A burst of NO is produced by activating nNOS and eNOS in a short order. However, when iNOS is activated, it generates NO in a persistent way because iNOS has calmodulin as a subunit (Bredt, 1999; Dawson & Snyder, 1994).

In the brain, NO derived from nNOS is the predominant source. nNOS is abundant in several brain areas, such as the cerebral cortex, the striatum, the amygdala, the hypothalamus, the thalamus, the cerebellum, and the hippocampus (Bredt et al., 1991; Dawson & Snyder, 1994; Rodrigo et al., 1994). In the hippocampus, nNOS-positive neurons exist in the CA1–CA3 region and the DG (Rodrigo et al., 1994). nNOS reactivity is detectable in the numerous pyramidal neurons and the presumed GABAergic interneurons throughout the pyramidal cell layer of CA1–CA3 region, and also in the granule cells and polymorphic layers of the DG and in the stratum oriens (Doyle & Slater, 1997). nNOS is expressed in both mature and immature neurons. A light eNOS is localized to the cell bodies of CA1–CA3 subfield pyramidal neurons by a small population, and the granule cells of the DG (Calabrese, Butterfield, Scapagnini, Stella, & Maines, 2006).

2.2. NO diffusion

Rapid diffusion is an important characteristic of NO and has a great influence on its local concentration and effects. NO is a free radical molecule which has a low molecular weight and is soluble in both water and lipid (Shaw & Vosper, 1977). The diffusion coefficient for NO is about 3.8×10^{-2} cm^2/s (Meulemans, 1994). So it is highly diffusible in the brain and can cross a few cells from its site of synthesis to influence the neighboring cells within few seconds. A first strong evidence for the transcellular diffusion of NO in the brain is that NO induces synaptic potentiation between paired neurons and synapses approximately 100 μm distant in hippocampal slice (Schuman & Madison, 1994).

The diffusion of NO is not only in a rapid manner but also in a multidimensional way. It has been predicted that if a single source of NO can diffuse to the distance of 200 μm for 1–10 s, it will influence a volume of brain containing 2 million synapses (Wood & Garthwaite, 1994). That is reason to say that one small molecule causes the extensive effects.

2.3. NO metabolism

NO undergoes various reactions in the brain resulting in the formation of nitrites (NO_2^-), nitrates (NO_3^-), and peroxynitrites ($ONOO^-$) according to the following formulas:

$$2NO + O_2 \rightarrow N_2O_4 \overset{H_2O}{\rightarrow} NO_2^- + NO_3^- + 2H^+$$

$$NO + NO_2^- \rightarrow N_2O_3 \overset{H_2O}{\rightarrow} 2NO_2^- + 2H^+$$

$$NO + O_2^{\bullet -} \rightarrow ONOO^- \rightarrow ONOOH \rightarrow \left[\frac{^{\bullet}NO_2}{^{\bullet}OH}\right] \rightarrow NO_3^- + H^+$$

During the formation of nitrates, there are intermediate products such as nitrite radical ($^{\bullet}NO_2$) and hydroxyl radical ($^{\bullet}OH$) that are highly reactive (Beckman & Koppenol, 1996). NO reacts with $O_2^{\bullet -}$ quickly enough to form peroxynitrite anion ($ONOO^-$) because the affinity of $O_2^{\bullet -}$ is higher for NO than for superoxide dismutase (SOD) (Beckman, Beckman, Chen, Marshall, & Freeman, 1990; Huie & Padmaja, 1993). The limiting factors in the reaction are the amount of NO and its diffusion coefficient (Saran, Michel, & Bors, 1990). Under physiological conditions, $ONOO^-$ has a half-life of 1–2 s and an action radius of 100 μm, being degraded into multiple toxic products (Beckman et al., 1990) or scavenged by the reaction with bicarbonate to produce nitrosoperoxycarbonate ($ONOOCO_2^-$) (Whiteman, Ketsawatsakul, & Halliwell, 2002). Thus, excessive NO makes the overload of toxic products and mediates the destructive consequences under the pathological conditions.

2.4. NO signaling

Soluble guanylate cyclase (sGC) is the known intracellular receptor for NO. A large number of NO functions are mediated by NO/sGC signaling pathway (Garthwaite, 2010). Activation of sGC leads to the conversion of GTP to GMP, which in turn can act on various downstream targets, such as cGMP-dependent protein kinase (cG-PK), phosphodiesterases, cyclic-nucleotide-gated ion channels, and transcriptional factors (Francis, Busch, Corbin, & Sibley, 2010), and initiate the subsequent changes intracellularly. NO at nanomolar levels binding tightly on sGC results in a 100- to 200-fold activation of the enzyme.

In pathological conditions, excessive NO would produce superoxide radical and peroxynitrite ($ONOO^-$) to modify the microenvironment of neural cells. These reactive nitrogen species mediate both oxidative and nitrosative stresses, which form the basis of the cytotoxicity generally attributed to NO, relevant to the pathophysiology of inflammation, circulatory shock, and ischemia–reperfusion injury (Liaudet, Soriano, &

Szabó, 2000). A major pathway of peroxynitrite-dependent cytotoxicity is represented by the activation of the nuclear enzyme poly(ADP-ribose) polymerase (PARP) in response to DNA damage, which may trigger cell necrosis by ATP depletion (Brown, 2010; Jagtap & Szabó, 2005).

Besides the toxicity, more recent evidences indicate that peroxynitrite also behaves as a potent modulator of various cell signal transduction pathways (Pacher, Beckman, & Liaudet, 2007; Szabo, Ischiropoulos, & Radi, 2007). Due to its ability to nitrate tyrosine residues, peroxynitrite affects cellular processes dependent on tyrosine phosphorylation. Peroxynitrite also exerts complex effects on the activity of various kinases and phosphatases, resulting in the up- or downregulation of signaling cascades, in a concentration- and cell-dependent manner. Such roles of peroxynitrite in the redox regulation, including protein kinases B and C (PKB and PKC), the MAP kinases, nuclear factor kappa B, as well as signaling dependent on insulin and the sympatho-adrenergic system, are well documented (Liaudet, Vassalli, & Pacher, 2009).

Although sGC/cGMP pathway is crucial for the signaling transduction of NO, S-nitrosylation representing covalent reaction of an NO group with a reactive cysteine thiol on target proteins also mediates the vast majority of NO signaling (Nakamura & Lipton, 2011). The formation of S-nitrosoproteins (SNO-Ps) generally regulates protein function either allosterically or by direct modification of an active site cysteine. Interestingly, S-nitrosylation can mediate either protective or neurotoxic effects depending on the action of the target protein. For example, NO protects against excitotoxicity by S-nitrosylating the NR1 and NR2 subunits of the NMDA receptor, thereby diminishing Ca^{2+} influx (Choi et al., 2000). On the other hand, the pathological consequence of SNO-Ps has been characterized. The overproduction of oxidative/nitrosative stress contributes to neuronal cell death through S-nitrosylation of a number of targets, including MMP-9, cyclooxygenase-2 (COX-2), *N*-ethylmaleimide sensitive factor, and GAPDH (Nakamura & Lipton, 2011). For instance, S-nitrosylation of GAPDH augments its binding to Siah, which possesses a nuclear localization signal, and thus precipitates the translocation of the GAPDH/Siah complex into the nucleus. In the nucleus, SNO-GAPDH influences many signaling molecules, such as p300/CBP, SIRT1, and HDAC2, to initiate apoptotic cascades (Kornberg et al., 2010; Sen et al., 2008). Thus, SNO-Ps may become candidate targets for future treatments of neurological disorders.

3. NO AND HIPPOCAMPAL PLASTICITY

3.1. NO and LTP

Long-term potentiation (LTP) is a persistent increase in synaptic strength following high-frequency stimulation of a chemical synapse. It is one of several phenomena underlying synaptic plasticity. As the process of memory is considered to be encoded by modification of synaptic strength, LTP is widely counted as one of the major cellular mechanisms underlying learning and memory (Eichenbaum, Otto, & Cohen, 1992; Squire & Zola-Morgan, 1991). When giving a repetitive stimulation to presynaptic terminals, the synaptic transmission is enhanced for a long term. A major consequence is Ca^{2+} influx through NMDA receptor located on the postsynaptic membrane. For nNOS/NO regulated by Ca^{2+}/calmodulin, there is a potential link to LTP induction events by NO (Huang, 1997).

However, it cannot explain how postsynaptic events result in presynaptic changes during the sustained increase of synaptic strength. The hypothesized answer involves a membrane-permeable retrograde messenger which released by the postsynaptic cell and diffused to the presynaptic terminal to trigger the chain of events associated with LTP. Admittedly, NO has been proposed as the retrograde messenger for the following reasons (Huang, 1997). In initial studies, NOS inhibitors were found to block LTP when applied to hippocampal slice, and this experiment worked in an arginine-reversible manner (Schuman & Madison, 1991). Furthermore, LTP can be blocked when NOS inhibitors are selectively injected into postsynaptic cells through intracellular recording electrodes, which strongly suggests that the requirement of NO during the LTP formation (O'Dell, Hawkins, Kandel, & Arancio, 1991; Schuman & Madison, 1991). These results are also supported by the studies, showing that hemoglobin which binds NO and is not taken by cells attenuates LTP when applied extracellularly (Schuman & Madison, 1991).

Since NO is diffusible signal, whether the individual synapse or a local ensemble of synapses is influenced during the LTP induction has been discussed. Schuman and Madison have observed that inducing LTP at the synapses onto an individual CA1 neuron can enhance synaptic transmission at the synapses onto a neighboring neuron (Schuman & Madison, 1994). For limited diffusive distance of NO, the spatial extent of its influence to synaptic transmission should be restricted in certain range. To test this hypothesis, Schuman and Madison compared the synaptic enhancement exhibited by near and far neighboring cells. They found that LTP was induced in one neuron by paring postsynaptic depolarization with low-frequency stimulation of the presynaptic fibers when the cells bodies were about 35 μm apart. However, no potentiation was detected at the spatially remote cell synapses which were on average 595 μm apart (Schuman, 1995). It reveals that potentiation of synaptic transmission at neighboring cell synapses appears to be limited to the synapses near the site of LTP generation.

These evidences above still could not affirm that NO participates in the distributed potentiation directly. The researchers found that interfering with NO production using NOS inhibitor L-methyl-arginine (L-Me-Arg) in the paired cell blocked the LTP formation. Moreover, the injection of L-Me-Arg in the paired cell prevented the enhancement of synaptic transmission at the nearby neighboring cell synapses (O'Dell et al., 1991; Schuman & Madison, 1991). These results suggest a role for NO in the production of distributed synaptic activity.

In addition to that NO itself mediates the potentiation at neighboring synapses, it is possible that downstream of NO signals is involved. The potential role of cGMP as a mediator of LTP has been suggested firstly by Haley and colleagues who have shown that cGMP analogues can partially reproduce the effects of NO on LTP induction (Haley, Wilcox, & Chapman, 1992). Furthermore, high-frequency stimulation results in an elevation of cGMP in the CA1 region of the hippocampus which could be blocked by NOS inhibitor (Chetkovich et al., 1993). And hippocampal slices perfused with selective inhibitors of cG-PK fail to show LTP after a high-frequency stimulus (Zhuo, Hu, Schultz, Kandel, & Hawkins, 1994). These findings suggest that cGMP mediates NO-dependent LTP.

LTP has multiple temporal domains. One domain is associated with an enhancement of the excitatory postsynaptic potentials that persists for about 90 min. This form of LTP is referred to as early LTP (E-LTP). A second domain referred to as late LTP (L-LTP) is associated with synaptic enhancement that persists for periods of time greater than about 90 min. Many studies about the role of NO in LTP have focused on E-LTP. However, behavioral studies have shown that NO is involved in learning tasks, some of which are remembered for days, reminding us that NO signaling might also be involved in L-LTP. E-LTP and L-LTP involve different signaling pathways. In the CA1 region of the hippocampus, L-LTP is reduced by inhibitors or genetic disruption of cAMP-dependent protein kinase (PKA), whereas E-LTP is usually not PKA-dependent (Abel et al., 1997; Frey, Huang, & Kandel, 1993). Further results suggest that NO contributes to L-LTP by stimulating GC and cGMP-dependent protein kinase, which acts in parallel with PKA to increase phosphorylation of the transcription factor cAMP response element-binding protein (CREB) (Lu, Kandel, & Hawkins, 1999).

However, there were some controversial points of view. For using dibutyrul cGMP only partially mimicked LTP in the study by Haley et al. and the cGMP analogue 8-bromo-cGMP (8-Br-cGMP) was too weak in itself to induce LTP in the hippocampal slice (Zhuo et al., 1994), NO-mediated synaptic strength regulation may also involves the non-cGMP mechanism. Agreeing with that, some studies found the cGMP does not enhance synaptic transmission, even in some cases, depresses postsynaptic currents and neuronal firing (Doerner & Alger, 1988; Kleppisch & Feil, 2009). Therefore, further researches are required before any conclusions can be drawn.

3.2. NO and neurogenesis

Contrary to the long hold standpoint, it has been confirmed that continuous neurogenesis occurs in adult brain over the last years (Bruel-Jungerman, Rampon, & Laroche, 2007). The SGZ of DG in the hippocampus is one of the main sites where the neurogenesis occurs (Ehninger & Kempermann, 2008).

Current knowledge of hippocampal function indicates that adult neurogenesis provides a mechanism through synaptic plasticity to the process of learning and memory (Drapeau et al., 2003; Shors et al., 2001), and the associated mood states, such as depression, schizophrenia, and drug addiction (Jun, Mohammed Qasim Hussaini, Rigby, & Jang, 2012; Santarelli et al., 2003; Zhou et al., 2007, 2011).

In the hippocampus, neuronal progenitor cells from the SGZ of the DG migrate into the granule cell layer where they differentiate to neurons or gliocytes. Newly generated neuronal cells can extend their axons to the CA3 region of the hippocampus a few days after their mitosis. Following that, these axons can form synapses with adjacent cells. Thus, the newly generated cells can differentiate to mature neurons and integrate into the original neural networks to participate the hippocampal function (Gould, Tanapat, Rydel, & Hastings, 2000; van Praag et al., 2002).

There are multiple regulatory factors involved in the process of hippocampal neurogenesis, such as all kinds of neural growth factors, neurotrophic factors, steroid hormones, and neurotransmitters (Gould et al., 2000).

Recent studies have shown that NO plays an important roles in regulating neurogenesis (Hindley et al., 1997; Hua et al., 2008; Luo et al., 2007; Zhu et al., 2003, 2006). However, it is hard to draw a simple conclusion about whether NO promote or inhibit hippocampal neurogenesis.

In the development brain, NO is essential for the neuronal differentiation, survival, and precursor proliferation. nNOS knockout mice exhibit a transient decrease in neural precursors proliferation at early postnatal ages (Chen et al., 2004). In the cultured embryonic neural stem cells (NSCs), L-VNIO, a potent and highly selective inhibitor of nNOS, decreases neurospheres formation and cell proliferation. Consistent with pharmacological manipulations, deleting nNOS gene causes a markedly decreased proliferation of the cultured NSCs (Luo et al., 2010). In PC12 cells model, it has been found that the expression of differentiation markers can be reduced by NOS inhibitors (Poluha et al., 1997).

In adult brain, NO donor DETA/NONOate has obvious effects on promoting neurogenesis in two brain disease models. In a traumatic brain injury model in rats, DETA/NONOate treatment significantly increases the proliferation, survival, migration, and differentiation of neural progenitor cells in the hippocampus (Lu, Mahmood, Zhang, & Copp, 2003). In a classic depression mouse model, intrahippocampal infusion of DETA/NONOate increases both cell proliferation and newborn cells survival in the DG in mice with chronic mild stress (CMS) (Hua et al., 2008). These results suggest a positive role of NO in the fate of newly born cells in the hippocampus.

However, a numerous studies have revealed "dark side" of brain NO on neurogenesis. These studies gave a strong evidence to show that NO is a negative regulator of cell proliferation in neurogenic regions of the adult brain. Several earlier studies show that NO inhibits proliferation of NSCs under physiological conditions (Matarredona, Murillo-Carretero, Moreno-López, & Estrada, 2005; Moreno-López et al., 2004; Packer et al., 2003). In isolated NSCs from the SVZ, supraphysiological concentration of NO inhibits NSCs proliferation and promotes differentiation of precursors into astrocytes (Covacu et al., 2006; Torroglosa et al., 2007).

Infusing 7-nitroindazole (7-NI), a nNOS inhibitor, directly into the mice hippocampus, causes significantly increases in both the proliferation of NSCs and their survival in the DG. The similar results are observed in the mice lacking nNOS gene (Zhou et al., 2007, 2011), suggesting that nNOS-derived NO is a negative regulator for the hippocampal neurogenesis. Actually, it has been well established that nNOS-derived NO exerts an antiproliferative effect on neurogenesis for a long time (Packer et al., 2003;

Zhu et al., 2006). And it has been associated with the pathophysiology of ischemia closely (Sun et al., 2005).

Although nNOS-derived NO exerts a negative control on the neurogenesis in the ischemic brain, iNOS may be necessary in ischemic hippocampal for neurogenesis-mediated self-recovery of the injured brain. NO produced by iNOS activation seems to have a positive role in the ischemia-induced neurogenesis, since inhibition of iNOS or deleting iNOS gene prevents ischemia-induced neurogenesis in the DG (Zhu et al., 2003). Similarly, another study suggests that NO derived from iNOS is important for increase in cell proliferation following a brain insult (Carreira et al., 2010). Luo et al. reported that inhibition of nNOS or deleting nNOS gene in the ischemic hippocampus upregulates iNOS expression, thereby promoting neurogenesis, which explains the different roles of nNOS-derived NO and iNOS-derived NO in regulating neurogenesis in ischemic brain (Luo et al., 2007).

Similar to the role of iNOS, eNOS is likely necessary for neurogenesis. Deletion of eNOS gene in mice causes a significant reduction in neuronal progenitor cell proliferation in the DG (Reif et al., 2004). Thus, it seems that the effects of NO on hippocampal neurogenesis depend on which NOS isoform it comes from.

Besides the origins of NO, the local concentration of NO is involved in the different regulation on neurogenesis. It has been shown that 0.1 and 0.4 μM DETA/NONOate promotes neuronal differentiation and neurite outgrowth (Chen et al., 2006), but 100 and 1000 μM DETA/NONOate diverts NSCs fate from neurons toward astrogliocytes (Covacu et al., 2006). It was also reported that NO donor NOC-18 promotes NSCs proliferation at 10 μM, whereas it inhibits proliferation at 100 μM (Carreira et al., 2010). CREB is a well-known nuclear transcription factor contributing to the hippocampal neurogenesis (Dworkin et al., 2009). The recent studies indicate that NO regulates CREB phosphorylation bidirectionally in the hippocampus. The lower dose of NO donors upregulates pCREB levels, while higher levels of NO downregulate CREB phosphorylation, both *in vitro* and *in vivo* (Hu et al., 2012).

Interestingly, besides the isoform of NOS and the concentration of NO, the cellular source of NO is implicated in different regulation of neurogenesis. It has been reported that NSCs-derived nNOS and neurons-derived nNOS regulate NSCs proliferation and differentiation bidirectionally. NSCs-derived nNOS is positive, whereas neurons-derived nNOS is negative to neurogenesis (Luo et al., 2010). The bidirectional regulations of neurogenesis by nNOS might result from different subcellular localizations of nNOS in NSCs and neurons, mainly in nuclei for NSCs and in cytoplasm for neurons. Thus, the negative role of nNOS in neurogenesis in the adult brain may be explained by that the inhibitory effect of neurons-derived nNOS overwhelmingly surpasses the facilitative effect of NSCs-derived nNOS because of predominant expression of nNOS in neurons in the brains. This point of view is supported by studies from $nNOS^{-/-}$ mice, in which they exhibit a transient decrease in neural precursors proliferation at early postnatal day (Chen et al., 2004) but a persistent increase in newborn cells proliferation in adult hippocampus (Parker et al., 2003; Zhou et al., 2007).

3.3. NO and synaptogenesis

Another implication of NO in synaptic plasticity has to be mentioned is that NO plays important roles in synaptogenesis. Synaptogenesis is critical for the formation of synapses between neurons in the nervous system. Although it occurs throughout a healthy person's life-span, an explosion of synapse formation occurs during early brain development.

In the early 1990s, several studies have implicated the effects of NO in synaptogenesis (Gally, Montague, Reeke, & Edelman, 1990; Hess, Patterson, Smith, & Skene, 1993). In light of this, researchers investigated the development regulation of nNOS. They demonstrated the physiologic postnatal induction of $nNOS_{155}$ during rodent brain development with maxima and temporal and spatial pattern that correlates with synaptogenesis. Since then, more studies focused on the regulation of central NO to synaptogenesis.

Multiple factors are involved in the process of synaptogenesis. It has been shown that activation of NMDA receptors initiates synaptogenesis through activation of downstream products. Recent studies show that induction of LTP triggers the formation of dendritic filopodia (Maletic-Savatic, Malinow, & Svoboda, 1999), considered precursors of spines (Ziv & Smith, 1996), as well as a direct formation of new spines or new types of synapses (Engert & Bonhoeffer, 1999). Remodeling of postsynaptic structures has thus been proposed to play a major role in activity-dependent synaptogenesis. However, there are other evidences for presynaptic morphological changes in synaptogenesis (Colicos, Collins, Sailor, & Goda, 2002; De Paola, Arber, & Caroni, 2003; Tashiro, Dunaevsky, Blazeski, Mason, & Yuste, 2003). In the organotypic hippocampal slice cultures, anoxia/hypoglycemia and theta burst stimulation, two conditions that were shown to induce morphological plasticity, also trigger synaptogenesis. In these processes, activation of postsynaptic NMDA receptors and subsequent release of NO, as a retrograde messenger, can stimulate a presynaptic plasticity, which in turn contributes to synaptogenesis (Nikonenko, Jourdain, & Muller, 2003). Blockade of NO production using NOS inhibitor L-NAME or selective nNOS inhibitor 3Br-7NI markedly prevents the structural presynaptic changes induced by anoxia/hypoglycemia and by theta burst stimulation, including growth of presynaptic filopodia-like protrusions, formation of multi-innervated spines (MISs), and formation of a postsynaptic density on the target cell. Moreover, the NO donor DETA/NONOate is able to reproduce the presynaptic remodeling triggered by those two conditions. Thus, NO is implicated in the mechanism of presynaptic changes in synaptogenesis.

As mentioned above, generally, remodeling of postsynaptic structure is proposed to play a major role in synaptogenesis. Postsynaptic density protein-95 (PSD-95) is a major regulator of synaptic structure and function. Knockdown of PSD-95 reduces the development of synaptic structures (Ehrlich, Klein, Rumpel, & Malinow, 2007), and PSD-95 mutant mice exhibit variations in spine densities in several brain regions (Vickers et al., 2006). Thus, PSD-95 is an important molecular that regulates the synaptogenesis. Using a quantitative electron microscopic analysis of serially sectioned dendritic segments of CA1 neurons in hippocampal slice cultures, it is shown that overexpression of PSD-95 not only affects the morphological characteristics of dendritic spines but also results in the formation of spines contacted by several different presynaptic partners. Evidence is provided to show that this effect depends on an interaction between PSD-95 and NOS and requires NO signaling. Downregulation of NOS by siRNA approach and treatment of slice cultures with the NOS inhibitor prevents the formation of MISs. To test whether the NO is indeed responsible for MISs formation, exogenous application of NO and activating signaling cascades associated with NO have been used. Using serial sections and 3D reconstructions, researchers analyzed more than 2000 spines in randomly chosen volumes in stratum radiatum taken from nine different hippocampal slice cultures and determined the proportion of MISs. This proportion increased about 10-fold after DEAT/NONOate treatment. Furthermore, NO donor treatment increased MISs not only on all cells of treated cultures but also on cells overexpressing the PDZ_2 mutant of PSD-95, indicating that the effect of NO is indeed downstream of PSD-95. Similarly, the application of 8-Br-cGMP increased the proportion of MISs in the entire tissue five times more than under control condition. Conversely, the blockade of cGMP by the inhibitor ODQ prevented formation of MISs on PSD-95-transfected neurons. Thus, NO production is sufficient for the synaptogenesis.

Directly, *in vivo* studies provide more sufficient evidences for this argument. nNOS knockout mice display defects in dendritic arborization (Inglis, Furia, Zuckerman, Strittmatter, & Kalb, 1998). In developing rat brain, NOS plays a key role during synaptic maturation by influencing synapsin I (Sánchez-Islas & León-Olea, 2004). In this study, rat pups were treated with a competitive NOS antagonist, *N*-nitro-L-arginine methyl ester, from postnatal days 3 to 23. NOS inhibition elicited a significant reduction in synapsin I immunoreactive density in the CA2, CA3, DG, and hilus of hippocampus, indicating synaptic maturation requires NOS signaling. Thus, functions that depend on the formation of synaptic connections such as learning and memory and normal mood status could be affected by NO deficiency.

Besides synapse formation, neurites outgrowth is also important substantial base. The outgrowth of neurites (axons and dendrites) is a fundamental morphological change associated with neural development and differentiation (Halegoua, Armstrong, & Kremer, 1991). Neurite sprouting may also play an important role in the formation of collateral circuits and functional recovery following damage to the brain, including hippocampus (Gage, Buzsáki, & Armstrong, 1990; Van der Zee, Fawcett, & Diamond, 1992). The primary cultures of E17 mouse hippocampal neurons cocultured with neopallial astrocytes treated by NO donors sodium nitrite or sodium nitroprusside have much longer neurites, and the effects of the differentiation of neurons were proved to be acted through a cGMP-dependent mechanism (Hindley et al., 1997). In cultures of PC12 cells, NO donors also enhance the neuritogenic effects of NGF. Similar neuritogenic effects can be observed when using 8-Br-cGMP and *para*(chlorophenylthio)-cGMP (*p*CPT-cGMP), but this neuritogenic activity is inhibited by LY83583 and methylene blue, inhibitors of guanylyl cyclase, indicating that NO may regulate synapse formation and maintenance by stimulating neurite outgrowth through sGC/cGMP pathway.

It is significant that several studies advocate a synaptogenic or neuritogenic role for NO, whereas a few other studies describe synaptotoxic NO-mediated regulation. Using a model of hypoglossal axonal injury, a group reported that axonal injury-induced disconnection of motor neurons from myocytes produces withdrawal of synaptic inputs to motor neurons and concomitant upregulation of the neuronal isoform of nNOS, and after recovery of the neuromuscular function, synaptic coverage is reestablished and nNOS is downregulated (Sunico, Portillo, González-Forero, & Moreno-López, 2005). They considered that NO acts as a signal for synaptic detachment and inhibits synapse formation by cGMP-dependent and S-nitrosylation-mediated mechanisms, respectively. However, no similar results are reported based on the hippocampal injury model.

4. NO AND THE RELATED CNS DISORDERS

4.1. NO and ischemia

Stroke is a major public health problem leading to high rates of death and disability in adults (Flynn, MacWalter, & Doney, 2008; Gállego, Muñoz, & Martínez-Vila, 2009). Sudden cessation of blood flow to the brain results in a series of events that either result in rapid loss of brain cells or delayed neuronal injury. One of the most vulnerable areas to ischemia or hypoglycemia is CA1 hippocampal region due to pyramidal neurons death. The involvement of NO in the pathology is complicated that it is hard to say it is beneficial or harmful in a simple way.

Confessedly, excitotoxicity is a well-accepted theory to elucidate the pathology of brain ischemia in the acute stages. It is defined as an excessive synaptic release of glutamate or overstimulation of its membrane receptors, resulting in neuronal injury or death (Sattler & Tymianski, 2000). Of the four glutamate receptors, NMDA receptor activation mediates excitotoxicity and neuronal damage after focal ischemia owning to their high calcium permeability (Gardoni & Di Luca, 2006). Overstimulation of NMDA receptors leads to Ca^{2+} overload and activation of Ca^{2+}-sensitive enzymes consequently. Thus, nNOS plays a prominent role in mediating excitotoxicity as a Ca^{2+}-sensitive enzyme (Eliasson et al., 1999). *In vitro*, NMDA-induced excitotoxicity is decreased by NOS inhibitors (Dawson, Dawson, London, Bredt, & Snyder, 1991; Uehara et al., 2006) and is decreased in neurons isolated from nNOS deficient mice (Dawson, Kizushi, Huang, Snyder, & Dawson, 1996). Brain damage following middle cerebral artery occlusion is diminished by treatment by nNOS inhibitors (Yoshida, Limmroth, Irikura, & Moskowitz, 1994) and in mice lacking nNOS gene (Huang et al., 1994).

The excessive NO resulted by nNOS activation may contribute to excitotoxicity via oxidation products of NO. Activation of NMDAR in neurons leads to the formation of superoxide ($O_2^{\bullet -}$) through oxidation of fatty acids (Lafon-Cazal, Pietri, Culcasi, & Bockaert, 1993). In ischemic brain, the superfluous superoxide produces toxicity directly and reacts with NO to produce peroxynitrite ($ONOO^-$). Normally, superoxide concentrations are kept at very low levels by the large amounts of endogenous SOD, an enzyme that rapidly scavenges superoxide. As we mentioned in Section 2.3, a majority of superoxide would react with the large amount NO rather than SOD, leading to the formation of peroxynitrite (Beckman et al., 1990; Huie & Padmaja, 1993). Peroxynitrite is a powerful oxidant which appears to be far more toxic than superoxide. The results with cultured neurons showed that it is lethal in a few seconds (Zhu, Gunn, & Beckman, 1992). NO generated during reperfusion after ischemic stroke can diffuse significant distances where it rapidly reacts with superoxide to form peroxynitrite. Because the rate of peroxynitrite formation depends on the product of superoxide and NO concentrations, each 10-fold increase in the concentrations of superoxide and NO results in about 100-fold augmentation in the rate of peroxynitrite formation (Varner & Beckman, 1995). Thus, in order to reducing the peroxynitrite production, it is logical that scavenge superoxide by increasing amount of SOD or reduce NO via different methods. Intravenously administered polyethylene glycol-conjugated SOD can reduce ischemic injury (Liu, Beckman, Freeman, Hogan, & Hsu, 1989). Transgenic mice overexpressing SOD are also quite resistant to cerebral insults (Klein, Chan, & Chang, 2003). As discussed above, it is clear that inhibiting nNOS activity or deleting nNOS gene can partially prevent glutamate-induced neuronal death and ischemia-induced brain damage in rodents *in vitro* and *in vivo*.

Activation of PARP may explain of the neuron death caused by peroxynitrite (Brown, 2010). PARP is a DNA-repair enzyme, which can also mediate cell death by NO. Peroxynitrite-mediated DNA damage results in PARP activation (Jagtap & Szabó, 2005). PARP inhibitors decrease neurotoxicity in the cultured neurons markedly (Zhang, Dawson, Dawson, & Snyder, 1994; Zhang, White & Iadecola, 1994). And stroke damage following MACO is reduced up to 80% in mice lacking PARP gene (Eliasson et al., 1997) or received treatment of PARP inhibitors (Abdelkarim et al., 2001). Moreover, PARP activation by brain ischemia is decreased in nNOS knockout mice. Thus, activation of nNOS by ischemia leads to peroxynitrite overproduction, in turn, activates PARP, causing neuronal death or energy depletion (Endres et al., 1998).

The truth is worthy to mention is that effect of nNOS inhibitors on ischemia depends on the treatment time course and dose setting. Some experiments have showed the inhibiting NOS enhances injury after cerebral ischemia (Yamamoto, Golanov, Berger, & Reis, 1992; Zhang, Dawson, et al., 1994; Zhang, White, et al., 1994). These reports remind us that the unexpected effects produced by simply inhibiting NOS may because of the inhibition of physiological effects of NO, such us disrupting collateral circulation, differentiation, and survival of new born neural cells.

For using NOS inhibitors or NO scavengers indiscriminately to treat ischemia would cause some severe side effects because of the physiological functions and even neuroprotective effects of NO, the better strategy focuses on the upstream signaling of excitotoxicity to prevent the harmful results. PSD-95 is a scaffolding protein that binds NMDAR and nNOS at excitatory synapses and assembles them into a macromolecular complex (Aarts et al., 2002; Cao et al., 2005). Immunoprecipitation showed that cerebral ischemia induces the interaction of nNOS with PSD-95 and disrupting nNOS–PSD-95 interaction via overexpressing the N-terminal amino acid residues 1–133 of nNOS (nNOS-N_{133}) prevented glutamate-induced excitotoxicity and ischemic damage. A blocker of nNOS–PSD-95 association ZL006 exhibits potent neuroprotective activity *in vitro*, ameliorates focal cerebral ischemic damage in rodents subjected to MACO, and has no undesirable effects of directly inhibiting nNOS (Zhou et al., 2010).

The reason that we should not inhibit NOS or scavenge endogenous NO is because the beneficial effects of NOS/NO involved in the pathophysiology of ischemia. NO has a crucial role in angiogenesis after ischemic stroke. It has been found that DEAT/NONOate increased angiogenesis after experimental cerebral ischemia (Chen et al., 2004; Zhang et al., 2003). eNOS-deficient mice display significantly impaired neovascularization after stroke, indicating that eNOS-derived NO mediates this effect (Cui et al., 2009). Moreover, eNOS-deficient mice have lower postischemic cerebral blood flow (CBF) levels, larger infarcts, and smaller penumbral regions than their wild-type littermates following MACO (Huang et al., 1996; Lo et al., 1996). Another restorative mechanism after stroke is NO-dependent neurogenesis. Although iNOS-derived NO was shown to be neurotoxic in the acute phase after stroke, several studies suggest that it plays an important role in cell proliferation after ischemia (Zhu et al., 2003; Zhang, Zhang, & Chopp, 2005).

Thus, scavenging NO or inhibiting NOS blindly or nonselectively could not be the good choice for treatment of stroke. Even several treatment strategies have been examined that increasing postischemic NO content specifically generated by eNOS activity to augment CBF could ameliorate the ischemic pathologic change (Endres, Laufs, Liao, & Moskowitz, 2004).

4.2. NO and mood disorders

Affective disorders such as depression and anxiety are life threatening and highly prevalent stress-related disorders. They rank among the top causes of worldwide disease burden and disability (Ressler & Mayberg, 2007). The hippocampus is one of several limbic brain regions implicated in the pathophysiology and treatment of affective disorders (Warner-Schmidt & Duman, 2006). As mentioned above, NO is involved in many hippocampal functions, including newly born cell proliferation, LTP, synaptogenesis, neurodegeneration, neuronal survival, and differentiation (Cho et al., 2009; Zhou & Zhu, 2009). A number of studies have shown that NO has significant implications for the psychiatric disorders characterized by depression (Hu et al., 2012; Zhou & Zhu, 2009; Zhou et al., 2007, 2011), anxiety (Zhang et al., 2010), schizophrenia (Reif et al., 2006; Shinkai, Ohmori, Hori, & Nakamura, 2002), and aggressiveness (Chiavegatto et al., 2001; Nelson et al., 1995).

Depression is a severe, recurring mental illness, which would be the primary cause of disability in the world by the year 2020 estimated by the World Health Organization (Murray & Lopez, 1997). It is commonly suggested that depression is highly correlated with hippocampal functions. Previous studies found that paroxetine and imipramine, the typical antidepressants, inhibited the enzymatic activity of NOS in animal and humans, and antidepressants withdraw resumes NOS activity (Angulo et al., 2001; Harvey, Retief, Korff, & Wegener, 2006). Moreover, plasma nitrate concentration, an index of NO production, is prominently higher in depressed patients (Suzuki, Yagi, Nakaki, Kanba, & Asai, 2001). These results suggest a possible involvement of NO in depression. Moreover, nNOS inhibitor 7-NI produces antidepressant-like effects, and these effects are reversed by L-arginine (Yildiz, Erden, Ulak, Utkan, & Gacar, 2000). Recently, it is reported that the CMS exposure upregulates nNOS expression in the hippocampus and induces depressive behaviors in male mice. nNOS inhibition or deletion reverses CMS-induced depressive syndromes (Zhou et al., 2007). The same group also gave a further evidence to show that nNOS contributes to chronic stress-induced depression by suppressing hippocampal neurogenesis, which involves the hippocampal telomerase activity (Zhou et al., 2011). Furthermore, they demonstrate that nNOS is a crucial factor mediating the behavioral effects of stress hormones, glucocorticoids. Hippocampal nNOS-derived NO significantly downregulates local Glucocorticoid receptor (GR) expression through both sGC/cGMP and peroxynitrite signal pathway and therefore elevates hypothalamic CRF, a peptide that governs the HPA axis (Zhou et al., 2011).

Epidemiological studies have indicated that about one-third of people will, at some periods in their lives, suffer from anxiety disorders (Hawgood & De Leo, 2008). Increasing evidence suggests that hippocampal NO is implicated in anxiety disorders (Miguel & Nunes-de-Souza, 2008; Workman, Trainor, Finy, & Nelson, 2008). NO donor sodium nitroprusside (SNP) facilitates an anxiogenic effect, whereas a nonselective NOS inhibitor L-NAME and nNOS inhibitor 7-NI produce the anxiolytic-like behaviors by infusing into the ventral hippocampus (Calixto et al., 2010). A recent work elucidates a model of signal pathway of anxiety. Postsynaptic 5-HT1A receptor stimulation downregulates nNOS expression in the hippocampus and leads to a decrease in NO that exerts a negative control on CREB phosphorylation, thus influencing anxiety-related behavior (Zhang et al., 2010). This report is in line with the viewpoints that hippocampal NO is involved in the effects of typical neurotransmitters, especially monoamine.

Interestingly, there are several evidences indicate that hippocampal NO plays a role in the sex difference of behaviors (Kant et al., 2000; Keser, Balkan, Gozen, Kanit, & Pogun, 2011; Taskiran, Kutay, Sozmen, & Pogun, 1997). It is commonly suggested that a female preponderance in depression and anxiety is universal and substantial. The prevalence of depression and anxiety for women is approximately twice that for men (Essau, Lewinsohn, Seeley, & Sasagawa, 2010; McLean, Asnaani, Litz, & Hofmann, 2011; Ressler & Mayberg, 2007). Female mice have substantially lower NO production in the hippocampus and are significantly more likely to display negative affective behaviors that their male littermates. Eliminating the difference in the basal hippocampal NO level between male and female mice mends the sex gap of affective behaviors (Hu et al., 2012). Interestingly, CMS upregulated nNOS in male mice, whereas it significantly downregulated nNOS in female mice (Hu et al., 2012; Palumbo et al., 2007; Zhou et al., 2007). For estradiol exerted a positive control on hippocampal NO production via ERβ-mediated nNOS expression, low estrogen in the female hippocampus (Mukai et al., 2010) accounts for lower local NO than in the male hippocampus. Although estrogen has important significance in modulating affective behaviors, it is not estrogen but NO in the hippocampus that mediates the gender difference of affective behaviors directly because hippocampal NO is necessary for the behavioral effects of estradiol and NO is an independent factor in modulating behaviors (Hu et al., 2012). Stress promoted hippocampal NO production in male due to glucocorticoid release, thus leading to local NO excess (Zhou et al., 2011). In contrast, stress suppresses NO production in female due to estrogen falling, thereby resulting in hippocampal NO shortage. While activating CREB rescues the depression-like effects of intrahippocampal NO donor DETA/NONOate, inactivating CREB abolishes the antidepressant-like effects of intrahippocampal NO donor DETA/NONOate. These findings suggest that NO is a critical molecular underlying the sex difference of affective disorders (Hu et al., 2012).

4.3. NO and Alzheimer's disease

Alzheimer's disease (AD) is the most common form of dementia, with progressive cognitive deficits being the primary symptom (Bekris, Yu, Bird, & Tsuang, 2010). AD affects approximately 35 million people worldwide (Querfurth & LaFerla, 2010). The prevalence of AD increases with age and is estimated to be about 1% in persons aged 60–64 but increases up to 33% in persons aged 85 or older (Mayeux, 2010).

It is neuropathologically characterized by amyloid and neurofibrillary tangle (NFT) depositions, basal forebrain cholinergic deficits, and extensive neuronal loss and synaptic changes in the cortex and hippocampus (Law, Gauthier, & Quirion, 2001). Besides the other brain regions, amygdalar and hippocampal atrophies are apparent in the AD brain (Kesslak, Nalcioglu, & Cotman, 1991). These pathological changes are considered to be associated with the neural cell damage caused by NO toxicity. The ability of NO to exert cellular damage due to its reactive oxidative properties is perhaps the primary neurotoxic mechanism (Law et al., 2001). The overproduction of NO caused by glutamatergic excitotoxicity will likely cause neuronal damage. Amyloid-β (Aβ), a peptide thought to play a key role in AD, has been shown to generating oxidative stress itself (Varadarajan, Yatin, Aksenova, & Butterfield, 2000). In addition, it also has synergistic action with glutamate to induce neuronal damage via an NO induction pathway (Aksenov et al., 1995; Noda, Nakanishi, & Akaike, 1999). It has been shown that Aβ stimulates microglial and astrocytic NO production (Akama, Albanese, Pestell, & Van Eldik, 1998). In the AD brain, Aβ activates CD4 + T cells, which produces various cytokines and induces microglial cell activation subsequently (Dickson, Lee, Mattiace, Yen, & Brosnan, 1993). Microglial activation induces iNOS-derived NO release excessively (Hu, Akama, Krafft, Chromy, & Van Eldik, 1998; Kröncke, Fehsel, & Kolb-Bachofen, 1995; Wallace, Geddes, Farquhar, & Masson, 1997). Moreover, activated microglial produces TNF-α which also potentiates NO production (Kröncke et al., 1995). The cytokines themselves can directly induce astrocytic iNOS-mediated NO production or in synergy with Aβ to induce iNOS expression (Akama et al., 1998). This resultant NO overproduction would most likely lead to increased peroxynitrite and possibly cell death.

Ca^{2+} homeostasis dysfunction is another mechanism that supports the link between NO and AD, since Ca^{2+} is a primary regulator of nNOS as discussed above. Aβ exposure or pretreatment could increase intracellular Ca^{2+} concentration and enhance the effects of glutamate and membrane depolarization postsynaptically (Hartmann, Eckert, & Müller, 1993; Mattson et al., 1992; Mattson, Tomaselli, & Rydel, 1993).

The involvements of NO in AD brain pathology also include the inflammatory process. For example, COX-2 expression has been shown to be upregulated in AD (Ho et al., 1999). The participation of NO in activating COX-2 and activation of pro-inflammatory has been well documented (del Zoppo et al., 2000). In addition to Aβ, NFT is another important AD marker (Law et al., 2001). There is evidence to show that NO may have a role in the development of NFT. NO is an activator of P21ras which has been implicated in the formation of NFT (Lander, Ogiste, Teng, & Novogrodsky, 1995; Lüth, Holzer, Gertz, & Arendt, 2000). Interestingly, colocalization of nNOS and P21ras in NFT-containing pyramidal neurons has been reported (Lüth et al., 2000). Apoptotic cell death may also be a feature of AD pathology, besides the necrotic death (Law et al., 2001). NO has been confirmed to participate in apoptotic events with low concentration being antiapoptotic, while higher concentration is proapoptotic (Dimmeler & Zeiher, 1997). It reminds us that NO is also likely to be involved in neuroprotection.

NO may exert its neuroprotectivity by means of inhibiting platelet aggregation or blocking leukocyte adhesion. And the participation of NO in LTP induction and maintaining could not be ignored because LTP abnormalities have been observed in AD brain (Bach et al., 1999). Aβ has been found to markedly impair hippocampal LTP (Bliss & Collingridge, 1993). Moreover, using NO donor, sGC stimulator, and cGMP analogs, researchers found that when modulating hippocampal synaptic plasticity, Aβ downregulates the NO/cGMP/cGK/CREB pathway (Puzzo et al., 2005). Thus, enhancement of the NO/cGMP signaling may provide a novel approach to the treatment of AD.

It is important to note that, when discussing the putative role of NO in neurological disorders, it is not sufficient to examine NO itself without consideration of NOS origins. However, it is still hardly to say which kind of NOS isoform is beneficial or harmful. Although it has been speculated that eNOS-derived NO may produce neuroprotective effects in AD brain (Law et al., 2001), just like in ischemic brain, data from animal studies demonstrated that the inactivation of iNOS increases both Aβ deposition and tau phosphorylation thus supporting a neuroprotective role of iNOS-derived NO (Colton et al., 2006). It is supported by studies from iNOS knockout mice (Colton et al., 2008). However, these hypotheses remain to be investigated fully.

5. CONCLUSION

Collectively, NO is implicated in a wide range of physiological functions and pathologies with multiple effects, due to its highly permeable and diffusible properties and derivation from different isoforms. Constructive NOS is distributed in the hippocampus abundantly, and iNOS can also express in many pathological situations. NO is produced in the hippocampus predominantly as a significant "neurotransmitter." NO participates in LTP and other forms of synaptic plasticity in the hippocampus. On applying LTP-inducing stimuli, NMDA receptor activation elicits a brief burst of NO. NO from nNOS is a good candidate for a retrograde messenger, diffusing rapidly to the presynaptic terminal and effecting changes in the neurotransmitter release machinery. NO signal transduction generally occurs through binding to sGC resulting in the formation of cGMP, with cGMP-dependent protein kinase being one of the downstream mechanisms leading to changes in synaptic strength. A growing evidences suggest that S-nitrosylation is another important signaling mediating the effects of NO on physical or pathological process in the CNS. The existence of two effector pathways has important functional implications, expanding and enriching the understanding that how NO modulates neuronal activities.

In the hippocampus, NO is a Janus molecular in the neuroprotection or neurotoxicity. It is important for neural cell survival and differentiation but also participates in the process of cell death or apoptosis. Interestingly, both NO donors and NOS inhibitors increase neurogenesis in the DG in different situations. Similarly, either complementing or scavenging NO could be used as an antistroke strategy depending on the time course and doses of treatments. In animal models, both NO donors and NOS inhibitors produce the antidepression-like behaviors. Thus, in the pathological situation, effects of NO are determined by the concentration, the time course, and the origins of NOS isoforms. To develop highly selective and specific new NO-related therapies for neurological diseases, further studies are required to clarify the precise changes in the different pathophysiology process.

REFERENCES

Aarts, M., Liu, Y., Liu, L., Besshoh, S., Arundine, M., Gurd, J. W., et al. (2002). Treatment of ischemic brain damage by perturbing NMDA receptor–PSD-95 protein interactions. *Science, 298*(5594), 846–850.

Abdelkarim, G. E., Gertz, K., Harms, C., Katchanov, J., Dirnagl, U., Szabó, C., et al. (2001). Protective effects of PJ34, a novel, potent inhibitor of poly(ADP-ribose) polymerase (PARP) in in vitro and in vivo models of stroke. *International Journal of Molecular Medicine,* 7(3), 255–260.

Abel, T., Nguyen, P. V., Barad, M., Deuel, T. A. S., Kandel, E. R., & Bourtchouladze, R. (1997). Genetic demonstration of a role for PKA in the late phase of LTP and in hippocampus-based long-term memory. *Cell, 88,* 615–626.

Akama, K. T., Albanese, C., Pestell, R. G., & Van Eldik, L. J. (1998). Amyloid beta-peptide stimulates nitric oxide production in astrocytes through an NFkappaB-dependent mechanism. *Proceedings of the National Academy of Sciences of the United States of America, 95*(10), 5795–5800.

Aksenov, M. Yu., Aksenova, M. V., Harris, M. E., Hensley, K., Butterfield, D. A., & Carney, J. M. (1995). Enhancement of beta-amyloid peptide A beta(1–40)-mediated neurotoxicity by glutamine synthetase. *Journal of Neurochemistry, 65*(4), 1899–1902.

Angulo, J., Peiró, C., Sanchez-Ferrer, C. F., Gabancho, S., Cuevas, P., Gupta, S., et al. (2001). Differential effects of serotonin reuptake inhibitors on erectile responses, NO-production, and neuronal NO synthase expression in rat corpus cavernosum tissue. *British Journal of Pharmacology, 134*(6), 1190–1194.

Bach, M. E., Barad, M., Son, H., Zhuo, M., Lu, Y. F., Shih, R., et al. (1999). Age-related defects in spatial memory are correlated with defects in the late phase of hippocampal long-term potentiation in vitro and are attenuated by drugs that enhance the cAMP signaling pathway. *Proceedings of the National Academy of Sciences of the United States of America, 96*(9), 5280–5285.

Beckman, J. S., Beckman, T. W., Chen, J., Marshall, P. A., & Freeman, B. A. (1990). Apparent hydroxyl radical production by peroxynitrite: Implications for endothelial injury from nitric oxide and superoxide. *Proceedings of the National Academy of Sciences of the United States of America, 87*(4), 1620–1624.

Beckman, J. S., & Koppenol, W. H. (1996). Nitric oxide, superoxide, and peroxynitrite: The good, the bad, and ugly. *The American Journal of Physiology, 271,* C1424–C1437.

Bekris, L. M., Yu, C. E., Bird, T. D., & Tsuang, D. W. (2010). Genetics of Alzheimer disease. *Journal of Geriatric Psychiatry and Neurology, 23*(4), 213–227.

Bliss, T. V., & Collingridge, G. L. (1993). A synaptic model of memory: Long-term potentiation in the hippocampus. *Nature, 361*(6407), 31–39.

Bowman, A., Gillespie, J. S., & Pollock, D. (1982). Oxyhaemoglobin blocks non-adrenergic non-cholinergic inhibition in the bovine retractor penis muscle. *European Journal of Pharmacology, 85*(2), 221–224.

Bredt, D. S. (1999). Endogenous nitric oxide synthesis: Biological functions and pathophysiology. *Free Radical Research, 31*(6), 577–596.

Bredt, D. S., Glatt, C. E., Hwang, P. M., Fotuhi, M., Dawson, T. M., & Snyder, S. H. (1991). Nitric oxide synthase protein and mRNA are discretely localized in neuronal populations of the mammalian CNS together with NADPH diaphorase. *Neuron,* 7(4), 615–624.

Brown, G. C. (2010). Nitric oxide and neuronal death. *Nitric Oxide, 23*(3), 153–165.
Bruel-Jungerman, E., Rampon, C., & Laroche, S. (2007). Adult hippocampal neurogenesis, synaptic plasticity and memory: Facts and hypotheses. *Reviews in the Neurosciences, 18*(2), 93–114.
Calabrese, V., Butterfield, D. A., Scapagnini, G., Stella, A. M., & Maines, M. D. (2006). Redox regulation of heat shock protein expression by signaling involving nitric oxide and carbon monoxide: Relevance to brain aging, neurodegenerative disorders, and longevity. *Antioxidants & Redox Signaling, 8*(3–4), 444–477.
Calixto, A. V., Duarte, F. S., Duzzioni, M., Nascimento Häckl, L. P., Faria, M. S., & De Lima, T. C. (2010). Role of ventral hippocampal nitric oxide/cGMP pathway in anxiety-related behaviors in rats submitted to the elevated T-maze. *Behavioural Brain Research, 207*(1), 112–117.
Cao, J., Viholainen, J. I., Dart, C., Warwick, H. K., Leyland, M. L., & Courtney, M. J. (2005). The PSD95–nNOS interface: A target for inhibition of excitotoxic p38 stress-activated protein kinase activation and cell death. *The Journal of Cell Biology, 168*(1), 117–126.
Carreira, B. P., Morte, M. I., Inácio, A., Costa, G., Rosmaninho-Salgado, J., Agasse, F., et al. (2010). Nitric oxide stimulates the proliferation of neural stem cells bypassing the epidermal growth factor receptor. *Stem Cells, 28*(7), 1219–1230.
Chen, J., Tu, Y., Moon, C., Matarazzo, V., Palmer, A. M., & Ronnett, G. V. (2004). The localization of neuronal nitric oxide synthase may influence its role in neuronal precursor proliferation and synaptic maintenance. *Developmental Biology, 269*(1), 165–182.
Chen, J., Zacharek, A., Li, Y., Li, A., Wang, L., Katakowski, M., et al. (2006). N-cadherin mediates nitric oxide-induced neurogenesis in young and retired breeder neurospheres. *Neuroscience, 140*(2), 377–388.
Chetkovich, D. M., Klann, E., & Sweatt, J. D. (1993). Nitric oxide synthase-independent long-term potentiation in area CA1 of hippocampus. *Neuroreport, 4*(7), 919–922.
Chiavegatto, S., Dawson, V. L., Mamounas, L. A., Koliatsos, V. E., Dawson, T. M., & Nelson, R. J. (2001). Brain serotonin dysfunction accounts for aggression in male mice lacking neuronal nitric oxide synthase. *Proceedings of the National Academy of Sciences of the United States of America, 98*(3), 1277–1281.
Cho, D. H., Nakamura, T., Fang, J., Cieplak, P., Godzik, A., Gu, Z., et al. (2009). S-nitrosylation of Drp1 mediates beta-amyloid-related mitochondrial fission and neuronal injury. *Science, 324*, 102–105.
Choi, Y. B., Tenneti, L., Le, D. A., Ortiz, J., Bai, G., Chen, H. S., et al. (2000). Molecular basis of NMDA receptor-coupled ion channel modulation by S-nitrosylation. *Nature Neuroscience, 3*(1), 15–21.
Colicos, M. A., Collins, B. E., Sailor, M. J., & Goda, Y. (2002). Remodeling of synaptic actin induced by photoconductive stimulation. *Cell, 107*, 605–616.
Colton, C. A., Vitek, M. P., Wink, D. A., Xu, Q., Cantillana, V., Previti, M. L., et al. (2006). NO synthase 2 (NOS2) deletion promotes multiple pathologies in a mouse model of Alzheimer's disease. *Proceedings of the National Academy of Sciences of the United States of America, 103*(34), 12867–12872.
Colton, C. A., Wilcock, D. M., Wink, D. A., Davis, J., Van Nostrand, W. E., & Vitek, M. P. (2008). The effects of NOS2 gene deletion on mice expressing mutated human AbetaPP. *Journal of Alzheimer's Disease, 15*(4), 571–587.
Covacu, R., Danilov, A. I., Rasmussen, B. S., Hallén, K., Moe, M. C., Lobell, A., et al. (2006). Nitric oxide exposure diverts neural stem cell fate from neurogenesis towards astrogliogenesis. *Stem Cells, 24*(12), 2792–2800.
Cui, X., Chopp, M., Zacharek, A., Zhang, C., Roberts, C., & Chen, J. (2009). Role of endothelial nitric oxide synthetase in arteriogenesis after stroke in mice. *Neuroscience, 159*(2), 744–750.

Dawson, V. L., Dawson, T. M., London, E. D., Bredt, D. S., & Snyder, S. H. (1991). Nitric oxide mediates glutamate neurotoxicity in primary cortical cultures. *Proceedings of the National Academy of Sciences of the United States of America, 88*(14), 6368–6371.

Dawson, V. L., Kizushi, V. M., Huang, P. L., Snyder, S. H., & Dawson, T. M. (1996). Resistance to neurotoxicity in cortical cultures from neuronal nitric oxide synthase-deficient mice. *The Journal of Neuroscience, 16*(8), 2479–2487.

Dawson, T. M., & Snyder, S. H. (1994). Gases as biological messengers: Nitric oxide and carbon monoxide in the brain. *The Journal of Neuroscience, 14*(9), 5147–5159.

del Zoppo, G., Ginis, I., Hallenbeck, J. M., Iadecola, C., Wang, X., & Feuerstein, G. Z. (2000). Inflammation and stroke: Putative role for cytokines, adhesion molecules and iNOS in brain response to ischemia. *Brain Pathology, 10*(1), 95–112.

De Paola, V., Arber, S., & Caroni, P. (2003). AMPA receptors regulate dynamic equilibrium of presynaptic terminals in mature hippocampal networks. *Nature Neuroscience, 6*, 491–500.

Dickson, D. W., Lee, S. C., Mattiace, L. A., Yen, S. H., & Brosnan, C. (1993). Microglia and cytokines in neurological disease, with special reference to AIDS and Alzheimer's disease. *Glia, 7*(1), 75–83.

Dimmeler, S., & Zeiher, A. M. (1997). Nitric oxide and apoptosis: Another paradigm for the double-edged role of nitric oxide. *Nitric Oxide, 1*(4), 275–281.

Doerner, D., & Alger, B. E. (1988). Cyclic GMP depresses hippocampal Ca^{2+} current through a mechanism independent of cGMP-dependent protein kinase. *Neuron, 1*(8), 693–699.

Doyle, C. A., & Slater, P. (1997). Localization of neuronal and endothelial nitric oxide synthase isoforms in human hippocampus. *Neuroscience, 76*(2), 387–395.

Drapeau, E., Mayo, W., Aurousseau, C., Le Moal, M., Piazza, P. V., & Abrous, D. N. (2003). Spatial memory performances of aged rats in the water maze predict levels of hippocampal neurogenesis. *Proceedings of the National Academy of Sciences of the United States of America, 100*(24), 14385–14390.

Dworkin, S., Malaterre, J., Hollande, F., Darcy, P. K., Ramsay, R. G., & Mantamadiotis, T. (2009). cAMP response element binding protein is required for mouse neural progenitor cell survival and expansion. *Stem Cells, 27*(6), 1347–1357.

Ehninger, D., & Kempermann, G. (2008). Neurogenesis in the adult hippocampus. *Cell and Tissue Research, 331*(1), 243–250.

Ehrlich, I., Klein, M., Rumpel, S., & Malinow, R. (2007). PSD-95 is required for activity-driven synapse stabilization. *Proceedings of the National Academy of Sciences of the United States of America, 104*(10), 4176–4181.

Eichenbaum, H., & Cohen, N. J. (1993). *Memory, amnesia, and the hippocampal system.* Cambridge: MIT Press.

Eichenbaum, H., Otto, T., & Cohen, N. J. (1992). The hippocampus—What does it do? *Behavioral and Neural Biology, 57*(1), 2–36.

Eliasson, M. J., Huang, Z., Ferrante, R. J., Sasamata, M., Molliver, M. E., Snyder, S. H., et al. (1999). Neuronal nitric oxide synthase activation and peroxynitrite formation in ischemic stroke linked to neural damage. *The Journal of Neuroscience, 19*(14), 5910–5918.

Eliasson, M. J., Sampei, K., Mandir, A. S., Hurn, P. D., Traystman, R. J., Bao, J., et al. (1997). Poly(ADP-ribose) polymerase gene disruption renders mice resistant to cerebral ischemia. *Nature Medicine, 3*(10), 1089–1095.

Endres, M., Laufs, U., Liao, J. K., & Moskowitz, M. A. (2004). Targeting eNOS for stroke protection. *Trends in Neurosciences, 27*(5), 283–289.

Endres, M., Scott, G., Namura, S., Salzman, A. L., Huang, P. L., Moskowitz, M. A., et al. (1998). Role of peroxynitrite and neuronal nitric oxide synthase in the activation of

poly(ADP-ribose) synthetase in a murine model of cerebral ischemia-reperfusion. *Neuroscience Letters*, *248*(1), 41–44.
Engert, F., & Bonhoeffer, T. (1999). Dendritic spine changes associated with hippocampal long-term synaptic plasticity. *Nature*, *399*, 66–70.
Essau, C. A., Lewinsohn, P. M., Seeley, J. R., & Sasagawa, S. (2010). Gender differences in the developmental course of depression. *Journal of Affective Disorders*, *127*(1–3), 185–190.
Fanselow, M. S., & Dong, H. W. (2010). Are the dorsal and ventral hippocampus functionally distinct structures? *Neuron*, *65*(1), 7–19.
Flynn, R. W., MacWalter, R. S., & Doney, A. S. (2008). The cost of cerebral ischaemia. *Neuropharmacology*, *55*, 250–256.
Francis, S. H., Busch, J. L., Corbin, J. D., & Sibley, D. (2010). cGMP-dependent protein kinases and cGMP phosphodiesterases in nitric oxide and cGMP action. *Pharmacological Reviews*, *62*(3), 525–563.
Frey, U., Huang, Y.-Y., & Kandel, E. R. (1993). Effects of cAMP simulate a late stage of LTP in hippocampal CA1 neurons. *Science*, *260*, 1661–1664.
Furchgott, R. F., & Zawadzki, J. V. (1980). The obligatory role of endothelial cells in the relaxation of arterial smooth muscle by acetylcholine. *Nature*, *288*(5789), 373–376.
Gage, F. H., Buzsáki, G., & Armstrong, D. M. (1990). NGF-dependent sprouting and regeneration in the hippocampus. *Progress in Brain Research*, *83*, 357–370.
Gállego, J., Muñoz, R., & Martínez-Vila, E. (2009). Emergent cerebrovascular disease risk factor weighting: Is transient ischemic attack an imminent threat? *Cerebrovascular Diseases*, *27*(Suppl. 1), 88–96.
Gally, J. A., Montague, P. R., Reeke, G. N., Jr., & Edelman, G. M. (1990). The NO hypothesis: Possible effects of a short-lived, rapidly diffusible signal in the development and function of the nervous system. *Proceedings of the National Academy of Sciences of the United States of America*, *87*(9), 3547–3551.
Gardoni, F., & Di Luca, M. (2006). New targets for pharmacological intervention in the glutamatergic synapse. *European Journal of Pharmacology*, *545*(1), 2–10.
Garthwaite, J. (2010). New insight into the functioning of nitric oxide-receptive guanylyl cyclase: Physiological and pharmacological implications. *Molecular and Cellular Biochemistry*, *334*(1–2), 221–232.
Garthwaite, J., Charles, S. L., & Chess-Williams, R. (1988). Endothelium-derived relaxing factor release on activation of NMDA receptors suggests role as intercellular messenger in the brain. *Nature*, *336*(6197), 385–388.
Gould, E., Tanapat, P., Rydel, T., & Hastings, N. (2000). Regulation of hippocampal neurogenesis in adulthood. *Biological Psychiatry*, *48*(8), 715–720.
Gruetter, C. A., Barry, B. K., McNamara, D. B., Gruetter, D. Y., Kadowitz, P. J., & Ignarro, J. L. (1979). Relaxation of bovine coronary artery and activation of coronary arterial guanylate cyclase by nitric oxide, nitroprusside and a carcinogenic nitrosoamine. *Journal of Cyclic Nucleotide Research*, *5*(3), 211–224.
Halegoua, S., Armstrong, R. C., & Kremer, N. E. (1991). Dissecting the mode of action of a neuronal growth factor. *Current Topics in Microbiology and Immunology*, *165*, 119–170.
Haley, J. E., Wilcox, G. L., & Chapman, P. F. (1992). The role of nitric oxide in hippocampal long-term potentiation. *Neuron*, *8*(2), 211–216.
Hartmann, H., Eckert, A., & Müller, W. E. (1993). beta-Amyloid protein amplifies calcium signalling in central neurons from the adult mouse. *Biochemical and Biophysical Research Communications*, *194*(3), 1216–1220.
Harvey, B. H., Retief, R., Korff, A., & Wegener, G. (2006). Increased hippocampal nitric oxide synthase activity and stress responsiveness after imipramine discontinuation: Role of 5HT 2A/C-receptors. *Metabolic Brain Disease*, *21*(2–3), 211–220.
Hawgood, J., & De Leo, D. (2008). Anxiety disorders and suicidal behaviour: An update. *Current Opinion in Psychiatry*, *21*, 51–64.

Herman, J. P., Ostrander, M. M., Mueller, N. K., & Figueiredo, H. (2005). Limbic system mechanisms of stress regulation: Hypothalamo-pituitary-adrenocortical axis. *Progress in Neuro-Psychopharmacology & Biological Psychiatry*, *29*(8), 1201–1213.

Hess, D. T., Patterson, S. I., Smith, D. S., & Skene, J. H. (1993). Neuronal growth cone collapse and inhibition of protein fatty acylation by nitric oxide. *Nature*, *366*(6455), 562–565.

Hindley, S., Juurlink, B. H., Gysbers, J. W., Middlemiss, P. J., Herman, M. A., & Rathbone, M. P. (1997). Nitric oxide donors enhance neurotrophin-induced neurite outgrowth through a cGMP-dependent mechanism. *Journal of Neuroscience Research*, *47*(4), 427–439.

Ho, L., Pieroni, C., Winger, D., Purohit, D. P., Aisen, P. S., & Pasinetti, G. M. (1999). Regional distribution of cyclooxygenase-2 in the hippocampal formation in Alzheimer's disease. *Journal of Neuroscience Research*, *57*(3), 295–303.

Hu, J., Akama, K. T., Krafft, G. A., Chromy, B. A., & Van Eldik, L. J. (1998). Amyloid-beta peptide activates cultured astrocytes: Morphological alterations, cytokine induction and nitric oxide release. *Brain Research*, *785*(2), 195–206.

Hu, Y., Wu, D. L., Luo, C. X., Zhu, L. J., Zhang, J., Wu, H. Y., et al. (2012). Hippocampal nitric oxide contributes to sex difference in affective behaviors. *Proceedings of the National Academy of Sciences of the United States of America*, *109*(35), 14224–14229.

Hua, Y., Huang, X. Y., Zhou, L., Zhou, Q. G., Hu, Y., Luo, C. X., et al. (2008). DETA/NONOate, a nitric oxide donor, produces antidepressant effects by promoting hippocampal neurogenesis. *Psychopharmacology*, *200*(2), 231–242.

Huang, E. P. (1997). Synaptic plasticity: A role for nitric oxide in LTP. *Current Biology*, 7(3), R141–R143.

Huang, Z., Huang, P. L., Ma, J., Meng, W., Ayata, C., Fishman, M. C., et al. (1996). Enlarged infarcts in endothelial nitric oxide synthase knockout mice are attenuated by nitro-L-arginine. *Journal of Cerebral Blood Flow and Metabolism*, *16*(5), 981–987.

Huang, Z., Huang, P. L., Panahian, N., Dalkara, T., Fishman, M. C., & Moskowitz, M. A. (1994). Effects of cerebral ischemia in mice deficient in neuronal nitric oxide synthase. *Science*, *265*(5180), 1883–1885.

Huie, R. E., & Padmaja, S. (1993). The reaction of no with superoxide. *Free Radical Research Communications*, *18*(4), 195–199.

Ignarro, L. J., Buga, G. M., Wood, K. S., Byrns, R. E., & Chaudhuri, G. (1987). Endothelium-derived relaxing factor produced and released from artery and vein is nitric oxide. *Proceedings of the National Academy of Sciences of the United States of America*, *84*(24), 9265–9269.

Inglis, F. M., Furia, F., Zuckerman, K. E., Strittmatter, S. M., & Kalb, R. G. (1998). The role of nitric oxide and NMDA receptors in the development of motor neuron dendrites. *The Journal of Neuroscience*, *18*(24), 10493–10501.

Jagtap, P., & Szabó, C. (2005). Poly(ADP-ribose) polymerase and the therapeutic effects of its inhibitors. *Nature Reviews. Drug Discovery*, *4*(5), 421–440.

Jun, H., Mohammed Qasim Hussaini, S., Rigby, M. J., & Jang, M. H. (2012). Functional role of adult hippocampal neurogenesis as a therapeutic strategy for mental disorders. *Neural Plasticity*, *2012*, 854285.

Kant, L., Yilmaz, O., Taskiran, D., Kulali, B., Furedy, J. J., Demirgören, S., et al. (2000). Sexually dimorphic cognitive style, female sex hormones, and cortical nitric oxide. *Physiology & Behavior*, *71*(3–4), 277–287.

Katsuki, S., Arnold, W. P., & Murad, F. (1977). Effects of sodium nitroprusside, nitroglycerin, and sodium azide on levels of cyclic nucleotides and mechanical activity of various tissues. *Journal of Cyclic Nucleotide Research*, *3*(4), 239–247.

Keser, A., Balkan, B., Gozen, O., Kanit, L., & Pogun, S. (2011). Hippocampal neuronal nitric oxide synthase (nNOS) is regulated by nicotine and stress in female but not in male rats. *Brain Research*, *1368*, 134–142.

Kesslak, J. P., Nalcioglu, O., & Cotman, C. W. (1991). Quantification of magnetic resonance scans for hippocampal and parahippocampal atrophy in Alzheimer's disease. *Neurology, 41*(1), 51–54.

Klein, M. B., Chan, P. H., & Chang, J. (2003). Protective effects of superoxide dismutase against ischemia-reperfusion injury: Development and application of a transgenic animal model. *Plastic and Reconstructive Surgery, 111*(1), 251–255, discussion 256–257.

Kleppisch, T., & Feil, R. (2009). cGMP signalling in the mammalian brain: Role in synaptic plasticity and behaviour. *Handbook of Experimental Pharmacology, 191*, 549–579.

Klüver, H., & Bucy, P. C. (1937). "Psychic blindness" and other symptoms following bilateral temporal lobectomy in Rhesus monkeys. *The American Journal of Physiology, 119*, 352–353.

Knowles, R. G. (1997). Nitric oxide biochemistry. *Biochemical Society Transactions, 25*(3), 895–901.

Kornberg, M. D., Sen, N., Hara, M. R., Juluri, K. R., Nguyen, J. V., Snowman, A. M., et al. (2010). GAPDH mediates nitrosylation of nuclear proteins. *Nature Cell Biology, 12*, 1094–1100.

Kröncke, K. D., Fehsel, K., & Kolb-Bachofen, V. (1995). Inducible nitric oxide synthase and its product nitric oxide, a small molecule with complex biological activities. *Biological Chemistry Hoppe-Seyler, 376*(6), 327–343.

Lafon-Cazal, M., Pietri, S., Culcasi, M., & Bockaert, J. (1993). NMDA-dependent superoxide production and neurotoxicity. *Nature, 364*(6437), 535–537.

Lander, H. M., Ogiste, J. S., Teng, K. K., & Novogrodsky, A. (1995). p21ras as a common signaling target of reactive free radicals and cellular redox stress. *The Journal of Biological Chemistry, 270*(36), 21195–21198.

Law, A., Gauthier, S., & Quirion, R. (2001). Say NO to Alzheimer's disease: The putative links between nitric oxide and dementia of the Alzheimer's type. *Brain Research. Brain Research Reviews, 35*(1), 73–96.

Liaudet, L., Soriano, F. G., & Szabó, C. (2000). Biology of nitric oxide signaling. *Critical Care Medicine, 28*(4 Suppl.), N37–N52.

Liaudet, L., Vassalli, G., & Pacher, P. (2009). Role of peroxynitrite in the redox regulation of cell signal transduction pathways. *Frontiers in Bioscience (Landmark Edition), 14*, 4809–4814.

Liu, T. H., Beckman, J. S., Freeman, B. A., Hogan, E. L., & Hsu, C. Y. (1989). Polyethylene glycol-conjugated superoxide dismutase and catalase reduce ischemic brain injury. *The American Journal of Physiology, 256*(2 Pt 2), H589–H593.

Lo, E. H., Hara, H., Rogowska, J., Trocha, M., Pierce, A. R., Huang, P. L., et al. (1996). Temporal correlation mapping analysis of the hemodynamic penumbra in mutant mice deficient in endothelial nitric oxide synthase gene expression. *Stroke, 27*(8), 1381–1385.

Lu, Y. F., Kandel, E. R., & Hawkins, R. D. (1999). Nitric oxide signaling contributes to late-phase LTP and CREB phosphorylation in the hippocampus. *The Journal of Neuroscience, 19*(23), 10250–10261.

Lu, D., Mahmood, A., Zhang, R., & Copp, M. (2003). Upregulation of neurogenesis and reduction in functional deficits following administration of DEtA/NONOate, a nitric oxide donor, after traumatic brain injury in rats. *Journal of Neurosurgery, 99*(2), 351–361.

Luo, C. X., Jin, X., Cao, C. C., Zhu, M. M., Wang, B., Chang, L., et al. (2010). Bidirectional regulation of neurogenesis by neuronal nitric oxide synthase derived from neurons and neural stem cells. *Stem Cells, 28*(11), 2041–2052.

Luo, C. X., Zhu, X. J., Zhou, Q. G., Wang, B., Wang, W., Cai, H. H., et al. (2007). Reduced neuronal nitric oxide synthase is involved in ischemia-induced hippocampal neurogenesis by up-regulating inducible nitric oxide synthase expression. *Journal of Neurochemistry, 103*(5), 1872–1882.

Lüth, H. J., Holzer, M., Gertz, H. J., & Arendt, T. (2000). Aberrant expression of nNOS in pyramidal neurons in Alzheimer's disease is highly co-localized with p21ras and p16INK4a. *Brain Research*, *852*(1), 45–55.
Maletic-Savatic, M., Malinow, R., & Svoboda, K. (1999). Rapid dendritic morphogenesis in CA1 hippocampal dendrites induced by synaptic activity. *Science*, *283*, 1923–1927.
Marletta, M. A., Yoon, P. S., Iyengar, R., Leaf, C. D., & Wishnok, J. S. (1988). Macrophage oxidation of L-arginine to nitrite and nitrate: Nitric oxide is an intermediate. *Biochemistry*, *27*(24), 8706–8711.
Matarredona, E. R., Murillo-Carretero, M., Moreno-López, B., & Estrada, C. (2005). Role of nitric oxide in subventricular zone neurogenesis. *Brain Research. Brain Research Reviews*, *49*(2), 355–366.
Mattson, M. P., Cheng, B., Davis, D., Bryant, K., Lieberburg, I., & Rydel, R. E. (1992). beta-Amyloid peptides destabilize calcium homeostasis and render human cortical neurons vulnerable to excitotoxicity. *The Journal of Neuroscience*, *12*(2), 376–389.
Mattson, M. P., Tomaselli, K. J., & Rydel, R. E. (1993). Calcium-destabilizing and neurodegenerative effects of aggregated beta-amyloid peptide are attenuated by basic FGF. *Brain Research*, *621*(1), 35–49.
Mayeux, R. (2010). Clinical practice. Early Alzheimer's disease. *The New England Journal of Medicine*, *362*(23), 2194–2201.
McEwen, B. S., Conrad, C. D., Kuroda, Y., Frankfurt, M., Magarinos, A. M., & McKittrick, C. (1997). Prevention of stress-induced morphological and cognitive consequences. *European Neuropsychopharmacology*, 7(Suppl. 3), S323–S328.
McLean, C. P., Asnaani, A., Litz, B. T., & Hofmann, S. G. (2011). Gender differences in anxiety disorders: Prevalence, course of illness, comorbidity and burden of illness. *Journal of Psychiatric Research*, *45*(8), 1027–1035.
Meulemans, A. (1994). Diffusion coefficients and half-lives of nitric oxide and N-nitroso-L-arginine in rat cortex. *Neuroscience Letters*, *171*(1–2), 89–93.
Miguel, T. T., & Nunes-de-Souza, R. L. (2008). Anxiogenic-like effects induced by NMDA receptor activation are prevented by inhibition of neuronal nitric oxide synthase in the periaqueductal gray in mice. *Brain Research*, *1240*, 39–46.
Mitchell, H. H., Shonle, H. A., & Grindley, H. S. (1916). The origins of nitrates in the urine. *The Journal of Biological Chemistry*, *24*, 461–490.
Moreno-López, B., Romero-Grimaldi, C., Noval, J. A., Murillo-Carretero, M., Matarredona, E. R., & Estrada, C. (2004). Nitric oxide is a physiological inhibitor of neurogenesis in the adult mouse subventricular zone and olfactory bulb. *The Journal of Neuroscience*, *24*(1), 85–95.
Mukai, H., Kimoto, T., Hojo, Y., Kawato, S., Murakami, G., Higo, S., et al. (2010). Modulation of synaptic plasticity by brain estrogen in the hippocampus. *Biochimica et Biophysica Acta*, *1800*(10), 1030–1044.
Murray, C. J., & Lopez, A. D. (1997). Alternative projections of mortality and disability by cause 1990–2020: Global burden of disease study. *Lancet*, *349*, 1498–1504.
Nakamura, T., & Lipton, S. A. (2011). Redox modulation by S-nitrosylation contributes to protein misfolding, mitochondrial dynamics, and neuronal synaptic damage in neurodegenerative diseases. *Cell Death and Differentiation*, *18*(9), 1478–1486.
Nelson, R. J., Demas, G. E., Huang, P. L., Fishman, M. C., Dawson, V. L., Dawson, T. M., et al. (1995). Behavioural abnormalities in male mice lacking neuronal nitric oxide synthase. *Nature*, *378*, 383–386.
Nikonenko, I., Jourdain, P., & Muller, D. (2003). Presynaptic remodeling contributes to activity-dependent synaptogenesis. *The Journal of Neuroscience*, *23*(24), 8498–8505.
Noda, M., Nakanishi, H., & Akaike, N. (1999). Glutamate release from microglia via glutamate transporter is enhanced by amyloid-beta peptide. *Neuroscience*, *92*(4), 1465–1474.
O'Dell, T. J., Hawkins, R. D., Kandel, E. R., & Arancio, O. (1991). Tests of the roles of two diffusible substances in long-term potentiation: Evidence for nitric oxide as a possible

early retrograde messenger. *Proceedings of the National Academy of Sciences of the United States of America, 88*(24), 11285–11289.
Pacher, P., Beckman, J. S., & Liaudet, L. (2007). Nitric oxide and peroxynitrite in health and disease. *Physiological Reviews, 87*(1), 315–424.
Packer, M. A., Stasiv, Y., Benraiss, A., Chmielnicki, E., Grinberg, A., Westphal, H., et al. (2003). Nitric oxide negatively regulates mammalian adult neurogenesis. *Proceedings of the National Academy of Sciences of the United States of America, 100*(16), 9566–9571.
Palmer, R. M., Ferrige, A. G., & Moncada, S. (1987). Nitric oxide release accounts for the biological activity of endothelium-derived relaxing factor. *Nature, 327*(6122), 524–526.
Palumbo, M. L., Fosser, N. S., Rios, H., Zorrilla Zubilete, M. A., Guelman, L. R., Cremaschi, G. A., et al. (2007). Loss of hippocampal neuronal nitric oxide synthase contributes to the stress-related deficit in learning and memory. *Journal of Neurochemistry, 102*(1), 261–274.
Poluha, W., Schonhoff, C. M., Harrington, K. S., Lachyankar, M. B., Crosbie, N. E., Bulseco, D. A., et al. (1997). A novel, nerve growth factor-activated pathway involving nitric oxide, p53, and p21WAF1 regulates neuronal differentiation of PC12 cells. *The Journal of Biological Chemistry, 272*(38), 24002–24007.
Puzzo, D., Vitolo, O., Trinchese, F., Jacob, J. P., Palmeri, A., & Arancio, O. (2005). Amyloid-beta peptide inhibits activation of the nitric oxide/cGMP/cAMP-responsive element-binding protein pathway during hippocampal synaptic plasticity. *The Journal of Neuroscience, 25*(29), 6887–6897.
Querfurth, H. W., & LaFerla, F. M. (2010). Alzheimer's disease. *The New England Journal of Medicine, 362*(4), 329–344.
Reif, A., Herterich, S., Strobel, A., Ehlis, A. C., Saur, D., Jacob, C. P., et al. (2006). A neuronal nitric oxide synthase (NOS-I) haplotype associated with schizophrenia modifies prefrontal cortex function. *Molecular Psychiatry, 11*, 286–300.
Reif, A., Schmitt, A., Fritzen, S., Chourbaji, S., Bartsch, C., Urani, A., et al. (2004). Differential effect of endothelial nitric oxide synthase (NOS-III) on the regulation of adult neurogenesis and behaviour. *The European Journal of Neuroscience, 20*(4), 885–895.
Ressler, K. J., & Mayberg, H. S. (2007). Targeting abnormal neural circuits in mood and anxiety disorders: From the laboratory to the clinic. *Nature Neuroscience, 10*(9), 1116–1124.
Rodrigo, J., Springall, D. R., Uttenthal, O., Bentura, M. L., Abadia-Molina, F., Riveros-Moreno, V., et al. (1994). Localization of nitric oxide synthase in the adult rat brain. *Philosophical Transactions of the Royal Society of London Series B, Biological Sciences, 345*(1312), 175–221.
Sánchez-Islas, E., & León-Olea, M. (2004). Nitric oxide synthase inhibition during synaptic maturation decreases synapsin I immunoreactivity in rat brain. *Nitric Oxide, 10*(3), 141–149.
Santarelli, L., Saxe, M., Gross, C., Surget, A., Battaglia, F., Dulawa, S., et al. (2003). Requirement of hippocampal neurogenesis for the behavioral effects of antidepressants. *Science, 301*(5634), 805–809.
Saran, M., Michel, C., & Bors, W. (1990). Reaction of NO with O_2^{-c} implications for the action of endothelium-derived relaxing factor (EDRF). *Free Radical Research Communications, 10*(4–5), 221–226.
Sattler, R., & Tymianski, M. (2000). Molecular mechanisms of calcium-dependent excitotoxicity. *Journal of Molecular Medicine (Berlin, Germany), 78*(1), 3–13.
Schmidt, H. H., Wilke, P., Evers, B., & Böhme, E. (1989). Enzymatic formation of nitrogen oxides from L-arginine in bovine brain cytosol. *Biochemical and Biophysical Research Communications, 165*(1), 284–291.
Schuman, E. M. (1995). Nitric oxide signaling, long term potentiation and long term depression. In S. R. Vincent (Ed.), *Nitric oxide in the central nervous system* (pp. 125–150). San Diego: Academic Press Inc.

Schuman, E. M., & Madison, D. V. (1991). A requirement for the intercellular messenger nitric oxide in long-term potentiation. *Science*, *254*(5037), 1503–1506.
Schuman, E. M., & Madison, D. V. (1994). Locally distributed synaptic potentiation in the hippocampus. *Science*, *263*, 532–536.
Sen, N., Hara, M. R., Kornberg, M. D., Cascio, M. B., Bae, B. I., Shahani, N., et al. (2008). Nitric oxide-induced nuclear GAPDH activates p300/CBP and mediates apoptosis. *Nature Cell Biology*, *10*, 866–873.
Shaw, A. W., & Vosper, A. J. (1977). Solubility of nitric oxide in aqueous and nonaqueous solvents. *Journal of the Chemical Society, Faraday Transactions*, *8*, 1239–1244.
Shinkai, T., Ohmori, O., Hori, H., & Nakamura, J. (2002). Allelic association of the neuronal nitric oxide synthase (NOS1) gene with schizophrenia. *Molecular Psychiatry*, *7*, 560–563.
Shors, T. J., Miesegaes, G., Beylin, A., Zhao, M., Rydel, T., & Gould, E. (2001). Neurogenesis in the adult is involved in the formation of trace memories. *Nature*, *410*(6826), 372–376.
Squire, L. R. (1992). Memory and the hippocampus: A synthesis from findings with rats, monkeys, and humans. *Psychological Review*, *99*(2), 195–231.
Squire, L. R., & Zola-Morgan, S. (1991). The medial temporal lobe memory system. *Science*, *253*(5026), 1380–1386.
Sun, Y., Jin, K., Childs, J. T., Xie, L., Mao, X. O., & Greenberg, D. A. (2005). Neuronal nitric oxide synthase and ischemia-induced neurogenesis. *Journal of Cerebral Blood Flow and Metabolism*, *25*(4), 485–492.
Sunico, C. R., Portillo, F., González-Forero, D., & Moreno-López, B. (2005). Nitric-oxide-directed synaptic remodeling in the adult mammal CNS. *The Journal of Neuroscience*, *25*(6), 1448–1458.
Suzuki, E., Yagi, G., Nakaki, T., Kanba, S., & Asai, M. (2001). Elevated plasma nitrate levels in depressive states. *Journal of Affective Disorders*, *63*(1–3), 221–224.
Szabo, C., Ischiropoulos, H., & Radi, R. (2007). Peroxynitrite: Biochemistry, pathophysiology and development of therapeutics. *Nature Reviews. Drug Discovery*, *6*(8), 662–680.
Tashiro, A., Dunaevsky, A., Blazeski, R., Mason, C. A., & Yuste, R. (2003). Bidirectional regulation of hippocampal mossy fiber filopodial motility by kainate receptors: A two-step model of synaptogenesis. *Neuron*, *38*, 773–784.
Taskiran, D., Kutay, F. Z., Sozmen, E., & Pogun, S. (1997). Sex differences in nitrite/nitrate levels and antioxidant defense in rat brain. *Neuroreport*, *8*(4), 881–884.
Torroglosa, A., Murillo-Carretero, M., Romero-Grimaldi, C., Matarredona, E. R., Campos-Caro, A., & Estrada, C. (2007). Nitric oxide decreases subventricular zone stem cell proliferation by inhibition of epidermal growth factor receptor and phosphoinositide-3-kinase/Akt pathway. *Stem Cells*, *25*(1), 88–97.
Uehara, T., Nakamura, T., Yao, D., Shi, Z. Q., Gu, Z., Ma, Y., et al. (2006). S-nitrosylated protein-disulphide isomerase links protein misfolding to neurodegeneration. *Nature*, *441*(7092), 513–517.
Van der Zee, C. E., Fawcett, J., & Diamond, J. (1992). Antibody to NGF inhibits collateral sprouting of septohippocampal fibers following entorhinal cortex lesion in adult rats. *The Journal of Comparative Neurology*, *326*(1), 91–100.
van Praag, H., Schinder, A. F., Christie, B. R., Toni, N., Palmer, T. D., & Gage, F. H. (2002). Functional neurogenesis in the adult hippocampus. *Nature*, *415*(6875), 1030–1034.
Varadarajan, S., Yatin, S., Aksenova, M., & Butterfield, D. A. (2000). Review: Alzheimer's amyloid beta-peptide-associated free radical oxidative stress and neurotoxicity. *Journal of Structural Biology*, *130*(2–3), 184–208.
Varner, Pamela D., & Beckman, Joseph S. (1995). Nitric oxide toxicity in neuronal injury and degeneration. In S. R. Vincent (Ed.), *Neuroscience perspectives: Nitric oxide in the nervous system* (pp. 191–204). San Diego: Academic Press Inc.

Vickers, C. A., Stephens, B., Bowen, J., Arbuthnott, G. W., Grant, S. G., & Ingham, C. A. (2006). Neurone specific regulation of dendritic spines in vivo by post synaptic density 95 protein (PSD-95). *Brain Research, 1090*(1), 89–98.

Wallace, M. N., Geddes, J. G., Farquhar, D. A., & Masson, M. R. (1997). Nitric oxide synthase in reactive astrocytes adjacent to beta-amyloid plaques. *Experimental Neurology, 144*(2), 266–272.

Warner-Schmidt, J. L., & Duman, R. S. (2006). Hippocampal neurogenesis: Opposing effects of stress and antidepressant treatment. *Hippocampus, 16*, 239–249.

Whiteman, M., Ketsawatsakul, U., & Halliwell, B. (2002). A reassessment of the peroxynitrite scavenging activity of uric acid. *Annals of the New York Academy of Sciences, 962*, 242–259.

Wood, J., & Garthwaite, J. (1994). Models of the diffusional spread of nitric oxide: Implications for neural nitric oxide signalling and its pharmacological properties. *Neuropharmacology, 33*, 1235–1244.

Workman, J. L., Trainor, B. C., Finy, M. S., & Nelson, R. J. (2008). Inhibition of neuronal nitric oxide reduces anxiety-like responses to pair housing. *Behavioural Brain Research, 187*, 109–115.

Yamamoto, S., Golanov, E. V., Berger, S. B., & Reis, D. J. (1992). Inhibition of nitric oxide synthesis increases focal ischemic infarction in rat. *Journal of Cerebral Blood Flow and Metabolism, 12*(5), 717–726.

Yildiz, F., Erden, B. F., Ulak, G., Utkan, T., & Gacar, N. (2000). Antidepressant-like effect of 7-nitroindazole in the forced swimming test in rats. *Psychopharmacology, 149*(1), 41–44.

Yoshida, T., Limmroth, V., Irikura, K., & Moskowitz, M. A. (1994). The NOS inhibitor, 7-nitroindazole, decreases focal infarct volume but not the response to topical acetylcholine in pial vessels. *Journal of Cerebral Blood Flow and Metabolism, 14*(6), 924–929.

Zhang, J., Dawson, V. L., Dawson, T. M., & Snyder, S. H. (1994). Nitric oxide activation of poly(ADP-ribose) synthetase in neurotoxicity. *Science, 263*(5147), 687–689.

Zhang, J., Huang, X. Y., Ye, M. L., Luo, C. X., Wu, H. Y., Hu, Y., et al. (2010). Neuronal nitric oxide synthase alteration accounts for the role of 5-HT1A-receptor in modulating anxiety-related behaviors. *The Journal of Neuroscience, 30*, 2433–2441.

Zhang, R., Wang, L., Zhang, L., Chen, J., Zhu, Z., Zhang, Z., et al. (2003). Nitric oxide enhances angiogenesis via the synthesis of vascular endothelial growth factor and cGMP after stroke in the rat. *Circulation Research, 92*(3), 308–313.

Zhang, F., White, J. G., & Iadecola, C. (1994). Nitric oxide donors increase blood flow and reduce brain damage in focal ischemia: Evidence that nitric oxide is beneficial in the early stages of cerebral ischemia. *Journal of Cerebral Blood Flow and Metabolism, 14*(2), 217–226.

Zhang, R. L., Zhang, Z. G., & Chopp, M. (2005). Neurogenesis in the adult ischemic brain: Generation, migration, survival, and restorative therapy. *The Neuroscientist, 11*(5), 408–416.

Zhou, Q. G., Hu, Y., Hua, Y., Hu, M., Luo, C. X., Han, X., et al. (2007). Neuronal nitric oxide synthase contributes to chronic stress-induced depression by suppressing hippocampal neurogenesis. *Journal of Neurochemistry, 103*(5), 1843–1854.

Zhou, L., Li, F., Xu, H. B., Luo, C. X., Wu, H. Y., Zhu, M. M., et al. (2010). Treatment of cerebral ischemia by disrupting ischemia-induced interaction of nNOS with PSD-95. *Nature Medicine, 16*(12), 1439–1443.

Zhou, L., & Zhu, D. Y. (2009). Neuronal nitric oxide synthase: Structure, subcellular localization, regulation, and clinical implications. *Nitric Oxide, 20*, 223–230.

Zhou, Q. G., Zhu, L. J., Chen, C., Wu, H. Y., Luo, C. X., Chang, L., et al. (2011). Hippocampal neuronal nitric oxide synthase mediates the stress-related depressive behaviors of glucocorticoids by downregulating glucocorticoid receptor. *The Journal of Neuroscience, 31*(21), 7579–7590.

Zhu, L., Gunn, C., & Beckman, J. S. (1992). Bactericidal activity of peroxynitrite. *Archives of Biochemistry and Biophysics*, *298*(2), 452–457.
Zhu, X. J., Hua, Y., Jiang, J., Zhou, Q. G., Luo, C. X., Han, X., et al. (2006). Neuronal nitric oxide synthase-derived nitric oxide inhibits neurogenesis in the adult dentate gyrus by down-regulating cyclic AMP response element binding protein phosphorylation. *Neuroscience*, *141*(2), 827–836.
Zhu, D. Y., Liu, S. H., Sun, H. S., & Lu, Y. M. (2003). Expression of inducible nitric oxide synthase after focal cerebral ischemia stimulates neurogenesis in the adult rodent dentate gyrus. *The Journal of Neuroscine*, *23*(1), 223–229.
Zhuo, M., Hu, Y., Schultz, C., Kandel, E. R., & Hawkins, R. D. (1994). Role of guanylyl cyclase and cGMP-dependent protein kinase in long-term potentiation. *Nature*, *368*(6472), 635–639.
Ziv, N. E., & Smith, S. J. (1996). Evidence for a role of dendritic filopodia in synaptogenesis and spine formation. *Neuron*, *17*, 91–102.
Zola-Morgan, S., & Squire, L. R. (1993). Neuroanatomy of memory. *Annual Review of Neuroscience*, *16*, 547–563.

FURTHER READING

Bauer, M. K., Lieb, K., Schulze-Osthoff, K., Berger, M., Gebicke-Haerter, P. J., Bauer, J., et al. (1997). Expression and regulation of cyclooxygenase-2 in rat microglia. *European Journal of Biochemistry*, *243*(3), 726–731.
Böhme, G. A., Bon, C., Stutzmann, J. M., Doble, A., & Blanchard, J. C. (1991). Possible involvement of nitric oxide in long-term potentiation. *European Journal of Pharmacology*, *199*(3), 379–381.
Chen, J., Li, Y., Zhang, R., Katakowski, M., Gautam, S. C., Xu, Y., et al. (1993). Nitric oxide synthase-independent long-term potentiation in area CA1 of hippocampus. *Neuroreport*, *4*(7), 919–922.
Chen, J., Zacharek, A., Li, Y., Li, A., Wang, L., Katakowski, M., et al. (2001). Brain serotonin dysfunction accounts for aggression in male mice lacking neuronal nitric oxide synthase. *Proceedings of the National Academy of Sciences of the United States of America*, *98*, 1277–1281.
Chopp, M. (2004). Combination therapy of stroke in rats with a nitric oxide donor and human bone marrow stromal cells enhances angiogenesis and neurogenesis. *Brain Research*, *1005*(1–2), 21–28.
Chopp, M. (2006). N-cadherin mediates nitric oxide-induced neurogenesis in young and retired breeder neurospheres. *Neuroscience*, *140*(2), 377–388.
Colasanti, M., Persichini, T., Fabrizi, C., Cavalieri, E., Venturini, G., Ascenzi, P., et al. (1998). Expression of a NOS-III-like protein in human astroglial cell culture. *Biochemical and Biophysical Research Communications*, *252*(3), 552–555.
Guix, F. X., Uribesalgo, I., Coma, M., & Muñoz, F. J. (2005). The physiology and pathophysiology of nitric oxide in the brain. *Progress in Neurobiology*, *76*(2), 126–152.
Malinski, T., Bailey, F., Zhang, Z. G., & Chopp, M. (1993). Nitric oxide measured by a porphyrinic microsensor in rat brain after transient middle cerebral artery occlusion. *Journal of Cerebral Blood Flow and Metabolism*, *13*(3), 355–358.
Son, H., Hawkins, R. D., Martin, K., Kiebler, M., Huang, P. L., Fishman, M. C., et al. (1996). Long-term potentiation is reduced in mice that are doubly mutant in endothelial and neuronal nitric oxide synthase. *Cell*, *87*(6), 1015–1023.
Zhang, J., & Zhu, D. Y. (2011). Hippocampal telomerase is involved in the modulation of depressive behaviors. *The Journal of Neuroscience*, *31*(34), 12258–12269.
Zhou, Q. G., Hu, Y., Wu, D. L., Zhu, L. J., Chen, C., Jin, X., et al. (2003). Expression of inducible nitric oxide synthase after focal cerebral ischemia stimulates neurogenesis in the adult rodent dentate gyrus. *The Journal of Neuroscience*, *23*(1), 223–229.

CHAPTER SEVEN

Nitric Oxide and Hypoxia Signaling

H.S. Jeffrey Man*,†,‡, Albert K.Y. Tsui†,§, Philip A. Marsden*,†,¶,1

*Institute of Medical Science, University of Toronto, Toronto, Ontario, Canada
†Li Ka Shing Knowledge Institute, St. Michael's Hospital, Toronto, Ontario, Canada
‡Divisions of Respirology and Critical Care Medicine, Department of Medicine, University of Toronto, Toronto, Ontario, Canada
§Department of Anesthesia, St. Michael's Hospital, Toronto, Ontario, Canada
¶Division of Nephrology, Department of Medicine, University of Toronto, Toronto, Ontario, Canada
[1]Corresponding author: e-mail address: p.marsden@utoronto.ca

Contents

Abstract

Nitric oxide (NO) production is catalyzed by three distinct enzymes, namely, neuronal nitric oxide synthase (nNOS), inducible NOS (iNOS), and endothelial NOS (eNOS). The production of NO by vascular endothelium relies mainly on eNOS. Curiously, iNOS and nNOS also are relevant for vascular NO production in certain settings. By relaxing vascular smooth muscle, the classical view is that NO participates in O_2 homeostasis by increasing local blood flow and O_2 delivery. It is now appreciated that NO has an even

Vitamins and Hormones, Volume 96
ISSN 0083-6729
http://dx.doi.org/10.1016/B978-0-12-800254-4.00007-6

more fundamental role in cellular oxygen sensing at the cellular and physiological level. A key component of cellular oxygen sensing is the hypoxia-inducible factor (HIF) that activates a transcriptional program to promote cellular survival under conditions of inadequate oxygen supply. Important new insights demonstrate that HIF protein is stabilized by two parallel pathways: (1) a decrease in the O_2-dependent prolyl hydroxylation of HIF and (2) NO-dependent S-nitrosylation of HIF pathway components including HIF-α. The need for these two complementary pathways to HIF activation arises because decreased oxygen delivery can occur not only by decreased ambient oxygen but also by decreased blood oxygen-carrying capacity, as with anemia. In turn, NO production is tightly linked to O_2 homeostasis. O_2 is a key substrate for the generation of NO and impacts the enzymatic activity and expression of the enzymes that catalyze the production of NO, the nitric oxide synthases. These relationships manifest in a variety of clinical settings ranging from the unique situation of humans living in hypoxic environments at high altitudes to the common scenario of anemia and the use of therapeutics that can bind or release NO.

1. INTRODUCTION

This chapter explores the relationship between nitric oxide (NO) biology and mechanisms by which cells sense and respond to a decrease in oxygen available for cellular metabolism. In most cells, oxygen (O_2) is responsible for a large proportion of energy production through oxidative phosphorylation. From a physiological perspective, blood flow, O_2-carrying capacity, and O_2 saturation are the factors that determine oxygen delivery to cells. Cells must be able to sense decreased oxygen delivery due to decreased O_2-carrying capacity and decreased partial pressure of oxygen. The intracellular sense and response pathway involves the hypoxia-inducible factor (HIF) that plays a major role in this cellular response to hypoxia (Semenza, 2012). We now know that NO plays a role in how cells sense and response to hypoxia, especially when oxygen delivery is reduced in the absence of hypoxemia (low oxygen saturation or partial pressure of oxygen) (Ho, Man, & Marsden, 2012). Traditionally, NO has been characterized as a paracrine molecule with the ability to modulate vasomotor tone locally and cardiovascular and respiratory responses centrally. Physiologically, this response results in greater blood flow to the tissues and cells. It is now appreciated that NO not only controls local blood flow, and hence O_2 bioavailability, but also signals and responds to decreased oxygen availability via interaction with the HIF pathway. We will discuss the oxygen-sensing pathways and their relation to NO signaling and provide examples where these pathways are clinically relevant.

1.1. Vignette—Tibetan highlanders and genetic evolution at the HIF genomic loci

The rise in atmospheric oxygen from very low levels to the current 21% is thought to have occurred approximately 500–600 million years ago (Taylor & McElwain, 2010). Concurrently, oxygen-sensing pathways have evolved to cope with these changes and promote survival of organisms. These mechanisms include nitric oxide synthases (NOSs), HIF, and prolyl hydroxylase domains (PHDs) (Taylor & McElwain, 2010). Interestingly, these ancient oxygen-sensing mechanisms arose prior to the appearance of hemoglobin (Hgb) (Feelisch & Martin, 1995). Hgb is found throughout the plant and animal kingdoms, including *Eubacteria* and unicellular eukaryotes. Its ubiquitous presence within organisms and conserved amino acid sequence suggested that the Hgb genes are descended from an ancient, common ancestor (Hardison, 1998). Although Hgb may have many different functions in the plant and animal kingdoms, the primary role of Hgb in complex organisms is to maximize oxygen-carrying capacity and to optimize oxygen delivery to tissues. The importance of Hgb in oxygen delivery is emphasized by the consequences of Hgb deficiency, such as anemia. Currently, anemia is a worldwide health problem that affects a quarter of the world's population (Worldwide prevalence of anaemia report 1993–2005. WHO global database on anaemia. Geneva, World Health Organization, 2008). Regardless of the types of anemia, a common observation is that anemia predicts morbidity and mortality. Although anemia can be treated with blood transfusion and erythropoietin-stimulating agents (ESAs), these treatment strategies require further study to understand their potential benefits. Interestingly, a large randomized clinical trial (TRICC) demonstrated that patients transfused to a higher Hgb threshold (liberal strategy) do not have survival advantage over those transfused at a lower Hgb threshold (restrictive strategy) (Hebert et al., 1999). In fact, younger and less ill patients have improved survival receiving a restrictive transfusion strategy. Additionally, anemic patients with chronic kidney disease given ESAs to correct Hgb concentration do not have better outcomes and survival (Drueke et al., 2006; Pfeffer et al., 2009; Singh et al., 2006). These puzzling observations suggest that simply correcting Hgb levels in anemia is not sufficient. Thus, there is a need to investigate the cellular mechanisms of anemia. Recently, others and we have demonstrated that anemia leads to reduced tissue oxygen tension and activates hypoxic mechanisms, such as HIF and NO (McLaren et al., 2007; Tsui et al., 2011). These observations remind us that the ancient oxygen-sensing mechanisms and oxygen delivery systems are tightly regulated. This chapter will explore the relationship of Hgb, NO, and HIF pathways in anemia and hypoxia.

First, we will present the case of the Tibetan highlanders. This experiment of nature illustrates that increased NO signaling and genetic evolution at HIF genomic loci can serve as a highly effective adaptive population response in adverse conditions, specifically an environment of chronic hypoxia. Tibetan highlanders have undergone over a 25,000-year period an amazing genetic adaptive response to living at high altitude under hypoxic conditions (Beall, 2007). Living at ~4000 meters above sea level, the oxygen saturation of Hgb is only ~84–85% at rest compared with ~97–98% for those living at the sea level. Despite lower oxygen content, Tibetan highlanders have work capacities similar to sea-level dwellers (Erzurum et al., 2007). Compared to sea-level dwellers, Tibetan highlanders have higher levels of circulating NO products including nitrates, nitrites, and nitroso-proteins (Erzurum et al., 2007). Tibetan highlanders have higher levels of forearm blood flow and oxygen delivery despite lower Hgb levels and lower Hgb saturation. How the highlanders have this remarkable ability is not completely characterized. However, compared to nearby lowland dwelling Han populations, genetic associate studies revealed that Tibetan highlanders have diverged from their neighbors near the EPAS1/HIF-2α locus (Beall et al., 2010). Phenotypically, these genetic alterations, along with others, are associated with a lower Hgb concentration in highlanders (Beall et al., 2010; Simonson et al., 2010). Curiously, Andean high-altitude dwellers who migrated to high altitude living in South America ~11,000 years ago do not have the same adaptations as their Tibetan counterparts. Rather, they have higher levels of Hgb and lower levels of exhaled NO and demonstrate hypoxic pulmonary vasoconstriction (Beall, 2007). While the molecular changes that underlie the differences in NO biology in Tibetan and Andean populations require further investigation, their contrasting physiology draws attention to the curious relationship between NO and hypoxia.

This chapter consists of two main themes. The first part focuses on the role of NO in the cellular response to hypoxia. The second part focuses on how oxygen availability influences NO production.

2. PART 1: NO BIOLOGY IN NORMOXIA, HYPOXIA, AND ANEMIA

This section briefly reviews basic NO biology and discusses its interaction with the classical hypoxia pathway involving the HIF.

2.1. NO biology, S-nitrosylation, and equilibrium of S-nitrosylated products (Fig. 7.1)

In humans, NO is a signaling molecule involved in many physiological and pathological conditions (Anand & Stamler, 2012; Ho et al., 2012). NO in the body is generated by a reaction catalyzed by a family of NOSs. NOS enzymes use L-arginine and O_2 as substrates to generate NO and L-citrulline (Moncada & Higgs, 1993). There are three distinct isoforms of NOS with distinct tissue- and cell-specific expression patterns (Michel & Feron, 1997): neuronal NOS (nNOS; NOS1) (Bredt et al., 1991), inducible NOS (iNOS; NOS2) (Lowenstein, Glatt, Bredt, & Snyder, 1992), and endothelial NOS (eNOS; NOS3) (Marsden et al., 1992). Initially identified as an endothelial-derived relaxing factor in 1984 (Griffith, Edwards, Lewis, Newby, & Henderson, 1984), NO binding to ferrous (Fe^{2+}) proteins is one of the most potent and well-known actions of NO (Furchgott & Zawadzki, 1980). In particular, NO is known for its role in cardiovascular homeostasis via interaction with soluble guanylate cyclase, resulting in the generation of cyclic GMP as a second messenger and vasodilation (Ignarro, 1999). However, recent evidence demonstrates that signal transduction and other cellular functions occur largely through

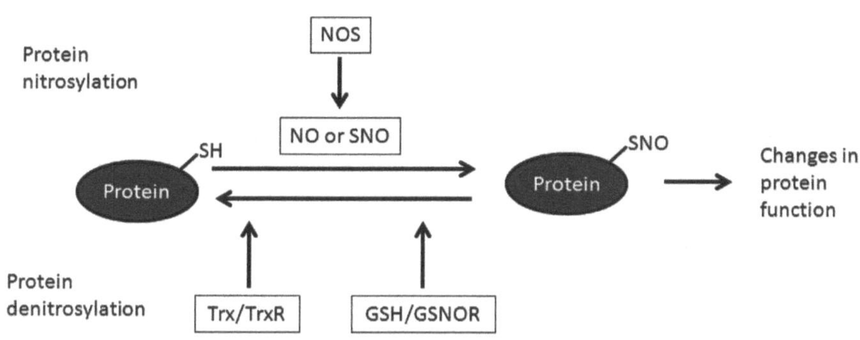

Figure 7.1 Equilibrium of nitrosylation and denitrosylation systems *in vivo*. Protein S-nitrosylation is dynamically regulated by "nitroso" synthases and denitrosylation systems. To induce S-nitrosylation, nitric oxide (NO) produced by nitric oxide synthases (NOSs) covalently binds to cysteine residues of target proteins, which leads to a range of effects including alterations in enzymatic activity, subcellular localization, protein–protein interaction, and stability. There are two denitrosylation systems identified *in vivo*. Both the thioredoxin reductase (TrxR) system and the S-nitrosoglutathione reductase (GSNOR) system have the ability to remove the NO group from the S-nitrolsylated cysteine residue. These pathways play important roles in physiological and pathophysiological conditions. (See the color plate.)

cGMP-independent mechanisms (Hess, Matsumoto, Kim, Marshall, & Stamler, 2005). Rather, formation of S-nitrosoproteins by the reversible, covalent addition of an NO moiety to the thiol of cysteine (Cys), known as protein S-nitrosylation, gives rise to S-nitrosothiols and accounts for a major mechanism by which NO regulates cellular targets including the HIF pathway. S-Nitrosylation can have a range of effects on target proteins, including changes in enzymatic activity, subcellular localization, protein–protein interactions, and protein stability, among others (Anand & Stamler, 2012). This process has such importance that NOSs may now be regarded as "nitrosothiol synthases" (Hess et al., 2005). NO, via NO-derived species such as N_2O_3, can react with Cys thiols to generate S-nitrosothiols, or S-nitrosothiols can transfer an NO group to an acceptor Cys thiol, a process known as transnitrosylation (Foster, McMahon, & Stamler, 2003; Hess et al., 2005). Formation of S-nitrosothiols is balanced by denitrosylation by denitrosylases (Fig. 7.1). The two major denitrosylation pathways are the thioredoxin system and the S-nitrosoglutathione reductase (GSNOR) system (Benhar, Forrester, & Stamler, 2009). S-Nitrosoglutathione (GSNO) is the most abundant low-molecular-weight S-nitrosothiol in mammalian cells and, as a key component of the denitrosylation pathway, plays an important role in setting levels of cellular nitrosylation. S-Nitrosylation and denitrosylation are associated with a variety of physiological and pathological conditions, including the cellular response to hypoxia, which will be discussed later.

2.2. SNO products and physiology

In addition to promoting blood flow, SNOs are also involved in many aspects of the integrative physiological responses in the body. Stamler and colleagues first explored this concept in the context of the Hgb oxygenation/deoxygenation cycle. Within the β-globin chain, a cysteine residue (Cysβ93) can function as a site for transnitrosylation from the heme group (Fe-NO) to form SNO-Cysβ93 in the R structure of oxyHgb. As O_2 concentration decreases and Hgb conformation switches to the T structure, this SNO-Cysβ93 is exposed, which allows the transfer of the NO group to other thiols such as glutathione. This process is proposed as means of NO transport to the microcirculation and is favored in the setting of deoxygenation when there are higher levels of deoxyhemoglobin. In this way, NO release from Hgb is coordinated with O_2 release and participates physiologically to increase blood flow in areas of relative hypoxia (Jia, Bonaventura, Bonaventura, & Stamler, 1996; Marsden, 2007). Although SNO-Hgb typically accounts for <1% of total Hgb, biochemical methods allow us to identify this phenomenon. Yet, measuring the specific physiological effects of SNO-Hgb cycling in real time remains challenging. Nevertheless, there are data that low-molecular-weight SNOs, such as GSNO, S-nitrosocysteinyl glycine, and S-nitroso-L-cysteine, can activate the respiratory centers in the brain to stimulate ventilation (Lipton et al., 2001). These SNOs are derived from SNO-Hgb and possess the ability to produce a ventilatory response as with hypoxia. While endogenous SNOs have important physiological effects, exogenously administered SNOs can have pathophysiological effects when they are involved in transnitrosylation reactions that interfere with endogenous SNO functions. For example, chronic exposure to a commonly used antioxidant, *N*-acetylcysteine (NAC), mirrors chronic hypoxia and leads to pulmonary hypertension with a phenotype similar to that seen with eNOS-deficient mice (Palmer et al., 2007). In this situation, NAC readily interacts with NO and is transnitrosylated by NO in red blood cells to form S-nitroso-acetylcysteine (SNOAC) that functionally sequesters NO, thereby preventing its normal activity. These findings demonstrate physiological and pathophysiological examples where NO, via S-nitrosylation, signals hypoxia.

2.3. HIF protein stabilization in hypoxia

Hypoxia refers to a pathologic state in an organism in which there is an imbalance between oxygen supply and demand. Conceptually, hypoxia may apply to the whole organism or to local deficiency of oxygen (tissue hypoxia). Nevertheless, O_2 availability is below the physiological requirement of the tissue. It is important to note that hypoxia can occur in the absence of hypoxemia, which refers to a decrease in oxygen concentration in the blood. A majority of the ATP generated for cellular metabolism depends on O_2; as a result, the mechanisms that respond to a decrease in O_2 availability are highly evolutionarily conserved (Brahimi-Horn, Chiche, & Pouyssegur, 2007; Brahimi-Horn & Pouyssegur, 2007). Within a cell, hypoxia leads to a global reduction in protein synthesis. Importantly, hypoxia regulates the expression of specific sets of genes (Koritzinsky et al., 2006; Shih & Claffey, 1998). Fundamental to these observed changes in gene expression is the response mediated by the transcription factor HIF (Wang,

Jiang, Rue, & Semenza, 1995). HIF is a heterodimeric transcription factor, which is composed of a hypoxia-inducible α subunit (HIF-α) and a constitutively expressed β subunit (HIF-β), also known as aryl hydrocarbon nuclear receptor (Wood, Gleadle, Pugh, Hankinson, & Ratcliffe, 1996). HIF is considered a master regulator of oxygen homeostasis, and under hypoxic conditions, HIF-α accumulates and translocates to the nucleus where it dimerizes with HIF-β and binds to hypoxia response elements (HREs; consensus sequence 5′-(R)CGTG-3′) in the enhancer/promoter regions of target genes (Semenza, 2012; Fig. 7.2). These target genes participate in an integrated physiological response to increase O_2 delivery to the tissues and optimize cellular metabolism. Vascular endothelial growth factor (VEGF) and erythropoietin function to promote blood vessel growth and Hgb production, respectively, while the glucose transporter GLUT1 and pyruvate dehydrogenase kinase (PDK1) function to increase glucose uptake and augment glycolysis, respectively (Forsythe et al., 1996; Ouiddir, Planes, Fernandes, VanHesse, & Clerici, 1999; Semenza & Wang, 1992). On the other hand, HIF mediates the transcriptional repression of other genes (Evans et al., 2007; Wang et al., 2009). There are two isoforms of HIF-α which are important in the regulation of oxygen homeostasis: HIF-1α and HIF-2α. While some target genes overlap, HIF-α isoforms also have unique targets and themselves have tissue-specific patterns of expression (Branco-Price et al., 2012; Hu, Wang, Chodosh, Keith, & Simon, 2003; Mole et al., 2009; Raval et al., 2005).

Dysregulated expression of HIFs contributes to the development of some cancers (Kaelin, 2007). As a result, HIF protein expression is tightly linked to hypoxia. Cells grown in normoxic conditions produce their respective HIF-α isoform mRNA and protein. However, under these normal conditions, HIF-α protein is rapidly degraded via the ubiquitin–proteasome pathway and does not function to activate the HIF transcriptional program (Kallio, Wilson, O'Brien, Makino, & Poellinger, 1999; Fig. 7.2). Rather, in an oxygen-dependent fashion, a family of enzymes known as PHD-containing proteins with three isoforms (PHD1, PHD2, and PHD3) catalyze the hydroxylation of two key proline residues in the oxygen-dependent degradation (ODD) domain of both HIF-α isoforms (Bruick & McKnight, 2001; Huang, Gu, Schau, & Bunn, 1998; O'Rourke, Tian, Ratcliffe, & Pugh, 1999). These hydroxyproline residues are then recognized by von Hippel–Lindau protein (pVHL), which forms E3-ubiquitin ligase complexes in association with other

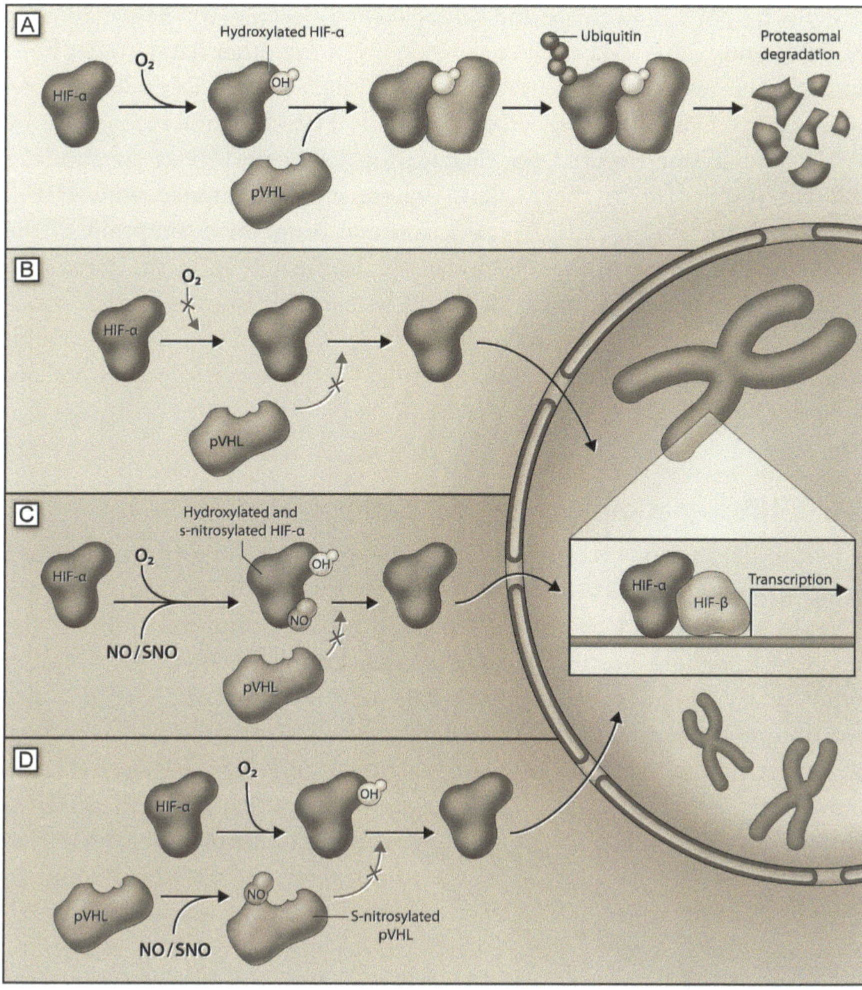

Figure 7.2 Regulation of hypoxia-inducible factor (HIF) by nitric oxide (NO) in the presence of oxygen. (A) In normoxia, the presence of oxygen leads to the activation of oxygen-dependent prolyl hydroxylase domains (PHDs), which hydroxylate proline residues of the HIF-α subunit. Von Hippel–Lindau protein (pVHL) recognizes and binds to the hydroxylated HIF-α, which is then subject to polyubiquitination and proteösomal degradation. Thus, HIF-α protein is normally produced and rapidly degraded in the presence of oxygen. (B) With hypoxia, oxygen-dependent PHDs do not hydroxylate HIF-α. HIF-α protein accumulates and translocates to the nucleus where it binds to the HIF-β subunit to activate target genes. (C–D) Even with enough O_2 for HIF-α hydroxylation, the presence of NO can result in HIF-α protein stabilization by S-nitrosylation of pVHL or HIF-α. S-Nitroslyation of pVHL or HIF-α prevents interaction of pVHL and HIF-α, allowing HIF-α to escape the degradation pathway. The accumulated HIF-α in the cytosol translocates to the nucleus and binds and dimerizes with HIF-1β, which can activate transcription of HIF-dependent genes. (See the color plate.)

proteins to polyubiquitinate and target HIF-α proteins for degradation in the 26S proteasome (Maxwell et al., 1999). On the other hand, under hypoxic conditions, there is a reduction of the oxygen-dependent prolyl hydroxylation of HIF-αs by PHDs. Hence, HIF-α is not recognized by pVHL and is not targeted for degradation. Rather, HIF-α accumulates and can translocate to the nucleus, where it will dimerize with HIF-β to mediate the canonical HIF transcriptional program. Disruption of the VHL gene in the von Hippel–Lindau syndrome results in the formation of highly vascularized tumors including hemangioblastomas and clear cell renal cell carcinoma (Ivan & Kaelin, 2001; Kaelin, 2008). Cells exhibit activation of the HIF transcription pathway even when the cells are not hypoxic.

2.4. S-Nitrosylation and the HIF pathway

The stabilization of HIF-α protein under hypoxic conditions described in the preceding text is the most well-known mechanism of HIF pathway activation. Key to this stabilization is that hypoxia inhibits the oxygen-dependent posttranslational modification of HIF-α. Thus, under hypoxia, pVHL does not bind the HIF-α ODD domain: HIF-α is not polyubiquitinated and is not degraded by the 26S proteasome. Hypoxia, however, is not the only regulator of the interaction between HIF-α and pVHL. S-Nitrosylation of the key molecules that destabilize HIF-α protein can prevent HIF-α degradation even under normoxia. HIF-α itself can be nitrosylated at Cys533 in the mouse HIF-α ODD (analogous to Cys520 in human HIF-α ODD), which prevents interaction with pVHL (Li et al., 2007). Alternatively, pVHL activity can be inhibited in normoxia by S-nitrosylation at Cys162, which disrupts the interaction of pVHL with Elongin C, a component of the E3 ubiquitin ligase complex (Palmer et al., 2007). Although PHDs also have potential sites for nitrosylation, the functional implications of PHD S-nitrosylation are less clear (Berchner-Pfannschmidt, Yamac, Trinidad, & Fandrey, 2007; Chowdhury et al., 2011). The stabilization and thus activation of HIF by NO/S-nitrosylation are functionally relevant to anemia, inflammation, and cancer (Li et al., 2007; Tsui et al., 2011). In this way, HIF-α stabilization and HIF pathway activation should be considered depending on two major mechanisms: (1) reduction in oxygen-dependent prolyl hydroxylation of HIF and (2) the S-nitrosylation of HIF pathway components.

2.5. NO inhibits HIF stabilization in hypoxia

While S-nitrosylation of pVHL or HIF-α inhibits HIF degradation in normoxia, intracellular NO can blunt HIF stabilization and activity in hypoxia (Marsden, 2007). NO is an inhibitor of the mitochondrial cytochrome *c* oxidase, which is the primary site of cellular oxygen consumption. Consequently, this leads to a reduction of oxygen utilization in the oxygen transport chain, changes in intracellular oxygen availability, and a redistribution of oxygen toward other oxygen-dependent targets, including prolyl hydroxylases. As a result, there are oxygen-dependent prolyl hydroxylation of HIF-α and a reduction of HIF-α stability (Hagen, Taylor, Lam, & Moncada, 2003). Thus, intracellular NO blunts the normal accumulation of HIF-α protein in hypoxia. Collectively, these observations suggest that oxygen availability and distribution dictate how NO interacts with HIF signaling pathways.

2.6. Divergent roles of nNOS in hypoxia versus anemia

Activation of specific NOS enzymes becomes important when examining cellular responses to pathologic states such as anemia and hypoxia. Both anemia and hypoxia impact oxygen delivery, yet recent studies have found that nNOS-mediated S-nitrosylation improves survival in anemia and not hypoxia (Tsui et al., 2011). These findings underscore the need to understand how the cellular response is different between these two conditions. Anemia is a common finding in both chronic conditions and with acute critical illness and is associated with increased mortality (Beattie, Karkouti, Wijeysundera, & Tait, 2009; Carson et al., 1996; Pfeffer et al., 2009). Treatment of anemia with either red blood cell transfusion or erythropoiesis-stimulating agents may not improve survival or reduce morbidity (Drueke et al., 2006; Hebert et al., 1999; Pfeffer et al., 2009; Singh et al., 2006). These data suggest that factors other than the absolute level of Hgb are important for cellular adaptation and survival. Because oxygen delivery is compromised with both anemia (low Hgb) and hypoxia (low pO_2), it is important to compare hypoxic cell signaling between these conditions.

Because HIF pathway proteins can be S-nitrosylated, mice lacking individual NOS isoforms were exposed to either hypoxia or anemia and physiological and cellular responses, including HIF activation, were examined. Only nNOS-deficient mice (nNOS (−/−)), but not eNOS- or iNOS-deficient mice, had an increase in mortality with anemia (Tsui et al.,

2011). These mice did not generate the same adaptive cardiovascular responses to anemia that wild-type (nNOS expressing) mice demonstrated. While they had comparable tissue hypoxia to wild-type mice, nNOS (−/−) mice were not able to increase cardiac output (CO) or reduce systemic vascular resistance (SVR) in response to anemia. Resistance vessels did not vasodilate in nNOS (−/−) mice compared to wild-type mice. In contrast with wild-type mice, nNOS (−/−) mice did not have increased expression of HIF-1α and HIF-responsive genes such as EPO, VEGF, GLUT1, and PDK1, especially in the brain. HIF-α(ODD)-luciferase transgenic mice, which express the reporter luciferase protein when HIF-α protein is stable, showed that nNOS was essential for HIF-α protein stability in anemia. This stabilization of HIF-α protein with anemia was mediated, in part, by the nNOS-dependent S-nitrosylation of pVHL (Fig. 7.2). The increase in bioavailable NO may also have resulted from a reduction in GSNO reductase, a key enzyme in the denitrosylation pathway (Fig. 7.1). On the other hand, these same nNOS (−/−) mice had a survival advantage in hypoxia. In response to hypoxic conditions, nNOS (−/−) mice were able to increase CO and decrease SVR. nNOS was not necessary to increase the expression of HIF-1α and HIF-responsive genes with hypoxia. Together, these results demonstrate a specific role for nNOS in the cellular hypoxic signaling pathways involving HIF-α and the physiological response to anemia.

Although blood oxygen content is reduced with both hypoxia and anemia, differences in distribution of O_2 may necessitate alternate pathways to activate HIF in the arterial vasculature (Fig. 7.3). Hypoxia results in a reduction of immediately available dissolved O_2 (reduced pO_2) in the vascular compartment and reduced O_2 bound to Hgb. Oxygen-dependent PHDs cannot hydroxylate HIF-α, and HIF-α protein is not degraded. Anemia results in a reduction in the amount of Hgb and, as a result, a decrease in the oxygen-carrying capacity of the blood. However, the pO_2 (dissolved oxygen) and Hgb O_2 saturation may remain the same or even slightly increase in the arterial vasculature due to eNOS-dependent changes in lung ventilation–perfusion matching (Deem et al., 1999). Thus, with anemia, there may be greater free O_2 levels, though total oxygen delivery will be reduced. As a result, classical O_2 sensing via oxygen-dependent PHDs may not be sufficient to stabilize HIF-α protein. Therefore, hypoxia signaling occurs through nNOS-mediated S-nitrosylation of pVHL. Compared to hypoxia, anemia may also allow Hgb-mediated NO signaling via release of NO in the oxy-deoxyhemoglobin cycling. These mechanisms are complementary in signaling decreased O_2 availability resulting from different

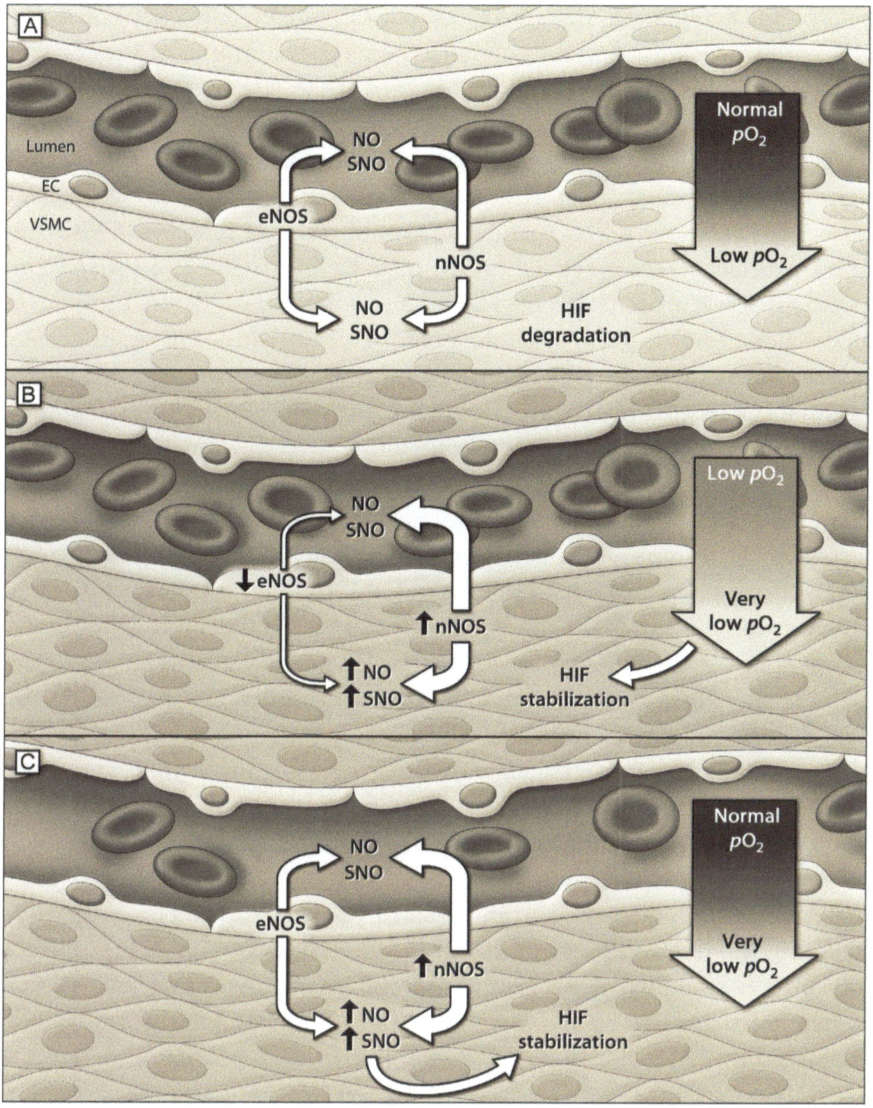

Figure 7.3 Mechanisms of vascular HIF stabilization in hypoxia versus anemia. (A) In normoxia (normal O_2 content, normal pO_2), constitutive expression of eNOS from the endothelium and nNOS from vascular smooth muscle contributes to the NO and SNO content in the vascular lumen and smooth muscle compartments. A normal O_2 gradient exists from blood to tissue. In anemia and hypoxia, HIF-α can be stabilized by different mechanisms despite similar reduction in blood oxygen content: (1) the classical prolyl hydroxylase pathway and (2) nNOS/NO-mediated S-nitroslyation of HIF pathway. (B) In hypoxia (low arterial pO_2, low O_2 saturation, normal hemoglobin levels), there is increased nNOS in smooth muscle but reduced eNOS expression in endothelial cells.

(Continued)

physiological mechanisms. The key issue is that the evolutionary conserved HIF pathway cannot classically signal hypoxia in the arteries of anemic animals. Rather, NO derived from nNOS primes the HIF pathways in arteries of anemic animals.

While GSNO reductase was decreased in anemia, nNOS-mediated NO signaling is the primary protective mechanism that promotes survival in anemia. A reduction in GSNO reductase would favor an increase in NO and SNO products. Decreased GSNOR may be a protective adaptation to improve SNO protein levels. The functional response is an increase in blood flow (Jia et al., 1996), myocardial protection, and angiogenesis (Lima et al., 2009). As GSNOR-deficient mice possess higher blood and tissue SNO levels, this mouse model serves as an excellent tool to study the physiological effects of S-nitrosylation (Liu et al., 2004). In a model of myocardial infarction, GSNOR-deficient mice demonstrated reduced infarct size, preserved ventricular function, as well as preserved myocardial tissue oxygen tension compared to GSNOR-replete mice (Lima et al., 2009). With respect to hypoxia signaling, these effects were also associated with increased S-nitrosylation of HIF-1α and HIF activity. These observations suggest that endogenous SNO level is an important physiological element in myocardial infarction. In contrast, higher SNO levels can be detrimental and contribute to pathophysiological effects. For example, GSNOR-deficient mice had increased mortality from endotoxic and septic shock models (Liu et al., 2004). Abnormally, high SNO levels in these mice result in tissue injury. Interestingly, iNOS inhibition in GSNOR-deficient mice abolished the detrimental effects. In this way, SNO levels are tightly regulated by both nitrosylation and denitrosylation pathways for integrative physiological responses in mammals. Nonetheless, GSNOR-deficient mice did not have significant survival advantage in anemia (Tsui et al., 2011), suggesting that specific nNOS-mediated NO signaling is protective in anemia.

Figure 7.3—Cont'd nNOS is the main source NO/SNO in hypoxia. Due to the reduced level of free oxygen (low arterial pO_2), increased HIF-α expression is predominantly regulated via the classic prolyl hydroxylase pathways due to oxygen deprivation. Also, the O_2 gradient from blood to tissue is greatly reduced in systemic hypoxia. (C) In anemia (normal arterial pO_2, normal O_2 saturation, low hemoglobin concentration), nNOS is increased, whereas eNOS level remains unchanged. There exists a greater blood to tissue O_2 gradient in anemia compared to hypoxia. HIF-α is not stabilized via the prolyl hydroxylase pathways; rather, HIF-α stabilization occurs via nNOS-mediated S-nitrosylation of pVHL. nNOS is the major source for tissue NO/SNO levels in anemia. (See the color plate.)

2.7. Physiological responses to anemia—Role of NO

Anemia is defined as a reduction in Hgb concentration or red blood cell mass. Because most of the blood oxygen content is carried by Hgb, anemia leads to a decrease in oxygen-carrying capacity in the blood, which is characterized by low blood oxygen content and normal arterial partial pressure of oxygen (PaO_2 or free oxygen). In response to anemia, the body activates a series of adaptive mechanisms to optimize oxygen delivery.

2.7.1 Cardiovascular adaptation

During acute anemia, the sudden drop in blood oxygen content may be sensed at multiple redundant levels. Once tissue hypoxia is sensed, the associated efferent signals may trigger a central neuronal response that activates the adrenergic nervous system and increases the sympathetic activity. This mechanism contributes to increased CO, stroke volume (SV), and heart rate (HR) during anemia, as demonstrated by cardiac denervation experiments (Glick, Plauth, & Braunwald, 1964). The composite effect is an increase in CO and maintenance of mean arterial pressure until very low Hgb levels. In addition, active and passive mechanisms lead to an associated reduction in SVR, which further promotes O_2 delivery to tissues (Eichelbronner, D'Almeida, Sielenkamper, Sibbald, & Chin-Yee, 2002; Ickx, Rigolet, & Van Der Linden, 2000; Ragoonanan et al., 2009). These cardiovascular responses are consistently observed in humans and animals and act to maintain a global balance of O_2 delivery and consumption (Murray & Rapaport, 1972; Weiskopf et al., 1998). Interestingly, S-nitrosylation can modulate cardiac function by regulating intracellular Ca^{2+} levels in cardiomyocytes (Gonzalez, Beigi, Treuer, & Hare, 2007), which may, in part, account for the enhanced CO response to acute anemia.

At the vascular level, an important characteristic response in acute anemia is blood flow that is redistributed to ensure adequate oxygen delivery to vital organs to support the survival of the organism during stress conditions. For example, cerebral (Hudetz, Wood, & Kampine, 2000; Ragoonanan et al., 2009) and coronary blood flow (Murray & Rapaport, 1972) increases during anemia, whereas intestinal blood flow decreases (van Bommel et al., 2002). Of interest, NO may be an important mediator of increased cerebral blood flow during acute anemia (Hudetz et al., 2000). Despite this "preferential perfusion," noninvasive measurements of brain tissue pO_2 demonstrate that the brain microvascular PO_2 decreases progressively as Hgb levels become reduced (Ragoonanan et al., 2009). The observation of blood flow redistribution in anemia suggests that blood viscosity may not be the primary driving force to increase vital organ blood flow. Rather, active, hypoxia-driven, organ-specific vasodilation may be one of the main contributors for the observed effect in preferential blood flow. Collectively, increased CO and vital organ blood flow in anemia ensure that tissue oxygen delivery is optimized to promote survival.

2.7.2 Respiratory adaptation

To compensate for the reduced Hgb level during anemia, oxygen loading of Hgb in the lungs is optimized. The mechanisms are thought to involve NO-mediated signals to improve ventilation and perfusion matching in the lungs (Deem et al., 1999; Sproule, Mitchell, & Miller, 1960). Anemia leads to an active redistribution of blood flow within the lungs, which shunts blood away from the poorly oxygenated alveoli and improves blood flow toward the better-oxygenated alveoli. Venous return and deoxyhemoglobin are directed towards ventilated alveoli where it can be better oxygenated. This reduces the amount of deoxyhemoglobin passing through poorly ventilated lung units and entering the arterial circulation still deoxygenated. These respiratory responses lead to a characteristic rise in P_aO_2 and S_aO_2 and ensure that optimal S_aO_2 is maintained at a time of reduced Hgb to maximize blood oxygen content.

2.7.3 Metabolic adaptation

In addition, alteration in the metabolic requirement helps to preserve the overall balance of systemic oxygen supply and demand during anemia (Murray & Rapaport, 1972; Weiskopf et al., 1998). For example, increase in myocardial oxygen utilization results in reduction in tissue O_2 availability for consumption in other organs (Murray & Rapaport, 1972; von Restorff, Hofling, Holtz, & Bassenge, 1975). While O_2 consumption is maintained in the brain (Johannsson & Siesjo, 1974), it decreases in the kidney during anemia (Johannes, Mik, Nohe, Unertl, & Ince, 2007). Such mechanisms are activated in hypoxia (Matsuoka, Saiki, & Mortola, 1994) and may contribute to adaptive changes to promote survival in acute anemia. These mechanisms are dependent, in part, on HIF-mediated metabolic responses (Semenza, 2009). The activation of these adaptive responses ensures that tissue O_2 homeostasis and the balance of O_2 supply and demand are optimized during anemic stress.

2.7.4 Increased tissue oxygen extraction during anemia

Experimental studies in humans and animals demonstrated that global and tissue-specific O_2 delivery is maintained during acute anemia, in part, through increased systemic oxygen extraction (van Bommel et al., 2002; Weiskopf et al., 1998). This mechanism may be less important to organs with a high basal oxygen extraction ratio, such as the heart (70%), which are more dependent on increased flow to match increased oxygen consumption during anemia (Wolff, 2007). By contrast, the oxygen extraction ratio in the brain increases from about 30% at baseline to near 50% during anemia under experimental conditions (Ragoonanan et al., 2009; van Bommel et al., 2002). Increase in tissue oxygen extraction likely depends on at least three important factors: (1) a right shift of the oxyhemoglobin dissociation curve to reduce Hgb-oxygen affinity (this may occur by a number of mediators including decreased pH, increased 2,3-DPG, and NO-mediated signaling events (Doctor et al., 2005; Sladen, 1981)); (2) increased tissue blood flow that favors increased oxygen diffusion into the tissue (Halperin, Cheema-Dhadli, Lin, & Kamel, 2006); and (3) increased capillary recruitment and density may limit the diffusion distance to cells during anemia (El Hasnaoui-Saadani et al., 2009). The primary goal of these mechanisms is to facilitate oxygen diffusion from the microcirculation to the tissues, thereby sustaining mitochondrial oxidative phosphorylation (aerobic respiration).

3. PART 2: EFFECTS OF HYPOXIA ON NOSs

The first part of this chapter focused on the regulation of the HIF pathway and response to hypoxia by NO. This next part will focus on how hypoxia affects the enzymes that generate NO. The relationship of NOS enzyme expression and NOS enzyme function in hypoxia is highly specific to each isoform.

3.1. Neuronal NOS

nNOS/NOS1 was the first human NOS gene to be cloned and characterized (Bredt et al., 1991). nNOS contributes to protective effects in myocardial infarction (Burger et al., 2009; Dawson et al., 2005; Saraiva et al., 2005), learning and memory (Weitzdoerfer et al., 2004; Xu, Russell, Bazner, & Hamilton, 2001), muscular dystrophic pathology (Kobayashi et al., 2008; Percival, Anderson, Gregorevic, Chamberlain, & Froehner, 2008; Percival, Anderson, Huang, Adams, & Froehner, 2010), and atherosclerosis

(Kuhlencordt et al., 2006; Morishita et al., 2002). From a pathophysiological perspective, nNOS is detrimental in models of ischemic stroke (Eliasson et al., 1999) and heart failure (Burkard et al., 2007). Although named as a "neuronal" NOS gene, nNOS is expressed in a variety of cell types (Michel & Feron, 1997). The tissue- and cell-specific expression of nNOS utilizes a greater number of distinct promoters and associated exon 1 variants than any other human gene (Newton et al., 2003; Wang, Newton, & Marsden, 1999; Wang et al., 2002). Most basally expressed nNOS variants have poor translational efficiency (Wang et al., 1999). Chronic hypoxia induces the expression of a translation-efficient 5′-untranslated region (5′UTR) variant of nNOS that leads to greater nNOS protein levels and greater nNOS activity (measured at 16–24 h) (Ward et al., 2005). Basally, nNOS expression occurs from a promoter that produces a long and translation-inefficient 5′UTR. In hypoxia, a promoter just upstream of exon 2 initiates transcription in intron 1 over 100 kb downstream from the constitutively expressed exon 1 and produces a short 5′UTR. This mRNA transcript is efficiently translated under hypoxic conditions such that nNOS

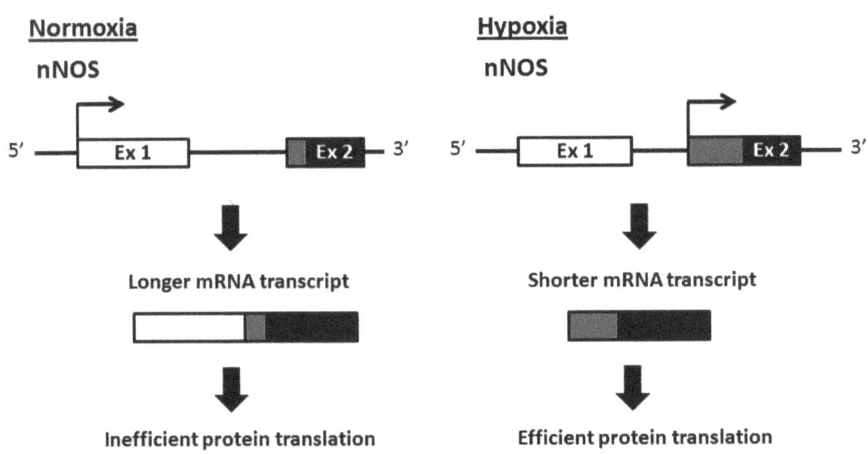

Figure 7.4 Neuronal nitric oxide synthase (nNOS) mRNA is efficiently translated under hypoxia. Left: Under normoxia, nNOS transcription (arrow) in exon 1 produces a long 5′UTR (white- and gray-boxed area). Translation begins in exon 2 (black-boxed area). The translational efficiency of basally expressed nNOS mRNA transcripts with this long 5′UTR is very poor. However, hypoxia (right) activates a novel promoter just downstream of exon 1. This results in the production of an nNOS mRNA containing alternative, shorter 5′UTR. Under hypoxic conditions, the translation of mRNA with exon 1 is inhibited. However, the nNOS mRNA containing this alternative 5′UTR is favored and protein translation becomes very efficient. In this way, nNOS expression can be highly regulated by oxygen concentrations.

from aortic smooth muscle cells accounts for a majority of NOS enzyme activity and NO production from the aorta under hypoxic conditions (Fig. 7.4). Interestingly, nNOS has a very high K_m for O_2 of 350 μM, and nNOS activity is highly dependent on O_2 being almost linearly related to O_2 availability in the physiological range (Elayan, Axley, Prasad, Ahlers, & Auker, 2000; Semenza, 2005). Ward et al. found that while nNOS protein increased ~10-fold, nNOS activity increased only 3.5-fold (Ward et al., 2005). Overall, compared to baseline, NO activity in the vessel wall increased by 1.25-fold. Nonetheless, this increase in NO activity will reduce vasoconstriction in chronic hypoxic conditions. The regulation of the nNOS gene with hypoxia and the biochemical dependence of nNOS on O_2 availability may serve to coordinate a physiologically favorable response to acute and chronic hypoxia. nNOS-derived NO may not be beneficial with acute hypoxia, and in fact, nNOS activity can increase neurotoxicity after acute ischemic stroke (Dawson & Dawson, 1996). Due to its high dependence on oxygen availability, NO production by nNOS will decrease with acute hypoxia. However, this initial decrease in NO is attenuated in chronic hypoxia by increase in nNOS expression, which may be beneficial to restore blood flow to chronically hypoxic tissue (Ho et al., 2012; Semenza, 2005). Nonetheless, the highly regulated nature of nNOS expression with hypoxia highlights the importance of NO biology in oxygen homeostasis.

3.2. Inducible NOS

Compared to the other NOS genes, NOS2/iNOS is different because human cells and tissues do not typically express iNOS mRNA and protein under basal conditions. Rather, iNOS expression is induced by pro-inflammatory cytokines, such as TNF-α and IL-1, which, in combination with IFN-γ, can increase the transcription of iNOS mRNA in a variety of cell types including macrophages and vascular smooth muscle cells. This inflammatory induction of iNOS often involves NF-κB (Robinson, Baumgardner, & Otto, 2011; Shaked et al., 2012). We have found that epigenetic regulation (DNA methylation and histone posttranslational modifications) is important for the differential inducibility of iNOS between cell types and between species. Models of iNOS activation are difficult to generalize as iNOS induction in human cells is more difficult than in rodents (Chan et al., 2005). These differences may reflect a greater advantage for iNOS induction to be limited to chronic inflammation and infection in humans compared to mice, in which iNOS plays a key role for a more responsive innate immune system. Nevertheless, evidence suggests that hypoxia, in certain settings, can regulate iNOS expression and that there is a complex relationship with HIFs and inflammatory mediators. Upregulation of iNOS expression in hypoxia often requires costimulation by cytokines. While the mouse iNOS promoter contains a HRE (the human iNOS promoter does not), hypoxia can only upregulate iNOS expression in most cells in the presence of costimulation by IFN-γ (Melillo et al., 1995). The HIFs themselves have a differential role in iNOS biology. While HIF-1α is induced by Th1 cytokines in macrophages and induces iNOS expression, HIF-2α is induced by Th2 cytokines and induces arginase-1, which competes with iNOS for L-arginine as a substrate (Takeda et al., 2010). In most situations, iNOS mRNA levels correlate with iNOS protein levels, suggesting that regulation of iNOS in hypoxia is dependent on mRNA regulation (Robinson et al., 2011). Functionally, iNOS activity, like nNOS, is highly dependent on oxygen availability with a K_m for O_2 of 130 μM (Semenza, 2005). In general, NO production by iNOS decreases across cell types with hypoxia (Robinson et al., 2011). Given the complex relationships of iNOS expression and function with hypoxia and the requirement for inflammatory stimulation, evaluation of iNOS activity requires an understanding of the specific setting under investigation.

3.3. Endothelial NOS

Endothelium-derived NO is dependent on the endothelial-specific isoform of NOS—the eNOS. eNOS is important in regulating vasomotor tone, hemostasis, and angiogenesis (Jamal, Man, & Marsden, 2012). It is expressed almost exclusively in the endothelium of medium- and large-sized vessels. Its expression is controlled by transcriptional, posttranscriptional, and posttranslational mechanisms (Matouk & Marsden, 2008).

3.3.1 Regulation of eNOS transcription in hypoxia

In most *in vitro* models, hypoxia reduces the expression of eNOS from endothelial cells (Fish et al., 2007, 2010; McQuillan, Leung, Marsden, Kostyk, & Kourembanas, 1994; Ostergaard et al., 2007). Decreases in eNOS mRNA occur due to both transcriptional and posttranscriptional mechanisms (Fish et al., 2010; McQuillan et al., 1994). Recently, we found that chromatin-based mechanisms contribute to the decreased transcription of eNOS mRNA. In endothelial cells cultured under normoxic conditions, the eNOS promoter contains activating chromatin marks such as methylation at

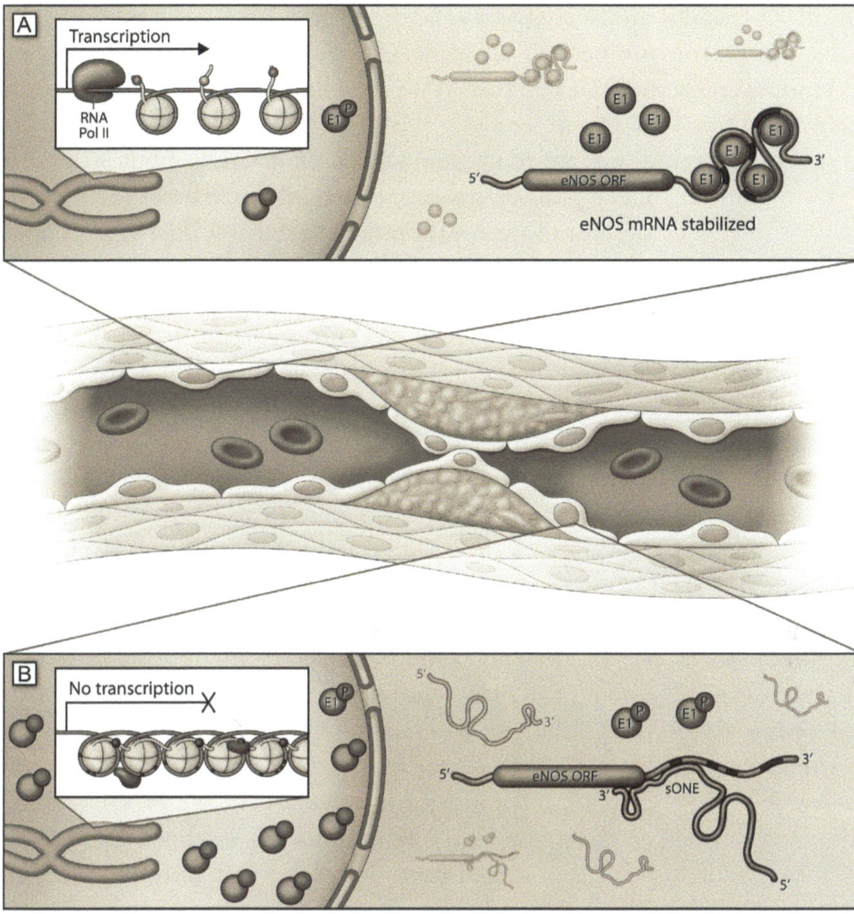

Figure 7.5 Effects of hypoxia on transcription and posttranscription regulation of eNOS mRNA. (A) Normoxia—with adequate oxygen availability within the cell, eNOS mRNA is actively transcribed. The chromatin adopts an open confirmation with epigenetic marks that favor transcription (green). RNA polymerase II (purple) engages at the promoter and eNOS mRNA is transcribed. In the cytoplasm, eNOS mRNA is bound and stabilized in the cytoplasm by heterogeneous ribonucleoprotein E1 (E1). (B) Hypoxia—chromatin at the eNOS promoter adopts a closed conformation with epigenetic marks that favor transcription repression (red and amber circles). RNA polymerase II cannot bind and eNOS mRNA is not transcribed. In addition, heterogeneous ribonucleoprotein E1 (E1) is phosphorylated and is more abundant in the nucleus. Phosphorylated E1 cannot bind to and stabilize eNOS mRNA. eNOS mRNA is vulnerable to destabilizing factors. At the same time, the expression of the natural *cis*-antisense transcript sONE is increased with hypoxia and can bind to eNOS mRNA. eNOS mRNA stability is reduced and eNOS mRNA and protein levels decrease with hypoxia. (See the color plate.)

histone-3 lysine-4 and acetylation at both histone-3 lysine-9 and histone-4 lysine-12. Chromatin bearing these marks allows RNA polymerase II (Pol II) binding and active transcription. With acute hypoxia (1 h), these histones are evicted from the promoter along with Pol II. With longer duration of hypoxia (24 h), histones are reincorporated at the promoter but lack these activating histone marks. Rather, transcription is inhibited because the chromatin is reset and does not allow Pol II binding to the eNOS promoter (Fish et al., 2010; Fig. 7.5).

3.3.2 Posttranscription regulation of eNOS in hypoxia

The decrease in transcription of eNOS mRNA in hypoxia is accompanied by a decrease in eNOS mRNA stability. The 3′UTR of eNOS mRNA contains multiple pyrimidine-rich sequences that are conserved across species (Ho et al., 2013). Deletion of these sequences decreases the stability of eNOS mRNA, and addition of these sequences to a firefly luciferase reporter increased the stability of reporter mRNA. Stabilizing complexes can form at these 3′UTR sequences involving RNA-binding proteins, specifically the heterogeneous nuclear ribonucleoprotein E1 (hnRNP E1). These complexes protect the eNOS mRNA from destabilization from natural antisense RNA, microRNAs, and exogenously administered short-interfering RNAs (siRNAs). As a result, eNOS mRNA is stable under normoxic conditions with a half-life of greater than 24 h. Hypoxia disrupts these stabilizing complexes and leads to a decrease in eNOS mRNA stability. Akt-mediated serine phosphorylation of hnRNP E1 prevents the nuclear localization of hnRNP E1 and disrupts interaction of hnRNP E1 with eNOS mRNA. Thus, the 3′UTR of eNOS becomes accessible to destabilizing elements with hypoxia. At the same time, hypoxia increases the expression of a natural antisense transcript, sONE, that has exon sequence complementarity to 662 nucleotides of eNOS protein coding and 3′UTR sequences (Fish et al., 2007; Robb et al., 2004; Fig. 7.5). Together, the loss of stabilization of eNOS mRNA by hnRNP E1 and increase in the destabilizing antisense RNA, sONE, lead to a reduction in eNOS expression and eNOS-derived NO in hypoxic human endothelial cells *in vitro*.

3.4. eNOS regulation in *in vivo* models of hypoxia

These *in vitro* models of endothelial cells provide good evidence of eNOS mRNA regulatory mechanisms in hypoxia. *In vivo*, regulation of eNOS expression and function in a given vascular bed may be subjected to influences other than the availability of O_2, which varies by tissue specificity and disease conditions. eNOS regulation is also subject to shear stress, VEGF, and inflammation (Ho et al., 2012; Teichert et al., 2008). Clinical phenotypes may also be influenced by NO generation by other NOS enzymes. Thus, while the effect of hypoxia in isolation on endothelial cells is well studied *in vitro*, a variety of results can be found *in vivo*. Nevertheless, there are data that hypoxic rats have decreased endothelial NO production selectively from lung endothelium in both acute and chronic hypoxia (Shaul, Wells, & Horning, 1993). This can occur due to changes in the cellular localization and function of eNOS with hypoxia, without significant decreases in eNOS protein levels (Murata, Sato, Hori, Ozaki, & Karaki, 2002). Thus, the changes in eNOS function are multifactorial. A reduction in eNOS phosphorylation at Ser1177 and increased phosphorylation at Threonine-495 are associated with decreased eNOS function in hypoxia. eNOS protein is also found toward the periphery of the cell and is more likely to be colocalized with Arginase-2 which competes with eNOS for L-arginine as a substrate (Prieto et al., 2011). Interestingly, eNOS has the lowest K_m for O_2 of the three NOS enzymes and is biochemically the least dependent on changes in O_2 availability (Semenza, 2005). Disease models of eNOS function in hypoxia are complicated but generally demonstrate that eNOS is important for the vascular phenotype. A drop in NO production in lung vasculature with hypoxia is traditionally thought to play a role in hypoxic pulmonary vasoconstriction to acutely match ventilation and perfusion in a heterogeneously aerated lung. Pulmonary vasoconstriction in poorly aerated and relatively hypoxic lung units diverts blood flow toward better aerated and oxygenated lung units. Despite a decrease in eNOS-derived NO, an increase in eNOS protein may be an attempt to protect the lung vasculature in chronic hypoxia. Fagan et al. found that the increase in pathologic pulmonary artery muscularization and right ventricular systolic pressure in eNOS-null mice (eNOS (−/−)) exposed to hypoxia was rescued with inhaled NO (Fagan et al., 1999). In these mice, there is a dose–response relationship with progressively hypoxic conditions producing more pulmonary hypertension than seen with eNOS deletion (eNOS (−/−)) at baseline. Similarly in rats, there is an increase in whole lung eNOS protein levels when animals are hypoxic (Le Cras, Xue, Rengasamy, & Johns, 1996) and this increase in eNOS persists at 2 weeks posthypoxia exposure (Resta, Chicoine, Omdahl, & Walker, 1999). Others have found that eNOS-null mice have a decrease in pulmonary artery muscularization when exposed to hypoxia (Quinlan et al., 2000). While Fagan et al. did find an increase in iNOS mRNA expression in their system, neither study measured overall NO generation in the pulmonary vasculature. Those results would be important in interpreting the differences in phenotype between the studies.

In the coronary circulation, both epicardial and microvascular coronary vessels demonstrate increased eNOS protein and NO production during hypoxia. The mechanisms of eNOS upregulation, however, were different. While there was an increase of eNOS mRNA in the epicardial endothelium, eNOS mRNA did not change in the microvascular endothelium. The heart, along with the brain, is one of the organs prioritized for oxygen supply in systemic hypoxia. eNOS expression is one mechanism by which blood flow to the heart is increased with hypoxia (Justice, Tanner, & Myers, 2000).

3.5. O_2 dependency of NOS enzymes

Compared to nNOS and iNOS, eNOS enzymatic activity varies less with changes in O_2 concentration. The K_m for O_2 of eNOS is relatively low at 4 μM (Table 7.1; Semenza, 2005; Stuehr, Santolini, Wang, Wei, & Adak, 2004). Functionally, eNOS enzymatic activity is relatively unchanged throughout the physiological range of O_2 concentration. Whereas nNOS and iNOS function more as O_2-dependent switches, changes in NO production by eNOS are more titratable with gradual reductions in NO production with decreasing O_2 availability. From a physiological standpoint, this property may be important because endothelial cells are the first cells exposed to changes in O_2 availability (Fig. 7.3) and NO generation by eNOS may have implications both for vascular homeostasis locally and NO signaling more distally. Work that focuses on individual stimuli that either activate or repress the NOS system is an important step to understanding NOS regulatory mechanisms and NO generation.

Nevertheless, when studying NO biology *in vivo*, even in hypoxia, other factors including blood flow and inflammation may contribute to NOS enzyme expression and activity. Similarly, the HIF pathway has evolved to sense decrease O_2 bioavailability in two settings: (1) in the setting of

Table 7.1 K_m of nitric oxide synthases (NOSs) for oxygen

Nitric oxide synthase	K_m for O_2 (μM)
nNOS	350
iNOS	140
eNOS	4

hypoxia where there is a low partial pressure of oxygen such that HIF is stabilized by a reduction oxygen-dependent prolyl hydroxylase activity; (2) in the setting of anemia during which NO signaling and S-nitrosylation of pVHL prevent HIF degradation.

4. SUMMARY

In this chapter, we highlight the role of NO as a molecule that signals hypoxia. NO interacts with the cellular oxygen-sensing pathway and can stabilize the HIF. NO is also important for physiological responses to hypoxia. nNOS-derived NO improves survival in anemia and highlights important differences between anemia and hypoxia. We discuss S-nitrosylation of multiple proteins at cysteine residues to be a major mechanism by which NO has biological activity and by which NO can be produced locally and have an effect systemically. While NO can signal hypoxia, hypoxia itself has a role in the regulation of NOS enzyme expression and has variable effects on NOS isoform enzymatic activity. These relationships are important for temporal responses in acute and chronic hypoxia. NO regulation and activity occur in the context of other complementary physiological conditions and molecular effectors.

REFERENCES

Anand, P., & Stamler, J. S. (2012). Enzymatic mechanisms regulating protein S-nitrosylation: Implications in health and disease. *Journal of Molecular Medicine*, *90*(3), 233–244.

Beall, C. M. (2007). Two routes to functional adaptation: Tibetan and Andean high-altitude natives. *Proceedings of the National Academy of Sciences of the United States of America*, *104*(Suppl. 1), 8655–8660.

Beall, C. M., Cavalleri, G. L., Deng, L., Elston, R. C., Gao, Y., Knight, J., et al. (2010). Natural selection on EPAS1 (HIF2alpha) associated with low hemoglobin concentration in Tibetan highlanders. *Proceedings of the National Academy of Sciences of the United States of America*, *107*(25), 11459–11464.

Beattie, W. S., Karkouti, K., Wijeysundera, D. N., & Tait, G. (2009). Risk associated with preoperative anemia in noncardiac surgery: A single-center cohort study. *Anesthesiology*, *110*(3), 574–581.

Benhar, M., Forrester, M. T., & Stamler, J. S. (2009). Protein denitrosylation: Enzymatic mechanisms and cellular functions. *Nature Reviews. Molecular Cell Biology*, *10*(10), 721–732.

Berchner-Pfannschmidt, U., Yamac, H., Trinidad, B., & Fandrey, J. (2007). Nitric oxide modulates oxygen sensing by hypoxia-inducible factor 1-dependent induction of prolyl hydroxylase 2. *The Journal of Biological Chemistry*, *282*(3), 1788–1796.

Brahimi-Horn, M. C., Chiche, J., & Pouyssegur, J. (2007). Hypoxia and cancer. *Journal of Molecular Medicine*, *85*(12), 1301–1307.

Brahimi-Horn, M. C., & Pouyssegur, J. (2007). Oxygen, a source of life and stress. *FEBS Letters*, *581*(19), 3582–3591.

Branco-Price, C., Zhang, N., Schnelle, M., Evans, C., Katschinski, D. M., Liao, D., et al. (2012). Endothelial cell HIF-1alpha and HIF-2alpha differentially regulate metastatic success. *Cancer Cell, 21*(1), 52–65.

Bredt, D. S., Hwang, P. M., Glatt, C. E., Lowenstein, C., Reed, R. R., & Snyder, S. H. (1991). Cloned and expressed nitric oxide synthase structurally resembles cytochrome P-450 reductase. *Nature, 351*(6329), 714–718.

Bruick, R. K., & McKnight, S. L. (2001). A conserved family of prolyl-4-hydroxylases that modify HIF. *Science, 294*(5545), 1337–1340.

Burger, D. E., Lu, X., Lei, M., Xiang, F. L., Hammoud, L., Jiang, M., et al. (2009). Neuronal nitric oxide synthase protects against myocardial infarction-induced ventricular arrhythmia and mortality in mice. *Circulation, 120*(14), 1345–1354.

Burkard, N., Rokita, A. G., Kaufmann, S. G., Hallhuber, M., Wu, R., Hu, K., et al. (2007). Conditional neuronal nitric oxide synthase overexpression impairs myocardial contractility. *Circulation Research, 100*(3), e32–e44.

Carson, J. L., Duff, A., Poses, R. M., Berlin, J. A., Spence, R. K., Trout, R., et al. (1996). Effect of anaemia and cardiovascular disease on surgical mortality and morbidity. *Lancet, 348*(9034), 1055–1060.

Chan, G. C., Fish, J. E., Mawji, I. A., Leung, D. D., Rachlis, A. C., & Marsden, P. A. (2005). Epigenetic basis for the transcriptional hyporesponsiveness of the human inducible nitric oxide synthase gene in vascular endothelial cells. *Journal of Immunology, 175*(6), 3846–3861.

Chowdhury, R., Flashman, E., Mecinovic, J., Kramer, H. B., Kessler, B. M., Frapart, Y. M., et al. (2011). Studies on the reaction of nitric oxide with the hypoxia-inducible factor prolyl hydroxylase domain 2 (EGLN1). *Journal of Molecular Biology, 410*(2), 268–279.

Dawson, V. L., & Dawson, T. M. (1996). Nitric oxide neurotoxicity. *Journal of Chemical Neuroanatomy, 10*(3–4), 179–190.

Dawson, D., Lygate, C. A., Zhang, M. H., Hulbert, K., Neubauer, S., & Casadei, B. (2005). nNOS gene deletion exacerbates pathological left ventricular remodeling and functional deterioration after myocardial infarction. *Circulation, 112*(24), 3729–3737.

Deem, S., Hedges, R. G., McKinney, S., Polissar, N. L., Alberts, M. K., & Swenson, E. R. (1999). Mechanisms of improvement in pulmonary gas exchange during isovolemic hemodilution. *Journal of Applied Physiology, 87*(1), 132–141.

Doctor, A., Platt, R., Sheram, M. L., Eischeid, A., McMahon, T., Maxey, T., et al. (2005). Hemoglobin conformation couples erythrocyte S-nitrosothiol content to O2 gradients. *Proceedings of the National Academy of Sciences of the United States of America, 102*(16), 5709–5714.

Drueke, T. B., Locatelli, F., Clyne, N., Eckardt, K. U., Macdougall, I. C., Tsakiris, D., et al. (2006). Normalization of hemoglobin level in patients with chronic kidney disease and anemia. *The New England Journal of Medicine, 355*(20), 2071–2084.

Eichelbronner, O., D'Almeida, M., Sielenkamper, A., Sibbald, W. J., & Chin-Yee, I. H. (2002). Increasing P(50) does not improve DO(2CRIT) or systemic VO(2) in severe anemia. *American Journal of Physiology. Heart and Circulatory Physiology, 283*(1), H92–H101.

El Hasnaoui-Saadani, R., Pichon, A., Marchant, D., Olivier, P., Launay, T., Quidu, P., et al. (2009). Cerebral adaptations to chronic anemia in a model of erythropoietin-deficient mice exposed to hypoxia. *American Journal of Physiology. Regulatory, Integrative and Comparative Physiology, 296*(3), R801–R811.

Elayan, I. M., Axley, M. J., Prasad, P. V., Ahlers, S. T., & Auker, C. R. (2000). Effect of hyperbaric oxygen treatment on nitric oxide and oxygen free radicals in rat brain. *Journal of Neurophysiology, 83*(4), 2022–2029.

Eliasson, M. J., Huang, Z., Ferrante, R. J., Sasamata, M., Molliver, M. E., Snyder, S. H., et al. (1999). Neuronal nitric oxide synthase activation and peroxynitrite formation in ischemic stroke linked to neural damage. *The Journal of Neuroscience, 19*(14), 5910–5918.

Erzurum, S. C., Ghosh, S., Janocha, A. J., Xu, W., Bauer, S., Bryan, N. S., et al. (2007). Higher blood flow and circulating NO products offset high-altitude hypoxia among Tibetans. *Proceedings of the National Academy of Sciences of the United States of America, 104*(45), 17593–17598.

Evans, A. J., Russell, R. C., Roche, O., Burry, T. N., Fish, J. E., Chow, V. W., et al. (2007). VHL promotes E2 box-dependent E-cadherin transcription by HIF-mediated regulation of SIP1 and snail. *Molecular and Cellular Biology, 27*(1), 157–169.

Fagan, K. A., Fouty, B. W., Tyler, R. C., Morris, K. G., Jr., Hepler, L. K., Sato, K., et al. (1999). The pulmonary circulation of homozygous or heterozygous eNOS-null mice is hyperresponsive to mild hypoxia. *The Journal of Clinical Investigation, 103*(2), 291–299.

Feelisch, M., & Martin, J. F. (1995). The early role of nitric oxide in evolution. *Trends in Ecology & Evolution, 10*(12), 496–499.

Fish, J. E., Matouk, C. C., Yeboah, E., Bevan, S. C., Khan, M., Patil, K., et al. (2007). Hypoxia-inducible expression of a natural cis-antisense transcript inhibits endothelial nitric-oxide synthase. *The Journal of Biological Chemistry, 282*(21), 15652–15666.

Fish, J. E., Yan, M. S., Matouk, C. C., St Bernard, R., Ho, J. J., Gavryushova, A., et al. (2010). Hypoxic repression of endothelial nitric-oxide synthase transcription is coupled with eviction of promoter histones. *The Journal of Biological Chemistry, 285*(2), 810–826.

Forsythe, J. A., Jiang, B. H., Iyer, N. V., Agani, F., Leung, S. W., Koos, R. D., et al. (1996). Activation of vascular endothelial growth factor gene transcription by hypoxia-inducible factor 1. *Molecular and Cellular Biology, 16*(9), 4604–4613.

Foster, M. W., McMahon, T. J., & Stamler, J. S. (2003). S-nitrosylation in health and disease. *Trends in Molecular Medicine, 9*(4), 160–168.

Furchgott, R. F., & Zawadzki, J. V. (1980). The obligatory role of endothelial cells in the relaxation of arterial smooth muscle by acetylcholine. *Nature, 288*(5789), 373–376.

Glick, G., Plauth, W. H., Jr., & Braunwald, E. (1964). Role of the autonomic nervous system in the circulatory response to acutely induced anemia in unanesthetized dogs. *The Journal of Clinical Investigation, 43*, 2112–2124.

Gonzalez, D. R., Beigi, F., Treuer, A. V., & Hare, J. M. (2007). Deficient ryanodine receptor S-nitrosylation increases sarcoplasmic reticulum calcium leak and arrhythmogenesis in cardiomyocytes. *Proceedings of the National Academy of Sciences of the United States of America, 104*(51), 20612–20617.

Griffith, T. M., Edwards, D. H., Lewis, M. J., Newby, A. C., & Henderson, A. H. (1984). The nature of endothelium-derived vascular relaxant factor. *Nature, 308*(5960), 645–647.

Hagen, T., Taylor, C. T., Lam, F., & Moncada, S. (2003). Redistribution of intracellular oxygen in hypoxia by nitric oxide: Effect on HIF1alpha. *Science, 302*(5652), 1975–1978.

Halperin, M. L., Cheema-Dhadli, S., Lin, S. H., & Kamel, K. S. (2006). Properties permitting the renal cortex to be the oxygen sensor for the release of erythropoietin: Clinical implications. *Clinical Journal of the American Society of Nephrology, 1*(5), 1049–1053.

Hardison, R. (1998). Hemoglobins from bacteria to man: Evolution of different patterns of gene expression. *The Journal of Experimental Biology, 201*(Pt 8), 1099–1117.

Hebert, P. C., Wells, G., Blajchman, M. A., Marshall, J., Martin, C., Pagliarello, G., et al. (1999). A multicenter, randomized, controlled clinical trial of transfusion requirements in critical care. Transfusion Requirements in Critical Care Investigators, Canadian Critical Care Trials Group. *The New England Journal of Medicine, 340*(6), 409–417.

Hess, D. T., Matsumoto, A., Kim, S. O., Marshall, H. E., & Stamler, J. S. (2005). Protein S-nitrosylation: Purview and parameters. *Nature Reviews. Molecular Cell Biology*, *6*(2), 150–166.

Ho, J. J., Man, H. S., & Marsden, P. A. (2012). Nitric oxide signaling in hypoxia. *Journal of Molecular Medicine*, *90*(3), 217–231.

Ho, J. J., Robb, G. B., Tai, S. C., Turgeon, P. J., Mawji, I. A., Man, H. S., et al. (2013). Active stabilization of human endothelial nitric oxide synthase mRNA by hnRNP E1 protects against antisense RNA and microRNAs. *Molecular and Cellular Biology*, *33*(10), 2029–2046.

Hu, C. J., Wang, L. Y., Chodosh, L. A., Keith, B., & Simon, M. C. (2003). Differential roles of hypoxia-inducible factor 1alpha (HIF-1alpha) and HIF-2alpha in hypoxic gene regulation. *Molecular and Cellular Biology*, *23*(24), 9361–9374.

Huang, L. E., Gu, J., Schau, M., & Bunn, H. F. (1998). Regulation of hypoxia-inducible factor 1alpha is mediated by an O2-dependent degradation domain via the ubiquitin-proteasome pathway. *Proceedings of the National Academy of Sciences of the United States of America*, *95*(14), 7987–7992.

Hudetz, A. G., Wood, J. D., & Kampine, J. P. (2000). 7-Nitroindazole impedes erythrocyte flow response to isovolemic hemodilution in the cerebral capillary circulation. *Journal of Cerebral Blood Flow and Metabolism*, *20*(2), 220–224.

Ickx, B. E., Rigolet, M., & Van Der Linden, P. J. (2000). Cardiovascular and metabolic response to acute normovolemic anemia. Effects of anesthesia. *Anesthesiology*, *93*(4), 1011–1016.

Ignarro, L. J. (1999). Nitric oxide: A unique endogenous signaling molecule in vascular biology. *Bioscience Reports*, *19*(2), 51–71.

Ivan, M., & Kaelin, W. G., Jr. (2001). The von Hippel-Lindau tumor suppressor protein. *Current Opinion in Genetics & Development*, *11*(1), 27–34.

Jamal, A., Man, H. S., & Marsden, P. A. (2012). Gene regulation in the vascular endothelium: Why epigenetics is important for the kidney. *Seminars in Nephrology*, *32*(2), 176–184.

Jia, L., Bonaventura, C., Bonaventura, J., & Stamler, J. S. (1996). S-nitrosohaemoglobin: A dynamic activity of blood involved in vascular control. *Nature*, *380*(6571), 221–226.

Johannes, T., Mik, E. G., Nohe, B., Unertl, K. E., & Ince, C. (2007). Acute decrease in renal microvascular PO_2 during acute normovolemic hemodilution. *American Journal of Physiology. Renal Physiology*, *292*(2), F796–F803.

Johannsson, H., & Siesjo, B. K. (1974). Blood flow and oxygen consumption in the rat brain in dilutional anemia. *Acta Physiologica Scandinavica*, *91*(1), 136–138.

Justice, J. M., Tanner, M. A., & Myers, P. R. (2000). Endothelial cell regulation of nitric oxide production during hypoxia in coronary microvessels and epicardial arteries. *Journal of Cellular Physiology*, *182*(3), 359–365.

Kaelin, W. G., Jr. (2007). The von Hippel-Lindau tumor suppressor protein and clear cell renal carcinoma. *Clinical Cancer Research*, *13*(2 Pt 2), 680s–684s.

Kaelin, W. G., Jr. (2008). The von Hippel-Lindau tumour suppressor protein: O2 sensing and cancer. *Nature Reviews. Cancer*, *8*(11), 865–873.

Kallio, P. J., Wilson, W. J., O'Brien, S., Makino, Y., & Poellinger, L. (1999). Regulation of the hypoxia-inducible transcription factor 1alpha by the ubiquitin-proteasome pathway. *The Journal of Biological Chemistry*, *274*(10), 6519–6525.

Kobayashi, Y. M., Rader, E. P., Crawford, R. W., Iyengar, N. K., Thedens, D. R., Faulkner, J. A., et al. (2008). Sarcolemma-localized nNOS is required to maintain activity after mild exercise. *Nature*, *456*(7221), 511–515.

Koritzinsky, M., Magagnin, M. G., van den Beucken, T., Seigneuric, R., Savelkouls, K., Dostie, J., et al. (2006). Gene expression during acute and prolonged hypoxia is regulated by distinct mechanisms of translational control. *The EMBO Journal*, *25*(5), 1114–1125.

Kuhlencordt, P. J., Hotten, S., Schodel, J., Rutzel, S., Hu, K., Widder, J., et al. (2006). Atheroprotective effects of neuronal nitric oxide synthase in apolipoprotein e knockout mice. *Arteriosclerosis, Thrombosis, and Vascular Biology*, *26*(7), 1539–1544.

Le Cras, T. D., Xue, C., Rengasamy, A., & Johns, R. A. (1996). Chronic hypoxia upregulates endothelial and inducible NO synthase gene and protein expression in rat lung. *The American Journal of Physiology*, *270*(1 Pt 1), L164–L170.

Li, F., Sonveaux, P., Rabbani, Z. N., Liu, S., Yan, B., Huang, Q., et al. (2007). Regulation of HIF-1alpha stability through S-nitrosylation. *Molecular Cell*, *26*(1), 63–74.

Lima, B., Lam, G. K., Xie, L., Diesen, D. L., Villamizar, N., Nienaber, J., et al. (2009). Endogenous S-nitrosothiols protect against myocardial injury. *Proceedings of the National Academy of Sciences of the United States of America*, *106*(15), 6297–6302.

Lipton, A. J., Johnson, M. A., Macdonald, T., Lieberman, M. W., Gozal, D., & Gaston, B. (2001). S-nitrosothiols signal the ventilatory response to hypoxia. *Nature*, *413*(6852), 171–174.

Liu, L., Yan, Y., Zeng, M., Zhang, J., Hanes, M. A., Ahearn, G., et al. (2004). Essential roles of S-nitrosothiols in vascular homeostasis and endotoxic shock. *Cell*, *116*(4), 617–628.

Lowenstein, C. J., Glatt, C. S., Bredt, D. S., & Snyder, S. H. (1992). Cloned and expressed macrophage nitric oxide synthase contrasts with the brain enzyme. *Proceedings of the National Academy of Sciences of the United States of America*, *89*(15), 6711–6715.

Marsden, P. A. (2007). Low-molecular-weight S-nitrosothiols and blood vessel injury. *The Journal of Clinical Investigation*, *117*(9), 2377–2380.

Marsden, P. A., Schappert, K. T., Chen, H. S., Flowers, M., Sundell, C. L., Wilcox, J. N., et al. (1992). Molecular cloning and characterization of human endothelial nitric oxide synthase. *FEBS Letters*, *307*(3), 287–293.

Matouk, C. C., & Marsden, P. A. (2008). Epigenetic regulation of vascular endothelial gene expression. *Circulation Research*, *102*(8), 873–887.

Matsuoka, T., Saiki, C., & Mortola, J. P. (1994). Metabolic and ventilatory responses to anemic hypoxia in conscious rats. *Journal of Applied Physiology*, *77*(3), 1067–1072.

Maxwell, P. H., Wiesener, M. S., Chang, G. W., Clifford, S. C., Vaux, E. C., Cockman, M. E., et al. (1999). The tumour suppressor protein VHL targets hypoxia-inducible factors for oxygen-dependent proteolysis. *Nature*, *399*(6733), 271–275.

McLaren, A. T., Marsden, P. A., Mazer, C. D., Baker, A. J., Stewart, D. J., Tsui, A. K., et al. (2007). Increased expression of HIF-1alpha, nNOS, and VEGF in the cerebral cortex of anemic rats. *American Journal of Physiology. Regulatory, Integrative and Comparative Physiology*, *292*(1), R403–R414.

McQuillan, L. P., Leung, G. K., Marsden, P. A., Kostyk, S. K., & Kourembanas, S. (1994). Hypoxia inhibits expression of eNOS via transcriptional and posttranscriptional mechanisms. *The American Journal of Physiology*, *267*(5 Pt 2), H1921–H1927.

Melillo, G., Musso, T., Sica, A., Taylor, L. S., Cox, G. W., & Varesio, L. (1995). A hypoxia-responsive element mediates a novel pathway of activation of the inducible nitric oxide synthase promoter. *The Journal of Experimental Medicine*, *182*(6), 1683–1693.

Michel, T., & Feron, O. (1997). Nitric oxide synthases: Which, where, how, and why? *The Journal of Clinical Investigation*, *100*(9), 2146–2152.

Mole, D. R., Blancher, C., Copley, R. R., Pollard, P. J., Gleadle, J. M., Ragoussis, J., et al. (2009). Genome-wide association of hypoxia-inducible factor (HIF)-1alpha and HIF-2alpha DNA binding with expression profiling of hypoxia-inducible transcripts. *The Journal of Biological Chemistry*, *284*(25), 16767–16775.

Moncada, S., & Higgs, A. (1993). The L-arginine-nitric oxide pathway. *The New England Journal of Medicine*, *329*(27), 2002–2012.

Morishita, T., Tsutsui, M., Shimokawa, H., Horiuchi, M., Tanimoto, A., Suda, O., et al. (2002). Vasculoprotective roles of neuronal nitric oxide synthase. *FASEB Journal*, *16*(14), 1994–1996.

Murata, T., Sato, K., Hori, M., Ozaki, H., & Karaki, H. (2002). Decreased endothelial nitric-oxide synthase (eNOS) activity resulting from abnormal interaction between eNOS and its regulatory proteins in hypoxia-induced pulmonary hypertension. *The Journal of Biological Chemistry, 277*(46), 44085–44092.

Murray, J. F., & Rapaport, E. (1972). Coronary blood flow and myocardial metabolism in acute experimental anaemia. *Cardiovascular Research, 6*(4), 360–367.

Newton, D. C., Bevan, S. C., Choi, S., Robb, G. B., Millar, A., Wang, Y., et al. (2003). Translational regulation of human neuronal nitric-oxide synthase by an alternatively spliced 5'-untranslated region leader exon. *The Journal of Biological Chemistry, 278*(1), 636–644.

O'Rourke, J. F., Tian, Y. M., Ratcliffe, P. J., & Pugh, C. W. (1999). Oxygen-regulated and transactivating domains in endothelial PAS protein 1: Comparison with hypoxia-inducible factor-1alpha. *The Journal of Biological Chemistry, 274*(4), 2060–2071.

Ostergaard, L., Stankevicius, E., Andersen, M. R., Eskildsen-Helmond, Y., Ledet, T., Mulvany, M. J., et al. (2007). Diminished NO release in chronic hypoxic human endothelial cells. *American Journal of Physiology Heart and Circulatory Physiology, 293*(5), H2894–H2903.

Ouiddir, A., Planes, C., Fernandes, I., VanHesse, A., & Clerici, C. (1999). Hypoxia upregulates activity and expression of the glucose transporter GLUT1 in alveolar epithelial cells. *American Journal of Respiratory Cell and Molecular Biology, 21*(6), 710–718.

Palmer, L. A., Doctor, A., Chhabra, P., Sheram, M. L., Laubach, V. E., Karlinsey, M. Z., et al. (2007). S-nitrosothiols signal hypoxia-mimetic vascular pathology. *The Journal of Clinical Investigation, 117*(9), 2592–2601.

Percival, J. M., Anderson, K. N., Gregorevic, P., Chamberlain, J. S., & Froehner, S. C. (2008). Functional deficits in nNOSmu-deficient skeletal muscle: Myopathy in nNOS knockout mice. *PLoS One, 3*(10), e3387.

Percival, J. M., Anderson, K. N., Huang, P., Adams, M. E., & Froehner, S. C. (2010). Golgi and sarcolemmal neuronal NOS differentially regulate contraction-induced fatigue and vasoconstriction in exercising mouse skeletal muscle. *The Journal of Clinical Investigation, 120*(3), 816–826.

Pfeffer, M. A., Burdmann, E. A., Chen, C. Y., Cooper, M. E., de Zeeuw, D., Eckardt, K. U., et al. (2009). A trial of darbepoetin alfa in type 2 diabetes and chronic kidney disease. *The New England Journal of Medicine, 361*(21), 2019–2032.

Prieto, C. P., Krause, B. J., Quezada, C., San Martin, R., Sobrevia, L., & Casanello, P. (2011). Hypoxia-reduced nitric oxide synthase activity is partially explained by higher arginase-2 activity and cellular redistribution in human umbilical vein endothelium. *Placenta, 32*(12), 932–940.

Quinlan, T. R., Li, D., Laubach, V. E., Shesely, E. G., Zhou, N., & Johns, R. A. (2000). eNOS-deficient mice show reduced pulmonary vascular proliferation and remodeling to chronic hypoxia. *American Journal of Physiology Lung Cellular and Molecular Physiology, 279*(4), L641–L650.

Ragoonanan, T. E., Beattie, W. S., Mazer, C. D., Tsui, A. K., Leong-Poi, H., Wilson, D. F., et al. (2009). Metoprolol reduces cerebral tissue oxygen tension after acute hemodilution in rats. *Anesthesiology, 111*(5), 988–1000.

Raval, R. R., Lau, K. W., Tran, M. G., Sowter, H. M., Mandriota, S. J., Li, J. L., et al. (2005). Contrasting properties of hypoxia-inducible factor 1 (HIF-1) and HIF-2 in von Hippel-Lindau-associated renal cell carcinoma. *Molecular and Cellular Biology, 25*(13), 5675–5686.

Resta, T. C., Chicoine, L. G., Omdahl, J. L., & Walker, B. R. (1999). Maintained upregulation of pulmonary eNOS gene and protein expression during recovery from chronic hypoxia. *The American Journal of Physiology, 276*(2 Pt 2), H699–H708.

Robb, G. B., Carson, A. R., Tai, S. C., Fish, J. E., Singh, S., Yamada, T., et al. (2004). Post-transcriptional regulation of endothelial nitric-oxide synthase by an overlapping antisense mRNA transcript. *The Journal of Biological Chemistry*, *279*(36), 37982–37996.

Robinson, M. A., Baumgardner, J. E., & Otto, C. M. (2011). Oxygen-dependent regulation of nitric oxide production by inducible nitric oxide synthase. *Free Radical Biology & Medicine*, *51*(11), 1952–1965.

Saraiva, R. M., Minhas, K. M., Raju, S. V., Barouch, L. A., Pitz, E., Schuleri, K. H., et al. (2005). Deficiency of neuronal nitric oxide synthase increases mortality and cardiac remodeling after myocardial infarction: Role of nitroso-redox equilibrium. *Circulation*, *112*(22), 3415–3422.

Semenza, G. L. (2005). New insights into nNOS regulation of vascular homeostasis. *The Journal of Clinical Investigation*, *115*(11), 2976–2978.

Semenza, G. L. (2009). Regulation of oxygen homeostasis by hypoxia-inducible factor 1. *Physiology (Bethesda)*, *24*, 97–106.

Semenza, G. L. (2012). Hypoxia-inducible factors in physiology and medicine. *Cell*, *148*(3), 399–408.

Semenza, G. L., & Wang, G. L. (1992). A nuclear factor induced by hypoxia via de novo protein synthesis binds to the human erythropoietin gene enhancer at a site required for transcriptional activation. *Molecular and Cellular Biology*, *12*(12), 5447–5454.

Shaked, H., Hofseth, L. J., Chumanevich, A., Chumanevich, A. A., Wang, J., Wang, Y., et al. (2012). Chronic epithelial NF-kappaB activation accelerates APC loss and intestinal tumor initiation through iNOS up-regulation. *Proceedings of the National Academy of Sciences of the United States of America*, *109*(35), 14007–14012.

Shaul, P. W., Wells, L. B., & Horning, K. M. (1993). Acute and prolonged hypoxia attenuate endothelial nitric oxide production in rat pulmonary arteries by different mechanisms. *Journal of Cardiovascular Pharmacology*, *22*(6), 819–827.

Shih, S. C., & Claffey, K. P. (1998). Hypoxia-mediated regulation of gene expression in mammalian cells. *International Journal of Experimental Pathology*, *79*(6), 347–357.

Simonson, T. S., Yang, Y., Huff, C. D., Yun, H., Qin, G., Witherspoon, D. J., et al. (2010). Genetic evidence for high-altitude adaptation in Tibet. *Science*, *329*(5987), 72–75.

Singh, A. K., Szczech, L., Tang, K. L., Barnhart, H., Sapp, S., Wolfson, M., et al. (2006). Correction of anemia with epoetin alfa in chronic kidney disease. *The New England Journal of Medicine*, *355*(20), 2085–2098.

Sladen, R. N. (1981). The oxyhemoglobin dissociation curve. *International Anesthesiology Clinics*, *19*(3), 39–70.

Sproule, B. J., Mitchell, J. H., & Miller, W. F. (1960). Cardiopulmonary physiological responses to heavy exercise in patients with anemia. *The Journal of Clinical Investigation*, *39*, 378–388.

Stuehr, D. J., Santolini, J., Wang, Z. Q., Wei, C. C., & Adak, S. (2004). Update on mechanism and catalytic regulation in the NO synthases. *The Journal of Biological Chemistry*, *279*(35), 36167–36170.

Takeda, N., O'Dea, E. L., Doedens, A., Kim, J. W., Weidemann, A., Stockmann, C., et al. (2010). Differential activation and antagonistic function of HIF-{alpha} isoforms in macrophages are essential for NO homeostasis. *Genes & Development*, *24*(5), 491–501.

Taylor, C. T., & McElwain, J. C. (2010). Ancient atmospheres and the evolution of oxygen sensing via the hypoxia-inducible factor in metazoans. *Physiology (Bethesda)*, *25*(5), 272–279.

Teichert, A. M., Scott, J. A., Robb, G. B., Zhou, Y. Q., Zhu, S. N., Lem, M., et al. (2008). Endothelial nitric oxide synthase gene expression during murine embryogenesis: Commencement of expression in the embryo occurs with the establishment of a unidirectional circulatory system. *Circulation Research*, *103*(1), 24–33.

Tsui, A. K., Marsden, P. A., Mazer, C. D., Adamson, S. L., Henkelman, R. M., Ho, J. J., et al. (2011). Priming of hypoxia-inducible factor by neuronal nitric oxide synthase is essential for adaptive responses to severe anemia. *Proceedings of the National Academy of Sciences of the United States of America, 108*(42), 17544–17549.
van Bommel, J., Trouwborst, A., Schwarte, L., Siegemund, M., Ince, C., & Henny Ch, P. (2002). Intestinal and cerebral oxygenation during severe isovolemic hemodilution and subsequent hyperoxic ventilation in a pig model. *Anesthesiology, 97*(3), 660–670.
von Restorff, W., Hofling, B., Holtz, J., & Bassenge, E. (1975). Effect of increased blood fluidity through hemodilution on coronary circulation at rest and during exercise in dogs. *Pflügers Archiv, 357*(1–2), 15–24.
Wang, G. L., Jiang, B. H., Rue, E. A., & Semenza, G. L. (1995). Hypoxia-inducible factor 1 is a basic-helix-loop-helix-PAS heterodimer regulated by cellular O2 tension. *Proceedings of the National Academy of Sciences of the United States of America, 92*(12), 5510–5514.
Wang, Y., Newton, D. C., & Marsden, P. A. (1999). Neuronal NOS: Gene structure, mRNA diversity, and functional relevance. *Critical Reviews in Neurobiology, 13*(1), 21–43.
Wang, Y., Newton, D. C., Miller, T. L., Teichert, A. M., Phillips, M. J., Davidoff, M. S., et al. (2002). An alternative promoter of the human neuronal nitric oxide synthase gene is expressed specifically in Leydig cells. *The American Journal of Pathology, 160*(1), 369–380.
Wang, Y., Roche, O., Yan, M. S., Finak, G., Evans, A. J., Metcalf, J. L., et al. (2009). Regulation of endocytosis via the oxygen-sensing pathway. *Nature Medicine, 15*(3), 319–324.
Ward, M. E., Toporsian, M., Scott, J. A., Teoh, H., Govindaraju, V., Quan, A., et al. (2005). Hypoxia induces a functionally significant and translationally efficient neuronal NO synthase mRNA variant. *The Journal of Clinical Investigation, 115*(11), 3128–3139.
Weiskopf, R. B., Viele, M. K., Feiner, J., Kelley, S., Lieberman, J., Noorani, M., et al. (1998). Human cardiovascular and metabolic response to acute, severe isovolemic anemia. *JAMA: The Journal of the American Medical Association, 279*(3), 217–221.
Weitzdoerfer, R., Hoeger, H., Engidawork, E., Engelmann, M., Singewald, N., Lubec, G., et al. (2004). Neuronal nitric oxide synthase knock-out mice show impaired cognitive performance. *Nitric Oxide, 10*(3), 130–140.
Wolff, C. B. (2007). Normal cardiac output, oxygen delivery and oxygen extraction. *Advances in Experimental Medicine and Biology, 599*, 169–182.
Wood, S. M., Gleadle, J. M., Pugh, C. W., Hankinson, O., & Ratcliffe, P. J. (1996). The role of the aryl hydrocarbon receptor nuclear translocator (ARNT) in hypoxic induction of gene expression. Studies in ARNT-deficient cells. *The Journal of Biological Chemistry, 271*(25), 15117–15123.
Worldwide prevalence of anaemia report 1993–2005. WHO global database on anaemia. (2008). Geneva: World Health Organization.
Xu, X., Russell, T., Bazner, J., & Hamilton, J. (2001). NMDA receptor antagonist AP5 and nitric oxide synthase inhibitor 7-NI affect different phases of learning and memory in goldfish. *Brain Research, 889*(1–2), 274–277.

FURTHER READING

Semenza, G. L. (2011). Oxygen sensing, homeostasis, and disease. *The New England Journal of Medicine, 365*(6), 537–547.

CHAPTER EIGHT

NO Binding to the Proapoptotic Cytochrome *c*–Cardiolipin Complex

Michael A. Hough, Gary Silkstone, J.A.R. Worrall, Michael T. Wilson[1]
School of Biological Sciences, University of Essex, Colchester, United Kingdom
[1]Corresponding author: e-mail address: wilsmt@essex.ac.uk

Contents

Abstract

Cytochrome *c* is a heme protein that is localized in the compartment between the inner and outer mitochondrial membranes where it functions to transfer electrons between complex III and complex IV of the respiratory chain. It can also form an intimate association with the mitochondrion-specific phospholipid cardiolipin that induces a conformational change in the protein enabling it to act as a peroxidase catalyzing the oxidation of cardiolipin and thereby instigating a chain of events that leads to apoptosis. Unlike the native protein, cytochrome *c* within the complex binds ligands rapidly; in particular, NO can coordinate to either the ferric or ferrous iron of the heme. Remarkably, in the ferrous form, NO binds preferentially to the proximal side of the heme and thus behaves in a way similar to cytochrome *c′*-type proteins and to guanylate cyclase. The implications of NO binding to the proapoptotic cytochrome *c*/cardiolipin complex are discussed in terms of modulating the apoptotic response and buffering NO concentrations. Insights into the structure of the complex are provided by comparison with cytochrome *c′* for which X-ray structures are available.

Vitamins and Hormones, Volume 96
ISSN 0083-6729
http://dx.doi.org/10.1016/B978-0-12-800254-4.00008-8

1. INTRODUCTION

Nitric oxide (NO), although a neutral gaseous molecule comprising only two atoms, has an important function in biological systems. It plays a key role in a variety of biological processes including vasodilation, neuronal function, inflammation, and immune function (Ignarro, 1990). For example, NO exerts its effects on blood pressure via the catalytic activity of the soluble heme-containing enzyme guanylate cyclase (sGC), to which NO binds as the activation signal (Poulos, 2006). Binding of NO induces a conformational transition in the enzyme, involving dissociation of a His residue coordinated to the iron, which leads to catalysis of the conversion of GTP to cyclic GMP. This in turn activates protein kinase G which causes epithelial relaxation by opening calcium-activated potassium channels (Robertson, Schubert, Hescheler, & Nelson, 1993).

It has also been demonstrated that NO is involved in the regulation of apoptosis (programmed cell death), having a dual role, stimulatory in some cells and inhibitory in others, through mechanisms not fully elucidated (Chung, Pae, Choi, Billiar, & Kim, 2001; Choi, Pae, Jang, Kim, & Chung, 2002). Due to its hydrophobic nature, which enables easy diffusion in biomembranes (Radi, Cassina, Hodara, Quijano, & Castro, 2002; Rubbo et al., 1994), NO can reach mitochondria, where it is known to have a complex inhibitory effect on cytochrome *c* (cyt *c*) oxidase (Mason, Nicholls, Wilson, & Cooper, 2006), the major oxygen-metabolizing enzyme of the cell and responsible for cellular respiration. It may also be produced in mitochondria by the recently reported mitochondrial NO synthase (Ghafourifar & Cadenas, 2005; Haynes, Elfering, Traaseth, & Giulivi, 2004).

A recent development in this complex story is the finding that NO can bind to the heme iron of cyt *c* when this protein is associated with certain phospholipids. As these complexes are now considered to play an important role in the initiation of apoptosis, this gives rise to the proposal that NO may play a significant role in one of the key apoptotic events, namely, the release of cyt *c* from the mitochondrion (Orrenius, Gogvadze, & Zhivotovsky, 2007).

This review focuses on the nature of the interactions between NO and specific complexes of cyt *c* with the mitochondrial phospholipid, cardiolipin (CL). Through comparison with other heme proteins that bind NO and for which X-ray structures exist, we are able to make proposals regarding the heme site in the cyt *c*–CL complex. Finally, we consider how NO may regulate the apoptotic process and how these complexes may act to buffer the NO concentration in the mitochondrion.

2. THE cyt *c*–CL COMPLEX

The importance of the interactions of phospholipids with water-soluble proteins has been largely understated, yet such interactions are critical in regulating many biological pathways (Cantley, 2002; Green & Kroemer, 2004). An example of such an interaction is that between mitochondrial cyt *c* and CL (Kagan et al., 2005; Rajagopal, Silkstone, Nicholls, Wilson, & Worrall, 2012).

Mitochondria are bounded by a double membrane, which creates two compartments: the intermembrane space and the matrix. The inner mitochondrial membrane contains a phospholipid, diphosphatidylglycerol, commonly referred to as CL which has a structure comprising two phosphatidylglycerols connected by a glycerol molecule to form a dimeric structure with four fatty acid chains (Fig. 8.1). This phospholipid is important in the function of multiple enzymes involved in mitochondrial energy metabolism (Hasan, Yamashita, Ryan, Whitelegge, & Cramer, 2011) and is central to triggering the early stages of apoptosis (Kagan et al., 2005).

Mitochondrial cyt *c* is a soluble multifunctional heme-containing protein that plays a key role in life and death decisions of the cell (Huttemann et al., 2011). For life-sustaining energy production, cyt *c* acts as an electron carrier (redox protein) in the mitochondrial electron transfer chain leading to the production of adenosine triphosphate. In this role, cyt *c* has a six-coordinate heme iron with His and Met residues as axial ligands (Fig. 8.2) and a Fe(III)/Fe(II) heme redox potential of +260 mV. When problems caused by stress, e.g., oxidative stress, outweigh the benefits of sustaining viability of a cell,

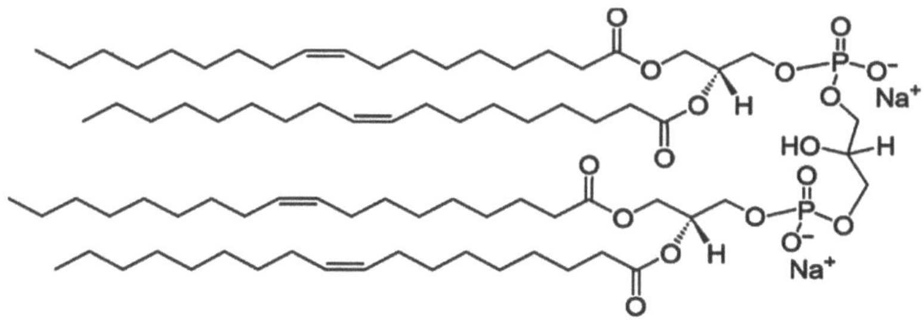

Figure 8.1 The molecular structure of the phospholipid, cardiolipin (CL).

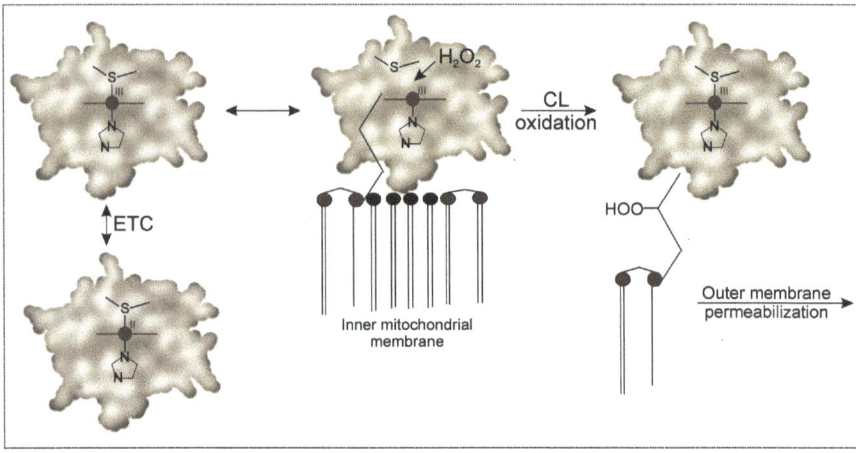

Figure 8.2 Schematic of the functional roles of human cyt *c* in the inner mitochondrial membrane. A small fraction of cyt *c* is considered to be bound to CL and transformed to a peroxidase by dissociation of the intrinsic methionine ligand through the insertion of one (or two) of the hydrocarbon chains. Under those oxidative stress conditions in which H_2O_2 is formed, this may oxidize the hydrocarbon catalyzed by the heme iron. The resulting lipid peroxide dissociates from the protein and eventually leads to opening of the transition pore initiating apoptosis. CL in the membrane is depicted in blue. *Adapted from figure 2B published in Journal of Biological Chemistry. Gary Silkstone, Sofia M. Kapetanaki, Ivan Husu, Marten H. Vos, Michael T. Wilson. Nitric Oxide Binds to the Proximal Heme Coordination Site of the Ferrocytochrome c/Cardiolipin Complex: FORMATION MECHANISM AND DYNAMICS. Journal of Biological Chemistry. 2010; Vol: pp.19785-19792. © the American Society for Biochemistry and Molecular Biology.* (See the color plate.)

various signals can initiate the programmed cell death process through intrinsic (mitochondrial) type II apoptosis. In a small fraction of the ferri-cyt *c*, the CL, by association with the protein, switches the heme protein's function from that of an electron transfer protein to a peroxidase, thus by creating a protein–phospholipid assembly cyt *c* is converted to a redox enzyme (Kagan et al., 2005; Fig. 8.2). CL achieves this by binding to the protein through electrostatic interactions involving the positively charged cyt *c* molecule and the anionic phosphate head groups of CL, and via a hydrophobic interaction which arises from a fatty acid acyl chain leaving the lipid bilayer and inserting into the interior of the cyt *c* molecule (Fig. 8.2; Kinnunen, 1992; Kinnunen, Koiv, Lehtonen, Rytomaa, & Mustonen, 1994; Rytomaa & Kinnunen, 1995; Stewart, Blakely, & Johnson, 2000; Tuominen, Wallace, & Kinnunen, 2002).

These interactions trigger the dissociation of the axial heme-iron Met ligand to create a five-coordinate heme iron, which readily reacts with the reactive oxygen species, H_2O_2, when this becomes available from, for example, the leakage of electrons from complex I to oxygen. The resulting interaction of peroxide with cyt *c*/CL complex leads to the oxidation of a fatty acid chain of CL (Kagan et al., 2005; Fig. 8.2). Oxidation of CL causes the cyt *c*/CL complex to dissociate and the modified CL then serves as a signal for downstream events, culminating in the permeabilization of the outer mitochondrial membrane and subsequent release of proapoptotic factors into the cytoplasm (Ow, Green, Hao, & Mak, 2008). These factors include cyt *c* itself, which is then required for the assembly of a multisubunit protease activation complex called the apoptosome (Riedl & Salvesen, 2007; Zou, Li, Liu, & Wang, 1999). Thus, the cyt *c*/CL complex, when activated by H_2O_2 produced, say, under conditions of oxidative stress, sets in motion a series of events that lead to the "point-of-no return" in the mitochondrial apoptotic pathway, mitochondrial outer membrane permeabilization (Ow et al., 2008).

Cyt *c* release from mitochondria into the cytosol of the cell is most notably associated with the assembly of the apoptosome (Liu, Kim, Yang, Jemmerson, & Wang, 1996). However, it has also been reported that peroxidase activity of cyt *c* in the cytosol can be induced by a phospholipid, phosphatidylserine (PS), located in the plasma membrane (Kagan et al., 2000; Tyurina et al., 2000). PS has a structure comprising one PS and two fatty acid chains as opposed to four in CL. PS has been observed to undergo oxidation in apoptotic cells that is temporally separated from CL oxidation (Arroyo et al., 2002; Kagan et al., 2002; Tyurina, Kawai, et al., 2004; Tyurina, Serinkan, et al., 2004). The resulting peroxidation of PS facilitates its externalization from the cytosolic side of the plasma membrane to the outer cell surface, essential for the recognition and clearance of apoptotic cells by macrophages (Fadok, de Cathelineau, Daleke, Henson, & Bratton, 2001; Kagan et al., 2002). Thus, cyt *c* phospholipid complexes appear to be functionally critical to the type II apoptosis pathway.

3. LIGAND BINDING TO THE cyt *c*/CL COMPLEX

An important consequence of the displacement of methionine from coordination to the heme iron (see Fig. 8.2) is that, unlike the CL-free protein, cyt *c* in association with the phospholipid now readily binds extraneous ligands. The cyt *c*/CL complex, once formed, may be reduced to the ferrous form and now behaves in a similar fashion to other penta-coordinate heme proteins that allow easy access to the iron. For example, the ferro-cyt *c*/CL complex binds CO rapidly and with high affinity to the iron (Kapetanaki et al., 2009; Rajagopal et al., 2012). The second-order rate constant for CO binding is $\sim 10^7\ M^{-1}\ s^{-1}$, larger than the comparable constant for myoglobin ($\sim 10^5\ M^{-1}\ s^{-1}$). This indicates an open and sterically unhindered heme pocket on the distal side, a conclusion that is supported by the high quantum efficiency of photolysis of CO from the heme into the bulk phase, very different from the Met80Ala mutant of cyt *c* which provides a penta-coordinate but sterically highly hindered route for ligand egress from the heme site (Kapetanaki et al., 2009; Silkstone, Jasaitis, Wilson, & Vos, 2007). In keeping with all heme proteins that bind CO the cyt *c*/CL complex also binds NO. However, in contrast to CO and O_2, that only have affinity for ferrous heme, NO binds to both the ferric and ferrous species.

3.1. NO binding to cyt *c*/CL complex

Although methionine is displaced from coordination to heme in the cyt *c*/CL complex, the sixth coordination site in the ferric form is not vacant. Through a combination of magnetic circular dichroism spectroscopy and potentiometric titrations, it has been demonstrated that the heme group of the cyt *c*/CL complex in the ferric form at pH 7.4 exists as two forms with either His/Lys- or His/OH^--coordination sets (Bradley, Silkstone, Wilson, Cheesman, & Butt, 2011).

Even though the sixth coordination site is filled either by a Lys nitrogen atom or by an OH^- ion, this does not preclude NO binding as this ligand has the highest affinity and can displace the others from coordination. However, the rate of NO binding is now limited by the rate of dissociation of the Lys/OH^- ligands. This rate has been determined to be ~5 s^{-1}; thus, NO binding is slow and the rate is independent of the NO concentration. The intrinsic rate constant for NO binding to the ferric heme, determined by laser photolysis, is ~10^6 M^{-1} s^{-1} and the dissociation rate constant, although not directly measured, may be estimated from these experiments to be ~0.01 s^{-1}, yielding an intrinsic affinity, typical for heme proteins of ~10^8 M^{-1} (Silkstone, Kapetanaki, Husu, Vos, & Wilson, 2012). As the Lys/OH^- ligands compete with NO for the site, the overall affinity for NO may be estimated as ~10^6 M^{-1}.

The mechanism of binding of NO to the ferrous form of the cyt *c*/CL is more complex than might have been expected from a knowledge of the CO-binding properties and of the behavior of the Met80 mutants of cyt *c*. First, the NO complex of ferro-cyt *c*/CL has spectroscopic properties that indicate that the heme is penta-coordinate and thus the histidine residue that binds to the proximal coordination site of the heme has dissociated. The Soret band in the optical spectrum is relatively broad and centered at 395 nm (see spectrum V in Fig. 8.3A), while hexa-coordinate NO adducts of *c*-type cytochromes have bands typically in the 415–420 nm range.

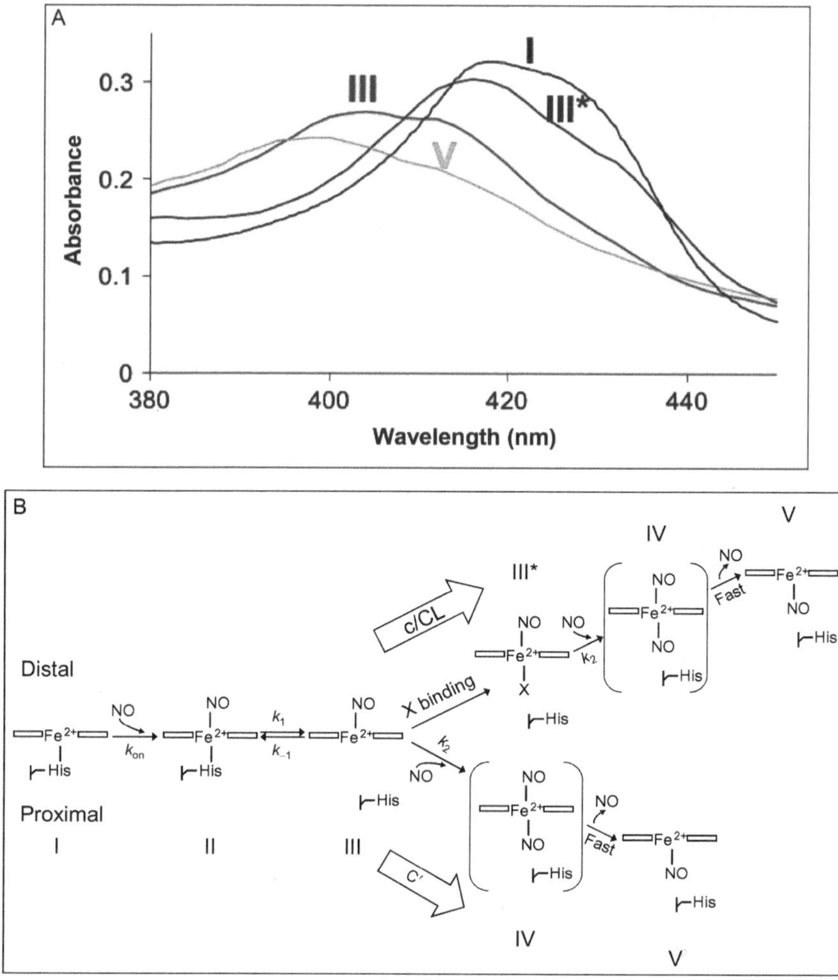

Figure 8.3 (A) Spectral intermediates in the reaction between the ferrous cyt *c*/CL complex (3 μM heme) and NO (60 μM) at temperature 20 °C (see also text for details and Silkstone et al., 2010). The spectra are shown in temporal sequence and color coded as follows: black, the initial spectrum of the high spin ferrous form (I); red, a penta-coordinate NO adduct populated some 20 ms after mixing the cyt *c*/CL complex with NO (III); blue, a hexa-coordinate NO species that accumulates after ~200 ms after mixing (III*); green, the final penta-coordinate NO species after 20 s (IV). The spectra are also labeled I, III, III*, and V to correspond with the species on the pathway given in Fig. 8.3B. (B) Comparison of the mechanism of NO binding to either the cyt *c*/CL complex or to Ax cyt *c*′. The steps in this pathway starting from the ferrous high spin form (I) are as follows: (a) an initial second-order association of NO with the heme at the vacant distal site to form the hexa-coordinate NO species (II). (b) The dissociation of the proximal histidine residue to form the penta-coordinate species (III). This step is slower than the initial NO-binding step in wild-type Ax cyt *c*′ allowing species (II) to be populated and observed. In the cyt *c*/CL complex, histidine dissociation is very rapid and thus species

(Continued)

In addition, the EPR spectrum shows a distinct three-line form, diagnostic for a penta-coordinate NO complex of ferrous heme. The comparable EPR spectrum for the Met80 mutants have a nine-line form that indicates that bound NO is coupled to a His residue *trans* to the NO, i.e., the proximal histidine remains coordinated to the heme iron. Dissociation of proximal histidine from ferrous heme on binding NO on the distal side has been observed in a number of heme proteins including *c*-type cytochromes and even hemoglobin under certain conditions, but has not previously been observed in mammalian cyt *c*. It is clear therefore that addition of CL to the protein does more to the structure than dissociate the methionine residue from the heme.

Experiments designed to study the kinetics of NO binding to ferro-cyt *c*/CL revealed further complexities. Time-resolved optical spectra, Fig. 8.3A, showed that following initial binding to the vacant sixth coordination site of the ferrous heme, a complex sequence of events ensued generating a number of distinct spectral forms on the pathway to the final penta-coordinate NO complex (Silkstone et al., 2010). The salient features of the mechanism (Fig. 8.3B) revealed by a combination of stopped-flow spectroscopy and laser photolysis experiments are that (1) there are two penta-coordinate NO forms and (2) there are two steps that have rates that are NO concentration dependent. With these details in mind, it has been proposed that NO first binds to the vacant distal position, dissociating the histidine from the *trans* position. Following further structural rearrangements of the molecule, a second NO binds to form a transient (hypothetical) hexa-coordinate species with NO coordinated to both distal and proximal sides. Thereafter, the distal NO dissociates from the distal side yielding the final penta-coordinate adduct with NO bound to the proximal side. The steps in this mechanism and the spectra associated with the species and the times at which they are populated are shown in Fig. 8.3A. The accompanying scheme (Fig. 8.3B) also shows the analogous mechanism for *Alcaligenes xylosoxidans* cytochrome *c*′ (Ax cyt *c*′) for which there are both kinetic evidence and definitive structural information, showing that NO binds to the proximal side of the heme in the final product (Lawson, Stevenson,

Figure 8.3—Cont'd (II) is not seen. (c) In the cyt *c*/CL complex, a rearrangement of the protein occurs that leads to a second hexa-coordinate form (III*) that is not seen in Ax cyt *c*′. (d) A second NO binds to the proximal side to form the postulated intermediate (IV) that has two-coordinated NO molecules. This is very unstable and has not been directly observed. (e) NO dissociates from the distal side to yield the final penta-coordinate NO form with NO bound to the proximal side of the heme. (See the color plate.)

Andrew, & Eady, 2000; Andrew, Green, Lawson, & Eady, 2001). The mechanisms of binding to cyt *c*/CL and cyt *c*′ are essentially the same, differing only in the value of rate constants, and thus species on the pathways are populated to different extents; e.g., the hexa-coordinate species, II of the scheme, is observed in wild-type Ax cyt *c*′ but is not seen in the cyt *c*/CL system because the rate of histidine dissociation is so fast that its concentration is vanishingly small.

3.2. NO binding to the proximal side of heme: Insights from cyt *c*′ and sGC

Binding of NO to the proximal side of heme is an unusual phenomenon and to date has been observed in only three proteins including cyt *c*/CL. The best-characterized proximal binding, and currently the only case for which structural information is available for the NO complex, occurs in Ax cyt *c*′ (Andrew, George, Lawson, & Eady, 2002; Andrew et al., 2001; Lawson et al., 2000). In this case, the ferrous heme is five-coordinate with a His proximal ligand, prior to an initial NO-binding event leading to the formation of a distal, six-coordinate His–Fe–NO complex. *trans* effects promote dissociation of the proximal histidine ligand, followed by binding of a second NO molecule in a second-order process to the now-vacant proximal coordination site to form a transient dinitrosyl intermediate (NO–Fe–NO) (Andrew et al., 2002). Dissociation of the distal NO from this intermediate forms the final 5c-NO proximal adduct (Fig. 8.4A).

This process leading to formation of the proximal NO adduct in Ax cyt *c*′ has been shown by mutagenesis studies to be critically dependent on the nature of the proximal and distal heme pockets. In the largely hydrophobic distal pocket, a hydrophobic or an aromatic residue is positioned above the distal binding site (Fig. 8.4). Residues naturally found in this position in different cytochromes *c*′ are Leu, Phe, Tyr, and Met. Mutation of Leu 16 in Ax cyt *c*′ to Ala resulted in formation of a stable six-coordinate distal His–Fe–NO complex (Garton et al., 2012) as well as an increase in CO-affinity of eight orders of magnitude (Antonyuk et al., 2011). A proximal NO species is not observed in this mutant, strongly suggesting that steric hindrance of distal NO binding is key to conversion to the 5c proximal NO state.

In the proximal pocket, a positively charged amino acid is maintained near to the proximal histidine, most commonly arginine but also lysine in some species. Mutations of the proximal Arg 124 in Ax cyt *c*′ showed increases in the rate of distal-to-proximal conversion, with the smallest change being in the R124K mutant where a positive charge is maintained and the largest in

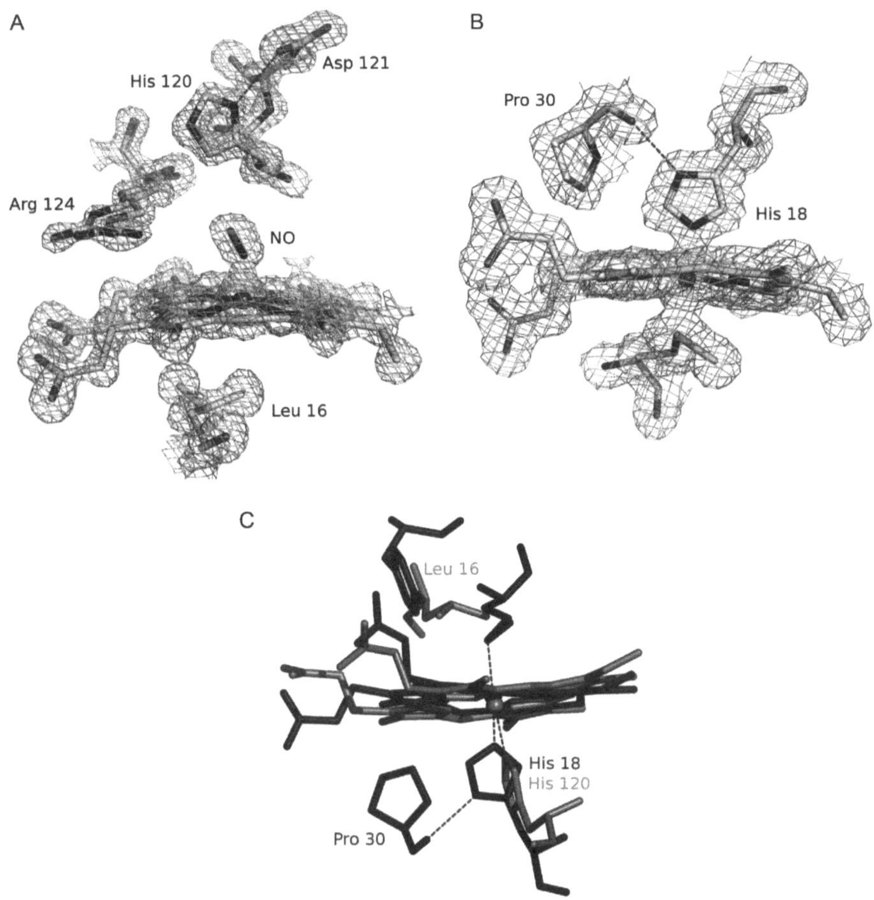

Figure 8.4 (A) Proximal NO binding to Ax cyt *c′* (Hough et al., 2011; PDB accession code 2xlm). The proximal His ligand has been displaced and is stabilized in this position by a H-bond to a nearby Asp residue. Note the presence of a positively charged Arg residue in the proximal pocket and a hydrophobic Leu residue in the distal pocket. (B) Heme environment in human cyt *c* (Rajagopal et al., 2013; PDB accession code 3zcf). Note that the proximal His 18 ligand is stabilized by a hydrogen bond to the carbonyl of Pro 30. (C) Least squares superposition of the heme group in ferrous cyt *c′* (salmon) and cyt *c* (blue). In cyt *c*, the heme is six coordinate with the Met ligand being dissociated upon formation of the cyt *c*/CL complex. (See the color plate.)

the R124E mutant where a negative charge was introduced (Barbieri, Murphy, Sawers, Eady, & Hasnain, 2008; Hough et al., 2011).

The existence of proximal NO binding in the eukaryotic NO-sensor sGC has been the subject of much debate. It has been established that two NO molecules, binding in two separate [NO] dependent steps, are required for full activation of this enzyme and that the first NO binds very rapidly [$k_{on} > 1.4 \times 10^8\ M^{-1}\ s^{-1}$] to the distal face of heme (Zhao, Brandish, Ballou, & Marletta, 1999). It is the second NO interaction, however, that is the dominant step in activation. Proposals have been made that this second NO interacts with a thiol group or another nonheme site (Fernhoff, Derbyshire, & Marletta, 2009). However, recent work (Martin, Berka, Sharina, & Tsai, 2012; Tsai, Berka, Sharina, & Martin, 2011) has confirmed earlier proposals (Zhao et al., 1999) (1) that heme is the binding site for the second NO and (2) that this NO binds to the proximal side of the heme. In sGC, based on the observation that binding of the second NO was not limited by the intrinsic dissociation constant of the proximal histidine, it is clear that proximal NO binding occurs in a concerted manner such that NO facilitates proximal His dissociation. From the resulting dinitrosyl complex, the distal NO is preferentially dissociated in a manner strongly analogous to that occurring in Ax cyt *c'*. In sGC, the activation signal results must be transmitted to the catalytic domain by as yet unresolved mechanisms that are likely to involve propagation of conformational changes following His dissociation.

4. INSIGHTS INTO THE NATURE AND FORMATION OF THE PROXIMAL NO COMPLEX IN cyt *c*–CL

Understanding the likely geometry and environment of the five-coordinate proximal NO species in cyt *c*/CL is challenging as the available crystal structures of cyt *c* are of the lipid-free form, Fig. 8.4B, and the protein is known to undergo significant structural rearrangements on lipid binding. Binding of CL to cyt *c* results in dissociation of the axial Met-heme ligand, although the resulting position of the dissociated methionine is unknown. Earlier work (Rajagopal et al., 2012) has shown that the acyl chain of CL occupies a near-heme position.

Despite similarities in the mode of NO binding, the near-heme environments in cyt *c* and Ax cyt *c'* are significantly different (Fig. 8.4B and C). Notably, while cyt *c* is six-coordinate with His-Met ligation, Ax cyt *c'* is five-coordinate with proximal His ligation, and the distal residue is a non-coordinated leucine. Common features likely to be important are the dissociation of the proximal His to allow binding of NO to the proximal side of heme and a mechanism to disfavor distal NO binding in a putative dinitrosyl intermediate.

By analogy to the mechanism established for Ax cyt *c'*, we can propose that *trans* effects following the initial (distal) NO-binding event will dissociate the proximal histidine ligand in cyt *c*/CL and that this ligand may then be required to reorientate in order to allow proximal NO binding. In contrast to the proximal His in Ax cyt *c*′, the equivalent residue in cyt *c* is stabilized by a strong hydrogen bond, to the carbonyl of Pro 30 (Fig. 8.4B). Provided that this H-bond remains in cyt *c*/CL, this might limit the conformational rearrangement the histidine is able to undergo following dissociation from Fe. However, it is possible that this H-bond is disrupted as part of the conformational change in cyt *c* on binding to CL and that this may promote proximal NO binding. In this context, the results of fluorescence (FRET) experiments to monitor the conformational transition brought about in cyt *c* on binding to CL membranes (Hanske et al., 2012) are relevant. These studies demonstrated that the protein could adopt a number of related conformations, the majority of which were characterized by being partially unfolded to render the heme accessible, consistent with the enhanced peroxidase activity and with the open heme-crevice structure found in ligand binding and photolysis investigations (see above). These fluorescence experiments also suggest that the Met residue is removed from the vicinity of the heme and that the proximal side of the heme group is toward the membrane and that the conformation on this side could be perturbed.

If a similar mechanism of distal-to-proximal NO conversion indeed occurs in Ax cyt *c*′ and cyt *c*/CL, we would expect to find a hydrophobic or aromatic moiety providing a degree of steric hindrance near to the distal face. In principle, this could arise either from a no longer ligated Met residue or from the acyl chain of CL itself. Data showing that the acyl chain of CL resides in close proximity to the distal side of the heme (Rajagopal et al., 2012) could suggest that it acts to lower the affinity for distal NO. If one assumes that a transient dinitrosyl intermediate is formed in cyt *c*/CL following dissociation of the proximal histidine (Fig. 8.3B), a mechanism must exist for the preferential dissociation of the distal NO to leave the observed proximal nitrosyl product. As the affinity of the distal site in cyt *c*/CL is very high, this requires either the affinity on the proximal side to be yet higher or conformational relaxation of the protein following His dissociation to alter the affinity on the distal side. Conceivably, such a conformational change could bring Met80 or the acyl group of CL closer to the distal NO-binding site, lowering its affinity below that of the proximal site. This could be related to loss of the H-bond between His and the carbonyl of Pro 30, although there is no immediate structural link between this residue and Met80.

5. CONCLUSIONS

The view we have advanced above is that a fraction, probably a small fraction, of cyt *c* is bound to CL in the inner mitochondrial membrane and is functionally altered by insertion of one (or two) of the hydrocarbon chains of this phospholipid into the interior of the protein. This cyt *c*/CL complex, while itself playing no role in the mitochondrion, acts as a receptor awaiting the arrival of H_2O_2 that can be an oxidative signal for apoptosis. If such a signal is received and H_2O_2 becomes bound to the heme in the cyt *c*/CL complex, this catalyzes the peroxidation of the hydrocarbon chain and in so doing acts as a transducer to change an oxidative signal into an electrophilic lipid signal, for which there are numerous examples (Higdon, Landar, Barnes, & Darley-Usmar, 2012; Levonen et al., 2004). Downstream of this, the oxidized CL is transported to the outer mitochondrial membrane where it initiates cyt *c* release into the cytoplasm by opening the transition pore.

This signal pathway is itself modulated by another signaling molecule, namely, NO. This molecule has affinity for both ferric and ferrous forms of the cyt *c*/CL complex, binding directly to the iron atom. In doing so, it prevents H_2O_2 from binding to the iron and thus NO acts to inhibit the peroxidative oxidation of the lipid chain. In this way, NO plays an antiapoptotic role. It may be supposed that the balance between the proapoptotic signal, H_2O_2, and the antiapoptotic signal, NO, is important in governing the fate of the cell.

The fact that the cyt *c*/CL complex binds NO may also be important in buffering/sequestering NO in the mitochondrion. The ferric protein has relatively low affinity for NO which means that it is unlikely to sequester NO efficiently and that any NO bound would readily be released. This is not so for the ferrous complex that binds NO tightly and thus acts to maintain NO at low concentration. An interesting feature of the ferrous protein is that it exhibits two modes of NO binding. At very low NO concentrations, binding to the distal side would predominate as binding to the proximal side requires two NO molecules to bind simultaneously to the heme, an event unlikely at low [NO]. It is possible, although not experimentally shown as yet, that the dissociation rates from the two sides of the heme are different. Comparison with cyt *c*′ suggests that dissociation from the distal side is relatively rapid allowing NO to return to the bulk phase, while dissociation from the proximal binding site, is slow and thus this mode sequesters

NO. In this way, the cyt *c*/CL complex would regulate NO concentrations, allowing rapid binding and release at low NO concentrations while sequestering NO at high concentrations.

Despite the structural differences between cyt *c* and Ax cyt *c*′ and the conformational changes in cyt *c* upon formation of the cyt *c*/CL complex, the similarity in NO-binding properties gives insights into the nature of proximal NO binding in cyt *c*/CL. First, *trans* effects will favor scission of the Fe–His 18 bond following distal NO binding. It is likely that the H-bond between His 18 and a nearby carbonyl (which is not present in cyt *c*′) is disrupted as part of the conformational change in cyt *c* on binding to CL and that this may promote proximal NO binding. Second, a mechanism may be presumed to exist in cyt *c*/CL to disfavor the distally bound NO of the transient dinitrosyl intermediate, even though k_{on} for distal binding is very high (in cyt *c*′, distal affinity is lower due to steric hindrance from Leu 16). Candidates are the displaced axial Met ligand or, perhaps more likely, the acyl group of CL itself, which may shift in response to His 18 dissociation from heme.

REFERENCES

Andrew, C. R., George, S. J., Lawson, D. M., & Eady, R. R. (2002). Six- to five-coordinate heme-nitrosyl conversion in cytochrome c' and its relevance to guanylate cyclase. *Biochemistry*, *41*(7), 2353–2360.

Andrew, C. R., Green, E. L., Lawson, D. M., & Eady, R. R. (2001). Resonance Raman studies of cytochrome c' support the binding of NO and CO to opposite sides of the heme: Implications for ligand discrimination in heme-based sensors. *Biochemistry*, *40*(13), 4115–4122.

Antonyuk, S. V., Rustage, N., Petersen, C. A., Arnst, J. L., Heyes, D. J., Sharma, R., et al. (2011). Carbon monoxide poisoning is prevented by the energy costs of conformational changes in gas-binding hemeproteins. *Proceedings of the National Academy of Sciences of the United States of America*, *108*(38), 15780–15785.

Arroyo, A., Modriansky, M., Serinkan, F. B., Bello, R. I., Matsura, T., Jiang, J., et al. (2002). NADPH oxidase-dependent oxidation and externalization of phosphatidylserine during apoptosis in me2so-differentiated hl-60 cells. Role in phagocytic clearance. *Journal of Biological Chemistry*, *277*(51), 49965–49975.

Barbieri, S., Murphy, L. M., Sawers, R. G., Eady, R. R., & Hasnain, S. S. (2008). Modulation of NO binding to cytochrome c' by distal and proximal heme pocket residues. *Journal of Biological Inorganic Chemistry*, *13*, 531–540.

Bradley, J. M., Silkstone, G., Wilson, M. T., Cheesman, M. R., & Butt, J. N. (2011). Probing a complex of cytochrome c and cardiolipin by magnetic circular dichroism spectroscopy: Implications for the initial events in apoptosis. *Journal of the American Chemical Society*, *133*(49), 19676–19679.

Cantley, L. C. (2002). The phosphoinositide 3-kinase pathway. *Science*, *296*(5573), 1655–1657.

Choi, B. M., Pae, H. O., Jang, S. I., Kim, Y. M., & Chung, H. T. (2002). Nitric oxide as a pro-apoptotic as well as anti-apoptotic modulator. *Journal of Biochemistry and Molecular Biology*, *35*(1), 116–126.

Chung, H. T., Pae, H. O., Choi, B. M., Billiar, T. R., & Kim, Y. M. (2001). Nitric oxide as a bioregulator of apoptosis. *Biochemical and Biophysical Research Communications*, *282*(5), 1075–1079.

Fadok, V. A., de Cathelineau, A., Daleke, D. L., Henson, P. M., & Bratton, D. L. (2001). Loss of phospholipid asymmetry and surface exposure of phosphatidylserine is required for phagocytosis of apoptotic cells by macrophages and fibroblasts. *The Journal of Biological Chemistry*, *276*(2), 1071–1077.

Fernhoff, N. B., Derbyshire, E. R., & Marletta, M. A. (2009). A nitric oxide/cysteine interaction mediates the activation of soluble guanylate cyclase. *Proceedings of the National Academy of Sciences of the United States of America*, *106*, 21602–21607.

Garton, E. M., Pixton, D. A., Petersen, C. A., Eady, R. R., Hasnain, S. S., & Andrew, C. R. (2012). A distal pocket Leu residue inhibits the binding of O_2 and NO at the distal heme site of cytochrome c'. *Journal of the American Chemical Society*, *134*(3), 1461–1463.

Ghafourifar, P., & Cadenas, E. (2005). Mitochondrial nitric oxide synthase. *Trends in Pharmacological Sciences*, *26*(4), 190–195.

Green, D. R., & Kroemer, G. (2004). The pathophysiology of mitochondrial cell death. *Science*, *305*(5684), 626–629.

Hanske, J., Toffey, J. R., Morenz, A. M., Bonilla, A. J., Schiavoni, K. H., & Pletneva, E. V. (2012). Conformational properties of cardiolipin-bound cytochrome. *Proceedings of the National Academy of Sciences of the United States of America*, *109*, 125–130.

Hasan, S. S., Yamashita, E., Ryan, C. M., Whitelegge, J. P., & Cramer, W. A. (2011). Conservation of lipid functions in cytochrome bc complexes. *Journal of Molecular Biology*, *414*(1), 145–162.

Haynes, V., Elfering, S., Traaseth, N., & Giulivi, C. (2004). Mitochondrial nitric-oxide synthase: Enzyme expression, characterization, and regulation. *Journal of Bioenergetics and Biomembranes*, *36*(4), 341–346.

Higdon, A. N., Landar, A., Barnes, S., & Darley-Usmar, V. M. (2012). The electrophile responsive proteome: Integrating proteomics and lipidomics with cellular function. *Antioxidants and Redox Signaling*, *17*(11), 1580–1589.

Hough, M. A., Antonyuk, S. V., Barbieri, S., Rustage, N., McKay, A. L., Servid, A. E., et al. (2011). Distal-to-proximal NO conversion in hemoproteins: The role of the proximal pocket. *Journal of Molecular Biology*, *405*(2), 395–409.

Huttemann, M., Pecina, P., Rainbolt, M., Sanderson, T. H., Kagan, V. E., Samavati, L., et al. (2011). The multiple functions of cytochrome c and their regulation in life and death decisions of the mammalian cell: From respiration to apoptosis. *Mitochondrion*, *11*(3), 369–381.

Ignarro, L. J. (1990). Heme-dependent activation of guanylate cyclase and cyclic GMP formation by endogenous nitric oxide: A unique transduction mechanism for transcellular signaling. *Pharmacology & Toxicology*, *67*(1), 1–7.

Kagan, V. E., Fabisiak, J. P., Shvedova, A. A., Tyurina, Y. Y., Tyurin, V. A., Schor, N. F., et al. (2000). Oxidative signaling pathway for externalization of plasma membrane phosphatidylserine during apoptosis. *FEBS Letters*, *477*(1–2), 1–7.

Kagan, V. E., Gleiss, B., Tyurina, Y. Y., Tyurin, V. A., Elenstrom-Magnusson, C., Liu, S. X., et al. (2002). A role for oxidative stress in apoptosis: Oxidation and externalization of phosphatidylserine is required for macrophage clearance of cells undergoing fas-mediated apoptosis. *Journal of Immunology*, *169*(1), 487–499.

Kagan, V. E., Tyurin, V. A., Jiang, J., Tyurina, Y. Y., Ritov, V. B., Amoscato, A. A., et al. (2005). Cytochrome c acts as a cardiolipin oxygenase required for release of proapoptotic factors. *Nature Chemical Biology*, *1*(4), 223–232.

Kapetanaki, S. M., Silkstone, G., Husu, I., Liebl, U., Wilson, M. T., & Vos, M. H. (2009). Interaction of carbon monoxide with the apoptosis-inducing cytochrome c-cardiolipin complex. *Biochemistry*, *48*(7), 1613–1619.

Kinnunen, P. K. (1992). Fusion of lipid bilayers: A model involving mechanistic connection to HII phase forming lipids. *Chemistry and Physics of Lipids*, *63*, 251–258.
Kinnunen, P. K., Koiv, A., Lehtonen, J. Y., Rytomaa, M., & Mustonen, P. (1994). Lipid dynamics and peripheral interactions of proteins with membrane surfaces. *Chemistry and Physics of Lipids*, *73*, 181–207.
Lawson, D. M., Stevenson, C. E., Andrew, C. R., & Eady, R. R. (2000). Unprecedented proximal binding of nitric oxide to heme: Implications for guanylate cyclase. *The EMBO Journal*, *19*(21), 5661–5671.
Levonen, A. L., Landar, A., Ramachandran, A., Ceaser, E. K., Dickinson, D. A., Zanoni, G., et al. (2004). Cellular mechanisms of redox cell signalling: Role of cysteine modification in controlling antioxidant defences in response to electrophilic lipid oxidation products. *Biochemistry Journal*, *378*, 373–382.
Liu, X., Kim, C. N., Yang, J., Jemmerson, R., & Wang, X. (1996). Induction of apoptotic program in cell-free extracts: Requirement for dATP and cytochrome c. *Cell*, *86*(1), 147–157.
Martin, E., Berka, V., Sharina, I., & Tsai, A. L. (2012). Mechanism of binding of no to soluble guanylyl cyclase: Implication for the second no binding to the heme proximal site. *Biochemistry*, *51*(13), 2737–2746.
Mason, M. G., Nicholls, P., Wilson, M. T., & Cooper, C. E. (2006). Nitric oxide inhibition of respiration involves both competitive (heme) and non-competitive (copper) binding to cytochrome c oxidase. *Proceedings of the National Academy of Sciences of the United States of America*, *103*, 708–713.
Orrenius, S., Gogvadze, V., & Zhivotovsky, B. (2007). Mitochondrial oxidative stress: Implications for cell death. *Annual Review of Pharmacology and Toxicology*, *47*, 143–183.
Ow, Y. P., Green, D. R., Hao, Z., & Mak, T. W. (2008). Cytochrome c: Functions beyond respiration. *Nature Reviews. Molecular Cell Biology*, *9*(7), 532–542.
Poulos, T. (2006). Soluble guanylate cyclase. *Current Opinion in Structural Biology*, *16*(6), 736–743.
Radi, R., Cassina, A., Hodara, R., Quijano, C., & Castro, L. (2002). Peroxynitrite reactions and formation in mitochondria. *Free Radical Biology & Medicine*, *33*(11), 1451–1464.
Rajagopal, B. S., Silkstone, G. G., Nicholls, P., Wilson, M. T., & Worrall, J. A. (2012). An investigation into a cardiolipin acyl chain insertion site in cytochrome c. *Biochimica et Biophysica Acta*, *1817*(5), 780–791.
Rajagopal, B. S., Edzuma, A. N., Hough, M. A., Blundell, K. L., Kagan, V. E., Kapralov, A. A., et al. (2013). The hydrogen-peroxide-induced radical behaviour in human cytochrome c-phospholipid complexes: implications for the enhanced pro-apoptotic activity of the G41S mutant. *Biochemical Journal*, *456*, 441–452.
Riedl, S. J., & Salvesen, G. S. (2007). The apoptosome: Signalling platform of cell death. *Nature Reviews. Molecular Cell Biology*, *8*(5), 405–413.
Robertson, B. E., Schubert, R., Hescheler, J., & Nelson, M. T. (1993). cGMP-dependent protein kinase activates Ca-activated K channels in cerebral artery smooth muscle cells. *American Journal of Physiology*, *265*, 299–303.
Rubbo, H., Radi, R., Trujillo, M., Telleri, R., Kalyanaraman, B., Barnes, S., et al. (1994). Nitric oxide regulation of superoxide and peroxynitrite-dependent lipid peroxidation. Formation of novel nitrogen-containing oxidized lipid derivatives. *The Journal of Biological Chemistry*, *269*(42), 26066–26075.
Rytomaa, M., & Kinnunen, P. K. (1995). Reversibility of the binding of cytochrome c to liposomes. Implications for lipid-protein interactions. *The Journal of Biological Chemistry*, *270*, 3197–3202.
Silkstone, G., Jasaitis, A., Wilson, M. T., & Vos, M. H. (2007). Ligand dynamics in an electron transfer protein. Picosecond geminate recombination of carbon monoxide to heme in mutant forms of cytochrome c. *The Journal of Biological Chemistry*, *282*(3), 1638–1649.

Silkstone, G., Kapetanaki, S. M., Husu, I., Vos, M. H., & Wilson, M. T. (2010). Nitric oxide binds to the proximal heme coordination site of the ferrocytochrome c/cardiolipin complex: Formation mechanism and dynamics. *The Journal of Biological Chemistry*, *285*(26), 19785–19792.

Silkstone, G., Kapetanaki, S. M., Husu, I., Vos, M. H., & Wilson, M. T. (2012). Nitric oxide binding to the cardiolipin complex of ferric cytochrome c. *Biochemistry*, *51*(34), 6760–6766.

Stewart, J. M., Blakely, J. A., & Johnson, M. D. (2000). The interaction of ferrocytochrome c with long-chain fatty acids and their CoA and carnitine esters. *Biochemistry and Cell Biology*, *78*, 675–681.

Tsai, A. L., Berka, V., Sharina, I., & Martin, E. (2011). Dynamic ligand exchange in soluble guanylyl cyclase (sgc): Implications for sgc regulation and desensitization. *The Journal of Biological Chemistry*, *286*(50), 43182–43192.

Tuominen, E. K., Wallace, C. J., & Kinnunen, P. K. (2002). Phospholipid-cytochrome c interaction: Evidence for the extended lipid anchorage. *The Journal of Biological Chemistry*, *277*, 8822–8826.

Tyurina, Y. Y., Kawai, K., Tyurin, V. A., Liu, S. X., Kagan, V. E., & Fabisiak, J. P. (2004). The plasma membrane is the site of selective phosphatidylserine oxidation during apoptosis: Role of cytochrome c. *Antioxidants and Redox Signaling*, *6*(2), 209–225.

Tyurina, Y. Y., Serinkan, F. B., Tyurin, V. A., Kini, V., Yalowich, J. C., Schroit, A. J., et al. (2004). Lipid antioxidant, etoposide, inhibits phosphatidylserine externalization and macrophage clearance of apoptotic cells by preventing phosphatidylserine oxidation. *The Journal of Biological Chemistry*, *279*(7), 6056–6064.

Tyurina, Y. Y., Shvedova, A. A., Kawai, K., Tyurin, V. A., Kommineni, C., Quinn, P. J., et al. (2000). Phospholipid signaling in apoptosis: Peroxidation and externalization of phosphatidylserine. *Toxicology*, *148*(2–3), 93–101.

Zhao, Y., Brandish, P. E., Ballou, D. P., & Marletta, M. A. (1999). A molecular basis for nitric oxide sensing by soluble guanylate cyclase. *Proceedings of the National Academy of Sciences of the United States of America*, *96*, 14753–14758.

Zou, H., Li, Y., Liu, X., & Wang, X. (1999). An APAF-1-cytochrome c multimeric complex is a functional apoptosome that activates procaspase-9. *Journal of Biological Chemistry*, *274*, 11549–11556.

CHAPTER NINE

The Nitric Oxide-Mediated Regulation of Prostaglandin Signaling in Medicine

Sangwon F. Kim[1]
Department of Psychiatry and Pharmacology, Center for Neurobiology and Behavior, The Perelman School of Medicine at University of Pennsylvania, Philadelphia, Pennsylvania, USA
[1]Corresponding author: e-mail address: sangwonk@mail.med.upenn.edu

Contents

Abstract

Nitric oxide synthase and cyclooxygenase pathways share a number of similarities. NOS produces nitric oxide (NO) and COX generates prostaglandins. Two major forms of NOS and COX have been discovered to date: constitutive versus inducible forms. The constitutive forms play a role in housekeeping and physiological states, while the inducible forms are overexpressed under pathological conditions such as inflammation in a variety of cells, producing a large amount of NO or prostaglandins. The cross talk between the COX and NOS pathways was initially reported in 1993 when they demonstrated in a series of *in vitro* and *in vivo* studies that NO activates the COX enzymes to produce increased amounts of prostaglandins. Those studies led to the concept that COX enzymes represent important endogenous "receptor" targets for amplifying or modulating the multifaceted roles of NO in physiology and pathology. Since then, numerous studies have been undertaken to delineate the functional consequences of this interaction as well as the potential mechanism by which each pathway interacts. This review focuses in particular on recent advances in this field that extend our understanding of these two pathways under various systems.

Vitamins and Hormones, Volume 96
ISSN 0083-6729
http://dx.doi.org/10.1016/B978-0-12-800254-4.00009-X

1. INTRODUCTION

Nitric oxide and prostaglandins are prominent local mediators that participate in the numerous homeostatic modulations of many cellular functions. Their constitutive biosynthesis and pulsed release in nanomolar concentrations occur via well-characterized enzymes, nitric oxide synthases (NOSs) and cyclooxygenases (COXs) utilizing specific substrates, L-arginine and arachidonic acid, respectively. This basal release of NO and PGs exerts a protective role in many pathophysiological conditions: vascular diseases (via enhanced vasodilatation and antiplatelet activity), gastric lesions (via activation of gastroprotective processes), erectile dysfunction (via enhanced vasodilatation), and learning and memory processing (via potentiation of neuronal plasticity). Moreover, in both acute and inflammatory settings, a large amount of NO and prostaglandins are released simultaneously due mainly to the activation of inducible enzymes possibly contributing to their cytotoxic effects (Fig. 9.1). Therefore, modulation of NOS or COX has been proposed to ameliorate various pathological conditions, and yet their possibly cross talk between two pathways supported by many studies for past 2 decades makes this task more challenging.

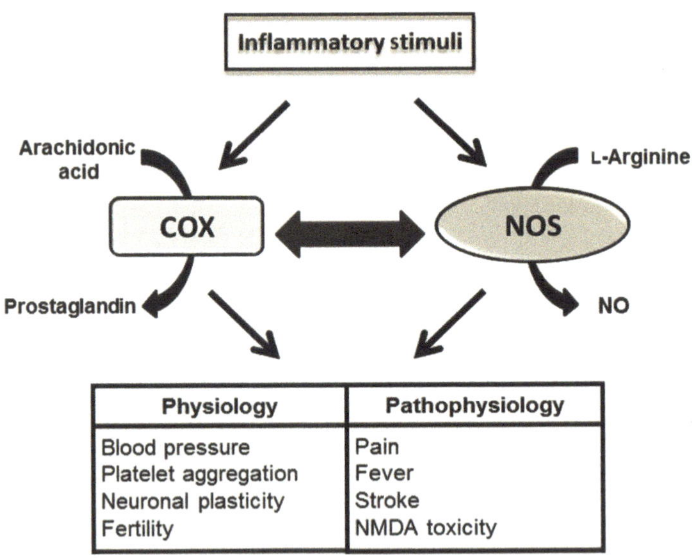

Figure 9.1 The role of nitric oxide synthase (NOS) and cyclooxygenase. Both enzymes share the similar pathway for activation and their physiological as well as pathophysiological implication. (See the color plate.)

2. NOS AND NO PRODUCTION

Three isozymes of NOS have been identified so far (Marletta, 1993; Stuehr, 1997). Isoform I, also known as neuronal or brain NO synthase (nNOS), is expressed in neuronal cells (Bredt et al., 1991; Marletta, 1994; Nathan, 1992), while isoform III or endothelium NOS was identified as the enzyme which produces "endothelium-derived relaxing factor" (Pollock et al., 1991) which turned out to be NO. Isoforms I and III are constitutively expressed and, hence, grouped together as constitutive NOS (cNOS). Activities of cNOS are regulated by intracellular calcium concentration via calmodulin (Stuehr, 1997). On the other hand, isoform II is expressed in a wide variety of host defense and other cells in response to inflammatory stimuli such as endogenous cytokines and bacterial lipopolysaccharide (LPS) endotoxin, resulting in a delayed (hours) but prolonged synthesis of high levels of NO (Hevel, White, & Marletta, 1991; Nathan & Xie, 1994). Isoform II is thus referred to as inducible NOS (iNOS). iNOS is bound to calmodulin even at resting intracellular calcium concentrations, and hence, the function of iNOS is not affected by intracellular calcium levels. Moreover, iNOS, compared to cNOS, produces a much higher amount of NO which is thought to be responsible for the cytotoxicity of NO (Moncada, Palmer, & Higgs, 1991; Nathan & Xie, 1994).

Despite differences in the mechanisms of NOS regulation, all NOS share a common structure and common cofactors. NOS is the only enzyme containing both an oxygenase and a reductase domain. The oxygenase domain includes terahydrobiopterin (BH4), heme, and calmodulin-binding sites, whereas the reductase domain harbors the binding sites for nicotinamide adenine dinucleotide phosphate, flavin adenine dinucleotide, and flavin mononucleotide (Stuehr, 1997). Later, mitochondrial NOS has been identified (Tatoyan & Giulivi, 1998), and it appears to be closely related to the regulation of mitochondrial respiration via interaction with superoxide ion (Giulivi, 1998). However, further investigation to characterize functions of mitochondrial NOS is required.

2.1. Different redox species of NO

NO is a highly reactive diatomic molecule that is generated via the oxidation of the terminal guanidino nitrogen atom of L-arginine by the enzyme NO synthase within mammalian cells (Abu-Soud, Presta, Mayer, & Stuehr, 1997; Moncada et al., 1991; Nathan, 1992; Stuehr et al., 1991). In biological systems, NO exists as three major redox-related species, and NO is used as a generic term to encompass all these species. Redox-related species of NO include $NO^{\bullet}$, which can modulate ferrous iron (Fe^{2+})-containing proteins by direct coordination to their Fe^{2+} centers; such $NO^{\bullet}$ can be cytotoxic by reacting with superoxide anion to produce peroxynitrite ($ONOO^{-}$). Further oxidation of $NO^{\bullet}$ species can generate nitrogen dioxide (NO_2), dinitrogen trioxide (N_2O_3), dinitrogen tetroxide (N_2O_4), nitrite (NO_2^{-}), nitrate (NO_3^{-}), and others (Szabo, Ischiropoulos, & Radi, 2007). The second important species of NO, the nitrosonium ion (NO^{+}), can nitrate thiol groups of proteins, a modification that may have important regulatory functions under both physiological and pathophysiological conditions (Foster, Hess, & Stamler, 2009). Nitrosonium ion has an extremely short half-life ($\approx 10^{-10}$ s) in solution at physiological pH (Stamler et al., 1992) and binds extremely rapidly to thiol groups. Following such binding, the resulting —SNO compounds maintain a "nitrosonium" characteristic and can transfer NO^{+} to other thiols, in a process called *trans*-nitrosylation (Kornberg et al., 2010; Lipton et al., 1993; Stamler, 1994). The final redox-related species of NO is the nitronyl (NO^{-}) anion, which was reported to have a similar reactivity to NO^{+} (Kim et al., 1999), leading to S-nitrosylation; however, the patho/physiological relevance of this species is unclear.

2.2. NO-mediated chemical reaction in biology

NO has two principal yet divergent functions in the biological system: homeostasis and cytotoxicity. For regulatory functions, NO is produced in small amounts under physiological conditions and mediates vasorelaxation. Nitric oxide released under these circumstances plays a crucial role in the cardiovascular and renal system where it controls organ blood flow distribution, inhibits the aggregation and adhesion of platelets to the vascular wall, limits leukocyte adhesion and smooth muscle cell proliferation, promotes diuresis and natriuresis with the kidney (Moncada & Higgs, 1995; Moncada et al., 1991), and is involved in neurotransmission (Bredt et al., 1991; Bredt & Snyder, 1994; Nathan, 1992). Most of these actions are mediated through the binding of NO to Fe in the heme prosthetic group of soluble guanylate cyclase which catalyzes the conversion of GTP to cyclic GMP (Ignarro, 1996). The importance of Fe in mediating the functions of NO is also apparent when examining the cytotoxic effects of the molecule, which are observed when NO produced from iNOS in much larger quantities by macrophages, hepatocytes, and other cells following their exposure to cytokines or microbial products. NO produced via such high-output systems inhibits proliferation of intracellular pathogens and tumor cells. These effects can be explained by the reactivity of $NO^{\bullet}$ with Fe in the [Fe–S] centers of several important macromolecules, including aconitase, and Complex I and Complex II of the electron transport chain (Drapier & Hibbs, 1988; Henry et al., 1993). The high affinity of $NO^{\bullet}$ for iron probably results in both the removal of Fe from [Fe–S] centers and the formation of nitrosyl Fe species within [Fe–S] proteins. Thus, $NO^{\bullet}$ exerts its function through its interaction with the Fe of heme and non-heme Fe proteins.

2.2.1 S-nitrosylation

The reactivity of NO with iron-containing moieties, such as heme or Fe–S clusters, adequately explains its immediate bioactivity but does not provide an explanation about the effects on cellular signaling pathways after post-translational modification of proteins by NO. In 1993, it was demonstrated that NO^{+} can interact with thiol groups of proteins to form *S*-nitrosothiols (Lipton et al., 1993). Importantly, S-nitrosylation of cysteine residues in proteins by the addition of NO^{+} results in the modification of their activities via affecting their stability or possible change in tertiary structure. S-nitrosylation can be achieved in a test tube by a reaction between the acidified nitrite and thiol groups.

It is highly unlikely that NO^{+} production can occur by this mechanism in biological systems because of the requirement for low pH. It is likely that biological production of NO^{+} requires iron-catalyzed oxidation of $NO^{\bullet}$ which is generated by NOS (Akaike et al., 1997; Inoue et al., 1999). Support for this argument has been found in the detection of iron–nitrosyl complexes and an increase of *S*-nitrosothiols in LPS/IFN-γ-treated macrophages, which produce a large quantity of NO (Lancaster & Hibbs, 1990).

Although NO^{+} can interact with thiol groups of proteins to form *S*-nitrosothiols in a nonenzymatically catalyzed S-nitrosylation reaction, there appears to be some specificity in this reaction. First, not all proteins containing free cysteines are likely to be S-nitrosylated, and not every available cysteine in a given target is S-nitrosylated. For example, the ryanodine receptor contains 87 cysteines but only 13 of them are S-nitrosylated (Xu, Eu, Meissner, & Stamler, 1998), suggesting a certain specificity of S-nitrosylation. Moreover, it has been proposed that there is a primary consensus sequence for S-nitrosylation requiring the presence of acidic or basic residues adjacent to target cysteines (Stamler, Toone, Lipton, & Sucher,

1997). However, this acid–base consensus motif can also arise as a result of three-dimensional protein structure, as shown by analysis of the crystal structure of caspase-3 whose active cysteine is S-nitrosylated by NO *in vivo* (Hess, Matsumoto, Nudelman, & Stamler, 2001) or as a form of protein–protein interaction (Kornberg et al., 2010). Hence, the conditions for S-nitrosylation appear to be far more complex than what may be suggested by a primary sequence or availability of a target cysteine.

More than a hundred proteins have been identified as targets of S-nitrosylation which is associated with modification of their functions or structures including COX2 (Foster et al., 2009; Stamler, 1994). Glyceraldehyde-3-phosphate dehydrogenase (GAPDH) is among the first proteins shown to be S-nitrosylated. It has been shown that NO production is correlated with inhibition of GAPDH in rat liver during chronic inflammation (Vedia et al., 1992); the S-nitrosylation of a cysteine residue in the active site of GAPDH not only causes reversible inactivation of the enzyme (Mohr, Hallak, de Boitte, Lapetina, & Brune, 1999) but also enhances its translocation to nucleus leading to cell death (Hara et al., 2005). Moreover, nitrosylated GAPDH functions as a carrier of NO and transnitrosylate nuclear proteins such as sirtuin-1 (SIRT1), histone deacetylase-2 (HDAC2), and DNA-activated protein kinase (DNA-PK) (Kornberg et al., 2010).

S-nitrosylation of hemoglobin was extensively studied, and Cys93 has been shown to be a site for this modification. The binding of NO to hemoglobin is dependent on the allosteric state of hemoglobin (Gow & Stamler, 1998). When oxygenated, hemoglobin is S-nitrosylated and the NO group is released during arterial–venous transit (Jia, Bonaventura, Bonaventura, & Stamler, 1996). Hence, NO released from hemoglobin plays an essential but previously unexpected role in the respiratory cycle by ensuring adequate blood flow to tissues.

S-nitrosylation of other proteins such as the *N*-methyl-D-aspartate (NMDA) receptor, $p21^{ras}$, caspase-3, AMPA receptor, and RASD1 affects their regulation or localization (Fang et al., 2000; Huang et al., 2005; Lander, Ogiste, Teng, & Novogrodsky, 1995; Mannick et al., 1999) and the list of S-nitrosylated proteins continues to grow (Foster et al., 2009). Interestingly, $p21^{ras}$ can be S-nitrosylated and activated by NO (Lander et al., 1995), with the concentration of NO required to obtain maximal activation of this protein *in vitro* being 1000-fold higher than that required for activation using intact cells, suggesting that the intracellular environment is more favorable for nitrosothiol formation (Lander et al., 1997).

2.2.2 Peroxynitrite-mediated reaction

Superoxide is generated by various sources under physiological conditions and its production is tightly regulated by superoxide dismutases (SODs) (McCord & Fridovich, 1969). However, during acute and chronic inflammation, superoxide production exceeds the capacity of endogenous SODs to remove it (Fridovich, 1999; Muscoli et al., 2003). Under the same conditions, NO production is also enhanced. Excess superoxide scavenges NO, generating peroxynitrite, thus destroying the biological activity of NO (Moncada et al., 1991). Peroxynitrite is highly reactive and exerts its action via various mechanisms; however, it is the interaction with tyrosine residues in proteins forming nitrotyrosine that has received the most attention. So far, numerous proteins have been identified as targets for nitrotyrosine modification but the specificity is not well understood. Nevertheless, nitrotyrosine formation within proteins leads to either gain or loss of function (Cassina et al., 2000; Nowak & Wachowicz, 2002; Szabo et al., 2007). MnSOD is a good example of an enzyme whose activity was lost after tyrosine nitration *in vivo* (McCord & Fridovich, 1969; Yamakura & Kawasaki, 2010). The functional consequence of nitrotyrosine formation of MnSOD is quite severe as loss of activity enhances the production of superoxide, leading to an increase in peroxynitrite, followed by additional modifications on other proteins such as aconitase, ATPase, and cytochrome *c*, further disrupting mitochondria respiratory function and consequently resulting in apoptosis (Cassina et al., 2000; Radi, Cassina, Hodara, Quijano, & Castro, 2002; Turko et al., 2003). This nitrotyrosine modification has been detected in various pathophysiological conditions such as ischemia and reperfusion injury, Alzheimer's disease (AD), and organ transplant (Ischiropoulos, 2009; Macmillan-Crow & Cruthirds, 2001; Mollace et al., 2001). Thus, peroxynitrite functions as a potent cytotoxic and anti-inflammatory molecule (Ischiropoulos, 2009; Radi, 2004; Szabo et al., 2007). Nitrotyrosine formation is similar to S-nitrosylation in that the NO moiety modifies a specific residue in the target proteins but many differences also exist (Foster et al., 2009; Szabo et al., 2007). Unlike the labile S-nitrosylation modification, once a tyrosine residue is modified by peroxynitrite, it appears to be long lasting. One way to remove the nitrotyrosine is by protein degradation (Ischiropoulos, 2009; Radi, 2004; Szabo et al., 2007). A yet unidentified protein, termed "denitrase," appears to be able to remove the nitro group without protein degradation and has been observed in tissues (Irie et al., 2003).

3. COX AND PROSTAGLANDIN IN BIOLOGY

3.1. COX1 and -2

COX was first purified in 1976 from sheep seminal vesicles a homodimer of molecular mass of 71 kDa containing both COX and hydroperoxidase activities (Hemler & Lands, 1976; Miyamoto, Ogino, Yamamoto, & Hayaishi, 1976) and led subsequently to COX being cloned from the same tissue (DeWitt, 1991; Merlie, Fagan, Mudd, & Needleman, 1988). With the availability of the cDNA encoding the protein, as well as COX-specific antibodies, numerous studies have evaluated the distribution, expression, and regulation of COX both *in vitro* and *in vivo*. Initially, COX was considered a single isoform that generated PGs in most tissues and cell types. However, several laboratories speculated an existence of COX isoenzymes as certain stimuli could induce the production of prostaglandin (Flower & Vane, 1972; Lin, Bienkowski, & Gorman, 1989). Later, Needleman's group reported production of prostaglandin generated from *de novo* synthesis of new COX protein in mouse peritoneal macrophage *in vivo* (Morrison, Moritz, & Needleman, 1978). Subsequently, COX was shown indeed to have two isoforms initially called prostaglandin H synthase-1 and -2 and now commonly referred to as COX1 and COX2 (Bailey, Muza, Hla, & Salata, 1985; Sano et al., 1992). These are encoded by two different genes; the human Ptgs2 (COX2) gene of 8.3 kb is a small immediate early gene on chromosome 1, while Ptgs1 (COX1) is a much longer 22 kb gene on chromosome 9. The main difference in gene of these two proteins can be also found in their promoter regions. The promoter region of the human COX1 gene lacks a TATA or CAAT box (Tanabe & Tohnai, 2002). These features lead COX1 to be a constitutive enzyme in the most of cells. In contrast, COX2 has various transcriptional regulatory sequences in its promoter region, including TATA box, an NF-IL6 motif, a CRE motif, an E-box, two NF-kB-sites, two AP-2 sites, and three Sp1 sites (Morita, 2002). Therefore, COX2 expression can be induced by multiple cytokines and growth factors via activation of transcriptional regulatory proteins that bind to these promoter sites. Hence, COX2 seems to be the primary COX-generating prostaglandin E2 (PGE2) during inflammation.

There is another interesting difference between COX1 and COX2. Previously, it has been shown that COX1 and COX2 may utilize different pools of arachidonic acid to produce prostaglandins. COX2 acts on arachidonic acid when its concentration is below 2.5 μM, which can be produced by endogenous source. However, COX1 preferentially utilize much higher concentration of arachidonic acid (>10 μM), which can be mostly derived by an exogenous source, or release during acute inflammation (Shitashige, Morita, & Murota, 1998).

Human COX1 and COX2 have about 600 amino acids with a molecular weight of around 70 kDa, sharing 60% identity in sequence (Otto & Smith, 1995; Rouzer & Marnett, 2009). Overall structure of COX1 and COX2 is similar, and those crystal structures can be almost superimposed. Both isoforms are composed of three domains: the epidermal growth factor domain, the membrane-bound domain, and the catalytic domain containing COX and peroxygenase active sites on the either side of heme prosthetic group which is important for both activities. Peroxidase active site, on the opposite side of membrane-binding domain, consists of heme prosthetic group (Garavito & Mulichak, 2003; Rouzer & Marnett, 2009), which from both isoforms has comparable affinities for cyanide as the sixth axial ligand, and yet sequence of side chain between two isoforms near peroxidase site is less conserved than any other catalytic domain area (Goodwin, Rowlinson, & Marnett, 2000). The most strongly conserved region in COX isoforms is around the COX active site, which is located on the top of a long hydrophobic channel originating from the membrane-binding domain near the heme. The binding site for fatty acids as well as competitive COX inhibitors resides above this channel (Kulmacz, van der Donk, & Tsai, 2003; Rouzer & Marnett, 2009). When arachidonic acid binds to the COX site, it sits at the end of channel and carbon-13 of arachidonic acid is placed near Tyr-385, which is a crucial amino acid for COX reaction. However, mutation of the Tyr-385 does not impact the midpoint potential of heme in COX2 further suggesting that peroxidase site is independent of COX site. Moreover, crystallographic as well as biochemical data showed that two active sites are physically separated as fatty acids do not interfere peroxidase catalysis (Kulmacz et al., 2003).

The conversion of arachidonic acid to the hydroperoxy-endoperoxide, prostaglandin G2 (PGG2), is initiated by abstraction of the pro-S hydrogen atom from carbon-13 and then dioxygen traps the pentadienyl radical at carbon 11. This reaction is followed by a series of radical reaction which is similar to polyunsaturated fatty acid autoxidation producing PGG2. The enzyme activation, which is dependent of peroxidase activity, is required for this COX reaction. In this process, an electron is transferred from heme to Tyr-385 generating tyrosyl radical in the cyclooxygenase active site and initiating a reaction. Moreover, tyrosyl radical is regenerated by reducing peroxyl radical to the hydroperoxide to form PGG2. After COX reaction, peroxidase reduces the 15-hyroperoxy of PGG2 to the corresponding alcohol of PGH2 (Fig. 9.2; Kulmacz et al., 2003; Rouzer & Marnett, 2009; Smith, DeWitt, & Garavito, 2000; Tsai & Kulmacz, 2010). Even though the two isoforms appear to follow the same basic mechanism and their three-dimensional crystal structures can be almost superimposed (Rouzer & Marnett, 2009), there is a minor difference within the substrate-binding site in catalytic domain which explains the specificity of nonsteroid anti-inflammatory drugs (NSAIDs) (Garavito & Mulichak,

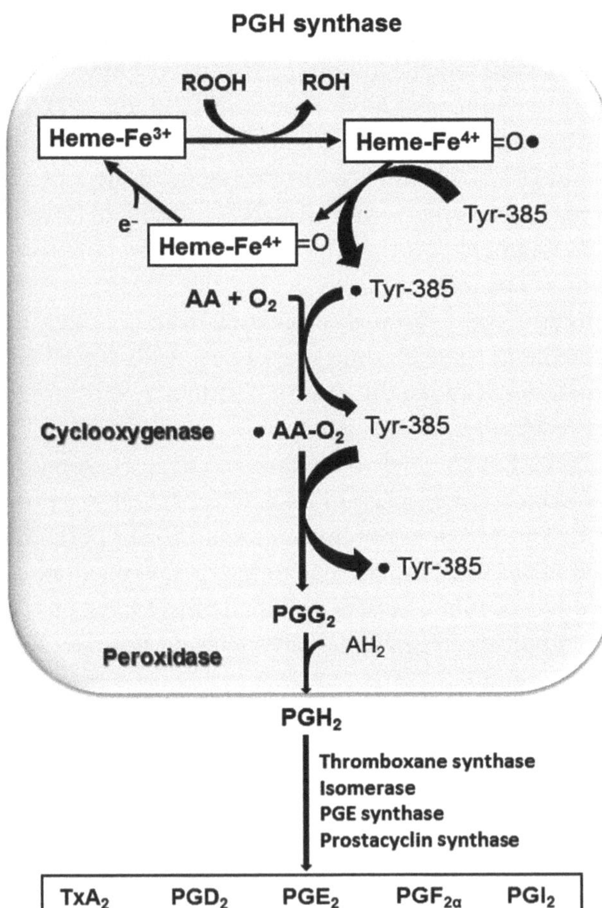

Figure 9.2 The scheme of COX enzyme reaction mechanism and prostaglandin synthesis. COX enzymes produce PGH2 by a two-step chemical reactions: cyclooxygenase and peroxidase. PGH2 is further converted to other prostanoids by tissue-specific enzymes.

2003). The active site of COX2 is larger due to a replacement of isoleucine-434 of COX1 with valine-434. The identification of structural difference between the active sites allowed the scientists to develop isoform-specific inhibitors.

Despite the structural similarity of the COX isoforms, their expression patterns and localization are very distinct. Constitutively expressed COX1 is present in tissues such as the stomach, gut, or kidney, where PGs play a cytoprotective role in maintaining normal physiological processes. By contrast, COX2 is expressed in many stimulated host defense cells at the site of inflammation, leading to robust production of proinflammatory PGs (Rouzer & Marnett, 2009). Considering this obvious difference in expression patterns, it had been speculated that COX1 produces prostaglandins which are required for homeostatic functions such as gastric protection and hemostasis, whereas COX2 participates in pathophysiological processes including inflammation and tumorigenesis. This idea drove pharmaceutical companies to invest massive resources to develop COX2-specific inhibitors as anti-inflammatory drugs to avoid common side effects of traditional NSAIDs such as gastrointestinal bleeding (DuBois et al., 1998; Masferrer et al., 1994). However, further extensive research has revealed that this path may not be ideal. First, genetic deletion of COX1 in mice did not show an increased susceptibility of gastric ulcer (Langenbach et al., 1997), while COX2 inhibitors have been shown to attenuate the healing process (Wallace, McKnight, Reuter, & Vergnolle, 2000). Second, COX1 is also expressed in inflammatory cells (Choi, Langenbach, & Bosetti, 2008), and COX2 is constitutively expressed in certain tissues including brain, and kidney where it exerts protective effects (Hao & Breyer, 2008; Langenbach et al., 1997; Lipsky et al., 2000; Yamagata, Andreasson, Kaufmann, Barnes, & Worley, 1993). Interestingly, it has been showed that COX1 knockout rather than COX2 deletion reduced the LPS-mediated inflammatory response in the brain (Aid, Langenbach, & Bosetti, 2008). Induced COX2 has also been observed in endothelial cells; its inhibition with COX2 inhibitors that spare COX1 function in platelets has been held responsible for certain cardiovascular side effects observed with the usage of COX2-specific inhibitors under chronic inflammatory conditions (Mukherjee, Nissen, & Topol, 2001; Solomon et al., 2004). Moreover, COX2 knockout mice showed unexpected results: disruption of postnatal kidney development, infertility in female mice due to failure to ovulate, and embryo implantation (Hao & Breyer, 2008; Langenbach, Loftin, Lee, & Tiano, 1999; Lipsky et al., 2000; McAdam et al., 2000). Hence, to ascribe such a dichotomy in function between the COX isoforms may be considered too general, though, there is clearly some differential functions between them. An elegant study performed by Funk's group showed that replacement of COX2 with COX1 cannot compensate for a loss of function in COX2 (Yu et al., 2007), strongly suggesting that there are specific roles assigned to each isoform. Moreover, utilizing various COX inhibitors and sophisticated structure–functional analyses, it has been shown that COX enzymes form structural homodimers, and yet exhibit functional heterodimers, as inhibition of only one catalytic site is sufficient to inhibit homodimer enzymatic activity, and binding of various fatty acids in a catalytic site can alter their enzyme activities (Prusakiewicz et al., 2009; Sidhu, Lee, Yuan, & Smith, 2010; Vecchio, Simmons, & Malkowski, 2010; Yu et al., 2006; Yuan et al., 2006). Furthermore, simultaneous expression of two isoforms of COXs led to an interesting observation by Yu and coworkers demonstrating that COX1 and COX2 can form a heterodimer (Yu et al., 2006) which appears to play a critical role in the transformation that occurs in the blood vessels of newly born mice, implicating more complex roles and regulation of COX enzymes in biology. Hence, it appears that the functions of COX1 and COX2 may be differentiated not only by their expression patterns or responses to stimuli but also by their intrinsic kinetic properties.

3.2. Function of prostaglandin signaling in biology

Prostaglandins and related compounds produced from arachidonic acids are some of the most prevalent compounds detected in every tissue and body fluid except for red blood cells. Prostaglandins are members of a lipid family which are enzymatically derived from polyunsaturated fatty acids with 20 carbon chains and 4 *cis*-double bonds (20:4) via the COX pathway. Prostaglandins are also known as "prostanoids," which include classical prostaglandin as well as specialized species such as thromboxane and prostacyclin (Miller, 2006; Smith et al., 1991).

The existence of biologically active PGs was first described in 1930 when Kurzrok and Lieb showed that fresh semen led to rhythmic contraction of human myometrium *in vitro* (Kurzrok & Lieb, 1930). Shortly thereafter, a couple of other groups independently confirmed that presence of an active substance responsible for muscle contraction and named it "prostaglandin," as an homage to the origin being the prostate gland (Goldblatt, 1935; Miller, 2006; von Euler, 1936). Continued research has revealed that prostaglandins are a mixture of biologically active substances such as prostaglandin E or prostaglandin F, but that ironically, the actual origin of prostaglandins is the seminal vesicle and not the prostate gland (Bergstrom & Sjovall, 1957).

3.2.1 Nervous system

The presence of prostaglandin in the brain was first reported in 1964 (Samuelsson, 1964). It has been speculated that prostaglandin production from COX2 plays a prominent role in fever, pain, or any pathological conditions (Choi, Aid, & Bosetti, 2009). In particular, it has been reported that there is an increase in inflammatory proteins in the brains of AD patients and some epidermiological studies indicated that NSAIDs decrease the risk of developing AD. Moreover, COX2 is upregulated in the brains of patients with AD, suggesting abnormal modulation of the COX pathway (Pasinetti & Aisen, 1998), and yet several clinical trials with NSAIDs or selective COX2 inhibitors showed little or no effect on clinical progression of AD (Launer, 2003; Townsend & Pratico, 2005).

Imbalance of glutamatergic system is associated with various neuropsychiatric conditions such as schizophrenia (Halene et al., 2009; Kantrowitz & Javitt, 2010). Moreover, this is closed related to the immune system. Indeed, it has been repeatedly shown that a change in immune response occurs in some of neuropsychiatric diseases. Interestingly, infection during pregnancy in mothers of offsprings later developing schizophrenia has been demonstrated by several laboratories (Brown et al., 2004; Buka et al., 2001; Westergaard, Mortensen, Pedersen, Wohlfahrt, & Melbye, 1999). COX2 inhibitors have been examined in animal models of schizophrenia showing a favorable effect compared to *placebo* (Kaiya et al., 1989). However, the role of prostaglandin in schizophrenia is not well characterized, even though an increase in PGE2 has been reported with schizophrenia patients (Kaiya et al., 1989).

It has been shown that a dramatic increase in COX2 mRNA and protein levels occurs under excitotoxic conditions, defined as a process by which high levels of glutamate or its analogs excite neurons and induce neuronal cell death (Chen, Magee, & Bazan, 2002). Glutamate enhances calcium influx, which can initiate multiple cascades of events including a stimulation of NO production by nNOS and activation of enzymes involved in generation of arachidonic acids and its metabolism, such as phospholipase A2, COX2, and lipoxygenase (Bosetti, Langenbach, & Weerasinghe, 2004; Piomelli et al., 1989). Hence, it is not surprising that glutamate-mediated excitotoxic conditions augment prostaglandin production. Indeed, our laboratory has shown that NMDA-mediated activation of COX2 requires nNOS and plays a crucial role in NMDA excitotoxicity which can be attenuated by COX2-specific inhibitors (Tian, Kim, Hester, & Snyder, 2008).

Ischemic insults also impact production or trafficking of prostaglandins and other arachidonic acid metabolites (Phillis & O'Regan, 2003). Interestingly, both COX1 and COX2 are increased in brain ischemia. Furthermore, selective inhibition of either COX1 or COX2 significantly reduces neuronal cell death, suggesting that both COX isoforms participate in the progression of neuronal damage following global cerebral ischemia and identifies a potential therapeutic use of COX inhibitors in cerebral ischemia (Candelario-Jalil et al., 2003).

One interesting aspect of neuronal COX2 is that unlike in other tissues, COX2 is constitutively expressed in neurons (Yamagata et al., 1993); thus, we can speculate that COX2 may play a role not only in pathological progression as described above but also in physiological neuronal function such as synaptic plasticity. Moreover, COX2 expression level is higher in hippocampal and cortical neurons involved in cognitive function than any other regions. Interestingly, Worley's group has shown that the basal expression of COX2 is regulated by NMDA receptor-dependent synaptic activity and especially upregulated by long-term potentiation (LTP), a form of synaptic plasticity and cellular model of learning and memory, suggesting that the expression and activity of COX2 may be associated with long-term synaptic plasticity (Kaufmann, Worley, Pegg, Bremer, & Isakson, 1996; Yamagata et al., 1993). This idea was further confirmed by showing that COX2-specific inhibitors or suppression of COX2 gene reduced high-frequency stimulation (HFS)-induced LTP implying that constitutively expressed COX2 plays a crucial role in synaptic plasticity (Chen et al., 2002; Yamagata et al., 1993). However, hippocampal slices prepared from COX2-deleted mice did not show any significant difference on HFS-induced LTP (Ballou, Botting, Goorha, Zhang, & Vane, 2000), implying that COX1 may be able to compensate for the loss of COX2 and its own role in synaptic plasticity (Chen & Bazan, 2005).

3.2.2 Cancer

The potential link between prostaglandin and tumorigenesis was initially proposed several decades ago (Sykes & Moddox, 1972; Williams, Karim, & Sandler, 1968). A large body of evidence from population-based studies, case control studies, and clinical trial indicates that regular use of NSAIDs over a 10–15-year period reduces the relative risk of developing a colon cancer (Giovannucci et al., 1995; Thun, Namboodiri, & Heath, 1991). The effects of prostaglandins on tumors are not limited to colon cancer and extend to cancers in other tissues such as breast (Ristimaki et al., 2002), pancreas (Tucker et al., 1999), esophagus (Zimmermann et al., 1999), lung (Hida et al., 1998), and gastric (Murata et al., 1999). Hence, COX2 expression is dysregulated in many types of cancer, and many laboratories investigated the possible role of COX2 in cancer as well as the potential mechanism of NSAID action as antitumor drug. In particular, COX2-derived PGE2 and its receptors play a predominant role in promoting cancer progression (Tian & Schiemann, 2010; Wang, Honn, & Nie, 2007). PGE2 elicits its cellular effects by binding to its cognate receptors (EP1–4) that belong to the family of seven-transmembrane G-protein-coupled receptors. A key role of PGE2 in tumorigenesis has been confirmed by genetic deletion of EP1 and EP4 receptors leading to a partial resistance to colon carcinogen-mediated induction of aberrant crypt foci (Greenhough et al., 2009; Mutoh et al., 2002; Watanabe et al., 1999). Moreover, an EP1-specific antagonist also reduced the incidence of crypt foci in wild-type mice, suggesting that PGE2-mediated activation of EP1 pathway plays a crucial role in tumorigenesis (Watanabe et al., 1999). Significant research efforts are currently being undertaken to identify more selective targets and elucidate the molecular pathway of prostaglandin-mediated oncogenesis. As the rate-limiting enzyme for this pathway, COX2 is garnering considerable attention in hopes that COX2-specific inhibitors may be used as anticancer drugs or for cancer prevention; however, further studies are necessary.

3.2.3 Cardiovascular

Prostaglandin E series are generally potent vasodilators of the vascular bed (Bergstrom, Carlson, & Weeks, 1968). They usually affect small vessels including arterioles and pre/postcapillary venules (Smyth, Grosser, Wang, Yu, & FitzGerald, 2009). Thromboxane and prostacyclin are more specialized prostaglandins that exert major impacts on the cardiovascular system (Bhagwat, Hamann, Still, Bunting, & Fitzpatrick, 1985; Smyth et al., 2009). Thromboxane is normally produced by platelets, acting as a very potent vasoconstrictor, and plays an important role on platelet aggregation and hemostasis. Its effect is counterbalanced by prostacyclin, which is a powerful vasodilator and inhibitor of platelet aggregation. Prostacyclin is the main product in the vascular system (Bunting, Gryglewski, Moncada, & Vane, 1976) and is produced in endothelial cells (MacIntyre, Pearson, & Gordon, 1978; Weksler, Marcus, & Jaffe, 1977). As prostaglandin production in the endothelial cell is dependent on COX2 induction (Schmedtje, Ji, Liu, DuBois, & Runge, 1997), the usage of COX2-specific inhibitors under chronic inflammatory conditions has raised concerns (Mukherjee et al., 2001; Solomon et al., 2004), leading to some COX2-specific inhibitors being withdrawn from the market due to an increased cardiovascular risk.

3.2.4 Thermogenesis and metabolism

Body temperature regulation is a fundamental homeostatic function that is regulated by the central nervous system in homeothermic animals (Morrison, Nakamura, & Madden, 2008). Moreover, the production of heat, thermogenesis, plays a crucial role in host defense mechanism to reduce pathogen viability as well as to stimulate immune cell responses. While prostaglandins are produced in virtually every system in the human body and play a role in cellular homeostasis, recent studies have uncovered an unexpected role for COX2-produced prostaglandins. Kristiansen's group first reported that uncoupling protein 1, a hallmark protein in brown adipocytes crucial for cold- or diet-induced thermogenesis, is regulated by COX2 (Madsen et al., 2010). Moreover, Herzig's group demonstrated that overexpression of COX2 in white adipose tissue, where excess fat is stored, can convert these cells to become brown adipose tissue, which burns fat to produce heat (Vegiopoulos et al., 2010). This brown fat is usually prominent in infants who require more energy to maintain their body temperature and continuously regress with age. A considerable amount of this brown fat still remains in adulthood. Since energy consumption during thermogenesis can involve oxidation of lipid molecules, regulation of thermogenesis in response to metabolic signals can also contributes to energy balance and regulation of body adipose store. Indeed, the amount of brown fat is correlated to leanness (Saito et al., 2009; van Marken Lichtenbelt et al., 2009). Hence, increasing brown fat content in humans has become an attractive target for antiobesity drugs and manipulation of COX2 activity may be a new therapeutic strategy, though a strong note of caution is warranted, as the triggering of the inflammatory response is a potential side effect of enhanced prostaglandin production.

3.2.5 Bone metabolism and fracture healing

Prostaglandins play an important role in bone metabolism (e.g., resorption and formation). Moreover, the production of various lipid mediators is increased in bone-resorptive disease such as osteoporosis, rheumatoid arthritis, and osteoarthritis (Hikiji, Takato, Shimizu, & Ishii, 2008). In particular, PGE2 is one of the most significant lipid mediators of bone remodeling and has been linked clinically to many bone-resorptive diseases (Kawaguchi, Pilbeam, Harrison, & Raisz, 1995). The bone-resorptive action of PGE2 was first documented *in vitro* more than 3 decades ago in a rat organ cultures (Klein & Raisz, 1970). Since then, numerous studies support the crucial effect of PGE2 on bone metabolism *in vitro* as well as *in vivo* (Hikiji et al., 2008; Kawaguchi et al., 1995). Osteoblasts is the primary site for the production of PGE2 which acts in an autocrine fashion (Kawaguchi et al., 1995). This action is medicated essentially mediated by PGE2 receptors (EP1–EP4). In rodents, all the isoforms were detected at transcript levels, while only EP3 and EP4 were observed immunohistochemically in human (Fortier, Gallant, Hackett, Patry, & de Brum-Fernandes, 2004; Miyaura et al., 2000). Genetic deletion study further supports the positive role of PGE2 in bone metabolism (Miyaura et al., 2000). Deletion of EP4-deficient mice exhibits reduced structural strength and imbalance in bone remodeling at the later stage while they appear normal at younger ages (Akhter, Cullen, & Pan, 2006; Li et al., 2005). EP2 knockout mice have unusually weak bone, whereas EP1-deficient seems normal (Akhter, Cullen, Gong, & Recker, 2001). Moreover, the effect of exogenously administered PGE2 was absent in EP4 knockout mice, suggesting that EP4 is the primary receptor mediating bone metabolism via PGE2 (Akhter et al., 2001, 2006; Li et al., 2005).

Bone healing is another complex cascades in our body to repair fractured bone without the formation of scar tissue (Sathyendra & Darowish, 2013). A bone fracture results in the disruption of local blood supply, hypoxia, and the formation of hematoma (Kwong & Harris, 2008; Sathyendra & Darowish, 2013). During this process, many cytokines are produced leading to elevation of PGE2 as well as COX2 expression (Su & O'Connor, 2013). Hence, the effects of NSAID have been studied extensively in various animal models. Several studies have shown that NSAIDs delay bone repair (Blackwell, Raisz, & Pilbeam, 2010; Goodman et al., 2002; Hochberg, Melin, & Reicin, 2003; Kurmis, Kurmis, O'Brien, & Dalen, 2012; Simon, Manigrasso, & O'Connor, 2002; Su & O'Connor, 2013). Moreover, fracture healing is slowed down in COX2-deficient mice compared to either wild-type or COX1-deficient mice. Importantly, PGE2 administration rescued the phenotypes observed in COX2-deficient mice (Pilbeam, Fall, Alander, & Raisz, 1997). However, it is prudent to point out that there are other numerous studies which have failed to reproduce positive correlation between prostaglandin and bone metabolism (Blackwell et al., 2010; Hochberg et al., 2003). It is partially due to the different strains of mice, dosages, or duration of drugs, cells isolated from various species and site. Furthermore, there are very few clinical studies firmly establishing this link (Pountos, Georgouli, Calori, & Giannoudis, 2012). Hence, it is necessary for the further studies defining an exact mechanism by which prostaglandin modulates bone metabolism along with well-randomized prospective clinical trials.

4. THE ROLE OF NITRIC OXIDE IN PROSTAGLANDIN REGULATION

NO and prostaglandin pathways share numerous similarities, and the two molecules can be generated simultaneously in the same tissues, as described in a number of models of inflammation such as endotoxin-induced septic shock, carrageenan-induced pouch, or paw inflammation (Fig. 9.2; Cuzzocrea & Salvemini, 2007). Hence, many studies have explored the potential link between NO and prostaglandin pathways and increasing evidence supports this possibility. However, the detailed mechanisms by which NO regulates prostaglandin production or vice versa is still controversial, partly because the interaction between these two pathways occurs at multiple levels along with the complexity of NO redox chemistry (Cuzzocrea & Salvemini, 2007; Mollace, Muscoli, Masini, Cuzzocrea, & Salvemini, 2005). An initial report detailing the cross talk between NO and prostaglandin was made by Needleman's group showing that NO can activate COX (Salvemini et al., 1993). In this study, NOS and COX2 activity were induced in a macrophage cell line, RAW264.7, treated with LPS to produce both NO and prostaglandin. The production of both was attenuated by NOS inhibitors. As the NOS inhibitors used in this study do not have any NSAID characteristics, the inhibition of prostaglandin production was likely due to the decrease in NO production and not as a direct effect on COX2 (Salvemini et al., 1993). This study was further supported by *in vitro* studies demonstrating that even exogenous NO gas or chemical donors can induce COX1 activity (Salvemini & Masferrer, 1996; Salvemini et al., 1994). Moreover, NO-mediated activation of COX does not seem to be limited to its enzyme activity. Liu et al. showed that NO can increase COX2 mRNA levels via the β-catenin/TCF pathway leading to activation of the polyoma enhancer activator 3 transcription factor (Liu, Borchert, & Phang, 2004).

Furthermore, NO also interacts with various other pathways which can influence COX expression such as the cAMP/PKA/CREB and JNK/Jun/ATF2 signaling cascades (Liu et al., 2004; Park et al., 2005).

In addition to this body of evidence showing an upregulation of COX activity by NO, numerous studies demonstrated that NO can inactivate COX under certain conditions. It has been shown that NOS inhibitors enhanced LPS-induced COX2 expression and production of prostaglandin in microglia, J774 macrophages cell line, or the rat Kupffer cells (Minghetti, Polazzi, Nicolini, Creminon, & Levi, 1996). Moreover, these results were confirmed by evidence that exogenous NO donors applied exogenously to LPS-treated microglial cells inhibit their PG release, an effect driven by reduced expression of COX2 (Stadler et al., 1993; Swierkosz, Mitchell, Warner, Botting, & Vane, 1995). Interestingly, the interaction between NO and prostaglandin is not unidirectional (Chen et al., 1998). Products of the COX pathway may inhibit or increase the release of NO. High levels of PGI_2 inhibit LPS-stimulated iNOS induction in J774.2 macrophages or murine peritoneal macrophages without affecting the enzyme once expressed (Kim, 2011). In rat Kupffer cells stimulated with endotoxin, inhibition of endogenously produced PG by indomethacin resulted in a concomitant inhibition of NO release (Kim, 2011), and conversely, indomethacin did not inhibit NO release in mouse RAW 264.7 cells nor J774.2 macrophages (Salvemini et al., 1993). Thus, there are clearly cell-specific responses to PG exposure. Other studies have found an inhibitory influence of PGE_2 on nNOS expression. Thus, in cultured macula densa cells, PGE_2 significantly reduced nNOS mRNA expression. In $COX2^{-/-}$ mice, nNOS mRNA expression in kidney was upregulated throughout the postnatal periods (Cheng et al., 2002).

COX activation in turn modulates L-arginine-NO pathway, and it has been shown that COX inhibition decreases NOS activity in human platelets (Chen, Salafranca, & Mehta, 1997). Both COX and NOS pathways regulate platelet function; thromboxane A_2 (TXA_2) induces platelet aggregation, whereas NO inhibits it. Inhibition of COX by two different agents, aspirin and indomethacin, markedly attenuates NOS activity as determined directly by the measurement of L-citrulline and nitrite formation and indirectly by the measurement of cGMP accumulation in platelets. These effects were not associated with a change in NOS protein expression in platelets, suggesting that the effects of COX inhibitors are mediated, at least in part, via TXA_2 inhibition and Ca^{2+} mobilization in platelets (Chen et al., 1997). It appears that these diverse experimental data on the inhibitory/stimulatory effects of PG on the NOS pathways are controversial and insufficient to draw any potential conclusion on functional relevance.

The mechanism of reciprocal regulation between the NOS and COX pathways interact is complicated by conflicting reports in the field. Currently, there are several proposed mechanisms by which NO modulates prostaglandin synthesis (Fig. 9.3). One is that NO and superoxide can be produced simultaneously and can react together to form peroxynitrite, which can eventually modify COX. Another possible mechanism is the direct S-nitrosylation of COX enzymes. These two posttranslational modifications can alter the function of these enzymes and are further discussed below.

4.1. Peroxynitrite-mediated regulation of prostaglandin synthesis

Superoxide and nitric oxide are readily produced in the cells under both physiologic and pathological conditions (Szabo et al., 2007). Peroxynitrite can be quickly generated upon production of these two compounds. Many laboratories have reported that peroxynitrite can interact with COX enzymes, but the consequence of the interaction differs depending on where this interaction occurs. So far, two peroxynitrite-binding sites in COX enzymes have been identified: within the endogenous heme moiety and at tyrosine residues on the protein.

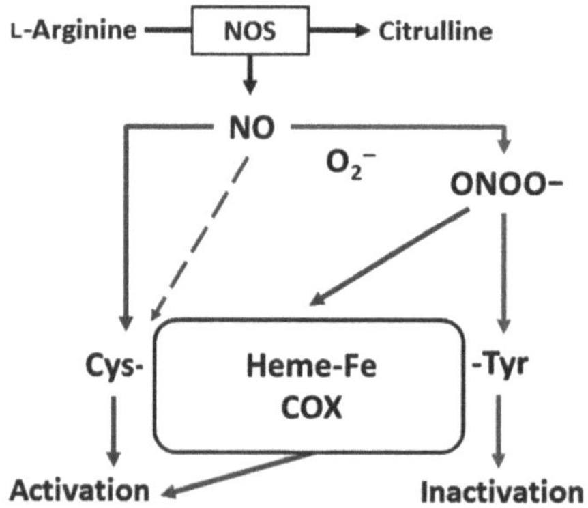

Figure 9.3 Proposed mechanisms of NO-mediated regulation of prostaglandin production. Nitric oxide produced by NOS can directly modify cysteine(s) in the COXs and activate it. Moreover, superoxide can be simultaneously generated under the same environment which can enhance NO production. Superoxide rapidly interacts with NO-producing peroxynitrite which can interact with either heme or tyrosine residue(s) in COX2. (See the color plate.)

Several groups have shown that peroxynitrite can activate both. In particular, treatment of cells with SIN-1, a chemical donor of peroxynitrite, increases prostaglandin synthesis and oxygen radical scavengers effectively prevented SIN-1-medicated COX activation, strongly implicating peroxynitrite as an activator of COX (Boulos, Jiang, & Balazy, 2000). The mechanism by which peroxynitrite activates COX activity still remains elusive; however, it has been speculated that peroxynitrite may interact with Fe in the heme group of COX forming a radical intermediate product which can facilitate the catalytic process (Karthein, Dietz, Nastainczyk, & Ruf, 1988; Marnett, Rowlinson, Goodwin, Kalgutkar, & Lanzo, 1999; Tsai, Palmer, & Kulmacz, 1992). This hypothesis is further supported by the fact that peroxynitrite is known to interact with the heme group of certain proteins such as myeloperoxidase and horseradish peroxidase to generate an intermediate product (Floris, Piersma, Yang, Jones, & Wever, 1993). However, Tsai et al. reported that it is unlikely that NO directly binds to heme group in COX to activate under anaerobic condition (Tsai, Wei, & Kulmacz, 1994).

Most of the studies about peroxynitrite and COX enzymes have been focused on tyrosine residue modification in the enzyme generating a nitrotyrosine residue in the enzymes. It appears that this modification, especially at Tyr385, leads to an inactivation of enzyme (Boulos et al., 2000; Deeb et al., 2002; Goodwin et al., 1998; Shi et al., 2000), as demonstrated by inhibition of COX1 enzyme activity when either purified enzyme COX1 or smooth muscle cells are treated with NO donors and arachidonic acid simultaneously (Upmacis, Deeb, & Hajjar, 1999). It is interesting to note that pure NO donors such as NOC-7 do not lead to formation of nitrotyrosine in the absence of arachidonic acid, while peroxynitrite, SIN-1, or tetranitromethane can readily carry out this reaction. Therefore, it appears that peroxynitrite has dual actions on COX depending on where it interacts in the enzyme.

Even though nitrotyrosine modification of COXs is detected in many pathological conditions and clearly inactivates its enzyme activity *in vitro*, the exact mechanism of nitration is poorly understood. There is no doubt that the levels of peroxynitrite is significantly elevated under pathological conditions, as inflammation induces NO production from iNOS and superoxide production from by NAD(P)H and xanthine oxidase and correlates to formation of nitrotyrosine in COXs. However, it is crucial to identify different conditions that determine the target selectivity of peroxynitrite: with a modification of tyrosine to inactivate an enzyme versus interaction with heme in COXs to activate it. Interestingly, Deeb et al. showed that holoCOX1 (heme-containing) can be nitrated at Tyr 385 by peroxynitrite, even in the absence of arachidonic acid, while apoCOX1 (heme-deficient) is modified at the different site, thus indicating a heme may be required to catalyze peroxynitrite-mediated Tyr385 nitration (Deeb et al., 2006). However, the identity of the nitrotyrosine residue in apoCOX1 and its downstream biological function are yet to be determined.

In other occasion, it has been proposed that peroxynitrite may interact with Fe in the heme moiety of COX to generate a short-lived intermediate which can facilitate the formation of nitrotyrosine (Upmacis, Deeb, & Hajjar, 2006). While heme is essential for this reaction within COX, the role of this cofactor is unclear, as there are other proteins which can be nitrated without heme moiety in the protein (Kanski, Behring, Pelling, & Schoneich, 2005). It is possible that Fe binding in the heme molecule in COX may expose the Tyr385 residue, which resides just below the heme moiety, making it more easily accessible to peroxynitrite. It will be interesting to examine whether recombinant COX with non-Fe protoporphyrin such as Zn-protoporphyrin or Sn-protoporphyrin can be targeted by peroxynitrite. This theory, however, conflicts in part with the aforementioned evidence that peroxynitrite binding to the heme moiety in COX leads to activation of the enzyme, while nitration of Tyr385 leads to enzyme inactivation. This may be partially explained by differing concentrations of arachidonic acid, which can be oxidized by peroxynitrite, with this oxidized substrate having different impacts on COX activities. Deeb et al. have indeed shown that a concentration of arachidonic acid and formation of nitrotyrosine in COX1 are inversely related (Deeb et al., 2010).

Finally, there are other enzymes which can be modified by peroxynitrite in prostaglandin synthesis pathway. Prostacyclin synthase (PGCI) is selectively inhibited by peroxynitrite at low concentration (IC50 = 50 nM), while TxA_2 synthase is activated (Zou, Martin, & Ullrich, 1997; Zou & Ullrich, 1996). It appears that nitration of tyrosine residue 430 in PGCI interferes the metal center of the active site and inhibits its activity (Zou, Shi, & Cohen, 2002; Zou & Ullrich, 1996).

Most importantly, this peroxynitration modification of COX enzymes has been detected in many pathological or chronic inflammatory conditions such as atherosclerosis, Parkinson's disease, or AD. However, it is not clear whether an inactivation of this enzyme by nitrotyrosine formation contributes to the pathogenesis of these diseases or simply an indication of nitrosative stress.

4.2. S-nitrosylation of COX

The recent development of a simple method called biotin switch assay to detect S-nitrosylation flourished identification of new targets and its functional consequence since this modification was first reported with albumin by Stamler et al. COX has also been identified as a target of this modification by various laboratories (Foster et al., 2009; Jaffrey, Erdjument-Bromage, Ferris, Tempst, & Snyder, 2001; Stamler et al., 1992). It was first proposed by Hajjar et al. showing that a formation of *S*-nitrosothiol may be responsible for NO-mediated activation of COX as this modification occurs in a heme-independent manner (Hajjar, Lander, Pearce, Upmacis, & Pomerantz, 1995). However, Marnett's group investigated the role of free cysteines in COX1 and showed that Cys313 or Cys540 mutation into serine alone inhibited its enzyme activity implying that NO-mediated Cys modification may interfere with COX1 activity (Kennedy, Smith, & Marnett, 1994). Moreover, Marnett's and Hajjar's groups independently reported that SNAP, an NO chemical donor, cannot activate COX1 (Landino, Crews, Timmons, Morrow, & Marnett, 1996; Upmacis et al., 1999); thus, confusion exists over whether S-nitrosylation of COX1 leads to its activation.

Our group has shown that COX2 can be activated by NO as a result of its S-nitrosylation as long as NO generated by iNOS is directly delivered to a target residue of COX2 by their protein–protein interaction (Kim, Huri, & Snyder, 2005). This interaction and modification are specific for COX2 as iNOS does not interact with COX1. In particular, we observed more than one S-nitrosylated cysteine residues in LPS/IFN-γ activated RAW264.7 cells and individually mutated each cysteine to serine. We have identified Cys523, located near the catalytic domain, as the only cysteine which does not affect base level enzyme activity, and when mutated, NO can no longer activate COX2. This cysteine is located near the catalytic site of COX2, and hence, we speculated that this modification leads to a slight conformational change in COX2 to affect its enzyme kinetics. Indeed, we performed viscosity studies examining enzyme activity in various concentrations of sucrose and showed that release of product is accelerated in the presence of NO donor. However, it is not possible to exclude a potential involvement of any other cysteines for NO-mediated activation of COX2 as mutation of other cysteines alone affected the basal level enzyme activity of COX2. Given that NO has a high affinity for heme, it is unclear how NO can specifically S-nitrosylate cysteines of COX2 rather than interacting with heme moiety. This may be partially explained by the fact that S-nitrosylation of

COX2 by iNOS requires physical interaction between the two proteins, specifically between the subunit-binding site of iNOS and catalytic domain of COX2. Hence, we postulated that NO generated from iNOS is simply in direct proximity of the S-nitrosylation target residue in COX2 for immediate delivery and modification. Importantly, disruption of iNOS/COX2 binding indeed diminishes S-nitrosylation and activation of COX2 in LPS-induced RAW264.7 cells. Interestingly, Goodwin et al. demonstrated that removal of superoxide in LPS-activated RAW264.7 cells inhibited prostaglandin formation to the same extent as NOS inhibitors (Goodwin et al., 1998). While the immediate speculation will be that nitrotyrosine is formed in COX2 by peroxynitrite, Cross' group showed that peroxynitrite can also generate *S*-nitrosothiol (van der Vliet, Hoen, Wong, Bast, & Cross, 1998), and therefore, further studies are necessary to identify the form of NO-mediated modification of COX2 in activated macrophages.

S-nitrosylation of COX2 does not seem to be limited to macrophages. Unlike other tissues, COX2 in the brain is constitutively expressed. We showed that COX2 is S-nitrosylated in neuronal cells by nNOS (Tian et al., 2008). To our surprise, nNOS binding is mediated by its PDZ domain, differing from the domain near the catalytic region that mediates iNOS binding to COX2. We also showed that a physical interaction between nNOS and COX2 is required for S-nitrosylation of COX2 and that disruption of binding attenuated NMDA-mediated excitotoxicity, as did a selective COX2 inhibitor.

5. PERSPECTIVE/CONCLUSION

The prostaglandin pathway has been the center of attention as a main target to treat numerous pathological conditions such as pain, inflammation, neurodegeneration, cancer, and metabolic deficiencies. It is clear that NO pathway is tightly coupled to prostaglandin signaling, and new evidence further supports that NO-conjugated COX inhibitors (e.g., nitroaspirin) may have a great potential for the treatment of both inflammatory as well as non-inflammatory conditions. Yet detailed mechanisms by which NO regulates prostaglandin production and vice versa remains poorly understood. Efforts from many outstanding laboratories have proposed multiple potential mechanisms, but some of them are quite contradictory. It is partly due to the multiple levels of interaction between these two pathways, the complexity of NO redox chemistry (Cuzzocrea & Salvemini, 2007; Mollace et al., 2005) and the lack of selective probes to decipher pathways *in vivo*.

Moreover, there is a considerable difference depending on which cell type or tissue is examined and the surrounding microenvironment may also be a crucial factor. Undoubtedly, further studies are necessary to understand the molecular basis of interaction between NO and prostaglandin and that these will unveil a crucial target for intervention in numerous pathological conditions.

REFERENCES

Abu-Soud, H. M., Presta, A., Mayer, B., & Stuehr, D. J. (1997). Analysis of neuronal NO synthase under single-turnover conditions: Conversion of Nomega-hydroxyarginine to nitric oxide and citrulline. *Biochemistry, 36*, 10811–10816.

Aid, S., Langenbach, R., & Bosetti, F. (2008). Neuroinflammatory response to lipopolysaccharide is exacerbated in mice genetically deficient in cyclooxygenase-2. *Journal of Neuroinflammation, 5*, 17.

Akaike, T., Inoue, K., Okamoto, T., Nishino, H., Otagiri, M., Fujii, S., et al. (1997). Nanomolar quantification and identification of various nitrosothiols by high performance liquid chromatography coupled with flow reactors of metals and Griess reagent. *Journal of Biochemistry, 122*, 459–466.

Akhter, M. P., Cullen, D. M., Gong, G., & Recker, R. R. (2001). Bone biomechanical properties in prostaglandin EP1 and EP2 knockout mice. *Bone, 29*, 121–125.

Akhter, M. P., Cullen, D. M., & Pan, L. C. (2006). Bone biomechanical properties in EP4 knockout mice. *Calcified Tissue International, 78*, 357–362.

Bailey, J. M., Muza, B., Hla, T., & Salata, K. (1985). Restoration of prostacyclin synthase in vascular smooth muscle cells after aspirin treatment: Regulation by epidermal growth factor. *The Journal of Lipid Research, 26*, 54–61.

Ballou, L. R., Botting, R. M., Goorha, S., Zhang, J., & Vane, J. R. (2000). Nociception in cyclooxygenase isozyme-deficient mice. *Proceedings of the National Academy of Sciences of the United States of America, 97*, 10272–10276.

Bergstrom, S., Carlson, L. A., & Weeks, J. R. (1968). The prostaglandins: A family of biologically active lipids. *Pharmacological Reviews, 20*, 1–48.

Bergstrom, S., & Sjovall, J. (1957). The isolation of prostaglandin. *Acta Chemica Scandinavica, 11*, 1086.

Bhagwat, S. S., Hamann, P. R., Still, W. C., Bunting, S., & Fitzpatrick, F. A. (1985). Synthesis and structure of the platelet aggregation factor thromboxane A2. *Nature, 315*, 511–513.

Blackwell, K. A., Raisz, L. G., & Pilbeam, C. C. (2010). Prostaglandins in bone: Bad cop, good cop? *Trends in Endocrinology and Metabolism, 21*, 294–301.

Bosetti, F., Langenbach, R., & Weerasinghe, G. R. (2004). Prostaglandin E2 and microsomal prostaglandin E synthase-2 expression are decreased in the cyclooxygenase-2-deficient mouse brain despite compensatory induction of cyclooxygenase-1 and Ca2+-dependent phospholipase A2. *Journal of Neurochemistry, 91*, 1389–1397.

Boulos, C., Jiang, H., & Balazy, M. (2000). Diffusion of peroxynitrite into the human platelet inhibits cyclooxygenase via nitration of tyrosine residues. *The Journal of Pharmacology and Experimental Therapeutics, 293*, 222–229.

Bredt, D. S., Hwang, P. M., Glatt, C. E., Lowenstein, C., Reed, R. R., & Snyder, S. H. (1991). Cloned and expressed nitric oxide synthase structurally resembles cytochrome P-450 reductase. *Nature, 351*, 714–718.

Bredt, D. S., & Snyder, S. H. (1994). Nitric oxide: A physiologic messenger molecule. *Annual Review of Biochemistry, 63*, 175–195.

Brown, A. S., Begg, M. D., Gravenstein, S., Schaefer, C. A., Wyatt, R. J., Bresnahan, M., et al. (2004). Serologic evidence of prenatal influenza in the etiology of schizophrenia. *Archives of General Psychiatry*, *61*, 774–780.

Buka, S. L., Tsuang, M. T., Torrey, E. F., Klebanoff, M. A., Bernstein, D., & Yolken, R. H. (2001). Maternal infections and subsequent psychosis among offspring. *Archives of General Psychiatry*, *58*, 1032–1037.

Bunting, S., Gryglewski, R., Moncada, S., & Vane, J. R. (1976). Arterial walls generate from prostaglandin endoperoxides a substance (prostaglandin X) which relaxes strips of mesenteric and coeliac arteries and inhibits platelet aggregation. *Prostaglandins*, *12*, 897–913.

Candelario-Jalil, E., Gonzalez-Falcon, A., Garcia-Cabrera, M., Alvarez, D., Al-Dalain, S., Martinez, G., et al. (2003). Assessment of the relative contribution of COX-1 and COX-2 isoforms to ischemia-induced oxidative damage and neurodegeneration following transient global cerebral ischemia. *Journal of Neurochemistry*, *86*, 545–555.

Cassina, A. M., Hodara, R., Souza, J. M., Thomson, L., Castro, L., Ischiropoulos, H., et al. (2000). Cytochrome *c* nitration by peroxynitrite. *The Journal of Biological Chemistry*, *275*, 21409–21415.

Chen, C., & Bazan, N. G. (2005). Lipid signaling: Sleep, synaptic plasticity, and neuroprotection. *Prostaglandins & Other Lipid Mediators*, *77*, 65–76.

Chen, K. D., Chen, L. Y., Huang, H. L., Lieu, C. H., Chang, Y. N., Chang, M. D., et al. (1998). Involvement of p38 mitogen-activated protein kinase signaling pathway in the rapid induction of the 78-kDa glucose-regulated protein in 9L rat brain tumor cells. *The Journal of Biological Chemistry*, *273*, 749–755.

Chen, C., Magee, J. C., & Bazan, N. G. (2002). Cyclooxygenase-2 regulates prostaglandin E2 signaling in hippocampal long-term synaptic plasticity. *Journal of Neurophysiology*, *87*, 2851–2857.

Chen, L., Salafranca, M. N., & Mehta, J. L. (1997). Cyclooxygenase inhibition decreases nitric oxide synthase activity in human platelets. *The American Journal of Physiology*, *273*, H1854–H1859.

Cheng, H. F., Wang, C. J., Moeckel, G. W., Zhang, M. Z., McKanna, J. A., & Harris, R. C. (2002). Cyclooxygenase-2 inhibitor blocks expression of mediators of renal injury in a model of diabetes and hypertension. *Kidney International*, *62*, 929–939.

Choi, S. H., Aid, S., & Bosetti, F. (2009). The distinct roles of cyclooxygenase-1 and -2 in neuroinflammation: Implications for translational research. *Trends in Pharmacological Sciences*, *30*, 174–181.

Choi, S. H., Langenbach, R., & Bosetti, F. (2008). Genetic deletion or pharmacological inhibition of cyclooxygenase-1 attenuate lipopolysaccharide-induced inflammatory response and brain injury. *FASEB Journal*, *22*, 1491–1501.

Cuzzocrea, S., & Salvemini, D. (2007). Molecular mechanisms involved in the reciprocal regulation of cyclooxygenase and nitric oxide synthase enzymes. *Kidney International*, *71*, 290–297.

Deeb, R. S., Cheung, C., Nuriel, T., Lamon, B. D., Upmacis, R. K., Gross, S. S., et al. (2010). Physical evidence for substrate binding in preventing cyclooxygenase inactivation under nitrative stress. *Journal of the American Chemical Society*, *132*, 3914–3922.

Deeb, R. S., Hao, G., Gross, S. S., Laine, M., Qiu, J. H., Resnick, B., et al. (2006). Heme catalyzes tyrosine 385 nitration and inactivation of prostaglandin H2 synthase-1 by peroxynitrite. *The Journal of Lipid Research*, *47*, 898–911.

Deeb, R. S., Resnick, M. J., Mittar, D., McCaffrey, T., Hajjar, D. P., & Upmacis, R. K. (2002). Tyrosine nitration in prostaglandin H(2) synthase. *The Journal of Lipid Research*, *43*, 1718–1726.

DeWitt, D. L. (1991). Prostaglandin endoperoxide synthase: Regulation of enzyme expression. *Biochimica et Biophysica Acta*, *1083*, 121–134.

Drapier, J. C., & Hibbs, J. B., Jr. (1988). Differentiation of murine macrophages to express nonspecific cytotoxicity for tumor cells results in L-arginine-dependent inhibition of mitochondrial iron-sulfur enzymes in the macrophage effector cells. *The Journal of Immunology, 140*, 2829–2838.

DuBois, R. N., Abramson, S. B., Crofford, L., Gupta, R. A., Simon, L. S., Van De Putte, L. B., et al. (1998). Cyclooxygenase in biology and disease. *FASEB Journal, 12*, 1063–1073.

Fang, M., Jaffrey, S. R., Sawa, A., Ye, K., Luo, X., & Snyder, S. H. (2000). Dexras1: A G protein specifically coupled to neuronal nitric oxide synthase via CAPON. *Neuron, 28*, 183–193.

Floris, R., Piersma, S. R., Yang, G., Jones, P., & Wever, R. (1993). Interaction of myeloperoxidase with peroxynitrite. A comparison with lactoperoxidase, horseradish peroxidase and catalase. *European Journal of Biochemistry, 215*, 767–775.

Flower, R. J., & Vane, J. R. (1972). Inhibition of prostaglandin synthetase in brain explains the anti-pyretic activity of paracetamol (4-acetamidophenol). *Nature, 240*, 410–411.

Fortier, I., Gallant, M. A., Hackett, J. A., Patry, C., & de Brum-Fernandes, A. J. (2004). Immunolocalization of the prostaglandin E2 receptor subtypes in human bone tissue: Differences in foetal, adult normal, osteoporotic and pagetic bone. *Prostaglandins, Leukotrienes, and Essential Fatty Acids, 70*, 431–439.

Foster, M. W., Hess, D. T., & Stamler, J. S. (2009). Protein S-nitrosylation in health and disease: A current perspective. *Trends in Molecular Medicine, 15*, 391–404.

Fridovich, I. (1999). Fundamental aspects of reactive oxygen species, or what's the matter with oxygen? *Annals of the New York Academy of Sciences, 893*, 13–18.

Garavito, R. M., & Mulichak, A. M. (2003). The structure of mammalian cyclooxygenases. *Annual Review of Biophysics and Biomolecular Structure, 32*, 183–206.

Giovannucci, E., Egan, K. M., Hunter, D. J., Stampfer, M. J., Colditz, G. A., Willett, W. C., et al. (1995). Aspirin and the risk of colorectal cancer in women. *The New England Journal of Medicine, 333*, 609–614.

Giulivi, C. (1998). Functional implications of nitric oxide produced by mitochondria in mitochondrial metabolism. *The Biochemical Journal, 332*(Pt. 3), 673–679.

Goldblatt, M. W. (1935). Properties of human seminal plasma. *The Journal of Physiology, 84*, 208–218.

Goodman, S., Ma, T., Trindade, M., Ikenoue, T., Matsuura, I., Wong, N., et al. (2002). COX-2 selective NSAID decreases bone ingrowth in vivo. *Journal of Orthopaedic Research, 20*, 1164–1169.

Goodwin, D. C., Gunther, M. R., Hsi, L. C., Crews, B. C., Eling, T. E., Mason, R. P., et al. (1998). Nitric oxide trapping of tyrosyl radicals generated during prostaglandin endoperoxide synthase turnover. Detection of the radical derivative of tyrosine 385. *The Journal of Biological Chemistry, 273*, 8903–8909.

Goodwin, D. C., Rowlinson, S. W., & Marnett, L. J. (2000). Substitution of tyrosine for the proximal histidine ligand to the heme of prostaglandin endoperoxide synthase 2: Implications for the mechanism of cyclooxygenase activation and catalysis. *Biochemistry, 39*, 5422–5432.

Gow, A. J., & Stamler, J. S. (1998). Reactions between nitric oxide and haemoglobin under physiological conditions. *Nature, 391*, 169–173.

Greenhough, A., Smartt, H. J., Moore, A. E., Roberts, H. R., Williams, A. C., Paraskeva, C., et al. (2009). The COX-2/PGE2 pathway: Key roles in the hallmarks of cancer and adaptation to the tumour microenvironment. *Carcinogenesis, 30*, 377–386.

Hajjar, D., Lander, H., Pearce, S., Upmacis, R., & Pomerantz, K. B. (1995). Nitric oxide enhances prostaglandin-H synthase-1 activity by a heme-independent mechanism: Evidence implicating nitrosothiols. *Journal of the American Chemical Society, 117*, 3340–3346.

Halene, T. B., Ehrlichman, R. S., Liang, Y., Christian, E. P., Jonak, G. J., Gur, T. L., et al. (2009). Assessment of NMDA receptor NR1 subunit hypofunction in mice as a model for schizophrenia. *Genes, Brain, and Behavior, 8*, 661–675.
Hao, C. M., & Breyer, M. D. (2008). Physiological regulation of prostaglandins in the kidney. *Annual Review of Physiology, 70*, 357–377.
Hara, M. R., Agrawal, N., Kim, S. F., Cascio, M. B., Fujimuro, M., Ozeki, Y., et al. (2005). S-nitrosylated GAPDH initiates apoptotic cell death by nuclear translocation following Siah1 binding. *Nature Cell Biology, 7*, 665–674.
Hemler, M., & Lands, W. E. (1976). Purification of the cyclooxygenase that forms prostaglandins. Demonstration of two forms of iron in the holoenzyme. *The Journal of Biological Chemistry, 251*, 5575–5579.
Henry, Y., Lepoivre, M., Drapier, J. C., Ducrocq, C., Boucher, J. L., & Guissani, A. (1993). EPR characterization of molecular targets for NO in mammalian cells and organelles. *FASEB Journal, 7*, 1124–1134.
Hess, D. T., Matsumoto, A., Nudelman, R., & Stamler, J. S. (2001). S-nitrosylation: Spectrum and specificity. *Nature Cell Biology, 3*, E46–E49.
Hevel, J. M., White, K. A., & Marletta, M. A. (1991). Purification of the inducible murine macrophage nitric oxide synthase. Identification as a flavoprotein. *The Journal of Biological Chemistry, 266*, 22789–22791.
Hida, T., Yatabe, Y., Achiwa, H., Muramatsu, H., Kozaki, K., Nakamura, S., et al. (1998). Increased expression of cyclooxygenase 2 occurs frequently in human lung cancers, specifically in adenocarcinomas. *Cancer Research, 58*, 3761–3764.
Hikiji, H., Takato, T., Shimizu, T., & Ishii, S. (2008). The roles of prostanoids, leukotrienes, and platelet-activating factor in bone metabolism and disease. *Progress in Lipid Research, 47*, 107–126.
Hochberg, M. C., Melin, J. M., & Reicin, A. (2003). Cox-2 inhibitors and fracture healing: An argument against such an effect. *Journal of Bone and Mineral Research, 18*, 583–587.
Huang, Y., Man, H. Y., Sekine-Aizawa, Y., Han, Y., Juluri, K., Luo, H., et al. (2005). S-nitrosylation of N-ethylmaleimide sensitive factor mediates surface expression of AMPA receptors. *Neuron, 46*, 533–540.
Ignarro, L. J. (1996). Physiology and pathophysiology of nitric oxide. *Kidney International. Supplement, 55*, S2–S5.
Inoue, K., Akaike, T., Miyamoto, Y., Okamoto, T., Sawa, T., Otagiri, M., et al. (1999). Nitrosothiol formation catalyzed by ceruloplasmin. Implication for cytoprotective mechanism in vivo. *The Journal of Biological Chemistry, 274*, 27069–27075.
Irie, Y., Saeki, M., Kamisaki, Y., Martin, E., & Murad, F. (2003). Histone H1.2 is a substrate for denitrase, an activity that reduces nitrotyrosine immunoreactivity in proteins. *Proceedings of the National Academy of Sciences of the United States of America, 100*, 5634–5639.
Ischiropoulos, H. (2009). Protein tyrosine nitration—An update. *Archives of Biochemistry and Biophysics, 484*, 117–121.
Jaffrey, S. R., Erdjument-Bromage, H., Ferris, C. D., Tempst, P., & Snyder, S. H. (2001). Protein S-nitrosylation: A physiological signal for neuronal nitric oxide. *Nature Cell Biology, 3*, 193–197.
Jia, L., Bonaventura, C., Bonaventura, J., & Stamler, J. S. (1996). S-nitrosohaemoglobin: A dynamic activity of blood involved in vascular control. *Nature, 380*, 221–226.
Kaiya, H., Uematsu, M., Ofuji, M., Nishida, A., Takeuchi, K., Nozaki, M., et al. (1989). Elevated plasma prostaglandin E2 levels in schizophrenia. *Journal of Neural Transmission, 77*, 39–46.
Kanski, J., Behring, A., Pelling, J., & Schoneich, C. (2005). Proteomic identification of 3-nitrotyrosine-containing rat cardiac proteins: Effects of biological aging. *American Journal of Physiology. Heart and Circulatory Physiology, 288*, H371–H381.

Kantrowitz, J. T., & Javitt, D. C. (2010). N-methyl-d-aspartate (NMDA) receptor dysfunction or dysregulation: The final common pathway on the road to schizophrenia? *Brain Research Bulletin*, *83*, 108–121.

Karthein, R., Dietz, R., Nastainczyk, W., & Ruf, H. H. (1988). Higher oxidation states of prostaglandin H synthase. EPR study of a transient tyrosyl radical in the enzyme during the peroxidase reaction. *European Journal of Biochemistry*, *171*, 313–320.

Kaufmann, W. E., Worley, P. F., Pegg, J., Bremer, M., & Isakson, P. (1996). COX-2, a synaptically induced enzyme, is expressed by excitatory neurons at postsynaptic sites in rat cerebral cortex. *Proceedings of the National Academy of Sciences of the United States of America*, *93*, 2317–2321.

Kawaguchi, H., Pilbeam, C. C., Harrison, J. R., & Raisz, L. G. (1995). The role of prostaglandins in the regulation of bone metabolism. *Clinical Orthopaedics and Related Research*, *313*, 36–46.

Kennedy, T. A., Smith, C. J., & Marnett, L. J. (1994). Investigation of the role of cysteines in catalysis by prostaglandin endoperoxide synthase. *The Journal of Biological Chemistry*, *269*, 27357–27364.

Kim, S. F. (2011). The role of nitric oxide in prostaglandin biology; update. *Nitric Oxide*, *25*, 255–264.

Kim, W. K., Choi, Y. B., Rayudu, P. V., Das, P., Asaad, W., Arnelle, D. R., et al. (1999). Attenuation of NMDA receptor activity and neurotoxicity by nitroxyl anion, NO^-. *Neuron*, *24*, 461–469.

Kim, S. F., Huri, D. A., & Snyder, S. H. (2005). Inducible nitric oxide synthase binds, S-nitrosylates, and activates cyclooxygenase-2. *Science*, *310*, 1966–1970.

Klein, D. C., & Raisz, L. G. (1970). Prostaglandins: Stimulation of bone resorption in tissue culture. *Endocrinology*, *86*, 1436–1440.

Kornberg, M. D., Sen, N., Hara, M. R., Juluri, K. R., Nguyen, J. V., Snowman, A. M., et al. (2010). GAPDH mediates nitrosylation of nuclear proteins. *Nature Cell Biology*, *12*, 1094–1100.

Kulmacz, R. J., van der Donk, W. A., & Tsai, A. L. (2003). Comparison of the properties of prostaglandin H synthase-1 and -2. *Progress in Lipid Research*, *42*, 377–404.

Kurmis, A. P., Kurmis, T. P., O'Brien, J. X., & Dalen, T. (2012). The effect of nonsteroidal anti-inflammatory drug administration on acute phase fracture-healing: A review. *The Journal of Bone and Joint Surgery. American Volume*, *94*, 815–823.

Kurzrok, R., & Lieb, C. (1930). Biochemical studies of human semen. II. *Proceedings of the Society for Experimental Biology and Medicine*, *28*, 268–272.

Kwong, F. N., & Harris, M. B. (2008). Recent developments in the biology of fracture repair. *The Journal of the American Academy of Orthopaedic Surgeons*, *16*, 619–625.

Lancaster, J. R., Jr., & Hibbs, J. B., Jr. (1990). EPR demonstration of iron-nitrosyl complex formation by cytotoxic activated macrophages. *Proceedings of the National Academy of Sciences of the United States of America*, *87*, 1223–1227.

Lander, H. M., Hajjar, D. P., Hempstead, B. L., Mirza, U. A., Chait, B. T., Campbell, S., et al. (1997). A molecular redox switch on p21(ras). Structural basis for the nitric oxide-p21(ras) interaction. *The Journal of Biological Chemistry*, *272*, 4323–4326.

Lander, H. M., Ogiste, J. S., Teng, K. K., & Novogrodsky, A. (1995). p21ras as a common signaling target of reactive free radicals and cellular redox stress. *The Journal of Biological Chemistry*, *270*, 21195–21198.

Landino, L. M., Crews, B. C., Timmons, M. D., Morrow, J. D., & Marnett, L. J. (1996). Peroxynitrite, the coupling product of nitric oxide and superoxide, activates prostaglandin biosynthesis. *Proceedings of the National Academy of Sciences of the United States of America*, *93*, 15069–15074.

Langenbach, R., Loftin, C., Lee, C., & Tiano, H. (1999). Cyclooxygenase knockout mice: Models for elucidating isoform-specific functions. *Biochemical Pharmacology, 58*, 1237–1246.

Langenbach, R., Morham, S. G., Tiano, H. F., Loftin, C. D., Ghanayem, B. I., Chulada, P. C., et al. (1997). Disruption of the mouse cyclooxygenase 1 gene. Characteristics of the mutant and areas of future study. *Advances in Experimental Medicine and Biology, 407*, 87–92.

Launer, L. (2003). Nonsteroidal anti-inflammatory drug use and the risk for Alzheimer's disease: Dissecting the epidemiological evidence. *Drugs, 63*, 731–739.

Li, M., Healy, D. R., Li, Y., Simmons, H. A., Crawford, D. T., Ke, H. Z., et al. (2005). Osteopenia and impaired fracture healing in aged EP4 receptor knockout mice. *Bone, 37*, 46–54.

Lin, A. H., Bienkowski, M. J., & Gorman, R. R. (1989). Regulation of prostaglandin H synthase mRNA levels and prostaglandin biosynthesis by platelet-derived growth factor. *The Journal of Biological Chemistry, 264*, 17379–17383.

Lipsky, P. E., Brooks, P., Crofford, L. J., DuBois, R., Graham, D., Simon, L. S., et al. (2000). Unresolved issues in the role of cyclooxygenase-2 in normal physiologic processes and disease. *Archives of Internal Medicine, 160*, 913–920.

Lipton, S. A., Choi, Y. B., Pan, Z. H., Lei, S. Z., Chen, H. S., Sucher, N. J., et al. (1993). A redox-based mechanism for the neuroprotective and neurodestructive effects of nitric oxide and related nitroso-compounds. *Nature, 364*, 626–632.

Liu, Y., Borchert, G. L., & Phang, J. M. (2004). Polyoma enhancer activator 3, an ets transcription factor, mediates the induction of cyclooxygenase-2 by nitric oxide in colorectal cancer cells. *The Journal of Biological Chemistry, 279*, 18694–18700.

MacIntyre, D. E., Pearson, J. D., & Gordon, J. L. (1978). Localisation and stimulation of prostacyclin production in vascular cells. *Nature, 271*, 549–551.

Macmillan-Crow, L. A., & Cruthirds, D. L. (2001). Invited review: Manganese superoxide dismutase in disease. *Free Radical Research, 34*, 325–336.

Madsen, L., Pedersen, L. M., Lillefosse, H. H., Fjaere, E., Bronstad, I., Hao, Q., et al. (2010). UCP1 induction during recruitment of brown adipocytes in white adipose tissue is dependent on cyclooxygenase activity. *PLoS One, 5*, e11391.

Mannick, J. B., Hausladen, A., Liu, L., Hess, D. T., Zeng, M., Miao, Q. X., et al. (1999). Fas-induced caspase denitrosylation. *Science, 284*, 651–654.

Marletta, M. A. (1993). Nitric oxide synthase structure and mechanism. *The Journal of Biological Chemistry, 268*, 12231–12234.

Marletta, M. A. (1994). Nitric oxide synthase: Aspects concerning structure and catalysis. *Cell, 78*, 927–930.

Marnett, L. J., Rowlinson, S. W., Goodwin, D. C., Kalgutkar, A. S., & Lanzo, C. A. (1999). Arachidonic acid oxygenation by COX-1 and COX-2. Mechanisms of catalysis and inhibition. *The Journal of Biological Chemistry, 274*, 22903–22906.

Masferrer, J. L., Zweifel, B. S., Manning, P. T., Hauser, S. D., Leahy, K. M., Smith, W. G., et al. (1994). Selective inhibition of inducible cyclooxygenase 2 in vivo is antiinflammatory and nonulcerogenic. *Proceedings of the National Academy of Sciences of the United States of America, 91*, 3228–3232.

McAdam, B. F., Mardini, I. A., Habib, A., Burke, A., Lawson, J. A., Kapoor, S., et al. (2000). Effect of regulated expression of human cyclooxygenase isoforms on eicosanoid and isoeicosanoid production in inflammation. *The Journal of Clinical Investigation, 105*, 1473–1482.

McCord, J. M., & Fridovich, I. (1969). The utility of superoxide dismutase in studying free radical reactions. I. Radicals generated by the interaction of sulfite, dimethyl sulfoxide, and oxygen. *The Journal of Biological Chemistry, 244*, 6056–6063.

Merlie, J. P., Fagan, D., Mudd, J., & Needleman, P. (1988). Isolation and characterization of the complementary DNA for sheep seminal vesicle prostaglandin endoperoxide synthase (cyclooxygenase). *The Journal of Biological Chemistry, 263*, 3550–3553.
Miller, S. B. (2006). Prostaglandins in health and disease: An overview. *Seminars in Arthritis and Rheumatism, 36*, 37–49.
Minghetti, L., Polazzi, E., Nicolini, A., Creminon, C., & Levi, G. (1996). Interferon-gamma and nitric oxide down-regulate lipopolysaccharide-induced prostanoid production in cultured rat microglial cells by inhibiting cyclooxygenase-2 expression. *Journal of Neurochemistry, 66*, 1963–1970.
Miyamoto, T., Ogino, N., Yamamoto, S., & Hayaishi, O. (1976). Purification of prostaglandin endoperoxide synthetase from bovine vesicular gland microsomes. *The Journal of Biological Chemistry, 251*, 2629–2636.
Miyaura, C., Inada, M., Suzawa, T., Sugimoto, Y., Ushikubi, F., Ichikawa, A., et al. (2000). Impaired bone resorption to prostaglandin E2 in prostaglandin E receptor EP4-knockout mice. *The Journal of Biological Chemistry, 275*, 19819–19823.
Mohr, S., Hallak, H., de Boitte, A., Lapetina, E. G., & Brune, B. (1999). Nitric oxide-induced S-glutathionylation and inactivation of glyceraldehyde-3-phosphate dehydrogenase. *The Journal of Biological Chemistry, 274*, 9427–9430.
Mollace, V., Muscoli, C., Masini, E., Cuzzocrea, S., & Salvemini, D. (2005). Modulation of prostaglandin biosynthesis by nitric oxide and nitric oxide donors. *Pharmacological Reviews, 57*, 217–252.
Mollace, V., Nottet, H. S., Clayette, P., Turco, M. C., Muscoli, C., Salvemini, D., et al. (2001). Oxidative stress and neuroAIDS: Triggers, modulators and novel antioxidants. *Trends in Neurosciences, 24*, 411–416.
Moncada, S., & Higgs, E. A. (1995). Molecular mechanisms and therapeutic strategies related to nitric oxide. *FASEB Journal, 9*, 1319–1330.
Moncada, S., Palmer, R. M., & Higgs, E. A. (1991). Nitric oxide: Physiology, pathophysiology, and pharmacology. *Pharmacological Reviews, 43*, 109–142.
Morita, I. (2002). Distinct functions of COX-1 and COX-2. *Prostaglandins & Other Lipid Mediators, 68–69*, 165–175.
Morrison, A. R., Moritz, H., & Needleman, P. (1978). Mechanism of enhanced renal prostaglandin biosynthesis in ureter obstruction. Role of de novo protein synthesis. *The Journal of Biological Chemistry, 253*, 8210–8212.
Morrison, S. F., Nakamura, K., & Madden, C. J. (2008). Central control of thermogenesis in mammals. *Experimental Physiology, 93*, 773–797.
Mukherjee, D., Nissen, S. E., & Topol, E. J. (2001). Risk of cardiovascular events associated with selective COX-2 inhibitors. *JAMA, 286*, 954–959.
Murata, H., Kawano, S., Tsuji, S., Tsuji, M., Sawaoka, H., Kimura, Y., et al. (1999). Cyclooxygenase-2 overexpression enhances lymphatic invasion and metastasis in human gastric carcinoma. *The American Journal of Gastroenterology, 94*, 451–455.
Muscoli, C., Cuzzocrea, S., Riley, D. P., Zweier, J. L., Thiemermann, C., Wang, Z. Q., et al. (2003). On the selectivity of superoxide dismutase mimetics and its importance in pharmacological studies. *British Journal of Pharmacology, 140*, 445–460.
Mutoh, M., Watanabe, K., Kitamura, T., Shoji, Y., Takahashi, M., Kawamori, T., et al. (2002). Involvement of prostaglandin E receptor subtype EP(4) in colon carcinogenesis. *Cancer Research, 62*, 28–32.
Nathan, C. (1992). Nitric oxide as a secretory product of mammalian cells. *FASEB Journal, 6*, 3051–3064.
Nathan, C., & Xie, Q. W. (1994). Regulation of biosynthesis of nitric oxide. *The Journal of Biological Chemistry, 269*, 13725–13728.
Nowak, P., & Wachowicz, B. (2002). Peroxynitrite-mediated modification of fibrinogen affects platelet aggregation and adhesion. *Platelets, 13*, 293–299.

Otto, J. C., & Smith, W. L. (1995). Prostaglandin endoperoxide synthases-1 and -2. *Journal of Lipid Mediators and Cell Signalling*, *12*, 139–156.

Park, S. W., Sung, M. W., Heo, D. S., Inoue, H., Shim, S. H., & Kim, K. H. (2005). Nitric oxide upregulates the cyclooxygenase-2 expression through the cAMP-response element in its promoter in several cancer cell lines. *Oncogene*, *24*, 6689–6698.

Pasinetti, G. M., & Aisen, P. S. (1998). Cyclooxygenase-2 expression is increased in frontal cortex of Alzheimer's disease brain. *Neuroscience*, *87*, 319–324.

Phillis, J. W., & O'Regan, M. H. (2003). The role of phospholipases, cyclooxygenases, and lipoxygenases in cerebral ischemic/traumatic injuries. *Critical Reviews in Neurobiology*, *15*, 61–90.

Pilbeam, C. C., Fall, P. M., Alander, C. B., & Raisz, L. G. (1997). Differential effects of nonsteroidal anti-inflammatory drugs on constitutive and inducible prostaglandin G/H synthase in cultured bone cells. *Journal of Bone and Mineral Research*, *12*, 1198–1203.

Piomelli, D., Wang, J. K., Sihra, T. S., Nairn, A. C., Czernik, A. J., & Greengard, P. (1989). Inhibition of Ca2+/calmodulin-dependent protein kinase II by arachidonic acid and its metabolites. *Proceedings of the National Academy of Sciences of the United States of America*, *86*, 8550–8554.

Pollock, J. S., Forstermann, U., Mitchell, J. A., Warner, T. D., Schmidt, H. H., Nakane, M., et al. (1991). Purification and characterization of particulate endothelium-derived relaxing factor synthase from cultured and native bovine aortic endothelial cells. *Proceedings of the National Academy of Sciences of the United States of America*, *88*, 10480–10484.

Pountos, I., Georgouli, T., Calori, G. M., & Giannoudis, P. V. (2012). Do nonsteroidal anti-inflammatory drugs affect bone healing? A critical analysis. *ScientificWorldJournal*, *2012*, 606404.

Prusakiewicz, J. J., Duggan, K. C., Rouzer, C. A., & Marnett, L. J. (2009). Differential sensitivity and mechanism of inhibition of COX-2 oxygenation of arachidonic acid and 2-arachidonoylglycerol by ibuprofen and mefenamic acid. *Biochemistry*, *48*, 7353–7355.

Radi, R. (2004). Nitric oxide, oxidants, and protein tyrosine nitration. *Proceedings of the National Academy of Sciences of the United States of America*, *101*, 4003–4008.

Radi, R., Cassina, A., Hodara, R., Quijano, C., & Castro, L. (2002). Peroxynitrite reactions and formation in mitochondria. *Free Radical Biology & Medicine*, *33*, 1451–1464.

Ristimaki, A., Sivula, A., Lundin, J., Lundin, M., Salminen, T., Haglund, C., et al. (2002). Prognostic significance of elevated cyclooxygenase-2 expression in breast cancer. *Cancer Research*, *62*, 632–635.

Rouzer, C. A., & Marnett, L. J. (2009). Cyclooxygenases: Structural and functional insights. *The Journal of Lipid Research*, *50*(Suppl.), S29–S34.

Saito, M., Okamatsu-Ogura, Y., Matsushita, M., Watanabe, K., Yoneshiro, T., Nio-Kobayashi, J., et al. (2009). High incidence of metabolically active brown adipose tissue in healthy adult humans: Effects of cold exposure and adiposity. *Diabetes*, *58*, 1526–1531.

Salvemini, D., & Masferrer, J. L. (1996). Interactions of nitric oxide with cyclooxygenase: In vitro, ex vivo, and in vivo studies. *Methods in Enzymology*, *269*, 12–25.

Salvemini, D., Misko, T. P., Masferrer, J. L., Seibert, K., Currie, M. G., & Needleman, P. (1993). Nitric oxide activates cyclooxygenase enzymes. *Proceedings of the National Academy of Sciences of the United States of America*, *90*, 7240–7244.

Salvemini, D., Seibert, K., Masferrer, J. L., Misko, T. P., Currie, M. G., & Needleman, P. (1994). Endogenous nitric oxide enhances prostaglandin production in a model of renal inflammation. *The Journal of Clinical Investigation*, *93*, 1940–1947.

Samuelsson, B. (1964). Identification of a smooth muscle-stimulating factor in bovine brain. Prostaglandins and related factors 25. *Biochimica et Biophysica Acta*, *84*, 218–219.

Sano, H., Hla, T., Maier, J. A., Crofford, L. J., Case, J. P., Maciag, T., et al. (1992). In vivo cyclooxygenase expression in synovial tissues of patients with rheumatoid arthritis and osteoarthritis and rats with adjuvant and streptococcal cell wall arthritis. *The Journal of Clinical Investigation*, *89*, 97–108.

Sathyendra, V., & Darowish, M. (2013). Basic science of bone healing. *Hand Clinics*, *29*, 473–481.

Schmedtje, J. F., Jr., Ji, Y. S., Liu, W. L., DuBois, R. N., & Runge, M. S. (1997). Hypoxia induces cyclooxygenase-2 via the NF-kappaB p65 transcription factor in human vascular endothelial cells. *The Journal of Biological Chemistry*, *272*, 601–608.

Shi, W., Hoganson, C. W., Espe, M., Bender, C. J., Babcock, G. T., Palmer, G., et al. (2000). Electron paramagnetic resonance and electron nuclear double resonance spectroscopic identification and characterization of the tyrosyl radicals in prostaglandin H synthase 1. *Biochemistry*, *39*, 4112–4121.

Shitashige, M., Morita, I., & Murota, S. (1998). Different substrate utilization between prostaglandin endoperoxide H synthase-1 and -2 in NIH3T3 fibroblasts. *Biochimica et Biophysica Acta*, *1389*, 57–66.

Sidhu, R. S., Lee, J. Y., Yuan, C., & Smith, W. L. (2010). Comparison of cyclooxygenase-1 crystal structures: Cross-talk between monomers comprising cyclooxygenase-1 homodimers. *Biochemistry*, *49*, 7069–7079.

Simon, A. M., Manigrasso, M. B., & O'Connor, J. P. (2002). Cyclo-oxygenase 2 function is essential for bone fracture healing. *Journal of Bone and Mineral Research*, *17*, 963–976.

Smith, W. L., DeWitt, D. L., & Garavito, R. M. (2000). Cyclooxygenases: Structural, cellular, and molecular biology. *Annual Review of Biochemistry*, *69*, 145–182.

Smith, W. L., Marnett, L. J., & DeWitt, D. L. (1991). Prostaglandin and thromboxane biosynthesis. *Pharmacology and Therapeutics*, *49*, 153–179.

Smyth, E. M., Grosser, T., Wang, M., Yu, Y., & FitzGerald, G. A. (2009). Prostanoids in health and disease. *The Journal of Lipid Research*, *50*(Suppl.), S423–S428.

Solomon, D. H., Schneeweiss, S., Glynn, R. J., Kiyota, Y., Levin, R., Mogun, H., et al. (2004). Relationship between selective cyclooxygenase-2 inhibitors and acute myocardial infarction in older adults. *Circulation*, *109*, 2068–2073.

Stadler, J., Harbrecht, B. G., Di, S. M., Curran, R. D., Jordan, M. L., Simmons, R. L., et al. (1993). Endogenous nitric oxide inhibits the synthesis of cyclooxygenase products and interleukin-6 by rat Kupffer cells. *Journal of Leukocyte Biology*, *53*, 165–172.

Stamler, J. S. (1994). Redox signaling: Nitrosylation and related target interactions of nitric oxide. *Cell*, *78*, 931–936.

Stamler, J. S., Simon, D. I., Osborne, J. A., Mullins, M. E., Jaraki, O., Michel, T., et al. (1992). S-nitrosylation of proteins with nitric oxide: Synthesis and characterization of biologically active compounds. *Proceedings of the National Academy of Sciences of the United States of America*, *89*, 444–448.

Stamler, J. S., Toone, E. J., Lipton, S. A., & Sucher, N. J. (1997). (S)NO signals: Translocation, regulation, and a consensus motif. *Neuron*, *18*, 691–696.

Stuehr, D. J. (1997). Structure-function aspects in the nitric oxide synthases. *Annual Review of Pharmacology and Toxicology*, *37*, 339–359.

Stuehr, D. J., Kwon, N. S., Nathan, C. F., Griffith, O. W., Feldman, P. L., & Wiseman, J. (1991). N omega-hydroxy-L-arginine is an intermediate in the biosynthesis of nitric oxide from L-arginine. *The Journal of Biological Chemistry*, *266*, 6259–6263.

Su, B., & O'Connor, J. P. (2013). NSAID therapy effects on healing of bone, tendon, and the enthesis. *Journal of Applied Physiology (Bethesda, Md: 1985)*, *115*, 892–899.

Swierkosz, T. A., Mitchell, J. A., Warner, T. D., Botting, R. M., & Vane, J. R. (1995). Co-induction of nitric oxide synthase and cyclo-oxygenase: Interactions between nitric oxide and prostanoids. *British Journal of Pharmacology*, *114*, 1335–1342.

Sykes, J. A., & Moddox, I. S. (1972). Prostaglandin production by experimental tumours and effects of anti-inflammatory compounds. *Nature: New Biology*, *237*, 59–61.

Szabo, C., Ischiropoulos, H., & Radi, R. (2007). Peroxynitrite: Biochemistry, pathophysiology and development of therapeutics. *Nature Reviews Drug Discovery, 6*, 662–680.
Tanabe, T., & Tohnai, N. (2002). Cyclooxygenase isozymes and their gene structures and expression. *Prostaglandins & Other Lipid Mediators, 68–69*, 95–114.
Tatoyan, A., & Giulivi, C. (1998). Purification and characterization of a nitric-oxide synthase from rat liver mitochondria. *The Journal of Biological Chemistry, 273*, 11044–11048.
Thun, M. J., Namboodiri, M. M., & Heath, C. W., Jr. (1991). Aspirin use and reduced risk of fatal colon cancer. *The New England Journal of Medicine, 325*, 1593–1596.
Tian, J., Kim, S. F., Hester, L., & Snyder, S. H. (2008). S-nitrosylation/activation of COX-2 mediates NMDA neurotoxicity. *Proceedings of the National Academy of Sciences of the United States of America, 105*, 10537–10540.
Tian, M., & Schiemann, W. P. (2010). PGE2 receptor EP2 mediates the antagonistic effect of COX-2 on TGF-beta signaling during mammary tumorigenesis. *FASEB Journal, 24*, 1105–1116.
Townsend, K. P., & Pratico, D. (2005). Novel therapeutic opportunities for Alzheimer's disease: Focus on nonsteroidal anti-inflammatory drugs. *FASEB Journal, 19*, 1592–1601.
Tsai, A. L., & Kulmacz, R. J. (2010). Prostaglandin H synthase: Resolved and unresolved mechanistic issues. *Archives of Biochemistry and Biophysics, 493*, 103–124.
Tsai, A. L., Palmer, G., & Kulmacz, R. J. (1992). Prostaglandin H synthase. Kinetics of tyrosyl radical formation and of cyclooxygenase catalysis. *The Journal of Biological Chemistry, 267*, 17753–17759.
Tsai, A. L., Wei, C., & Kulmacz, R. J. (1994). Interaction between nitric oxide and prostaglandin H synthase. *Archives of Biochemistry and Biophysics, 313*, 367–372.
Tucker, O. N., Dannenberg, A. J., Yang, E. K., Zhang, F., Teng, L., Daly, J. M., et al. (1999). Cyclooxygenase-2 expression is up-regulated in human pancreatic cancer. *Cancer Research, 59*, 987–990.
Turko, I. V., Li, L., Aulak, K. S., Stuehr, D. J., Chang, J. Y., & Murad, F. (2003). Protein tyrosine nitration in the mitochondria from diabetic mouse heart. Implications to dysfunctional mitochondria in diabetes. *The Journal of Biological Chemistry, 278*, 33972–33977.
Upmacis, R. K., Deeb, R. S., & Hajjar, D. P. (1999). Regulation of prostaglandin H2 synthase activity by nitrogen oxides. *Biochemistry, 38*, 12505–12513.
Upmacis, R. K., Deeb, R. S., & Hajjar, D. P. (2006). Oxidative alterations of cyclooxygenase during atherogenesis. *Prostaglandins & Other Lipid Mediators, 80*, 1–14.
van der Vliet, A., Hoen, P. A., Wong, P. S., Bast, A., & Cross, C. E. (1998). Formation of S-nitrosothiols via direct nucleophilic nitrosation of thiols by peroxynitrite with elimination of hydrogen peroxide. *The Journal of Biological Chemistry, 273*, 30255–30262.
van Marken Lichtenbelt, W. D., Vanhommerig, J. W., Smulders, N. M., Drossaerts, J. M., Kemerink, G. J., Bouvy, N. D., et al. (2009). Cold-activated brown adipose tissue in healthy men. *The New England Journal of Medicine, 360*, 1500–1508.
Vecchio, A. J., Simmons, D. M., & Malkowski, M. G. (2010). Structural basis of fatty acid substrate binding to cyclooxygenase-2. *The Journal of Biological Chemistry, 285*, 22152–22163.
Vedia, L., McDonald, B., Reep, B., Brune, B., Di, S. M., Billiar, T. R., et al. (1992). Nitric oxide-induced S-nitrosylation of glyceraldehyde-3-phosphate dehydrogenase inhibits enzymatic activity and increases endogenous ADP-ribosylation. *The Journal of Biological Chemistry, 267*, 24929–24932.
Vegiopoulos, A., Muller-Decker, K., Strzoda, D., Schmitt, I., Chichelnitskiy, E., Ostertag, A., et al. (2010). Cyclooxygenase-2 controls energy homeostasis in mice by de novo recruitment of brown adipocytes. *Science, 328*, 1158–1161.

von Euler, U. S. (1936). On the specific vaso-dilating and plain muscle stimulating substances from accessory genital glands in man and certain animals (prostaglandin and vesiglandin). *The Journal of Physiology, 88*, 213–234.

Wallace, J. L., McKnight, W., Reuter, B. K., & Vergnolle, N. (2000). NSAID-induced gastric damage in rats: Requirement for inhibition of both cyclooxygenase 1 and 2. *Gastroenterology, 119*, 706–714.

Wang, M. T., Honn, K. V., & Nie, D. (2007). Cyclooxygenases, prostanoids, and tumor progression. *Cancer and Metastasis Reviews, 26*, 525–534.

Watanabe, K., Kawamori, T., Nakatsugi, S., Ohta, T., Ohuchida, S., Yamamoto, H., et al. (1999). Role of the prostaglandin E receptor subtype EP1 in colon carcinogenesis. *Cancer Research, 59*, 5093–5096.

Weksler, B. B., Marcus, A. J., & Jaffe, E. A. (1977). Synthesis of prostaglandin I2 (prostacyclin) by cultured human and bovine endothelial cells. *Proceedings of the National Academy of Sciences of the United States of America, 74*, 3922–3926.

Westergaard, T., Mortensen, P. B., Pedersen, C. B., Wohlfahrt, J., & Melbye, M. (1999). Exposure to prenatal and childhood infections and the risk of schizophrenia: Suggestions from a study of sibship characteristics and influenza prevalence. *Archives of General Psychiatry, 56*, 993–998.

Williams, E. D., Karim, S. M., & Sandler, M. (1968). Prostaglandin secretion by medullary carcinoma of the thyroid. A possible cause of the associated diarrhoea. *Lancet, 1*, 22–23.

Xu, L., Eu, J. P., Meissner, G., & Stamler, J. S. (1998). Activation of the cardiac calcium release channel (ryanodine receptor) by poly-S-nitrosylation. *Science, 279*, 234–237.

Yamagata, K., Andreasson, K. I., Kaufmann, W. E., Barnes, C. A., & Worley, P. F. (1993). Expression of a mitogen-inducible cyclooxygenase in brain neurons: Regulation by synaptic activity and glucocorticoids. *Neuron, 11*, 371–386.

Yamakura, F., & Kawasaki, H. (2010). Post-translational modifications of superoxide dismutase. *Biochimica et Biophysica Acta, 1804*, 318–325.

Yu, Y., Fan, J., Chen, X. S., Wang, D., Klein-Szanto, A. J., Campbell, R. L., et al. (2006). Genetic model of selective COX2 inhibition reveals novel heterodimer signaling. *Nature Medicine, 12*, 699–704.

Yu, Y., Fan, J., Hui, Y., Rouzer, C. A., Marnett, L. J., Klein-Szanto, A. J., et al. (2007). Targeted cyclooxygenase gene (ptgs) exchange reveals discriminant isoform functionality. *The Journal of Biological Chemistry, 282*, 1498–1506.

Yuan, C., Rieke, C. J., Rimon, G., Wingerd, B. A., & Smith, W. L. (2006). Partnering between monomers of cyclooxygenase-2 homodimers. *Proceedings of the National Academy of Sciences of the United States of America, 103*, 6142–6147.

Zimmermann, K. C., Sarbia, M., Weber, A. A., Borchard, F., Gabbert, H. E., & Schror, K. (1999). Cyclooxygenase-2 expression in human esophageal carcinoma. *Cancer Research, 59*, 198–204.

Zou, M., Martin, C., & Ullrich, V. (1997). Tyrosine nitration as a mechanism of selective inactivation of prostacyclin synthase by peroxynitrite. *Biological Chemistry, 378*, 707–713.

Zou, M. H., Shi, C., & Cohen, R. A. (2002). Oxidation of the zinc-thiolate complex and uncoupling of endothelial nitric oxide synthase by peroxynitrite. *The Journal of Clinical Investigation, 109*, 817–826.

Zou, M. H., & Ullrich, V. (1996). Peroxynitrite formed by simultaneous generation of nitric oxide and superoxide selectively inhibits bovine aortic prostacyclin synthase. *FEBS Letters, 382*, 101–104.

CHAPTER TEN

Nitric Oxide as a Mediator of Estrogen Effects in Osteocytes

Jisha Joshua, Hema Kalyanaraman, Nisha Marathe, Renate B. Pilz[1]
Department of Medicine, University of California, San Diego, California, USA
[1]Corresponding author: e-mail address: rpilz@ucsd.edu

Contents

Abstract

Postmenopausal osteoporosis due to estrogen deficiency is a major health problem, and available therapies rely largely on the inhibition of bone resorption, because estrogen replacement is associated with risks. Estrogen promotes bone health in large part by increasing osteocyte survival, but the molecular mechanisms involved are only partly understood. We showed that estradiol stimulates nitric oxide (NO) production in osteocytes, leading to increased cGMP synthesis and activation of cGMP-dependent protein kinases (PKGs). Moreover, we found that 17β-estradiol protects osteocytes against apoptosis via the NO/cGMP signaling pathway: type II PKG mediates estradiol-induced activation of the prosurvival kinases Erk and Akt, whereas type I PKG contributes to prosurvival signaling by directly phosphorylating and inactivating the cell death protein BAD. Preclinical data support an important role of NO in bone biology, and clinical trials suggest that NO donors may prevent bone loss in postmenopausal women. Our data provide novel insights into estrogen signaling through the NO/cGMP/PKG pathway and a rationale for using NO donors and other cGMP-elevating agents for treating postmenopausal osteoporosis.

Vitamins and Hormones, Volume 96
ISSN 0083-6729
http://dx.doi.org/10.1016/B978-0-12-800254-4.00010-6

1. INTRODUCTION

The mammalian skeleton undergoes constant remodeling, with osteoclasts and matrix-secreting osteoblasts working sequentially to replace old bone with new (Raisz, 2005). Osteoblast and osteoclast activities are coupled, in part because cells of the osteoblastic lineage produce receptor of activated nuclear factor-κB (RANKL) and its antagonist osteoprotegerin (OPG) (Edwards & Mundy, 2011). Osteoblasts become entombed in mineralized matrix and develop into long-living, postmitotic osteocytes. Osteocytes form a syncytial network that communicates with neighboring osteocytes and osteoblasts via cell processes (Dallas, Prideaux, & Bonewald, 2013). Because of their location and extensive communication, osteocytes are considered to be the major sensors of bone turnover and transmitters of local signals to both osteoblasts and osteoclasts (Dallas et al., 2013).

During menopause, decreased ovarian estrogen production leads to a period of rapid bone loss; this is followed by a more gradual decrease in bone mass associated with aging, which is only partly due to sex hormone deficiency (Manolagas, 2010; Watts et al., 2010). The loss of bone mass and strength leading to postmenopausal osteoporosis is a significant health problem: about 8 million American women ≥50 years old have osteoporosis by WHO criteria (i.e., their bone mineral density is more than 2.5 standard deviations below the mean peak gender-specific BMD, referred to a *T*-score of −2.5); by age 60, one-half of U.S. Caucasian women have osteopenia (*T*-score between −1 and −2.5) or osteoporosis (*T*-score <−2.5) (Watts et al., 2010). Nearly 50% of postmenopausal women will have an osteoporosis-related fracture during their lifetime; these fractures are associated with chronic pain, loss of independence, and increased mortality (Chrischilles, Butler, Davis, & Wallace, 1991; Khosla, Bellido, et al., 2011; Khosla, Melton, & Riggs, 2011; Watts et al., 2010).

The Woman's Health Initiative showed that estrogen replacement prevents bone fractures, but the routine use of hormone replacement therapy has diminished significantly. Estrogens are no longer recommended for treatment of osteoporosis, because risks of breast cancer and thromboembolic disease outweigh the benefits (Barrett-Connor, Grady, & Stefanick, 2005; Raisz, 2005). Selective estrogen receptor (ER) modulators decrease breast cancer risk, but still cause thromboembolic complications (Hopkins et al., 2011; Watts et al., 2010). Treatment of postmenopausal osteoporosis relies heavily on antiresorptive agents, including bisphosphonates and denosumab (an antibody to RANKL) (Hopkins et al., 2011; Watts et al., 2010). Currently, teriparatide (PTH) is the only FDA-approved anabolic therapy for osteoporosis; PTH increases not only bone formation but also bone resorption, and its use is limited to 2 years, in part due to concerns about osteosarcomas in rats (Watts et al., 2010). New anabolic agents for the treatment of osteoporosis are needed, and nitric oxide (NO)-generating agents may represent a new class of bone-anabolic agents alternative to estrogen therapy (Jamal & Hamilton, 2012). Here, we will review the effects of NO in bone and describe how estrogen utilizes the NO/cGMP pathway to exert some of its anabolic effects in bone cells.

2. BONE-PROTECTIVE EFFECTS OF ESTROGEN

High bone turnover bone loss, with increased osteoclastic bone resorption outpacing osteoblastic bone formation, is the hallmark of osteoporosis resulting from estrogen deficiency due to ovariectomy or postmenopausal ovarian involution (Khosla, Melton, et al., 2011; Krum, 2011; Manolagas, 2010). Estrogens prevent bone loss by shortening osteoclast survival, while prolonging the life span of osteoblasts and osteocytes (Almeida et al., 2007; Manolagas, 2010; Tomkinson, Reeve, Shaw, & Noble, 1997). Estrogens induce osteoclast apoptosis directly, by inducing transcription of Fas ligand via the nuclear ER-α (Krum, 2011; Krum et al., 2008). In addition, estrogens inhibit osteoclast differentiation and activity indirectly, by repressing interleukin-6 production in immune cells and by decreasing the ratio of RANKL/OPG production in cells of osteoblastic lineage (Dallas et al., 2013; Jilka et al., 1992; Khosla, Melton, et al., 2011).

Osteocyte viability plays an important role in the maintenance of bone homeostasis and integrity, as demonstrated by rapid bone loss after ablation of mature osteocytes through activation of a suicide gene in a transgenic mouse model (Tatsumi et al., 2007). The diphtheria toxin receptor was expressed under an osteocyte-specific promoter, and injection of diphtheria toxin killed >70% of mature osteocytes in these mice. There is a tight spatial and temporal coupling of osteocyte apoptosis to the accumulation of microdamage and subsequent induction of bone remodeling; i.e., regions of osteocyte apoptosis coincide with the location of subsequent osteoclast response. Thus, osteocyte apoptosis is a necessary and controlling factor for the resultant osteoclastic resorptive response that leads to unbalanced bone remodeling and ultimately bone loss in states of estrogen deficiency

(Emerton et al., 2010). Osteocyte apoptosis is not only induced by estrogen deprivation but also occurs during normal aging (Manolagas, 2010).

Previous work has shown that 17β-estradiol protects osteoblasts and osteocytes from apoptosis by activating c-Src and the extracellular signal-responsive protein kinases (Erk-1/2) via a plasma membrane-bound ER-α; these effects do not require nuclear localization or DNA binding of the ER, but do require nuclear translocation of Erk (Almeida et al., 2010; Chen et al., 2005; Kousteni et al., 2001). These estrogen effects, which do not involve DNA binding and transcriptional regulation by the classic nuclear ERs, are referred to as "nongenomic" effects and play an important role in mediating the bone-protective effects of estrogens (Kousteni et al., 2002). We recently elucidated the mechanism of estrogen-induced Erk/Akt activation in bone cells and established a signaling pathway through which estrogen protects osteocytes from apoptosis (discussed below) (Marathe, Rangaswami, Zhuang, Boss, & Pilz, 2012).

3. INDUCTION OF NO SYNTHESIS BY ESTROGEN

Rapid effects of estrogen in the cardiovascular system, i.e., vasodilation occurring within minutes after estradiol administration, led to the discovery that estrogen induces the release of NO from endothelial cells within minutes of hormone addition (Russell, Haynes, Sinha, Clerisme, & Bender, 2000). Estradiol binding to an N-terminally truncated ER-α variant induces plasma membrane recruitment of the receptor and assembly of a membrane signaling complex, which includes endothelial NO synthase (NOS3) (Li et al., 2007). Consequently, NOS3 is activated by phosphorylation and by association with heat-shock protein-90. Others have shown that administration of an estrogen-dendrimer conjugate, which is excluded from the nucleus, activates a Gαi-dependent increase of NO synthesis via binding to ER-α in endothelial cells (Chambliss et al., 2010). Consistent with the effects of estrogens on endothelial NOS3, estrogen replacement increases the plasma concentration of stable NO oxidation products (nitrite and nitrate) in postmenopausal women (Hayashi, Ito, Kano, Endo, & Iguchi, 2000). We found that estrogens induce NO production in osteoblasts and osteocytes via NOS3 activation (Figs. 10.1, 10.2A, and unpublished results) (Marathe et al., 2012). Estradiol can also activate the NOS1 isoform in neuronal cells (Lee, Andoh, Murphy, & Chiueh, 2003).

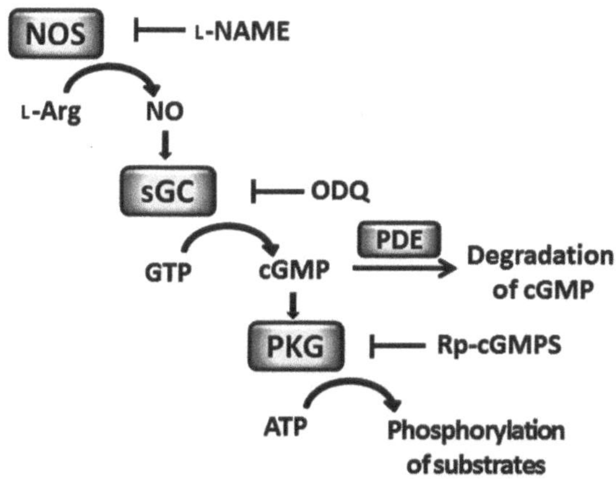

Figure 10.1 NO/cGMP/PKG signaling. Estrogens can activate NO synthase (NOS)-3, leading to increased NO synthesis, which stimulates soluble guanylate cyclase (sGC), leading to increased intracellular cGMP concentrations. cGMP is either degraded by phosphodiesterases (PDEs) or binds to various effectors, including cGMP-dependent protein kinases (PKGs). The following inhibitors are mentioned in the text: The NOS inhibitor L-NAME, the sGC inhibitor ODQ, and the PKG inhibitor Rp-cGMPS (abbreviated for Rp-8-pCPT-PET-cGMPS). (See the color plate.)

4. NO/cGMP SIGNALING IN BONE CELLS

An important role of NO in bone biology is supported by multiple studies in rodents: *First*, NOS3-deficient mice have reduced bone mass due to defects in osteoblast number and maturation; they have exaggerated bone loss after ovariectomy, and a blunted response to estrogens (Afzal, Polak, & Buttery, 2004; Aguirre et al., 2001; Armour et al., 2001; Grassi et al., 2006). *Second*, NO-generating nitrates prevent bone loss from estrogen deficiency in rats (Hukkanen et al., 2003; Wimalawansa, De, Gangula, & Yallampalli, 1996), whereas NOS inhibitors block *in vivo* bone formation induced by estrogens (Samuels, Perry, Gibson, Colley, & Tobias, 2001). *Third*, low concentrations of NO donors can enhance osteoblast proliferation and differentiation *in vitro* (Chae et al., 1997; Mancini, Moradi-Bidhendi, Becherini, Martineti, & MacIntyre, 2000; Otsuka et al., 1998). *Fourth*, low NO concentrations are necessary for osteoclast differentiation and survival, and NOS1-deficient mice have increased bone mass due to reduced osteoclast number and decreased bone resorption (Chae et al., 1998; Jung, Lin, Ramos, Faddis, & Chole, 2003; Lee et al., 2004; Nicolin et al., 2005; van't Hof, Macphee, Libouban, Helfrich, & Ralston, 2004);

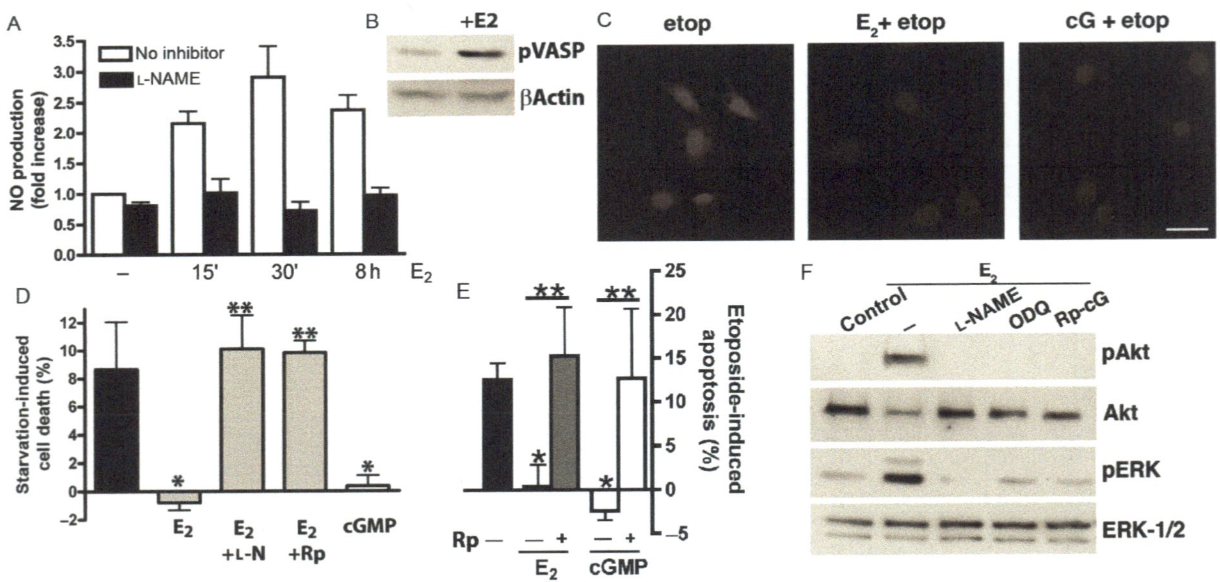

Figure 10.2 17β-Estradiol signals through NO/cGMP/PKG to protect osteoblasts and osteocytes from cell death. (A) MLO-Y4 osteocyte-like cells were treated with vehicle (white bars) or the NOS inhibitor L-NAME (black bars) for 1 h and received 10^{-7} M 17β-estradiol (E_2) for the indicated times. NO production was measured using the Griess reagent. (B) Cells were treated as in (A) and analyzed by Western blotting with an antibody specific for VASP phosphorylated on Ser^{259}, a preferred PKG phosphorylation site. (C) Cells were treated with etoposide (etop) to induce cell death and received 10^{-7} M E_2 or 10^{-4} M 8-pCPT-cGMP (cG) as indicated. Apoptotic cells were detected by immunofluorescence staining with an antibody specific for cleaved caspase-3 (red), and nuclei were counterstained with Hoechst 33342 (blue). (D, E) Murine primary osteoblasts (D) and MLO-Y4 osteocyte-like cells (E) received vehicle (black bar), 10^{-7} M E_2 (gray bars), or 10^{-4} M 8-pCPT-cGMP (white bars) and were serum-starved for 18 h (D) or treated with etoposide for 8 h (E) to induce apoptosis. Some cells were pretreated with L-NAME (L-N) or Rp-8-pCPT-PET-cGMPS (Rp) to inhibit NOS or PKG, respectively. Cell death was quantified by trypan blue staining (D) or TUNEL assay (E). (F) Serum-starved MLO-Y4 cells were treated with L-NAME, ODQ, or Rp-8-pCPT-PET-cGMPS to inhibit NOS, sGC, or PKG, and treated with vehicle (control) or 10^{-7} M E_2 for 5 min as indicated. Akt and Erk activation were assessed by Western blotting with phospho-specific antibodies (pAkt, pErk); loading was assessed by blotting with antibodies recognizing total Akt or Erk irrespective of their phosphorylation state. *This research was originally published in Marathe et al. (2012). ©The American Society for Biochemistry and Molecular Biology.* (See the color plate.)

however, NO inhibits bone resorption by mature osteoclasts (Dong, Williams, Jordan, Cornwell, & Blair, 1999; Fuller, Kirstein, & Chambers, 2006; Yaroslavskiy, Turkova, Wang, Robinson, & Blair, 2010).

NO has many different biological effects, including generation of reactive oxygen/nitrogen species and nitrosylation of heme- or sulfur-containing proteins (Nathan, 2004). One of its main cellular targets is the enzyme-soluble guanylate cyclase, which is activated by NO binding to the enzyme's heme prosthetic groups (Fig. 10.1). Soluble guanylate cyclase generates cGMP, which in turn regulates the activity of cGMP-dependent protein kinases (PKGs), phosphodiesterases, and cyclic nucleotide-gated ion channels (Francis, Busch, Corbin, & Sibley, 2010). The *PRKG1* gene encodes two splice variants differing in the N-terminal ~100 amino acids, PKG Iα and Iβ, that are largely cytosolic enzymes, while the *PRKG2* gene encodes a membrane-bound enzyme (Hofmann, Bernhard, Lukowski, & Weinmeister, 2009). PKG I and II differ in their tissue distribution, but both genes are expressed in osteoblasts and osteocytes (Hofmann et al., 2009; Rangaswami et al., 2009). We found that estrogen-induced NO synthesis in osteoblasts and MLO-Y4 osteocyte-like cells is sufficient for stimulation of cGMP synthesis and PKG activation, as demonstrated by increased phosphorylation of the PKG substrate vasodilator-stimulated phosphoprotein (VASP) (Fig. 10.2B) (Marathe et al., 2012). We also showed that stimulation of osteoblasts and osteocytes by fluid shear stress increases NO and cGMP levels, activating PKG II, which leads to Src and Erk activation, induction of *fos* family genes, and increased osteoblastic proliferation (Rangaswami et al., 2009, 2010).

5. BONE-ANABOLIC EFFECTS OF NO IN HUMANS—LESSONS FROM CLINICAL TRIALS

The organic nitrates nitroglycerin (NTG) and isosorbide mononitrate (ISMN) generate NO *in vivo* after mitochondrial biotransformation; they are currently approved for the management of hypertension, coronary artery insufficiency, and heart failure. Epidemiological studies indicate a protective effect of nitrates on bone, with a reduction in fracture risk (Jamal, Browner, Bauer, & Cummings, 1998; Pouwels et al., 2010; Rejnmark, Vestergaard, & Mosekilde, 2006). In the largest study, the use of nitrates was associated with an 11% reduction in the risk of any fracture and a 15% reduction in the risk of hip fractures (Rejnmark et al., 2006).

Four prospective, randomized trials have shown a positive effect of nitrates on bone density, at doses lower than those used for vasodilation (Jamal & Hamilton, 2012). *First*, in a small open-label study in young women who underwent oophorectomy, NTG ointment was as effective as standard estrogen replacement in preventing bone loss at 1 year (spine and hip) (Wimalawansa, 2000). *Second*, in a study of 60 postmenopausal women with established osteoporosis, individuals randomly assigned to receive ISMN showed similar improvement in BMD at 1 year as subjects randomized to standard bisphosphonate treatment (+10.8% and 12.1%) (Nabhan & Rabie, 2008). *Third*, in a short, 3-month study of 144 healthy postmenopausal women (aged 50–80), ISMN decreased a bone resorption marker (N-telopeptide, Ntx) and increased a bone formation marker (alkaline phosphatase, AlkP) compared to placebo (Jamal, Cummings, & Hawker, 2004). *Fourth*, in a 2-year trial of 243 healthy, but osteopenic, postmenopausal women (mean age 62, lumbar spine and total hip *T*-score of 0 to −2.0), NTG significantly increased BMD at the lumbar spine and hip (+6.7% and 7.0%, respectively) compared to placebo; NTG also increased cortical thickness at the radius and tibia, decreased Ntx, and increased AlkP (Jamal, Hamilton, Eastell, & Cummings, 2011). NTG increased indices of bone strength, suggesting that it should have the potential to reduce fractures, but the study was underpowered to assess fracture risk (Jamal et al., 2011; Khosla, 2011). In contrast, a similarly designed, 3-year double-blind, randomized, placebo-controlled trial of 186 healthy, early postmenopausal women (mean age 56, lumbar spine *T*-score of 0 to −2.5) failed to show any differences in BMD between treatment with NTG ointment or placebo, but treatment adherence was poor (Wimalawansa, Grimes, Wilson, & Hoover, 2009). Thus, consistent with the preclinical studies described earlier, clinical studies indicate that NO donors can act as anabolic agents in bone (Khosla, 2011).

6. ESTROGEN PROMOTION OF OSTEOCYTE SURVIVAL VIA THE NO/cGMP/PKG PATHWAY

Osteocyte survival is a key mechanism through which estrogen protects bone (Emerton et al., 2010; Kousteni et al., 2001; Manolagas & Parfitt, 2013; Tomkinson et al., 1997). NO/cGMP signaling has been implicated in regulating apoptosis in different cell types, including endothelial and neuronal cells (Blaise, Gauvin, Gangal, & Authier, 2005; Fiscus, 2002). Fluid shear stress-induced NO production or treatment with NO donors protect osteocytes and osteoblasts from apoptosis induced by exposure to tumor necrosis factor (TNF-α), but the downstream targets of NO were not studied (Chae et al., 2006; Tan et al., 2006). As preclinical and clinical data suggest that NO mediates at least some of estrogen's bone-protective effects, we examined the role of NO/cGMP signaling in the regulation of osteocyte survival by estrogen.

6.1. Antiapoptotic effects of estrogen in osteocytes and osteoblasts require NO/cGMP/PKG signaling

As described earlier, 100 nM 17β-estradiol (estradiol, E2) increased NO production by two- to threefold and induced phosphorylation of the PKG substrate VASP in osteocyte-like MLO-Y4 cells, indicating that estradiol activates the NO/cGMP/PKG pathway (Fig. 10.2A and B). We hypothesized that estrogen protects cells of the osteoblastic lineage from apoptosis via cGMP activation of PKG I and/or II.

We studied effects of estrogen and cGMP in primary osteoblasts exposed to serum starvation and in MLO-Y4 cells exposed to the topoisomerase inhibitor etoposide; both treatments induced apoptosis in 10–15% of the cells (Marathe et al., 2012). Treatment with either 100 nM estradiol or 100 μM of the membrane-permeable cGMP analog 8-*para*-chlorophenylthio-cGMP (8-pCPT-cGMP, cGMP) protected cells from starvation- or etoposide-induced cell death, as assessed by trypan blue staining and TUNEL assay (Fig. 10.2C–E). Estradiol protection from etoposide-induced death was lost, when NO synthesis was inhibited by L-NG-nitroarginine methyl ester (L-NAME) (Fig. 10.2A and D); however, the NO donor DETA-NONOate provided protection in the presence of L-NAME (Marathe et al., 2012). Inhibition of soluble guanylate cyclase by ODQ also blocked the protective effects of estradiol, but 8-pCPT-cGMP protected osteocytes from etoposide-induced cell death in the presence of either L-NAME or ODQ (Marathe et al., 2012). These data indicate estradiol signals through NO/cGMP to protect osteocytes from apoptosis and are consistent with L-NAME and ODQ specifically inhibiting NO and cGMP synthesis, respectively, rather than enhancing osteocyte death nonspecifically. Inhibition of PKGs with Rp-8-pCPT-PET-cGMPS prevented the prosurvival effects of both estradiol and 8-pCPT-cGMP (Fig. 10.2D and E), suggesting that osteocyte protection by estradiol and cGMP requires PKG. Similar results were obtained in murine primary osteoblasts and immortalized MC3T3 preosteoblast-like cells when subjected to injury from serum starvation or TNF-α, respectively (Marathe et al., 2012).

6.2. PKG Iα and PKG II are independently necessary for estradiol- and cGMP-mediated protection from apoptosis

We used an siRNA approach to investigate which PKG isoform was involved in the antiapoptotic effects of estradiol and 8-pCPT-cGMP. MLO-Y4 cells were transfected with siRNAs targeting either the common C-terminal region of PKG Iα and Iβ or PKG II, with an siRNA-targeting green fluorescent protein serving as a control. Cell death was quantified with trypan blue staining and confirmed by Western blotting and immunofluorescence staining for cleaved caspase-3. In MLO-Y4 cells transfected with either PKG I- or PKG II-specific siRNA, etoposide increased cell death regardless of whether estradiol or 8-pCPT-cGMP was present, indicating necessary and nonredundant roles for both PKG isoforms in estradiol- and cGMP-induced protection from cell death (Marathe et al., 2012).

Reconstituting PKG Iα in PKG I-depleted cells and PKG II in PKG II-depleted cells with adenoviral vectors encoding siRNA-resistant PKG Iα or PKG II, respectively, restored the ability of estradiol and 8-pCPT-cGMP to reverse etoposide-induced cell death (Marathe et al., 2012). Surprisingly, reconstituting PKG Iβ in PKG I-depleted cells had no effect despite similar expression levels of both kinases and comparable activities toward the shared substrate VASP (Marathe et al., 2012). There are few examples of PKG Iα-specific functions for which PKG Iβ cannot substitute. In PKG I-null mice, it appears that either of the PKG I isoforms can rescue basic vascular and intestinal smooth muscle functions (Weber et al., 2007). However, due to their unique N-terminal dimerization/leucine zipper domain, PKG Iα and Iβ bind different, isoform-specific G-kinase interaction proteins, which may target the kinases to unique subcellular localizations and substrates. For example, while PKG Iα specifically binds to the regulator of G-protein signaling and the myosin-targeting subunit of myosin phosphatase (Surks et al., 1999; Tang et al., 2003), PKG Iβ specifically binds to the inositol 1,4,5-trisphosphate receptor-associated G-kinase substrate IRAG and the transcriptional regulator TFII-I (Casteel et al., 2002; Schlossmann et al., 2000). Based on our findings, PKG Iα may be targeted to a subcellular compartment that favors downstream signaling that leads to cell survival.

6.3. Estradiol-induced Akt and Erk activation in osteocytes is mediated by NO/cGMP/PKG II and necessary for the hormone's antiapoptotic effects

In MLO-Y4 osteocyte-like cells, MC3T3 osteoblast-like cells, and in primary human and murine osteoblasts, we found that estradiol and 8-pCPT-cGMP rapidly induced Akt and Erk phosphorylation on sites that are known to cause activation of the kinases, and the estrogen effects were prevented by pharmacological inhibition of NO synthase, soluble guanylate cyclase, or PKG (Fig. 10.2F for MLO-Y4 cells) (Marathe et al., 2012, and unpublished results). Using the siRNA approach described earlier, we determined that depleting only PKG II, but not PKG I, resulted in loss of Akt and Erk activation by estradiol and 8-pCPT-cGMP, and the activation was restored by adenoviral reconstitution of PKG II (Marathe et al., 2012).

Previous work has shown that the antiapoptotic effects of estradiol require Erk activation and transient nuclear localization of the kinase (Chen et al., 2005). By Western blotting of nuclear extracts and immunofluorescence staining of MLO-Y4 cells, we showed that 8-pCPT-cGMP induces nuclear translocation of Erk in a PKG II-dependent manner, again mimicking the effect of estradiol (Marathe et al., 2012).

Pharmacological inhibition of phosphatidylinositol 3-kinase (PI3K) with LY294002 blocked estradiol- and 8-pCPT-cGMP-induced Akt phosphorylation and prevented estradiol from protecting MLO-Y4 cells from etoposide-induced apoptosis (Marathe et al., 2012). Thus, in addition to Erk activation and nuclear localization (Chen et al., 2005), the PI3K/Akt pathway is also necessary for the prosurvival effects of estradiol.

In endothelial cells, estradiol-induced Akt and Erk activation is dependent on Src (Cheskis, Greger, Nagpal, & Freedman, 2007; Levin, 2011). We previously reported that PKG II activates Erk and Akt in osteoblasts and osteocytes via Src activation, and that the membrane-bound PKG II enzyme is uniquely situated to stimulate Src through activation of a Shp-1/2 phosphatase complex (Rangaswami et al., 2010).

6.4. Antiapoptotic effects of estradiol and cGMP require BAD phosphorylation

The proapoptotic family member BAD sensitizes cells to apoptosis by binding and inactivating antiapoptotic Bcl-2 family members. BAD function is regulated by phosphorylation: Ser112 or Ser136 phosphorylation enables binding of 14-3-3 proteins and subsequent phosphorylation of Ser155 to fully disassociate BAD from the antiapoptotic Bcl-2 family member (Danial, 2008; Datta et al., 2000). We used mutant BAD constructs containing serine to alanine substitutions in each of the three phosphorylation sites (Zhou, Liu, Payne, Lutz, & Chittenden, 2000) to determine the role of BAD phosphorylation in the antiapoptotic effects of estradiol and cGMP. Etoposide induced similar levels of cell death in cells expressing wild-type or mutant BAD S112A (12.7% and 13.6%, respectively), and apoptosis was reduced to basal levels by estradiol or 8-pCPT-cGMP (Marathe et al., 2012). However, etoposide-treated cells expressing BAD S136A or BAD S155A continued to show a high percentage of cell death despite treatment with estradiol or 8-pCPT-cGMP. Thus, BAD phosphorylation on Ser136 and Ser155 is crucial for the effects of estradiol and cGMP on osteocyte apoptosis. The mutant proteins appear to act in a dominant-negative fashion, likely by dimerizing with endogenous Bcl-2 or Bcl-XL.

Previous studies suggest that PKG Iα phosphorylates BAD on Ser155 (Datta et al., 2000; Johlfs & Fiscus, 2010; Zhou et al., 2000). Using wild-type and mutant BAD constructs isolated from 293T cells, we confirmed that PKG I directly phosphorylates BAD Ser155, as alanine substitution for Ser155 completely prevented *in vitro* phosphorylation of BAD by PKG. A phospho-specific BAD antibody was used to examine the kinetics of BAD Ser155 phosphorylation in intact MLO-Y4 cells after cotransfection with wild-type BAD and PKG 1α (the antibody is not sensitive enough to detect phosphorylation of endogenous BAD). BAD Ser155 phosphorylation was found to increase within 10 min of adding 8-CPT-cGMP, and remained elevated at 2 h. Only PKG Iα, but not PKG Iβ, expression promoted cGMP-induced Ser155 phosphorylation in BAD-transfected cells, consistent with the antiapoptotic effect of PKG Iα, but not PKG Iβ, described earlier.

Treating MLO-Y4 cells and primary murine osteoblasts with 8-pCPT-cGMP also increased BAD phosphorylation on Ser112 and Ser136. This was most likely through cGMP/PKG II-mediated activation of Erk and Akt, as BAD Ser112 and Ser136 are targets of ribosomal S6 kinase (acting downstream of Erk) and Akt, respectively (Danial, 2008). Consistent with this hypothesis, siRNA-mediated depletion of PKG II prevented BAD phosphorylation at Ser112 and Ser136, and treatment with the PI3K inhibitor LY294002 blocked Akt phosphorylation and prevented BAD phosphorylation at Ser136. Thus, PKG Iα directly phosphorylates BAD on Ser155, while PKG II indirectly increases BAD phosphorylation on Ser136 by activating Akt. Akt is upstream of several prosurvival pathways, but one of its most prominent prosurvival effects is phosphorylation of BAD on Ser136, which is required for binding of 14-3-3 proteins and a prerequisite for BAD phosphorylation on Ser155 and dissociation from Bcl-2 (Datta et al., 2000).

In conclusion, PKG Iα and PKG II are important mediators of the anti-apoptotic effects of estradiol in osteocytes and their signaling converges on BAD phosphorylation. These results help explain why NOS3-deficient ovarectomized mice have a defective anabolic response to exogenous estrogens, how L-NAME reduces estradiol-induced bone formation in intact mice, and how NO donors alleviate ovariectomy-induced bone loss in rats (Armour et al., 2001; Grassi et al., 2006; Samuels et al., 2001; Wimalawansa et al., 1996). Our results provide a rationale for using NO donors or direct soluble guanylate cyclase activators as bone-protective agents.

ACKNOWLEDGMENT

We thank Dr. L. Bonewald for providing us MLO-Y4 cells.

REFERENCES

Afzal, F., Polak, J., & Buttery, L. (2004). Endothelial nitric oxide synthase in the control of osteoblastic mineralizing activity and bone integrity. *Journal of Pathology*, *202*, 503–510.

Aguirre, J., Buttery, L., O'Shaughnessy, M., Afzal, F., de Marticorena, I. F., Hukkanen, M., et al. (2001). Endothelial nitric oxide synthase gene-deficient mice demonstrate marked retardation in postnatal bone formation, reduced bone volume, and defects in osteoblast maturation and activity. *American Journal of Pathology*, *158*, 247–257.

Almeida, M., Han, L., Martin-Millan, M., Plotkin, L. I., Stewart, S. A., Roberson, P. K., et al. (2007). Skeletal involution by age-associated oxidative stress and its acceleration by loss of sex steroids. *Journal of Biological Chemistry*, *282*, 27285–27297.

Almeida, M., Martin-Millan, M., Ambrogini, E., Bradsher, R., III, Han, L., Chen, X. D., et al. (2010). Estrogens attenuate oxidative stress and the differentiation and apoptosis of osteoblasts by DNA-binding-independent actions of the ERalpha. *Journal of Bone and Mineral Research*, *25*, 769–781.

Armour, K. E., Armour, K. J., Gallagher, M. E., Godecke, A., Helfrich, M. H., Reid, D. M., et al. (2001). Defective bone formation and anabolic response to exogenous estrogen in mice with targeted disruption of endothelial nitric oxide synthase. *Endocrinology*, *142*, 760–766.

Barrett-Connor, E., Grady, D., & Stefanick, M. L. (2005). The rise and fall of menopausal hormone therapy. *Annual Review of Public Health*, *26*, 115–140.

Blaise, G. A., Gauvin, D., Gangal, M., & Authier, S. (2005). Nitric oxide, cell signaling and cell death. *Toxicology*, *208*, 177–192.

Casteel, D. E., Zhuang, S., Gudi, T., Tang, J., Vuica, M., Desiderio, S., et al. (2002). cGMP-dependent protein kinase I β physically and functionally interacts with the transcriptional regulator TFII-I. *Journal of Biological Chemistry*, *277*, 32003–32014.

Chae, H. J., Chin, H. Y., Lee, G. Y., Park, H. R., Yang, S. K., Chung, H. T., et al. (2006). Carbon monoxide and nitric oxide protect against tumor necrosis factor-alpha-induced apoptosis in osteoblasts: HO-1 is necessary to mediate the protection. *Clinica Chimica Acta*, *365*, 270–278.

Chae, H. J., Park, R. K., Chung, H. T., Kang, J. S., Kim, M. S., Choi, D. Y., et al. (1997). Nitric oxide is a regulator of bone remodelling. *Journal of Pharmacy and Pharmacology*, *49*, 897–902.

Chae, H. J., Park, R. K., Kang, J. S., Shin, H. S., Kim, S. C., Chung, H. T., et al. (1998). Effect of stem cell factor, interleukin-6, nitric oxide and transforming growth factor-beta on the osteoclast differentiation induced by 1 alpha, 25-(OH)2D3 in primary murine bone marrow cultures. *Pharmacology and Toxicology*, *82*, 223–229.

Chambliss, K. L., Wu, Q., Oltmann, S., Konaniah, E. S., Umetani, M., Korach, K. S., et al. (2010). Non-nuclear estrogen receptor alpha signaling promotes cardiovascular protection but not uterine or breast cancer growth in mice. *Journal of Clinical Investigation*, *120*, 2319–2330.

Chen, J. R., Plotkin, L. I., Aguirre, J. I., Han, L., Jilka, R. L., Kousteni, S., et al. (2005). Transient versus sustained phosphorylation and nuclear accumulation of ERKs underlie anti-versus pro-apoptotic effects of estrogens. *Journal of Biological Chemistry*, *280*, 4632–4638.

Cheskis, B. J., Greger, J. G., Nagpal, S., & Freedman, L. P. (2007). Signaling by estrogens. *Journal of Cellular Physiology*, *213*, 610–617.

Chrischilles, E. A., Butler, C. D., Davis, C. S., & Wallace, R. B. (1991). A model of lifetime osteoporosis impact. *Archives of Internal Medicine*, *151*, 2026–2032.

Dallas, S. L., Prideaux, M., & Bonewald, L. F. (2013). The osteocyte: An endocrine cell . . . and more. *Endocrine Reviews*, *34*, 658–690.

Danial, N. N. (2008). BAD: Undertaker by night, candyman by day. *Oncogene*, *27*(Suppl. 1), S53–S70.

Datta, S. R., Katsov, A., Hu, L., Petros, A., Fesik, S. W., Yaffe, M. B., et al. (2000). 14-3-3 proteins and survival kinases cooperate to inactivate BAD by BH3 domain phosphorylation. *Molecular Cell*, *6*, 41–51.

Dong, S. S., Williams, J. P., Jordan, S. E., Cornwell, T., & Blair, H. C. (1999). Nitric oxide regulation of cGMP production in osteoclasts. *Journal of Cellular Biochemistry*, *73*, 478–487.

Edwards, J. R., & Mundy, G. R. (2011). Advances in osteoclast biology: Old findings and new insights from mouse models. *Nature Reviews. Rheumatology*, *7*, 235–243.

Emerton, K. B., Hu, B., Woo, A. A., Sinofsky, A., Hernandez, C., Majeska, R. J., et al. (2010). Osteocyte apoptosis and control of bone resorption following ovariectomy in mice. *Bone*, *46*, 577–583.

Fiscus, R. R. (2002). Involvement of cyclic GMP and protein kinase G in the regulation of apoptosis and survival in neural cells. *Neurosignals*, *11*, 175–190.

Francis, S. H., Busch, J. L., Corbin, J. D., & Sibley, D. (2010). cGMP-dependent protein kinases and cGMP phosphodiesterases in nitric oxide and cGMP action. *Pharmacological Reviews*, *62*, 525–563.

Fuller, K., Kirstein, B., & Chambers, T. J. (2006). Murine osteoclast formation and function: Differential regulation by humoral agents. *Endocrinology*, *147*, 1979–1985.

Grassi, F., Fan, X., Rahnert, J., Weitzmann, M. N., Pacifici, R., Nanes, M. S., et al. (2006). Bone re/modeling is more dynamic in the endothelial nitric oxide synthase(−/−) mouse. *Endocrinology*, *147*, 4392–4399.

Hayashi, T., Ito, I., Kano, H., Endo, H., & Iguchi, A. (2000). Estriol (E3) replacement improves endothelial function and bone mineral density in very elderly women. *Journals of Gerontology. Series A, Biological Sciences and Medical Sciences*, *55*, B183–B190.

Hofmann, F., Bernhard, D., Lukowski, R., & Weinmeister, P. (2009). cGMP regulated protein kinases (cGK). *Handbook of Experimental Pharmacology*, *191*, 137–162.

Hopkins, R. B., Goeree, R., Pullenayegum, E., Adachi, J. D., Papaioannou, A., Xie, F., et al. (2011). The relative efficacy of nine osteoporosis medications for reducing the rate of fractures in post-menopausal women. *BMC Musculoskeletal Disorders*, *12*, 209.

Hukkanen, M., Platts, L. A., Lawes, T., Girgis, S. I., Konttinen, Y. T., Goodship, A. E., et al. (2003). Effect of nitric oxide donor nitroglycerin on bone mineral density in a rat model of estrogen deficiency-induced osteopenia. *Bone, 32*, 142–149.

Jamal, S. A., Browner, W. S., Bauer, D. C., & Cummings, S. R. (1998). Intermittent use of nitrates increases bone mineral density: The study of osteoporotic fractures. *Journal of Bone and Mineral Research, 13*, 1755–1759.

Jamal, S. A., Cummings, S. R., & Hawker, G. A. (2004). Isosorbide mononitrate increases bone formation and decreases bone resorption in postmenopausal women: A randomized trial. *Journal of Bone and Mineral Research, 19*, 1512–1517.

Jamal, S. A., & Hamilton, C. J. (2012). Nitric oxide donors for the treatment of osteoporosis. *Current Osteoporosis Reports, 10*, 86–92.

Jamal, S. A., Hamilton, C. J., Eastell, R., & Cummings, S. R. (2011). Effect of nitroglycerin ointment on bone density and strength in postmenopausal women: A randomized trial. *Journal of the American Medical Association, 305*, 800–807.

Jilka, R. L., Hangoc, G., Girasole, G., Passeri, G., Williams, D. C., & Abrams, J. S. (1992). Increased osteoclast development after estrogen loss: Mediation by interleukin-6. *Science, 257*, 88–91.

Johlfs, M. G., & Fiscus, R. R. (2010). Protein kinase G type-Ialpha phosphorylates the apoptosis-regulating protein Bad at serine 155 and protects against apoptosis in N1E-115 cells. *Neurochemistry International, 56*, 546–553.

Jung, J. Y., Lin, A. C., Ramos, L. M., Faddis, B. T., & Chole, R. A. (2003). Nitric oxide synthase I mediates osteoclast activity in vitro and in vivo. *Journal of Cellular Biochemistry, 89*, 613–621.

Khosla, S. (2011). Is nitroglycerin a novel and inexpensive treatment for osteoporosis? *Journal of the American Medical Association, 305*, 826–827.

Khosla, S., Bellido, T. M., Drezner, M. K., Gordon, C. M., Harris, T. B., Kiel, D. P., et al. (2011). Forum on aging and skeletal health: Summary of the proceedings of an ASBMR workshop. *Journal of Bone and Mineral Research, 26*, 2565–2578.

Khosla, S., Melton, L. J., III, & Riggs, B. L. (2011). The unitary model for estrogen deficiency and the pathogenesis of osteoporosis: Is a revision needed? *Journal of Bone and Mineral Research, 26*, 441–451.

Kousteni, S., Bellido, T., Plotkin, L. I., O'Brien, C. A., Bodenner, D. L., Han, L., et al. (2001). Nongenotropic, sex-nonspecific signaling through the estrogen or androgen receptors: dissociation from transcriptional activity. *Cell, 104*, 719–730.

Kousteni, S., Chen, J. R., Bellido, T., Han, L., Ali, A. A., O'Brien, C. A., et al. (2002). Reversal of bone loss in mice by nongenotropic signaling of sex steroids. *Science, 298*, 843–846.

Krum, S. A. (2011). Direct transcriptional targets of sex steroid hormones in bone. *Journal of Cellular Biochemistry, 112*, 401–408.

Krum, S. A., Miranda-Carboni, G. A., Hauschka, P. V., Carroll, J. S., Lane, T. F., Freedman, L. P., et al. (2008). Estrogen protects bone by inducing Fas ligand in osteoblasts to regulate osteoclast survival. *EMBO Journal, 27*, 535–545.

Lee, S. Y., Andoh, T., Murphy, D. L., & Chiueh, C. C. (2003). 17beta-estradiol activates ICI 182,780-sensitive estrogen receptors and cyclic GMP-dependent thioredoxin expression for neuroprotection. *FASEB Journal, 17*, 947–948.

Lee, S. K., Huang, H., Lee, S. W., Kim, K. H., Kim, K. K., Kim, H. M., et al. (2004). Involvement of iNOS-dependent NO production in the stimulation of osteoclast survival by TNF-α. *Experimental Cell Research, 298*, 359–368.

Levin, E. R. (2011). Minireview: Extranuclear steroid receptors: Roles in modulation of cell functions. *Molecular Endocrinology, 25*, 377–384.

Li, L., Hisamoto, K., Kim, K. H., Haynes, M. P., Bauer, P. M., Sanjay, A., et al. (2007). Variant estrogen receptor-c-Src molecular interdependence and c-Src structural requirements for endothelial NO synthase activation. *Proceedings of the National Academy of Sciences of the United States of America, 104*, 16468–16473.

Mancini, L., Moradi-Bidhendi, N., Becherini, L., Martineti, V., & MacIntyre, I. (2000). The biphasic effects of nitric oxide in primary rat osteoblasts are cGMP dependent. *Biochemical and Biophysical Research Communications, 274*, 477–481.

Manolagas, S. C. (2010). From estrogen-centric to aging and oxidative stress: A revised perspective of the pathogenesis of osteoporosis. *Endocrine Reviews, 31*, 266–300.

Manolagas, S. C., & Parfitt, A. M. (2013). For whom the bell tolls: distress signals from long-lived osteocytes and the pathogenesis of metabolic bone diseases. *Bone, 54*, 272–278.

Marathe, N., Rangaswami, H., Zhuang, S., Boss, G. R., & Pilz, R. B. (2012). Pro-survival effects of 17beta-estradiol on osteocytes are mediated by nitric Oxide/cGMP via differential actions of cGMP-dependent protein kinases I and II. *Journal of Biological Chemistry, 287*, 978–988.

Nabhan, A. F., & Rabie, N. H. (2008). Isosorbide mononitrate versus alendronate for post-menopausal osteoporosis. *International Journal of Gynaecology and Obstetrics, 103*, 213–216.

Nathan, C. (2004). The moving frontier in nitric oxide-dependent signaling. *Science's STKE, 2004*, e52.

Nicolin, V., Ponti, C., Narducci, P., Grill, V., Bortul, R., Zweyer, M., et al. (2005). Different levels of the neuronal nitric oxide synthase isoform modulate the rate of osteoclastic differentiation of TIB-71 and CRL-2278 RAW 264.7 murine cell clones. *The Anatomical Record Part A: Discoveries in Molecular, Cellular, and Evolutionary Biology, 286*, 945–954.

Otsuka, E., Hirano, K., Matsushita, S., Inoue, A., Hirose, S., Yamaguchi, A., et al. (1998). Effects of nitric oxide from exogenous nitric oxide donors on osteoblastic metabolism. *European Journal of Pharmacology, 349*, 345–350.

Pouwels, S., Lalmohamed, A., van Staa, T., Cooper, C., Souverein, P., Leufkens, H. G., et al. (2010). Use of organic nitrates and the risk of hip fracture: A population-based case–control study. *Journal of Clinical Endocrinology and Metabolism, 95*, 1924–1931.

Raisz, L. G. (2005). Pathogenesis of osteoporosis: Concepts, conflicts, and prospects. *Journal of Clinical Investigation, 115*, 3318–3325.

Rangaswami, H., Marathe, N., Zhuang, S., Chen, Y., Yeh, J. C., Frangos, J. A., et al. (2009). Type II cGMP-dependent protein kinase mediates osteoblast mechanotransduction. *Journal of Biological Chemistry, 284*, 14796–14808.

Rangaswami, H., Schwappacher, R., Marathe, N., Zhuang, S., Casteel, D. E., Haas, B., et al. (2010). Cyclic GMP and protein kinase G control a Src-containing mechanosome in osteoblasts. *Science Signaling, 3*, ra91.

Rejnmark, L., Vestergaard, P., & Mosekilde, L. (2006). Decreased fracture risk in users of organic nitrates: A nationwide case–control study. *Journal of Bone and Mineral Research, 21*, 1811–1817.

Russell, K. S., Haynes, M. P., Sinha, D., Clerisme, E., & Bender, J. R. (2000). Human vascular endothelial cells contain membrane binding sites for estradiol, which mediate rapid intracellular signaling. *Proceedings of the National Academy of Sciences of the United States of America, 97*, 5930–5935.

Samuels, A., Perry, M. J., Gibson, R. L., Colley, S., & Tobias, J. H. (2001). Role of endothelial nitric oxide synthase in estrogen-induced osteogenesis. *Bone, 29*, 24–29.

Schlossmann, J., Ammendola, A., Ashman, K., Zong, X., Huber, A., Neubauer, G., et al. (2000). Regulation of intracellular calcium by a signalling complex of IRAG, IP3 receptor and cGMP-dependent protein kinase. *Nature, 404*, 197–200.

Surks, H. K., Mochizuki, N., Kasai, Y., Georgescu, S. P., Tang, K. M., Ito, M., et al. (1999). Regulation of myosin phosphatase by a specific interaction with cGMP-dependent protein kinase Iα. *Science, 286*, 1583–1587.

Tan, S. D., Kuijpers-Jagtman, A. M., Semeins, C. M., Bronckers, A. L., Maltha, J. C., & Von den Hoff, J. W. (2006). Fluid shear stress inhibits TNFalpha-induced osteocyte apoptosis. *Journal of Dental Research*, *85*, 905–909.

Tang, K. M., Wang, G., Lu, P., Karas, R. H., Aronovitz, M., Heximer, S. P., et al. (2003). Regulator of G-protein signaling-2 mediates vascular smooth muscle relaxation and blood pressure. *Nature Medicine*, *9*, 1506–1512.

Tatsumi, S., Ishii, K., Amizuka, N., Li, M., Kobayashi, T., Kohno, K., et al. (2007). Targeted ablation of osteocytes induces osteoporosis with defective mechanotransduction. *Cell Metabolism*, *5*, 464–475.

Tomkinson, A., Reeve, J., Shaw, R. W., & Noble, B. S. (1997). The death of osteocytes via apoptosis accompanies estrogen withdrawal in human bone. *Journal of Clinical Endocrinology and Metabolism*, *82*, 3128–3135.

van't Hof, R. J., Macphee, J., Libouban, H., Helfrich, M. H., & Ralston, S. H. (2004). Regulation of bone mass and bone turnover by neuronal nitric oxide synthase. *Endocrinology*, *145*, 5068–5074.

Watts, N. B., Bilezikian, J. P., Camacho, P. M., Greenspan, S. L., Harris, S. T., Hodgson, S. F., et al. (2010). American Association of Clinical Endocrinologists Medical Guidelines for Clinical Practice for the diagnosis and treatment of postmenopausal osteoporosis. *Endocrine Practice*, *16*(Suppl. 3), 1–37.

Weber, S., Bernhard, D., Lukowski, R., Weinmeister, P., Worner, R., Wegener, J. W., et al. (2007). Rescue of cGMP kinase I knockout mice by smooth muscle specific expression of either isozyme. *Circulation Research*, *101*, 1096–1103.

Wimalawansa, S. J. (2000). Nitroglycerin therapy is as efficacious as standard estrogen replacement therapy (Premarin) in prevention of oophorectomy-induced bone loss: a human pilot clinical study. *Journal of Bone and Mineral Research*, *15*, 2240–2244.

Wimalawansa, S. J., De Marco, G., Gangula, P., & Yallampalli, C. (1996). Nitric oxide donor alleviates ovariectomy-induced bone loss. *Bone*, *18*, 301–304.

Wimalawansa, S. J., Grimes, J. P., Wilson, A. C., & Hoover, D. R. (2009). Transdermal nitroglycerin therapy may not prevent early postmenopausal bone loss. *Journal of Clinical Endocrinology and Metabolism*, *94*, 3356–3364.

Yaroslavskiy, B. B., Turkova, I., Wang, Y., Robinson, L. J., & Blair, H. C. (2010). Functional osteoclast attachment requires inositol-1,4,5-trisphosphate receptor-associated cGMP-dependent kinase substrate. *Laboratory Investigation*, *90*, 1533–1542.

Zhou, X. M., Liu, Y., Payne, G., Lutz, R. J., & Chittenden, T. (2000). Growth factors inactivate the cell death promoter BAD by phosphorylation of its BH3 domain on Ser155. *Journal of Biological Chemistry*, *275*, 25046–25051.

CHAPTER ELEVEN

Insights into the Diverse Effects of Nitric Oxide on Tumor Biology

Divya Vasudevan, Douglas D. Thomas[1]

Department of Medicinal Chemistry and Pharmacognosy, University of Illinois at Chicago, Chicago, Illinois, USA

[1]Corresponding author: e-mail address: ddthomas@uic.edu

Contents

Abstract

Among its many roles in cellular biology, nitric oxide (·NO) has long been associated with cancers both as a protumorigenic and as an antitumorigenic agent. The dual nature of this signaling molecule in varied settings is attributable to its temporal and concentration-dependent effects that produce different phenotypes. The steady-state ·NO concentration within the cell is a balance between its rate of enzymatic synthesis from the three nitric oxide synthase (NOS) isoforms and consumption via numerous metabolic pathways and demonstrates strong dependence on the tissue oxygen concentration. NOS expression and ·NO production are often deregulated and associated with numerous types of cancers with dissimilar prognostic outcomes. ·NO influences several facets of tumor initiation and progression including DNA damage, chronic inflammation, angiogenesis, epithelial-mesenchymal transition, and metastasis, to name a few. The role of ·NO as an epigenetic modulator has also recently emerged and has potentially important mechanistic implications in regulating transcription of oncogenes and tumor-suppressor genes. ·NO-derived cellular adducts such as dinitrosyliron complexes and the formation of higher nitrogen oxides further alter its cellular behavior. Among anticancer strategies, the use of NOS as a prognostic biomarker and modulation of ·NO production for therapeutic benefit have gained importance over the past decade. Numerous ·NO-releasing drugs and NOS inhibitors have been evaluated in preclinical and clinical settings to arrest tumor growth. Taken together, ·NO affects various arms of cancer signaling networks. An overview of this complex interplay is provided in this chapter.

Vitamins and Hormones, Volume 96
ISSN 0083-6729
http://dx.doi.org/10.1016/B978-0-12-800254-4.00011-8

1. INTRODUCTION

Nitric oxide (nitrogen monoxide, $^{\bullet}NO$) is a small, biologically short-lived free radical with numerous roles in health and disease. Since its discovery in the human body as the endothelium-derived relaxation factor in 1987 (Ignarro, Buga, Wood, Byrns, & Chaudhuri, 1987), $^{\bullet}NO$ has emerged as an important player in numerous signal transduction pathways. Although there are precursors for $^{\bullet}NO$ production that are derived from dietary sources, such as nitrate and nitrite (NO_2^-/NO_3^-), it is primarily synthesized enzymatically. Using arginine and oxygen (O_2) as cosubstrates, three isoforms of nitric oxide synthase (NOS): neuronal (nNOS), inducible (iNOS), and endothelial (eNOS), generate $^{\bullet}NO$ at various amounts (Stuehr, Santolini, Wang, Wei, & Adak, 2004). This is important because it is becoming increasingly clear that both the biochemical and phenotypic consequences of $^{\bullet}NO$ production are largely a function of its rate and duration of synthesis as well as the resulting concentration (Thomas et al., 2008). These concentration and time-dependent effects result in dramatic differences in the distribution of $^{\bullet}NO$-derived cellular adducts as well as the induction of specific signal transduction pathways (Thomas et al., 2004). Part of understanding the biological impact of $^{\bullet}NO$ is sorting out the context-specific factors that contribute to its overall steady-state concentration [$^{\bullet}NO_{ss}$]. As the steady-state concentration of $^{\bullet}NO$ is a function of both its rate of production as well as its rate of metabolism, there are numerous contributing factors. Variables include substrate and cofactor availability, the NOS isoform generating $^{\bullet}NO$, the amount of NOS, tissue/cell type, and especially the oxygen concentration. It is important to stress the important contribution of oxygen toward $^{\bullet}NO$ chemical biology, not only because is it a substrate for enzymatic $^{\bullet}NO$ synthesis but also because all dominant modes of $^{\bullet}NO$ metabolism are O_2-dependent (Hickok, Vasudevan, Jablonski, & Thomas, 2013).

Both deregulation in $^{\bullet}NO$ production and variations in NOS expression have been implicated in several types of cancer, although with contradicting outcomes. In some instances, they are correlated with tumor progression and metastasis, while in other cases they are associated with tumor suppression (Hickok & Thomas, 2010). To help understand the behavior of $^{\bullet}NO$ within a cell, we provide an overview of its basic reactions, resulting products, and metabolic pathways. This chapter will specifically focus on the roles of $^{\bullet}NO$ in the context of cancer etiology and advancement with an emphasis on explaining its dichotomous behavior in the tumor microenvironment.

New approaches toward understanding the pleiotropic nature of ˙NO including its ability to modulate iron homeostasis and function as an epigenetic modulator will also be highlighted. Finally, we discuss the therapeutic potential of ˙NO-releasing drugs, ˙NO-mimetics, and NOS inhibitors.

2. CELLULAR REACTIONS OF ˙NO

Before functional roles for ˙NO can be elucidated in cancer etiology, it is important to understand the diversity in its chemical biology. Physiological and pathological effects of ˙NO are driven by the chemical reactions in its environment rather than via classical protein receptor–ligand interactions. Although a multitude of reaction mechanisms, intermediates, and products have been cataloged for ˙NO, it only has two primary biological targets: transition metals and other free radicals. The reason for this is that ˙NO is a free radical and these two reaction types lead to stabilization of the unpaired electron on ˙NO. The biological consequences of these reactions, however, are a function of their products. Products formed from the reactions of ˙NO fall into two main categories: adducts containing nitrogen oxide functional groups and the formation of other nitrogen oxides (Fig. 11.1; Lancaster, 2003). Other nitrogen oxides including N_2O_3, $^{\cdot}NO_2$, $ONOO^-$, NO_2^-, NO_3^- arise from the reaction of ˙NO with oxygen or other free radicals such as superoxide O_2^-. These nitrogen oxides can either remain as stable end products (NO_2^-/NO_3^-) or go on to react with cellular targets to form adducts containing ˙NO functional groups. The addition of ˙NO functional groups to proteins has various consequences ranging from altering their catalytic activity to changing redox status. Among the biologically important ˙NO-derived cellular adducts are *S*-nitrosothiols (RSNO) and 3-nitrotyrosine (3-NT) (Gow, Farkouh, Munson, Posencheg, & Ischiropoulos, 2004; Thomas et al., 2008).

Various mechanisms for general RSNO formation have been proposed but explanations for site-specific targeting of RSNO remains an area of intense debate. RSNO arise from nitrosation of key cysteine residues in thiol-containing proteins (Stamler et al., 1992). Much like phosphorylation, S-nitrosation has been proposed as a protein posttranslational modification that regulates physiological and pathophysiological events. Although it is accepted that RSNO formation occurs *in vivo*, many questions still surround its viability as a novel ˙NO signaling mechanism. While kinases and phosphatases tightly regulate protein phosphorylation, no enzymatic equivalent mechanism(s) of regulated nitrosation or denitrosation have been reported.

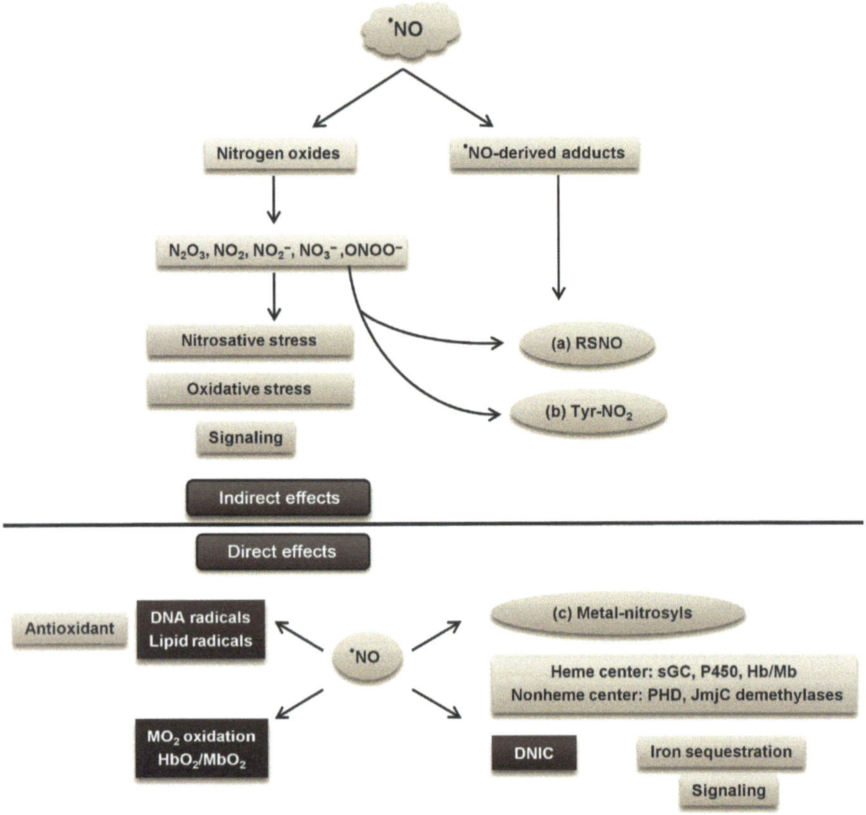

Figure 11.1 *The chemical biology of •NO signaling.* The biological effects of •NO can be broken down into two categories. Direct effects are those wherein •NO directly reacts with target molecules such as metal centers, DNA, and lipid radicals. Indirect effects are those wherein •NO first reacts with other molecules such as O_2 or O_2^- to form nitrogen oxides that then go on to react with target molecules. Both reaction types give rise to •NO-containing cellular adducts that can modulate signaling networks. (See the color plate.)

RSNO formation cannot be mediated by the direct reaction of •NO with free thiols and requires the formation of nitrosating species like N_2O_3 (Broniowska & Hogg, 2012). Formation of N_2O_3 is dependent on the reaction of •NO with O_2 or O_2^- and it is favored at high •NO concentrations in hydrophobic environments such as cell membranes (Liu, Miller, Joshi, Thomas, & Lancaster, 1998; Thomas et al., 2008). Therefore, other mechanisms must be invoked under alternate conditions to account for RSNO formation. One interesting possibility is that RSNO formation is a result of dinitrosyliron complex (DNIC) formation. Chelatable iron and the presence of DNIC have been correlated to RSNO formation (Bosworth, Toledo, Zmijewski, Li, & Lancaster, 2009). It is thought that DNIC act as storehouses of ·NO and mediate its oxidation to the nitrosonium (NO^+) moiety, which can go on to react with thiolates (RS^-) to form nitrosothiols (Vanin, Malenkova, & Serezhenkov, 1997). While the magnitude of cellular RSNO formation from free ·NO is substantially lower than amounts of DNIC formed, it appears that both iron-dependent and -independent mechanisms for RSNO formation exist (Hickok, Vasudevan, Thatcher, & Thomas, 2012). Nonetheless, S-nitrosation constitutes a potentially important arm of the cellular biology of ·NO and is implicated in several diseases such as neurodegeneration and cancer (Anand & Stamler, 2012).

The other dominant cellular ·NO adduct, 3-NT, is primarily derived from reactions of peroxynitrite ($ONOO^-$) generated in the local environment. Peroxynitrite is the product of the near diffusion-controlled reaction of ·NO with superoxide (O_2^-). Peroxynitrite can react to form several end products such as its reaction with tyrosine residues to form 3-NT (Beckman, Beckman, Chen, Marshall, & Freeman, 1990; Lymar, Khairutdinov, & Hurst, 2003; Thomas et al., 2008). Other possibilities include substrate oxidation and nitrate formation. Due to the rapid nature of the reaction that forms peroxynitrite, one consequence is the scavenging of ·NO which lowers the steady-state concentration ([$\cdot NO_{ss}$]). This can actually be quite important because decreases in [$\cdot NO_{ss}$] can preclude the reaction of ·NO with specific targets leading to altered phenotypes (Thomas et al., 2006). Another means of 3-NT formation is via nitrogen dioxide ($\cdot NO_2$). There are several pathways for the formation of $\cdot NO_2$. It is an important intermediate in the autooxidation reaction (see below) but it may also be formed in the absence of ·NO due to enzymatic or metal-catalyzed oxidation of nitrite (Thomas, Espey, Vitek, Miranda, & Wink, 2002). $\cdot NO_2$ is a strong oxidant which can react with proteins to either oxidize or nitrate functional groups (Thomas et al., 2002). Alternately, as shown below, $\cdot NO_2$ can react with ·NO to form N_2O_3, which as we have indicated is a strong nitrosating agent. Therefore, the formation of $\cdot NO_2$ can result in oxidative, nitrative, or nitrosative chemistry (Espey, Thomas, Miranda, & Wink, 2002; Espey, Xavier, Thomas, Miranda, & Wink, 2002). Given that $\cdot NO_2$ is uncharged, in the absence of ·NO, its lifetime is extended and it can diffuse several cell lengths. In this manner, $\cdot NO_2$ has been proposed to be a vital messenger molecule involved in signal transduction pathways (Espey, Xavier, et al., 2002).

Another important reaction of ·NO that has gained attention recently is the reaction of ·NO with the chelatable iron pool (CIP). The CIP is a methodologically defined population of redox active Fe^{2+}/Fe^{3+} and constitutes <5% of total cellular iron (Petrat, de Groot, Sustmann, & Rauen, 2002). Many enzymes utilize the iron (Fe^{2+}) from the CIP as a cofactor for catalytic activity. These include HIF prolyl hydroxylases, which mediate degradation of HIF-1α, and Jumonji C (JmjC) domain-containing histone demethylases, which have critical roles in cellular signaling networks (Cyr & Domann, 2011; Jaakkola et al., 2001; Knowles, Mole, Ratcliffe, & Harris, 2006). We and others have shown that when cells are exposed to ·NO in the presence of thiols, the CIP is quantitatively converted into paramagnetic DNIC with thiol-containing ligands (Hickok, Sahni, Shen, et al., 2011; Toledo et al., 2008). This means that iron sequestration which leads to an iron-starved phenotype is a potentially important corollary of DNIC formation. Although DNIC were detected in living systems over 50 years ago (Vithayathil, Ternberg, & Commoner, 1965), little is known regarding their biological function(s). Notwithstanding, DNIC are quantitatively the most abundant cellular adduct formed following ·NO exposure (Hickok, Sahni, Shen, et al., 2011). Recently, however, we demonstrated that DNIC formation resulted in inhibition of JmjC demethylase activity and altered histone methylation states, which in turn can modulate transcriptional machinery (Hickok, Vasudevan, Antholine, & Thomas, 2013; Fig. 11.2). Others have shown that certain proteins associated with drug resistance such as GST P1-1 and MRP regulate iron and ·NO trafficking in cells via DNIC formation. This machinery potentially modulates cellular ·NO levels and might have important effects in altering tumor cell signaling and proliferation (Lok et al., 2012). Deregulation of iron homeostasis itself has also been associated with the development and advancement of various types of cancers (Torti & Torti, 2011; Toyokuni, 2009). It is important to note that DNIC are formed rapidly after ·NO exposure and are stable even after removal of the ·NO source (Hickok, Sahni, Shen, et al., 2011; Hickok et al., 2012). Thus, iron sequestration by DNIC formation in ·NO-associated tumors can have both short term and long-term implications on gene transcription, ultimately dictating phenotypes.

Another way to classify biological reactions of ·NO is into *direct* and *indirect* reactions (Fig. 11.1). Direct reactions are those that occur fast enough for ·NO to be able to directly react with the target molecule. These reactions are prominent at low ·NO concentrations such as the binding of ·NO to the reduced iron in heme-containing enzymes. The most important of these is

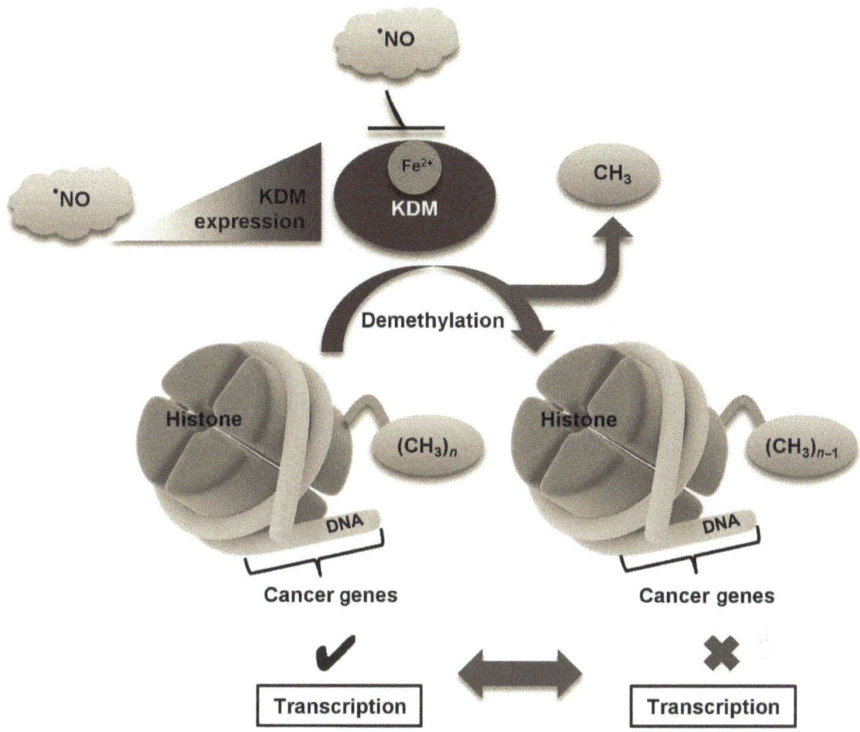

Figure 11.2 *Mechanisms of epigenetic regulation by •NO.* Nitric oxide can significantly alter the epigenetic landscape and regulate gene expression through a variety of mechanisms. It can directly inhibit the activity of JmjC histone demethylases (KDM) by binding to the non-heme iron atom. It can also change the expression level of a variety of histone-modifying enzymes. By changing both the degree and the location of post-translational histone modifications, •NO has potential to significantly alter gene transcription. This is a novel mechanism of •NO-mediated regulation of tumor phenotype. (See the color plate.)

the binding to and activation of soluble guanylyl cyclase (sGC) by •NO. The formation of DNIC is also an example of a key direct reaction of •NO. Conversely, indirect effects of •NO are the result of its reaction with other free radicals to form other nitrogen oxides as described above. These nitrogen oxides go on to react with various biological targets to have phenotypic effects. Indirect reactions are normally generated at higher •NO concentrations (Thomas et al., 2008; Wink et al., 1996).

One of the most important factors to consider when assessing the biological implications of •NO synthesis is the local oxygen concentration. In fact, the tissue oxygen gradient is a very important factor that dictates cellular responses to •NO (Hickok, Vasudevan, Jablonski, et al., 2013). In general, the O_2 availability in tissues is a function of its rate of delivery from the vasculature and its uptake via mitochondrial respiration. Although, normal O_2 tensions range from 10 to 60 μM (Buerk, 2007), tumors are often severely hypoxic. This becomes important because O_2 is a substrate for NOS enzymes and therefore, the local O_2 concentration is a key determinant of the rate of $^{\bullet}NO$ synthesis. The Km's for O_2 for each NOS isoform differ considerably; 23 μM for eNOS, 135 μM for iNOS, and 350 μM for nNOS. Therefore, changes in O_2 concentration affect $^{\bullet}NO$ production from eNOS to a lesser extent than that from nNOS (Stuehr et al., 2004). Additionally, the tissue distribution of NOS isoforms varies, making the local O_2 concentration even more critical in affecting $^{\bullet}NO$ output in a given microenvironment. Moreover, $^{\bullet}NO$ itself can extend the O_2 gradient in tissues by virtue of its ability to inhibit mitochondrial respiration (Buerk, 2007; Mason, Nicholls, Wilson, & Cooper, 2006; Thomas, Liu, Kantrow, & Lancaster, 2001).

In addition to its effect on $^{\bullet}NO$ synthesis, O_2 also affects $[^{\bullet}NO_{ss}]$ by modulating the rate of cellular $^{\bullet}NO$ consumption. One of the most studied reactions of $^{\bullet}NO$ is its reaction with O_2; the autooxidation reaction given by (Ford, Wink, & Stanbury, 1993):

$$^{\bullet}NO + O_2 \rightarrow [ONOO^{\bullet}]$$
$$[ONOO^{\bullet}] + {}^{\bullet}NO \rightarrow [N_2O_4]$$
$$^{\bullet}NO + {}^{\bullet}NO \rightarrow [N_2O_2]$$
$$[N_2O_2] + O_2 \rightarrow [N_2O_4]$$
$$[N_2O_4] \rightarrow 2\,^{\bullet}NO_2$$
$$^{\bullet}NO + {}^{\bullet}NO_2 \rightarrow N_2O_3$$
$$N_2O_3 + H_2O \rightarrow 2HNO_2$$

The overall stoichiometry of this reaction is:

$$4\,^{\bullet}NO + O_2 + 2H_2O \rightarrow 4H^+ + 4NO_2^-$$

Therefore, we have kinetic equation:

$$-\frac{d[NO]}{dt} = 4k[NO]^2[O_2]$$

where $k = 2 \times 10^6\ M^{-2}\ s^{-1}$.

It can be seen that several key intermediates that we have already discussed are formed during this reaction; namely, N_2O_3 and $^{\bullet}NO_2$. Due to the ubiquitous nature of O_2 within the most mammalian systems, this reaction undoubtedly occurs. It is important to note, however, that this reaction is second-order in $^{\cdot}NO$ concentration. At biological $^{\cdot}NO$ levels (<1 μM), the rate of autooxidation drops proportional to the square of the $^{\cdot}NO$ concentration. Therefore, the direct reaction of $^{\cdot}NO$ with O_2 would be too slow to compete with the reaction of $^{\cdot}NO$ with other cellular targets (radicals and metals). Proportionally, of all biological $^{\cdot}NO$ reactions, autooxidation does not account for a substantial percentage.

To date, the dominant mode of $^{\cdot}NO$ metabolism is not known. Interestingly, $^{\cdot}NO$ metabolism is an oxygen-dependent process that does not rely on the direct reaction of $^{\cdot}NO$ with oxygen. The greater the oxygen concentration, the faster $^{\cdot}NO$ is metabolized. In this manner, the rate of $^{\cdot}NO$ disappearance bears a direct correlation to the oxygen concentration (Thomas et al., 2001). Although the rapid nature of cellular $^{\cdot}NO$ metabolism overrides the autooxidation reaction, O_2 tensions are key in regulating [$^{\cdot}NO_{ss}$] (Hickok, Vasudevan, Jablonski, et al., 2013). $^{\cdot}NO$ metabolism coupled with O_2-dependent enzymatic synthesis of $^{\cdot}NO$ further highlight the importance of local oxygen concentrations in determining steady-state $^{\cdot}NO$ concentrations. In a hypoxic tumor, for example, the rate of $^{\cdot}NO$ disappearance will be slow but so will its rate of synthesis. In a well-oxygenated tumor, $^{\cdot}NO$ will be consumed more rapidly but it will also be synthesized at a higher rate. Complicating things further, it has been shown that the O_2 concentration affects iNOS expression, which is increased under hypoxia (Hickok, Sahni, Shen, et al., 2011; Hickok, Vasudevan, Jablonski, et al., 2013). As O_2 is limiting under hypoxia, increases in the amount of iNOS protein does not necessarily translate into the amount of $^{\cdot}NO$ produced. Interestingly, we found that there is an optimal O_2 concentration (~5–8%) where $^{\cdot}NO$ production and consumption pathways are ideally poised to achieve maximal [$^{\cdot}NO_{ss}$] and signaling responses (Hickok, Vasudevan, Jablonski, et al., 2013). This means that oxygen ultimately controls the [$^{\cdot}NO_{ss}$], but not in a manner that is necessarily predictable or linear. In summary, the main reason that the oxygen concentration is so important is because it influences local $^{\cdot}NO$ concentrations through numerous mechanisms.

This leads into the next topic of why the concentration of $^{\cdot}NO$ is so important. The main reason is that most biological responses to $^{\cdot}NO$ occur in a concentration-dependent manner. Almost all proteins respond to $^{\cdot}NO$ at specific threshold [$^{\cdot}NO_{ss}$] concentrations. For example, sGC activation requires <30 nM [$^{\cdot}NO$], Akt phosphorylation occurs between 30 and

100 nM [$^{\cdot}$NO], HIF-1α is stabilized in the presence of 100–300 nM [$^{\cdot}$NO], while p53 is phosphorylated at [$^{\cdot}$NO] > 500 nM (Thomas et al., 2004). These concentration-dependent $^{\cdot}$NO responses provide insight into its dichotomous behavior under various biological conditions. Not surprisingly, the oxygen concentration can indirectly have a huge influence on $^{\cdot}$NO signaling simply by regulating the $^{\cdot}$NO steady-state concentration (Hickok, Vasudevan, Jablonski, et al., 2013).

In addition to the $^{\cdot}$NO flux and the local redox environment, the duration of $^{\cdot}$NO exposure is also paramount. Some proteins like sGC and HIF-1α respond immediately to $^{\cdot}$NO (immediate responders), while others such as p53 require hours to be activated (delayed responders). Furthermore, sustained exposure to $^{\cdot}$NO is required for HIF-1α stabilization, which disappears when $^{\cdot}$NO levels drop below the required threshold concentration (transient responder). In contrast, p53 remains phosphorylated long after the $^{\cdot}$NO source is removed (sustained responder). It has long been wondered how $^{\cdot}$NO can produce dramatically different phenotypic outcomes under seemingly similar cellular environments. If one considers, however, that both the concentration and duration of $^{\cdot}$NO exposure bear different outcomes on cellular responses, many of the discrepancies associated with $^{\cdot}$NO signaling can be realized (Thomas et al., 2004, 2008).

2.1. Nitric Oxide and Cancer

The role of $^{\cdot}$NO on tumor biology is broad and often ambiguous. In some cases, it is protumorigenic and other times it is cytotoxic. Low $^{\cdot}$NO fluxes promote tumor cell survival by increasing angiogenesis and endothelial function. On the contrary, high $^{\cdot}$NO concentrations may induce apoptosis through numerous mechanisms such as activation of p53 phosphorylation and inhibition of mitochondrial respiration (Ridnour et al., 2006). Paradoxically, increased levels of $^{\cdot}$NO can also be tumor-initiating by introducing genotoxic DNA lesions (Wink et al., 1998). Regardless, whether or not $^{\cdot}$NO enhances or suppresses tumor cell metastasis or promotes differentiation into neoplastic phenotypes appears to be cell type and context specific.

Our lab recently demonstrated that $^{\cdot}$NO is an endogenously produced epigenetic regulator in tumor cells. It was shown that $^{\cdot}$NO could alter both histone methyl marks and the expression of histone methyl-modifying enzymes. Our findings also revealed that $^{\cdot}$NO directly inhibits JmjC domain-containing histone demethylases, potentially altering gene transcription (Hickok, Vasudevan, Antholine, et al., 2013; Fig. 11.2). This is important because the deregulation of both histone methyl marks and histone-modifying enzymes have been linked to the development or progression of cancers (Kooistra & Helin, 2012). This epigenetic model of ·NO signaling is novel and has the potential to explain much of the pleiotropic nature of this molecule in tumorigenesis. Furthermore, histone demethylases represent an emerging class of drug targets in chemotherapy (Hojfeldt, Agger, & Helin, 2013) and the design of specific ·NO-releasing drugs might be potentially attractive. Alternately, regulating ·NO signaling via targeting NOS isoforms with inhibitors is also an active area of cancer research.

2.2. ·NO and cancer etiology

There are numerous modes of action that have been proposed wherein ·NO can act as a tumor-initiating agent. Chronic inflammation is well documented to be a risk factor for the development and progression of various cancers. Activated macrophages in inflamed tissue can produce high levels of iNOS-derived ·NO which can go on to react with O_2 or O_2^- to form RNS such as N_2O_3, NO_2, and $ONOO^-$. These intermediates have been shown to induce DNA damage or alter oncogenic signaling pathways (Liu & Hotchkiss, 1995).

DNA damage in response to ·NO can occur via several mechanisms including the formation of carcinogenic nitrosamines, direct modification of DNA bases, or inhibition of DNA repair proteins (Wink, Ridnour, Hussain, & Harris, 2008). Nitrosamines can be formed by the nitrosation of amines either on protein residues or directly on DNA bases. Intermediates generated during ·NO autooxidation or the reduction of nitrite (which occurs in the acidic environment of the stomach) can form N_2O_3, which is a powerful nitrosating agent. This molecule can react with secondary amines on proteins or DNA to form carcinogenic nitrosamines (Wink et al., 1998). The nitrosation of secondary amines on DNA bases results in stable nitrosamines whereas nitrosation of primary amines on guanine, adenine, and cytosine can led to their deamination which results in transitions such as the conversion of cytosine to uracil (Nguyen et al., 1992; Wink et al., 1991). As mentioned earlier, another mechanism of nitrosative stress is via peroxynitrite formation. Peroxynitrite can either nitrate or oxidize DNA bases leading to the formation of 8-nitroguanine (Yermilov, Rubio, Becchi, et al., 1995; Yermilov, Rubio, & Ohshima, 1995), or cause DNA strand breaks *in vitro* (Salgo, Stone, Squadrito, Battista, & Pryor, 1995).

Modification of DNA repair proteins by ˙NO could also contribute to tumorigenesis by opposing apoptotic events. Several proteins such as ligases, Fgp, and alkyltransferases have been shown to be susceptible to inactivation under conditions of nitrosative stress (Laval, Wink, & Laval, 1997). Many of these proteins contain zinc fingers and nitrosation of cysteine residues in this motif leads to ejection of zinc and loss of structural integrity. This compromises their ability to repair damaged DNA, allowing genotoxic lesions to propagate and contribute toward cellular transformation to a neoplastic phenotype (Kroncke, 2001; Sidorkina, Espey, Miranda, Wink, & Laval, 2003; Wink & Laval, 1994). Before these results can be extrapolated from an *in vitro* setting, however, it is important to note that the feasibility of the above reactions *in vivo* depends on the relative fluxes the radical species involved (i.e., ˙NO and O_2^-). In fact, in many cases ˙NO has been shown to have antioxidant properties which may actually abate DNA damage. Although the reaction of ˙NO with O_2^- forms the powerful oxidant peroxynitrite, peroxynitrite itself can further react with ˙NO or O_2^- which tends to abolish much of its deleterious chemistry (Jourd'heuil et al., 1999; Thomas et al., 2006). Furthermore, O_2^- can reduce metals or oxidize macromolecules and therefore, its rapid reaction with ˙NO precludes this chemistry. As reduced metals can be strong oxidants, such as the via the Fenton reaction, removal of O_2^- in this case would be advantageous. Furthermore, there are numerous other cellular targets that can scavenge radicals such as GSH and ascorbate which may prevent their accumulation to high enough concentrations required for direct DNA modification (Wink et al., 1998). Thus, as with other aspects of its chemical biology, the ability of ˙NO to act as a cancer-initiating agent is also two-sided and dependent on numerous contributing factors.

Interactions of ˙NO with p53 have also been purported to play a role in cancer etiology. On one hand, ˙NO can induce phosphorylation of p53 at a number of residues which triggers programmed cell death (Hofseth et al., 2003). Accumulation of p53 can repress iNOS expression via a feedback loop and further protect against genotoxicity (Forrester et al., 1996). In this manner, ˙NO can counter DNA damage during chronic inflammation by activating apoptotic pathways. On the other hand, high iNOS levels have been shown to positively correlate with point mutations in the *p53* gene (Ambs et al., 1999) and increased risk of colon cancers. Thus, ˙NO–p53 interactions may bear different outcomes in different environments. A high mutation load of *p53* was seen in inflamed colon tissue obtained from patients suffering from ulcerative colitis (UC) (Hussain et al., 2000). Another study from the same group demonstrated UC samples lacked detectable levels of p53 phosphorylation at serine 46. They also noted that induction of p53-mediated apoptosis was also potentially inhibited due to the presence of high levels of macrophage inhibitory factor in UC cells compared to normal colon tissue. This study proposed that these somatic mutations would generate a cell population that is resistant to apoptosis and favor their clonal expansion. These cells would also be more tolerant to the genotoxic effects of ˙NO. Together, these processes can increase overall mutation rate and also confer protection against apoptosis, ultimately promoting carcinogenesis (Hofseth et al., 2003).

2.3. ˙NO and cancer progression

Besides its potential role as a causal agent in cancers, ˙NO has also been shown to facilitate conditions that influence tumor progression and metastasis. It has been demonstrated that ˙NO can promote tumor growth at low concentrations and exert antiproliferative effects at higher concentrations. When trying to understand the bimodal behavior of ˙NO in tumor biology, however, it is important to assess the study model being employed as it can substantially influence the measured responses.

Because of the controlled nature of *in vitro* cell culture experiments, the expression of NOS usually translates into the generation of ˙NO. An often overlooked caveat, however, is that *in vivo* NOS expression does not always correlate to ˙NO production. Nitric oxide synthesis requires the presence of substrates such as arginine and oxygen, as well as an abundance of cofactors. However, hypoxic tumors have poorly developed vasculature and restricted oxygen supply. Therefore, while hypoxia itself might elevate NOS expression (Hickok, Sahni, Shen, et al., 2011; Hickok, Vasudevan, Jablonski, et al., 2013), these tumors may be unable to synthesize substantial amounts of ˙NO. On the other hand, under these conditions, ˙NO metabolism is reduced (Hickok, Sahni, Shen, et al., 2011; Hickok, Vasudevan, Jablonski, et al., 2013) and its lifetime is increased. This would allow ˙NO to diffuse greater distances and differentially interact with additional cellular targets (Thomas et al., 2001). We have shown that NOS expression and the oxygen concentration do not linearly translate into the overall amount of ˙NO synthesized. There appears to be an optimal oxygen concentration where ˙NO-consumptive pathways and ˙NO synthesis is balanced to produce maximal amounts of ˙NO necessary for signaling (Hickok, Vasudevan, Jablonski, et al., 2013). Thus, NOS expression in a well-oxygenated tumor will produce a different ·NO microenvironment than in hypoxic tissue. Interestingly, regardless of whether NOS is producing bioactive amounts of ·NO, the presence of NOS has still been linked to more aggressive tumor phenotypes and, therefore, may serve a useful predictive marker and prognostic indicator (Hickok & Thomas, 2010).

2.3.1 NOS and cancer

·NO production and iNOS expression are generally considered hallmarks of aggressive tumor phenotypes and are often negative prognostic indicators (Jenkins et al., 1995; Vakkala et al., 2000; Wei et al., 2003). For example, iNOS upregulation and ·NO production have been associated with inflammatory lung carcinogenesis (Okayama et al., 2013), colon cancer pathogenesis (Ambs et al., 1998), as well as poor survival of ER-negative breast cancer patients (Glynn et al., 2010) and melanoma patients (Grimm, Ellerhorst, Tang, & Ekmekcioglu, 2008; Tanese, Grimm, & Ekmekcioglu, 2012). In ovarian cancer, iNOS positivity correlated with shorter disease-free survival and relapse of the tumor (Raspollini et al., 2004). Increased iNOS expression was found in almost 50% of head and neck carcinomas (Connelly, Macabeo-Ong, Dekker, Jordan, & Schmidt, 2005) and was also associated with the spread of metastases to lymph nodes (Brennan et al., 2008). Tumor-associated macrophages found in infiltrates of breast, cervical, and bladder cancers were found to express iNOS, and indicated poor prognosis (Bingle, Brown, & Lewis, 2002). The significance of iNOS as a molecular marker in cancers has therefore, gained considerable attention over the past two decades.

Animal models of iNOS-mediated tumor development have conflicting results. *Rag-2* deficient mice infected with *Helicobacter hepaticus* had increased leukocyte infiltration in the colon which correlated with increased iNOS expression at the site of infection as well as elevated urinary nitrite levels (Erdman et al., 2009). Genetic deletion of *iNOS* was found to decrease incidence of *Helicobacter pylori* induced gastritis (Nam et al., 2004) and urethane induced lung tumor growth (Kisley et al., 2002). PyV-mT mice lacking *iNOS* showed increased latency for developing mammary lesions with a delay period of 2–4 weeks (Ellies et al., 2003). On the other hand, when the highly metastatic K-1735 C4 murine melanoma cells were transfected with *iNOS*, they demonstrated increased apoptosis *in vitro*. After injection into Balb/c nude mice, these tumors were slow growing and less metastatic compared to injection with the parental cell line or cells lacking functional iNOS (Xie et al., 1995). Similarly, transfection of *iNOS* into human DLD-1 colon adenocarcinoma cells reduced their proliferation *in vitro*. However, when these cells were injected into nude mice, the resulting tumors were more vascularized and invasive compared to controls. A plausible explanation for this observation is that higher ·NO concentrations were generated by these cells *in vitro* as evidenced by the level of accumulated metabolites, causing apototic pathways to dominate. On the other hand, ·NO production in xenografted tumors was reported to be lower, thereby favoring progrowth and angiogenic pathways (Jenkins et al., 1995). While this is in line with the dual behavior of ·NO in cancer, it also highlights the underlying complexities of the purported functions of this molecule in tumor biology.

Adding to the difficulty, iNOS expression has varying outcomes in seemingly analogous *in vivo* tumor models. A few studies in *Apc* (*Min*/+) background mice capable of spontaneously developing multiple polyps in the intestine enumerate this paradox. In one study, deletion of *iNOS* in *Apc*(*Min*/+) mice led to a higher small intestine tumor number compared to *Apc*(*Min*/+)*iNOS*(+/+) littermates (Scott et al., 2001). In contrast, another group demonstrated that *Apc*(*Min*/+) mice developed fewer adenomas in the small intestine when their diet lacked arginine required for ·NO synthesis or contained the iNOS inhibitor, L-aminoguanidine. *iNOS* gene-knockout also had a similar effect on these mice (Ahn & Ohshima, 2001). A more recent study published that genetic ablation of *iNOS* or the administration of the iNOS inhibitor, L-NIL, did not significantly alter polyp formation in *Apc*(*Min*/+)$MSH2^{-/-}$ mice and eluted to the lack of a functional role for the ·NO/iNOS machinery in cancer development in these mice (Belcheva, Green, Weiss, Streutker, & Martin, 2013). Together, these studies illustrate the complexity of correlating ·NO production and iNOS expression with tumor development.

The role of eNOS and nNOS in tumor etiology and progression are understood to a much lesser extent. While it is well known that eNOS is associated with tumor proliferation by enhancing angiogenesis, mobility, and invasion (Jadeski, Hum, Chakraborty, & Lala, 2000; Jadeski & Lala, 1999), some findings suggest a broader role for this isoform in neoplastic transformation. For example, one study demonstrated that mice lacking *eNOS* were more susceptible to tumor development in the liver in response to carcinogens. eNOS overexpression correlated with downregulation of the matrix protease, cathepsin B, and led to decreased invasive potential of tumor cells (Decker et al., 2008). On the other hand, eNOS has also been shown to contribute to tumor initiation and advancement by activating oncogenic Ras proteins (Lim, Ancrile, Kashatus, & Counter, 2008). A more recent finding brought to light eNOS as a regulator of key pathways involved in the progression of prostate cancer. This study demonstrated that association of eNOS with chromatin-remodeling complexes mediated transcription of disease-related genes (Nanni et al., 2013).

Studies linking nNOS and cancer are rather scarce. In general, nNOS expression has been shown to be elevated with increasing histological grade of gliomas (Broholm et al., 2003; Tanriover et al., 2008). However, there was lack of correlation between nNOS expression and enzymatic ˙NO synthesis in these tumors. Evidence linking nNOS expression to metastatic potential is limited although one study observed higher levels of this isoform in melanoma metastases (Broholm et al., 2003).

2.3.2 Angiogenesis

Angiogenesis is a key event in the transformation of a benign tumor to its malignant and metastatic state. Several angiogenic factors such as the vascular endothelial growth factor (VEGF) and prostaglandins are recruited by the tumor to enhance blood supply required to secure nutrients and eliminate waste products (Hanahan & Folkman, 1996; Nishida, Yano, Nishida, Kamura, & Kojiro, 2006). The majority of ˙NO-driven angiogenic signals are attributable to low ˙NO fluxes derived from eNOS. To this end, HIF-1α stabilization by ˙NO causes upregulation of VEGF, which phosphorylates eNOS and enhances its activity (Papapetropoulos, Garcia-Cardena, Madri, & Sessa, 1997; Zhou, Schmid, & Brune, 2004). VEGF-mediated increases in angiogenesis were seen in both *iNOS*(+/+) and *iNOS*(−/−) mice but not in *eNOS*(−/−) mice, supporting the role of this isoform in VEGF-induced vascular permeability (Duda, Fukumura, & Jain, 2004; Ying & Hofseth, 2007). Prostaglandin E_2 (PGE_2) has been shown to support angiogenesis by increasing eNOS phosphorylation and activating the ˙NO/cGMP pathway. Increased endothelial sprouting in response to PGE_2 was seen in normal murine aortic segments but not in *eNOS* deficient types and the angiogenic effects were inhibited in the presence of inhibitors of eNOS or sGC (Namkoong et al., 2005). Along similar lines, Prostaglandin E_1-induced angiogenesis in rabbit cornea was potentiated by sodium nitroprusside (SNP) and abrogated by the eNOS inhibitor, L-NAME (Ziche et al., 1994).

Tumor neovascularization is also impacted by the interaction of ˙NO or eNOS with antiangiogenic factors. For example, it has been shown that very low (<1 nM), sustained fluxes of ˙NO can inhibit thrombospondin-1

(TSP1) and induce proliferation of endothelial cells. This progrowth response is dependent on sGC-driven ERK phosphorylation and is lost at higher ˙NO concentrations, where ERK dephosphorylation and reaccumulation of TSP1 occurs (Isenberg et al., 2005; Ridnour et al., 2005). The antiangiogenic agent, endostatin, decreases eNOS phosphorylation and subsequent ˙NO release by increasing PP2A phosphatase levels (Roberts, Isenberg, Ridnour, & Wink, 2007; Urbich et al., 2002). In addition to the above mentioned interactions, ˙NO also facilitates maturation and dilation of tumor blood vessels (Fukumura, Kashiwagi, & Jain, 2006).

Just like the disparate nature of ˙NO on other cellular responses, there are reports of ˙NO decreasing angiogenesis. One study showed that ˙NO donors inhibited VEGF upregulation by interfering with the binding of transcription factor activator protein (AP-1) to the *VEGF* promoter (Tsurumi et al., 1997). Another group reported that the use of SNP caused a dose-dependent inhibition of angiogenesis in the chorioallantoic membrane of chick embryos, while NOS inhibitors such as L-NAME and L-NMMA stimulated angiogenesis (Pipili-Synetos et al., 1994). However, these studies likely suffer from disparities in the dose of ˙NO delivered from dissimilar donor compounds which could potentially affect vascular growth. For example, lower concentrations (0.1–0.3 mM) of the ˙NO-mimetic, *S*-nitro-*N*-acetyl-penicillamine (SNAP), increased angiogenesis in microvascular endothelial cells and this effect was reversed in a dose-dependent manner when the concentration of donor rose above 0.5 mM (Jones, Tsugawa, Tarnawski, & Baatar, 2004). Regardless, these findings highlight the multifaceted functionality of this signaling molecule in different cellular milieu.

2.3.3 Epithelial-mesenchymal transition and metastasis

Epithelial-mesenchymal transition (EMT) is the process during which cells lose their epithelial adhesive properties and acquire mesenchymal properties of migration and invasion. EMT facilitates the conversion of a benign tumor to a more aggressive and metastatic phenotype; cells intravasate into the blood and lymph nodes and migrate to distant organs (Thiery, 2002, 2003). EMT is characterized by the functional loss of epithelial markers such as E-cadherin and is regulated by specific transcription factors including Snail. Activation of Snail, downstream of NF-κβ induces metastasis in part by repressing transcription of Raf-1 kinase inhibitor protein (RKIP). High levels of ˙NO have been shown to suppress EMT both *in vitro* and *in vivo* by inhibiting NF-κβ and Snail, thereby upregulating RKIP and E-cadherin protein levels (Baritaki et al., 2010). Modulation of this circuitry by ˙NO has also been shown to sensitize resistant tumor cells to apoptotic stimuli from chemotherapeutic agents (Bonavida & Baritaki, 2011).

We recently demonstrated another mechanism of ˙NO-mediated metastatic suppression *in vitro* using HCC 1806 triple negative breast cancer cells. This phenotype is attributable in part, to upregulation of N-myc downstream-regulated gene 1 (NDRG1) by ˙NO in a time and dose-dependent manner. We observed that these antimetastatic effects were reliant on sequestration of chelatable iron by ˙NO, such that incorporation of labile iron into DNIC correlated with decreased migratory and invasive potential of these cells (Hickok, Sahni, Mikhed, Bonini, & Thomas, 2011). These results highlight the important consequences of cellular ˙NO–iron interactions and also assign a functional role for DNIC in tumor settings.

There are several other reports of ˙NO-driven suppression of tumor cell migration and invasion that can be found in the literature. ˙NO has been shown to attenuate TGF-β-induced EMT in alveolar epithelial cells wherein cells maintained epithelial morphology and E-cadherin expression (Vyas-Read, Shaul, Yuhanna, & Willis, 2007). Matrix metalloproteinases (MMP) are tissue-remodeling proteins that have been implicated in the development of metastatic malignancies and are modulated by ˙NO. MMP-9 is an important member of this family and its overexpression has been associated with invasive tumor phenotypes (Deryugina & Quigley, 2006). One study showed that ˙NO decreased extracellular MMP-9 content and steady-state mRNA levels in MCF-7 cells and could, therefore, potentially contribute to antimetastatic outcomes (Jespersen et al., 2009).

˙NO has also been reported to decrease metastasis in *in vivo* model systems. ˙NO-producing tumor cells in ectopic or orthotopic xenograft nude mouse models induced apoptosis and failed to grow or form metastases (Le, Wei, Huang, Lancaster, & Xie, 2005). When B88 oral squamous carcinoma cells isolated from a patient were cultured and transfected with *iNOS in vitro*, they showed decreased migratory potential in Boyden chamber assays compared to nontransfected control cells. These cells formed smaller tumors when injected orthotopically into nude mice, and demonstrated no metastases in cervical lymph nodes. On the other hand, mice that received B88 control cells developed bigger tumors that metastasized to cervical lymph nodes (Harada et al., 2004). iNOS null tumor cell lines grew faster and produced more lung metastases when injected into *iNOS*(−/−) C57BL/6 mice compared to their enzymatically active *iNOS*(+/+) counterparts (Wei et al., 2003).

However, as with other facets of tumor progression, the role of ˙NO in metastasis is often contradictory and promigratory effects of ˙NO have been reported. Treatment of ER-negative breast cancer cells with the ˙NO-donor, DETA/NO, resulted in EMT by reducing adhesiveness, decreasing expression of E-cadherin, and upregulating expression of the mesenchymal marker, vimentin (Switzer et al., 2012). HRT-18 colorectal cancer cells constitutively expressing iNOS showed greater invasiveness *in vitro* than HT-29 cells of the same tumor type lacking iNOS (Siegert, Rosenberg, Schmitt, Denkert, & Hauptmann, 2002). Injection of B16-BL6 murine melanoma cells produced more and larger metastases in syngenic *iNOS* (+/+) mice than *iNOS*(−/−) mice. However, injection of M5076 murine ovarian sarcoma cells produced smaller and fewer metastases in the *iNOS* (+/+) mice than their *iNOS*(−/−) counterparts. This change was attributed to the differential sensitivity of these cell lines toward ˙NO-mediated cytotoxicity, arising from the disruption of host *iNOS* gene expression (Shi et al., 2000). Endogenous ˙NO production in EMT-6 murine breast cancer cells increased apoptosis *in vitro*, however, when these cells were injected into Balb/c mice, a twofold increase in pulmonary metastases was observed after 2 weeks (Edwards et al., 1996). Therefore, while cell culture experiments are an essential first step toward studying molecular pathways, care must be taken while extrapolating *in vitro* results to *in vivo* models. Currently, no reliable methods exist for accurately measuring ˙NO concentrations *in vivo* and ˙NO fluxes generated could be substantially different from those produced *in vitro*, giving rise to discrepancies in cellular responses. Local ˙NO effects might become significantly important while governing overall phenotype *in vivo* as ˙NO-mediated responses become limited by the prevalence of increased ˙NO-consumptive pathways and decreased diffusional distances in tissues.

There are also reports of ˙NO-mediated increases in MMP levels, which can lead to promigratory responses in cells. For example, use of the nitrosating agent SNAP increased mRNA and proteins levels of several MMPs including MMP-2 and -9 in WiDr colorectal cancer cells. This upregulation was partially mediated by a drop in tissue inhibitor of metalloproteinase-1/2 (TIMP-1/2) expression and correlated with increased cell migration/invasion in a cGMP-dependent manner (Babykutty et al., 2012). C3-L5 murine mammary adenocarcinoma cells expressed eNOS protein *in vitro* as well as *in vivo*, both at the primary site and lung metastases. Constitutively produced ˙NO increased invasiveness *in vitro* by down-regulating *TIMP-2* and *-3* mRNA expression, and this effect was diminished in the presence of inhibitors such as L-NAME and NMMA. When iNOS expression was induced *in vitro* in the presence of IFN-γ and LPS, these cells demonstrated a more invasive phenotype which paralleled upregulation of *MMP-2* mRNA and downregulation of *TIMP-2* and *-3* mRNA levels (Orucevic et al., 1999). In another study, *MMP-1* was elevated at the transcriptional level using both endogenous and exogenous sources of ·NO in human melanoma cell lines. This was also suggested to influence tumor progression (Ishii et al., 2003).

2.3.4 Epigenetics

Epigenetics is the study of heritable phenotypic changes caused by mechanisms other than altering the underlying DNA sequence. DNA methylation as well as histone modifications such as methylation, acetylation, phosphorylation, and ubiquitination can modulate transcriptional machinery and control gene expression (Kouzarides, 2007). Researchers have shown previously that free radicals can play an important role in epigenetic regulation via a variety of mechanisms including oxidation of nucleotides, oxidation of DNA and histone-modifying enzymes, and depletion of cofactor pools (Case, Madsen, Motto, Meyerholz, & Domann, 2013; Cyr & Domann, 2011; Hitchler & Domann, 2009, 2012; Hitchler et al., 2006). Building on these findings, our lab recently identified ·NO as an endogenously produced epigenetic regulator. We demonstrated that ·NO could inhibit JmjC domain-containing Fe (II) α-ketoglutarate-dependent histone demethylases, significantly altering global histone methylation patterns. This inhibition is direct, and occurs via binding of ·NO to the Fe (II) atom in the catalytic site of these enzymes leading to the formation of an iron–nitrosyl complex. The formation of this complex potentially precludes oxygen binding required for enzyme activation and thereby inhibits the demethylation reaction. Our findings also suggest that enzyme inhibition is potentiated indirectly due to the ability of ·NO to sequester iron during DNIC formation. The CIP is the source of iron cofactor required by JmjC demethylases, and DNIC formation depletes its availability thereby indirectly decreasing enzymatic activity. The relative amounts of cellular free iron versus DNIC impact overall histone methylation levels, which can consequently modify gene expression patterns and induce phenotypic changes. In addition, ·NO also alters the expression of numerous histone methyltransferases and demethylases at both the transcriptional and translational levels. Taken together, ·NO significantly changes the epigenetic landscape and in this manner can differentially regulate gene expression under a variety of physiological and pathophysiological conditions (Hickok, Vasudevan, Antholine, et al, 2013; Fig. 11.2).

JmjC demethylases and perturbations in cellular iron levels have independently been purported to play vital roles in tumor etiology and progression (Hojfeldt et al., 2013; Kooistra & Helin, 2012; Torti & Torti, 2011; Toyokuni, 2009). For example, cancer cells uptake more iron for growth and division, and the use of iron chelators has been shown to be efficacious in inducing antiproliferative effects (Buss, Torti, & Torti, 2003; Yu et al., 2012). These iron chelators target the same population of iron that is bound by ·NO during DNIC formation, i.e., the CIP (Petrat et al., 2002). Similarly, there are numerous reports of JmjC demethylases driving tumor development (Hojfeldt et al., 2013; Kooistra & Helin, 2012). Therefore, epigenetic regulatory effects of ·NO might have important underlying implications in the cause and advancement of several types of cancers.

Interestingly, we noted that ·NO exposure does not lead to S-nitrosation of thiol residues on the enzyme or destroy its structural integrity. In fact, ·NO-mediated JmjC demethylase inhibition is reversible and enzymatic activity is restored when the ·NO source is removed (Hickok, Vasudevan, Antholine, et al, 2013). As these enzymes are ubiquitously expressed, their activity can be modulated anywhere in the vicinity of sustained ·NO synthesis such as in ·NO-associated tumors. This can induce subtle changes in methylation at specific genomic loci in the local environment and differentially alter the tumor phenotype. Moreover, this mechanism holds true over a range of ·NO concentrations from nanomolar to micromolar and may overcome barriers associated with varying ·NO outputs across tissue zones arising from heterogeneous oxygen distribution. The pathway through which ·NO alters histone methylation is mechanistically simple. ·NO is synthesized, it then diffuses and binds to the non-heme iron atom in the active site of demethylases, thereby inhibiting their activity (Fig. 11.2).

$$^{\cdot}\text{NO synthesis} \rightarrow {}^{\cdot}\text{NO diffusion} \rightarrow \text{Demethylase inhibition}$$

This mechanism is perhaps the simplest that can be constructed for explaining the effects of ·NO that do not involve its interactions with heme-containing enzymes. It also does not invoke the complex chemistry of ·NO or the formation of specific adducts (RSNO or 3-NT formation). This mechanism does not eliminate or contradict currently accepted, well-defined paths of ·NO signaling but may turn out to be proportionally more probable under a multitude of diverse cellular conditions. Furthermore, it solves some fundamental problems in ·NO chemical biology by providing a simple and plausible alternate hypothesis for explaining the pleiotropic nature of a seemingly complex signaling molecule. Although time will tell, we suspect that epigenetic regulation by ·NO will prove to be equally if not more important than classical cGMP or RSNO-mediated signaling pathways for determining phenotypic outcome (Fig. 11.3).

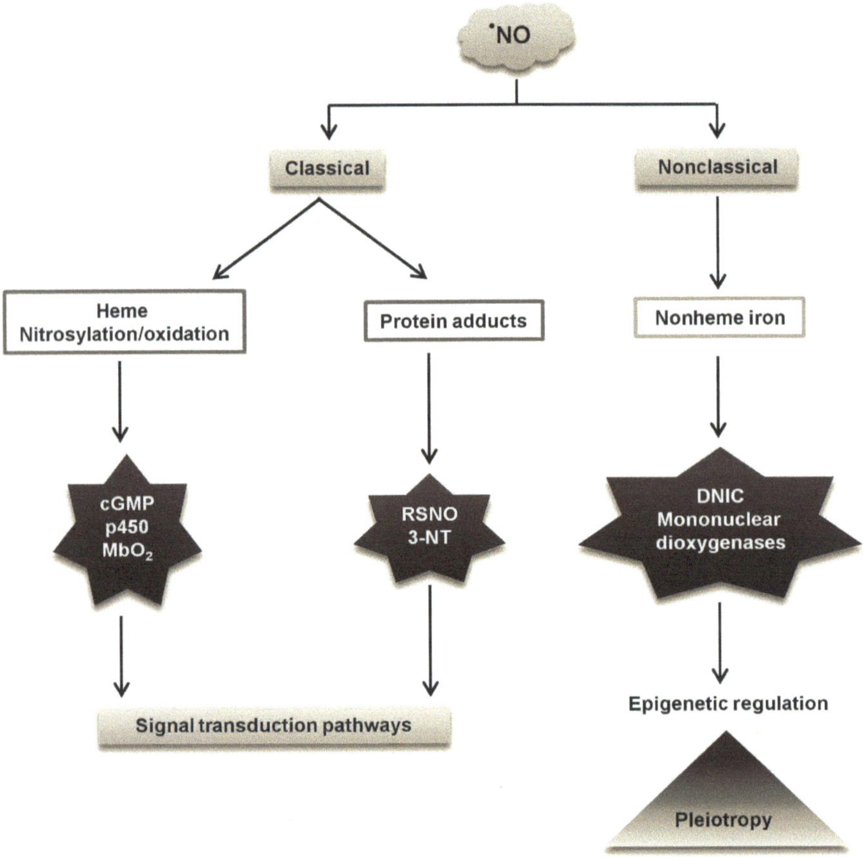

Figure 11.3 *The pleiotropic nature ·NO signaling.* Classical ·NO signaling mechanisms such as via sGC activation, RSNO, or 3-NT formation have long been studied and account for a portion of the observed cellular responses to ·NO. Inhibition of JmjC histone demethylases and DNIC formation are consequences of the interaction of ·NO with non-heme iron and represent novel, emerging mechanisms to explain much of its pleiotropic behavior. (See the color plate.)

2.4. ˙NO and cancer treatments

Given the various roles of ˙NO in cancers, it is no surprise that modulation of ˙NO production has been investigated for its therapeutic potential. Both ˙NO-releasing drugs and NOS inhibitors have been explored as therapies in numerous types of cancers. In some cases, NOS inhibitors have been tested to halt tumor progression. Inhibitors such as L-NAME, L-NIL, and 1400 W have been tested in several preclinical *in vivo* animal models for their antitumorigenic properties by blocking iNOS and eNOS-derived ˙NO production (Bian et al., 2012; Burke, Sullivan, Giles, & Glynn, 2013). On the other hand, a whole spectrum of ˙NO-releasing compounds have been synthesized and evaluated for use as antiproliferative agents in numerous settings. While NOS inhibitors and ˙NO donors have opposing effects on ˙NO generation, they can both serve as anticancer agents depending on the environment in which they are administered. This may once again be tied to the dichotomous nature of ˙NO in tumor etiology (Hickok & Thomas, 2010). Additionally, ˙NO donors have also been utilized for their radiosensitizing properties to enhance the efficaciousness of radiation therapies. Tumors frequently have poorly developed vasculature and for this reason, drugs administered into the blood often do not reach their intended target. Administration of ˙NO donors has been found to enhance tumor blood flow and oxygen supply, thereby resensitizing it to chemotherapeutic agents or radiation therapy (Wink et al., 2008).

There are several types of ˙NO-releasing drugs that have been developed for clinical evaluation. JS-K is a prodrug that releases ˙NO upon bioactivation and has been shown to inhibit cancer growth both *in vitro* and *in vivo*. JS-K has demonstrated efficacy as an anticancer agent in leukemia and several melanoma models concurrent with induction of DNA damage (Kiziltepe et al., 2007). A recent study demonstrated that another possible mode of action of JS-K is not via ˙NO generation but due to its ability to deplete cellular glutathione levels, which impacts the redox status leading to cellular stress and initiation of apoptosis (Maciag et al., 2013). NO–NSAIDs represent another class of derivative compounds where a traditional NSAID (like aspirin) is conjugated to an ˙NO-releasing moiety. These derivatives have been shown to be over 100 times more potent than their parent compounds in models of pancreatic and colon cancer by inducing apoptosis (Gao, Liu, & Rigas, 2005). GT-094 is an ˙NO chimera containing an NSAID that demonstrated reduction in aberrant crypt foci as well as iNOS expression levels in a rat model of colorectal cancer

(Hagos et al., 2007). NO–aspirin (NO–ASA) entered clinical trials to test its inhibitory effects on colon carcinogenesis among high-risk subjects. Whether these therapeutic effects are mediated by ˙NO itself or by generation of the compound backbone, remains to be determined (Rigas, 2007). Clinical trials were also conducted using low doses of nitroglycerin (GTN) administered via a patch to prostate cancer patients. Although GTN releases only minute quantities of free ˙NO, it did result in reduced tumor growth which was accompanied by protracted doubling time of the prostate-specific antigen (Siemens, Heaton, Adams, Kawakami, & Graham, 2009).

While these efforts are promising, there are several challenges to the practical use of ˙NO therapies that one must bear in mind. Among the disadvantages include risk of hypotensive shock, potential toxicity of the ˙NO-donor backbone, and other off-target effects (Miller & Megson, 2007). Achieving the desired ˙NO concentration consistently *in vivo* is also problematic, given the complex metabolic interplay that exists and the lack of tools to measure ˙NO levels under these conditions. Some compounds require metabolic activation to generate ˙NO, while others spontaneously release ˙NO. ˙NO delivery to the intended target must be precisely controlled in order to achieve maximal therapeutic efficacy and minimal side effects, which is complicated by the biochemical properties of this gaseous signaling molecule. Efforts to encapsulate the ˙NO-donor with suitable scaffolds, such as nanoparticles, to improve targeted delivery, and to prevent its release while in circulation have been attempted (Carpenter & Schoenfisch, 2012). Although modulation of tumor-˙NO pathways have promise as anticancer treatments for a variety of tumor types, further development is necessary before these options are of clinical relevance.

3. CONCLUSION/DISCUSSION

Unlike most signaling molecules, the chemical biology and signaling properties of ˙NO are immensely complex. There are few certainties surrounding the functional roles of ˙NO in tumor biology. As we have pointed out, in some cases ˙NO is inhibitory and in other cases it is tumor promoting. In the end, the main determinants of the downstream effects of ˙NO are its steady-state concentration and the duration of its exposure. Reasons being that these two properties not only determine the type of chemistry of ˙NO but also the distribution of cellular targets that are affected. The concentration of ˙NO determines much of its chemical reactions with other molecules. The types of reactions with other molecules (oxidation, nitration, nitrosation, nitrosylation) will influence their functionality. Ultimately, the functionality of ˙NO-target molecules will impact cell phenotype and have profound positive or negative effects on cancer etiology (Thomas et al., 2004, 2008). Despite the complexities surrounding ˙NO and cancer, one thing is clear; it is a key factor influencing numerous aspects of cancer etiology. What is important to realize is that reactions of ˙NO follow the laws of chemistry and physics and therefore, the behavior of ˙NO is not at all contradictory. Deciphering the influence of ˙NO on tumor biology is simply a matter of understanding the tumor microenvironmental conditions, which dictate the specific chemical biology of ˙NO.

REFERENCES

Ahn, B., & Ohshima, H. (2001). Suppression of intestinal polyposis in Apc(Min/+) mice by inhibiting nitric oxide production. *Cancer Research*, *61*(23), 8357–8360.

Ambs, S., Bennett, W. P., Merriam, W. G., Ogunfusika, M. O., Oser, S. M., Harrington, A. M., et al. (1999). Relationship between p53 mutations and inducible nitric oxide synthase expression in human colorectal cancer. *Journal of the National Cancer Institute*, *91*(1), 86–88.

Ambs, S., Merriam, W. G., Bennett, W. P., Felley-Bosco, E., Ogunfusika, M. O., Oser, S. M., et al. (1998). Frequent nitric oxide synthase-2 expression in human colon adenomas: Implication for tumor angiogenesis and colon cancer progression. *Cancer Research*, *58*(2), 334–341.

Anand, P., & Stamler, J. S. (2012). Enzymatic mechanisms regulating protein S-nitrosylation: Implications in health and disease. *Journal of Molecular Medicine*, *90*(3), 233–244. http://dx.doi.org/10.1007/s00109-012-0878-z.

Babykutty, S., Suboj, P., Srinivas, P., Nair, A. S., Chandramohan, K., & Gopala, S. (2012). Insidious role of nitric oxide in migration/invasion of colon cancer cells by upregulating MMP-2/9 via activation of cGMP-PKG-ERK signaling pathways. *Clinical and Experimental Metastasis*, *29*(5), 471–492. http://dx.doi.org/10.1007/s10585-012-9464-6.

Baritaki, S., Huerta-Yepez, S., Sahakyan, A., Karagiannides, I., Bakirtzi, K., Jazirehi, A., et al. (2010). Mechanisms of nitric oxide-mediated inhibition of EMT in cancer: Inhibition of the metastasis-inducer snail and induction of the metastasis-suppressor RKIP. *Cell Cycle*, *9*(24), 4931–4940. http://dx.doi.org/10.4161/cc.9.24.14229.

Beckman, J. S., Beckman, T. W., Chen, J., Marshall, P. A., & Freeman, B. A. (1990). Apparent hydroxyl radical production by peroxynitrite: Implications for endothelial injury from nitric oxide and superoxide. *Proceedings of the National Academy of Sciences of the United States of America*, *87*(4), 1620–1624.

Belcheva, A., Green, B., Weiss, A., Streutker, C., & Martin, A. (2013). Elevated incidence of polyp formation in APC(Min/(+))Msh2(−)/(−) mice is independent of nitric oxide-induced DNA mutations. *PLoS ONE*, *8*(5), e65204. http://dx.doi.org/10.1371/journal.pone.0065204.

Bian, K., Ghassemi, F., Sotolongo, A., Siu, A., Shauger, L., Kots, A., et al. (2012). NOS-2 signaling and cancer therapy. *IUBMB Life*, *64*(8), 676–683. http://dx.doi.org/10.1002/iub.1057.

Bingle, L., Brown, N. J., & Lewis, C. E. (2002). The role of tumour-associated macrophages in tumour progression: Implications for new anticancer therapies. *Journal of Pathology*, *196*(3), 254–265. http://dx.doi.org/10.1002/path.1027.

Bonavida, B., & Baritaki, S. (2011). Dual role of NO donors in the reversal of tumor cell resistance and EMT: Downregulation of the NF-kappaB/Snail/YY1/RKIP circuitry. *Nitric Oxide*, *24*(1), 1–7. http://dx.doi.org/10.1016/j.niox.2010.10.001.

Bosworth, C. A., Toledo, J. C., Jr., Zmijewski, J. W., Li, Q., & Lancaster, J. R., Jr. (2009). Dinitrosyliron complexes and the mechanism(s) of cellular protein nitrosothiol formation from nitric oxide. *Proceedings of the National Academy of Sciences of the United States of America*, *106*(12), 4671–4676. http://dx.doi.org/10.1073/pnas.0710416106.

Brennan, P. A., Dennis, S., Poller, D., Quintero, M., Puxeddu, R., & Thomas, G. J. (2008). Inducible nitric oxide synthase: Correlation with extracapsular spread and enhancement of tumor cell invasion in head and neck squamous cell carcinoma. *Head and Neck*, *30*(2), 208–214. http://dx.doi.org/10.1002/hed.20675.

Broholm, H., Rubin, I., Kruse, A., Braendstrup, O., Schmidt, K., Skriver, E. B., et al. (2003). Nitric oxide synthase expression and enzymatic activity in human brain tumors. *Clinical Neuropathology*, *22*(6), 273–281.

Broniowska, K. A., & Hogg, N. (2012). The chemical biology of *S*-nitrosothiols. *Antioxidants and Redox Signaling*, *17*(7), 969–980. http://dx.doi.org/10.1089/ars.2012.4590.

Buerk, D. G. (2007). Nitric oxide regulation of microvascular oxygen. *Antioxidants and Redox Signaling*, *9*(7), 829–843. http://dx.doi.org/10.1089/ars.2007.1551.

Burke, A. J., Sullivan, F. J., Giles, F. J., & Glynn, S. A. (2013). The yin and yang of nitric oxide in cancer progression. *Carcinogenesis*, *34*(3), 503–512. http://dx.doi.org/10.1093/carcin/bgt034.

Buss, J. L., Torti, F. M., & Torti, S. V. (2003). The role of iron chelation in cancer therapy. *Current Medicinal Chemistry*, *10*(12), 1021–1034.

Carpenter, A. W., & Schoenfisch, M. H. (2012). Nitric oxide release: Part II. Therapeutic applications. *Chemical Society Reviews*, *41*(10), 3742–3752. http://dx.doi.org/10.1039/c2cs15273h.

Case, A. J., Madsen, J. M., Motto, D. G., Meyerholz, D. K., & Domann, F. E. (2013). Manganese superoxide dismutase depletion in murine hematopoietic stem cells perturbs iron homeostasis, globin switching, and epigenetic control in erythrocyte precursor cells. *Free Radical Biology and Medicine*, *56*, 17–27. http://dx.doi.org/10.1016/j.freeradbiomed.2012.11.018.

Connelly, S. T., Macabeo-Ong, M., Dekker, N., Jordan, R. C., & Schmidt, B. L. (2005). Increased nitric oxide levels and iNOS over-expression in oral squamous cell carcinoma. *Oral Oncology*, *41*(3), 261–267. http://dx.doi.org/10.1016/j.oraloncology.2004.09.007.

Cyr, A. R., & Domann, F. E. (2011). The redox basis of epigenetic modifications: From mechanisms to functional consequences. *Antioxidants and Redox Signaling*, *15*(2), 551–589. http://dx.doi.org/10.1089/ars.2010.3492.

Decker, N. K., Abdelmoneim, S. S., Yaqoob, U., Hendrickson, H., Hormes, J., Bentley, M., et al. (2008). Nitric oxide regulates tumor cell cross-talk with stromal cells in the tumor microenvironment of the liver. *American Journal of Pathology*, *173*(4), 1002–1012. http://dx.doi.org/10.2353/ajpath.2008.080158.

Deryugina, E. I., & Quigley, J. P. (2006). Matrix metalloproteinases and tumor metastasis. *Cancer Metastasis Reviews*, *25*(1), 9–34. http://dx.doi.org/10.1007/s10555-006-7886-9.

Duda, D. G., Fukumura, D., & Jain, R. K. (2004). Role of eNOS in neovascularization: NO for endothelial progenitor cells. *Trends in Molecular Medicine*, *10*(4), 143–145.

Edwards, P., Cendan, J. C., Topping, D. B., Moldawer, L. L., MacKay, S., Copeland, E., et al. (1996). Tumor cell nitric oxide inhibits cell growth in vitro, but stimulates tumorigenesis and experimental lung metastasis in vivo. *Journal of Surgical Research*, *63*(1), 49–52. http://dx.doi.org/10.1006/jsre.1996.0221.

Ellies, L. G., Fishman, M., Hardison, J., Kleeman, J., Maglione, J. E., Manner, C. K., et al. (2003). Mammary tumor latency is increased in mice lacking the inducible nitric oxide

synthase. *International Journal of Cancer, 106*(1), 1–7. http://dx.doi.org/10.1002/ijc.11178.
Erdman, S. E., Rao, V. P., Poutahidis, T., Rogers, A. B., Taylor, C. L., Jackson, E. A., et al. (2009). Nitric oxide and TNF-alpha trigger colonic inflammation and carcinogenesis in Helicobacter hepaticus-infected, Rag2-deficient mice. *Proceedings of the National Academy of Sciences of the United States of America, 106*(4), 1027–1032. http://dx.doi.org/10.1073/pnas.0812347106.
Espey, M. G., Thomas, D. D., Miranda, K. M., & Wink, D. A. (2002). Focusing of nitric oxide mediated nitrosation and oxidative nitrosylation as a consequence of reaction with superoxide. *Proceedings of the National Academy of Sciences of the United States of America, 99*(17), 11127–11132. http://dx.doi.org/10.1073/pnas.152157599.
Espey, M. G., Xavier, S., Thomas, D. D., Miranda, K. M., & Wink, D. A. (2002). Direct real-time evaluation of nitration with green fluorescent protein in solution and within human cells reveals the impact of nitrogen dioxide vs. peroxynitrite mechanisms. *Proceedings of the National Academy of Sciences of the United States of America, 99*(6), 3481–3486. http://dx.doi.org/10.1073/pnas.062604199.
Ford, P. C., Wink, D. A., & Stanbury, D. M. (1993). Autoxidation kinetics of aqueous nitric oxide. *FEBS Letters, 326*(1–3), 1–3. http://dx.doi.org/10.1016/0014-5793(93)81748-O.
Forrester, K., Ambs, S., Lupold, S. E., Kapust, R. B., Spillare, E. A., Weinberg, W. C., et al. (1996). Nitric oxide-induced p53 accumulation and regulation of inducible nitric oxide synthase expression by wild-type p53. *Proceedings of the National Academy of Sciences of the United States of America, 93*(6), 2442–2447.
Fukumura, D., Kashiwagi, S., & Jain, R. K. (2006). The role of nitric oxide in tumour progression. *Nature Reviews. Cancer, 6*(7), 521–534. http://dx.doi.org/10.1038/nrc1910.
Gao, J., Liu, X., & Rigas, B. (2005). Nitric oxide-donating aspirin induces apoptosis in human colon cancer cells through induction of oxidative stress. *Proceedings of the National Academy of Sciences of the United States of America, 102*(47), 17207–17212. http://dx.doi.org/10.1073/pnas.0506893102.
Glynn, S. A., Boersma, B. J., Dorsey, T. H., Yi, M., Yfantis, H. G., Ridnour, L. A., et al. (2010). Increased NOS2 predicts poor survival in estrogen receptor-negative breast cancer patients. *Journal of Clinical Investigation, 120*(11), 3843–3854. http://dx.doi.org/10.1172/JCI42059.
Gow, A. J., Farkouh, C. R., Munson, D. A., Posencheg, M. A., & Ischiropoulos, H. (2004). Biological significance of nitric oxide-mediated protein modifications. *American Journal of Physiology Lung Cellular and Molecular Physiology, 287*(2), L262–L268. http://dx.doi.org/10.1152/ajplung.00295.2003.
Grimm, E. A., Ellerhorst, J., Tang, C. H., & Ekmekcioglu, S. (2008). Constitutive intracellular production of iNOS and NO in human melanoma: Possible role in regulation of growth and resistance to apoptosis. *Nitric Oxide, 19*(2), 133–137. http://dx.doi.org/10.1016/j.niox.2008.04.009.
Hagos, G. K., Carroll, R. E., Kouznetsova, T., Li, Q., Toader, V., Fernandez, P. A., et al. (2007). Colon cancer chemoprevention by a novel NO chimera that shows anti-inflammatory and antiproliferative activity in vitro and in vivo. *Molecular Cancer Therapeutics, 6*(8), 2230–2239. http://dx.doi.org/10.1158/1535-7163.MCT-07-0069.
Hanahan, D., & Folkman, J. (1996). Patterns and emerging mechanisms of the angiogenic switch during tumorigenesis. *Cell, 86*(3), 353–364. http://dx.doi.org/10.1016/S0092-8674(00)80108-7.
Harada, K., Supriatno, Kawaguchi, S., Tomitaro, O., Yoshida, H., & Sato, M. (2004). Overexpression of iNOS gene suppresses the tumorigenicity and metastasis of oral cancer cells. *In Vivo, 18*(4), 449–455.

Hickok, J. R., Sahni, S., Mikhed, Y., Bonini, M. G., & Thomas, D. D. (2011). Nitric oxide suppresses tumor cell migration through N-Myc downstream-regulated gene-1 (NDRG1) expression: Role of chelatable iron. *Journal of Biological Chemistry*, *286*(48), 41413–41424. http://dx.doi.org/10.1074/jbc.M111.287052.

Hickok, J. R., Sahni, S., Shen, H., Arvind, A., Antoniou, C., Fung, L. W., et al. (2011). Dinitrosyliron complexes are the most abundant nitric oxide-derived cellular adduct: Biological parameters of assembly and disappearance. *Free Radical Biology and Medicine*, *51*(8), 1558–1566. http://dx.doi.org/10.1016/j.freeradbiomed.2011.06.030.

Hickok, J. R., & Thomas, D. D. (2010). Nitric oxide and cancer therapy: The emperor has NO clothes. *Current Pharmaceutical Design*, *16*(4), 381–391.

Hickok, J. R., Vasudevan, D., Antholine, W. E., & Thomas, D. D. (2013). Nitric oxide modifies global histone methylation by inhibiting Jumonji C domain-containing demethylases. *Journal of Biological Chemistry*, *288*(22), 16004–16015. http://dx.doi.org/10.1074/jbc.M112.432294.

Hickok, J. R., Vasudevan, D., Jablonski, K., & Thomas, D. D. (2013). Oxygen dependence of nitric oxide-mediated signaling. *Redox Biology*, *1*(1), 203–209. http://dx.doi.org/10.1016/j.redox.2012.11.002.

Hickok, J. R., Vasudevan, D., Thatcher, G. R., & Thomas, D. D. (2012). Is S-nitrosocysteine a true surrogate for nitric oxide? *Antioxidants and Redox Signaling*, *17*(7), 962–968. http://dx.doi.org/10.1089/ars.2012.4543.

Hitchler, M. J., & Domann, F. E. (2009). Metabolic defects provide a spark for the epigenetic switch in cancer. *Free Radical Biology and Medicine*, *47*(2), 115–127. http://dx.doi.org/10.1016/j.freeradbiomed.2009.04.010.

Hitchler, M. J., & Domann, F. E. (2012). Redox regulation of the epigenetic landscape in cancer: A role for metabolic reprogramming in remodeling the epigenome. *Free Radical Biology and Medicine*, *53*(11), 2178–2187. http://dx.doi.org/10.1016/j.freeradbiomed.2012.09.028.

Hitchler, M. J., Wikainapakul, K., Yu, L., Powers, K., Attatippaholkun, W., & Domann, F. E. (2006). Epigenetic regulation of manganese superoxide dismutase expression in human breast cancer cells. *Epigenetics*, *1*(4), 163–171. http://dx.doi.org/10.4161/epi.1.4.3401.

Hofseth, L. J., Saito, S., Hussain, S. P., Espey, M. G., Miranda, K. M., Araki, Y., et al. (2003). Nitric oxide-induced cellular stress and p53 activation in chronic inflammation. *Proceedings of the National Academy of Sciences of the United States of America*, *100*(1), 143–148. http://dx.doi.org/10.1073/pnas.0237083100.

Hojfeldt, J. W., Agger, K., & Helin, K. (2013). Histone lysine demethylases as targets for anticancer therapy. *Nature Reviews. Drug Discovery*, *12*(12), 917–930. http://dx.doi.org/10.1038/nrd4154.

Hussain, S. P., Amstad, P., Raja, K., Ambs, S., Nagashima, M., Bennett, W. P., et al. (2000). Increased p53 mutation load in noncancerous colon tissue from ulcerative colitis: A cancer-prone chronic inflammatory disease. *Cancer Research*, *60*(13), 3333–3337.

Ignarro, L. J., Buga, G. M., Wood, K. S., Byrns, R. E., & Chaudhuri, G. (1987). Endothelium-derived relaxing factor produced and released from artery and vein is nitric oxide. *Proceedings of the National Academy of Sciences of the United States of America*, *84*(24), 9265–9269.

Isenberg, J. S., Ridnour, L. A., Perruccio, E. M., Espey, M. G., Wink, D. A., & Roberts, D. D. (2005). Thrombospondin-1 inhibits endothelial cell responses to nitric oxide in a cGMP-dependent manner. *Proceedings of the National Academy of Sciences of the United States of America*, *102*(37), 13141–13146. http://dx.doi.org/10.1073/pnas.0502977102.

Ishii, Y., Ogura, T., Tatemichi, M., Fujisawa, H., Otsuka, F., & Esumi, H. (2003). Induction of matrix metalloproteinase gene transcription by nitric oxide and mechanisms of

MMP-1 gene induction in human melanoma cell lines. *International Journal of Cancer, 103*(2), 161–168. http://dx.doi.org/10.1002/ijc.10808.

Jaakkola, P., Mole, D. R., Tian, Y. M., Wilson, M. I., Gielbert, J., Gaskell, S. J., et al. (2001). Targeting of HIF-alpha to the von Hippel-Lindau ubiquitylation complex by O2-regulated prolyl hydroxylation. *Science, 292*(5516), 468–472. http://dx.doi.org/10.1126/science.1059796.

Jadeski, L. C., Hum, K. O., Chakraborty, C., & Lala, P. K. (2000). Nitric oxide promotes murine mammary tumour growth and metastasis by stimulating tumour cell migration, invasiveness and angiogenesis. *International Journal of Cancer, 86*(1), 30–39. http://dx.doi.org/10.1002/(SICI)1097-0215(20000401)86:1<30::AID-IJC5>3.0.CO;2-I.

Jadeski, L. C., & Lala, P. K. (1999). Nitric oxide synthase inhibition by N(G)-nitro-L-arginine methyl ester inhibits tumor-induced angiogenesis in mammary tumors. *American Journal of Pathology, 155*(4), 1381–1390. http://dx.doi.org/10.1016/S0002-9440(10)65240-6.

Jenkins, D. C., Charles, I. G., Thomsen, L. L., Moss, D. W., Holmes, L. S., Baylis, S. A., et al. (1995). Roles of nitric oxide in tumor growth. *Proceedings of the National Academy of Sciences of the United States of America, 92*(10), 4392–4396.

Jespersen, C., Doller, A., Akool el, S., Bachmann, M., Muller, R., Gutwein, P., et al. (2009). Molecular mechanisms of nitric oxide-dependent inhibition of TPA-induced matrix metalloproteinase-9 (MMP-9) in MCF-7 cells. *Journal of Cellular Physiology, 219*(2), 276–287. http://dx.doi.org/10.1002/jcp.21658.

Jones, M. K., Tsugawa, K., Tarnawski, A. S., & Baatar, D. (2004). Dual actions of nitric oxide on angiogenesis: Possible roles of PKC, ERK, and AP-1. *Biochemical and Biophysical Research Communications, 318*(2), 520–528. http://dx.doi.org/10.1016/j.bbrc.2004.04.055.

Jourd'heuil, D., Miranda, K. M., Kim, S. M., Espey, M. G., Vodovotz, Y., Laroux, S., et al. (1999). The oxidative and nitrosative chemistry of the nitric oxide/superoxide reaction in the presence of bicarbonate. *Archives of Biochemistry and Biophysics, 365*(1), 92–100. http://dx.doi.org/10.1006/abbi.1999.1143.

Kisley, L. R., Barrett, B. S., Bauer, A. K., Dwyer-Nield, L. D., Barthel, B., Meyer, A. M., et al. (2002). Genetic ablation of inducible nitric oxide synthase decreases mouse lung tumorigenesis. *Cancer Research, 62*(23), 6850–6856.

Kiziltepe, T., Hideshima, T., Ishitsuka, K., Ocio, E. M., Raje, N., Catley, L., et al. (2007). JS-K, a GST-activated nitric oxide generator, induces DNA double-strand breaks, activates DNA damage response pathways, and induces apoptosis in vitro and in vivo in human multiple myeloma cells. *Blood, 110*(2), 709–718. http://dx.doi.org/10.1182/blood-2006-10-052845.

Knowles, H. J., Mole, D. R., Ratcliffe, P. J., & Harris, A. L. (2006). Normoxic stabilization of hypoxia-inducible factor-1alpha by modulation of the labile iron pool in differentiating U937 macrophages: Effect of natural resistance-associated macrophage protein 1. *Cancer Research, 66*(5), 2600–2607. http://dx.doi.org/10.1158/0008-5472.CAN-05-2351.

Kooistra, S. M., & Helin, K. (2012). Molecular mechanisms and potential functions of histone demethylases. *Nature Reviews. Molecular Cell Biology, 13*(5), 297–311. http://dx.doi.org/10.1038/nrm3327.

Kouzarides, T. (2007). Chromatin modifications and their function. *Cell, 128*(4), 693–705. http://dx.doi.org/10.1016/j.cell.2007.02.005.

Kroncke, K. D. (2001). Zinc finger proteins as molecular targets for nitric oxide-mediated gene regulation. *Antioxidants and Redox Signaling, 3*(4), 565–575. http://dx.doi.org/10.1089/15230860152542934.

Lancaster, J., Jr. (2003). Reactivity and diffusivity of nitrogen oxides in mammalian biology. In H. J. Forman, J. Fukuto, & M. Torres (Eds.), *Signal transduction by reactive oxygen and nitrogen species: pathways and chemical principles* (pp. 53–79). Netherlands: Springer.

Laval, F., Wink, D. A., & Laval, J. (1997). A discussion of mechanisms of NO genotoxicity: Implication of inhibition of DNA repair proteins. *Reviews of Physiology, Biochemistry and Pharmacology*, *131*, 175–191.

Le, X., Wei, D., Huang, S., Lancaster, J. R., Jr., & Xie, K. (2005). Nitric oxide synthase II suppresses the growth and metastasis of human cancer regardless of its up-regulation of protumor factors. *Proceedings of the National Academy of Sciences of the United States of America*, *102*(24), 8758–8763. http://dx.doi.org/10.1073/pnas.0409581102.

Lim, K. H., Ancrile, B. B., Kashatus, D. F., & Counter, C. M. (2008). Tumour maintenance is mediated by eNOS. *Nature*, *452*(7187), 646–649. http://dx.doi.org/10.1038/nature06778.

Liu, R. H., & Hotchkiss, J. H. (1995). Potential genotoxicity of chronically elevated nitric oxide: A review. *Mutational Research*, *339*(2), 73–89. http://dx.doi.org/10.1016/0165-1110(95)90004-7.

Liu, X., Miller, M. J., Joshi, M. S., Thomas, D. D., & Lancaster, J. R., Jr. (1998). Accelerated reaction of nitric oxide with O2 within the hydrophobic interior of biological membranes. *Proceedings of the National Academy of Sciences of the United States of America*, *95*(5), 2175–2179.

Lok, H. C., Suryo Rahmanto, Y., Hawkins, C. L., Kalinowski, D. S., Morrow, C. S., Townsend, A. J., et al. (2012). Nitric oxide storage and transport in cells are mediated by glutathione S-transferase P1-1 and multidrug resistance protein 1 via dinitrosyl iron complexes. *Journal of Biological Chemistry*, *287*(1), 607–618. http://dx.doi.org/10.1074/jbc.M111.310987.

Lymar, S. V., Khairutdinov, R. F., & Hurst, J. K. (2003). Hydroxyl radical formation by O-O bond homolysis in peroxynitrous acid. *Inorganic Chemistry*, *42*(17), 5259–5266. http://dx.doi.org/10.1021/ic030104l.

Maciag, A. E., Holland, R. J., Robert Cheng, Y. S., Rodriguez, L. G., Saavedra, J. E., Anderson, L. M., et al. (2013). Nitric oxide-releasing prodrug triggers cancer cell death through deregulation of cellular redox balance. *Redox Biology*, *1*(1), 115–124. http://dx.doi.org/10.1016/j.redox.2012.12.002.

Mason, M. G., Nicholls, P., Wilson, M. T., & Cooper, C. E. (2006). Nitric oxide inhibition of respiration involves both competitive (heme) and noncompetitive (copper) binding to cytochrome c oxidase. *Proceedings of the National Academy of Sciences of the United States of America*, *103*(3), 708–713. http://dx.doi.org/10.1073/pnas.0506562103.

Miller, M. R., & Megson, I. L. (2007). Recent developments in nitric oxide donor drugs. *British Journal of Pharmacology*, *151*(3), 305–321. http://dx.doi.org/10.1038/sj.bjp.0707224.

Nam, K. T., Oh, S. Y., Ahn, B., Kim, Y. B., Jang, D. D., Yang, K. H., et al. (2004). Decreased Helicobacter pylori associated gastric carcinogenesis in mice lacking inducible nitric oxide synthase. *Gut*, *53*(9), 1250–1255. http://dx.doi.org/10.1136/gut.2003.030684.

Namkoong, S., Lee, S. J., Kim, C. K., Kim, Y. M., Chung, H. T., Lee, H., et al. (2005). Prostaglandin E2 stimulates angiogenesis by activating the nitric oxide/cGMP pathway in human umbilical vein endothelial cells. *Experimental and Molecular Medicine*, *37*(6), 588–600. http://dx.doi.org/10.1038/emm.2005.72.

Nanni, S., Aiello, A., Re, A., Guffanti, A., Benvenuti, V., Colussi, C., et al. (2013). Estrogen-dependent dynamic profile of eNOS-DNA associations in prostate cancer. *PLoS ONE*, *8*(5), e62522. http://dx.doi.org/10.1371/journal.pone.0062522.

Nguyen, T., Brunson, D., Crespi, C. L., Penman, B. W., Wishnok, J. S., & Tannenbaum, S. R. (1992). DNA damage and mutation in human cells exposed to nitric oxide in vitro. *Proceedings of the National Academy of Sciences of the United States of America*, *89*(7), 3030–3034.

Nishida, N., Yano, H., Nishida, T., Kamura, T., & Kojiro, M. (2006). Angiogenesis in cancer. *Vascular Health and Risk Management*, *2*(3), 213–219.

Okayama, H., Saito, M., Oue, N., Weiss, J. M., Stauffer, J., Takenoshita, S., et al. (2013). NOS2 enhances KRAS-induced lung carcinogenesis, inflammation and microRNA-21 expression. *International Journal of Cancer*, *132*(1), 9–18. http://dx.doi.org/10.1002/ijc.27644.

Orucevic, A., Bechberger, J., Green, A. M., Shapiro, R. A., Billiar, T. R., & Lala, P. K. (1999). Nitric-oxide production by murine mammary adenocarcinoma cells promotes tumor-cell invasiveness. *International Journal of Cancer*, *81*(6), 889–896. http://dx.doi.org/10.1002/(SICI)1097-0215(19990611)81:6<889::AID-IJC9>3.0.CO;2-2.

Papapetropoulos, A., Garcia-Cardena, G., Madri, J. A., & Sessa, W. C. (1997). Nitric oxide production contributes to the angiogenic properties of vascular endothelial growth factor in human endothelial cells. *Journal of Clinical Investigation*, *100*(12), 3131–3139. http://dx.doi.org/10.1172/JCI119868.

Petrat, F., de Groot, H., Sustmann, R., & Rauen, U. (2002). The chelatable iron pool in living cells: A methodically defined quantity. *Biological Chemistry*, *383*(3–4), 489–502. http://dx.doi.org/10.1515/BC.2002.051.

Pipili-Synetos, E., Sakkoula, E., Haralabopoulos, G., Andriopoulou, P., Peristeris, P., & Maragoudakis, M. E. (1994). Evidence that nitric oxide is an endogenous antiangiogenic mediator. *British Journal of Pharmacology*, *111*(3), 894–902.

Raspollini, M. R., Amunni, G., Villanucci, A., Boddi, V., Baroni, G., Taddei, A., et al. (2004). Expression of inducible nitric oxide synthase and cyclooxygenase-2 in ovarian cancer: Correlation with clinical outcome. *Gynecologic Oncology*, *92*(3), 806–812. http://dx.doi.org/10.1016/j.ygyno.2003.12.023.

Ridnour, L. A., Isenberg, J. S., Espey, M. G., Thomas, D. D., Roberts, D. D., & Wink, D. A. (2005). Nitric oxide regulates angiogenesis through a functional switch involving thrombospondin-1. *Proceedings of the National Academy of Sciences of the United States of America*, *102*(37), 13147–13152. http://dx.doi.org/10.1073/pnas.0502979102.

Ridnour, L. A., Thomas, D. D., Donzelli, S., Espey, M. G., Roberts, D. D., Wink, D. A., et al. (2006). The biphasic nature of nitric oxide responses in tumor biology. *Antioxidants and Redox Signaling*, *8*(7–8), 1329–1337. http://dx.doi.org/10.1089/ars.2006.8.1329.

Rigas, B. (2007). Novel agents for cancer prevention based on nitric oxide. *Biochemical Society Transactions*, *35*(Pt 5), 1364–1368. http://dx.doi.org/10.1042/BST0351364.

Roberts, D. D., Isenberg, J. S., Ridnour, L. A., & Wink, D. A. (2007). Nitric oxide and its gatekeeper thrombospondin-1 in tumor angiogenesis. *Clinical Cancer Research*, *13*(3), 795–798. http://dx.doi.org/10.1158/1078-0432.CCR-06-1758.

Salgo, M. G., Stone, K., Squadrito, G. L., Battista, J. R., & Pryor, W. A. (1995). Peroxynitrite causes DNA nicks in plasmid pBR322. *Biochemical and Biophysical Research Communications*, *210*(3), 1025–1030. http://dx.doi.org/10.1006/bbrc.1995.1759.

Scott, D. J., Hull, M. A., Cartwright, E. J., Lam, W. K., Tisbury, A., Poulsom, R., et al. (2001). Lack of inducible nitric oxide synthase promotes intestinal tumorigenesis in the Apc(Min/+) mouse. *Gastroenterology*, *121*(4), 889–899. http://dx.doi.org/10.1053/gast.2001.27994.

Shi, Q., Xiong, Q., Wang, B., Le, X., Khan, N. A., & Xie, K. (2000). Influence of nitric oxide synthase II gene disruption on tumor growth and metastasis. *Cancer Research*, *60*(10), 2579–2583.

Sidorkina, O., Espey, M. G., Miranda, K. M., Wink, D. A., & Laval, J. (2003). Inhibition of poly(ADP-RIBOSE) polymerase (PARP) by nitric oxide and reactive nitrogen oxide species. *Free Radical Biology and Medicine*, *35*(11), 1431–1438. http://dx.doi.org/10.1016/j.freeradbiomed.2003.08.015.

Siegert, A., Rosenberg, C., Schmitt, W. D., Denkert, C., & Hauptmann, S. (2002). Nitric oxide of human colorectal adenocarcinoma cell lines promotes tumour cell invasion. *British Journal of Cancer, 86*(8), 1310–1315. http://dx.doi.org/10.1038/sj.bjc.6600224.

Siemens, D. R., Heaton, J. P., Adams, M. A., Kawakami, J., & Graham, C. H. (2009). Phase II study of nitric oxide donor for men with increasing prostate-specific antigen level after surgery or radiotherapy for prostate cancer. *Urology, 74*(4), 878–883. http://dx.doi.org/10.1016/j.urology.2009.03.004.

Stamler, J. S., Jaraki, O., Osborne, J., Simon, D. I., Keaney, J., Vita, J., et al. (1992). Nitric oxide circulates in mammalian plasma primarily as an S-nitroso adduct of serum albumin. *Proceedings of the National Academy of Sciences of the United States of America, 89*(16), 7674–7677.

Stuehr, D. J., Santolini, J., Wang, Z. Q., Wei, C. C., & Adak, S. (2004). Update on mechanism and catalytic regulation in the NO synthases. *Journal of Biological Chemistry, 279*(35), 36167–36170. http://dx.doi.org/10.1074/jbc.R400017200.

Switzer, C. H., Glynn, S. A., Cheng, R. Y., Ridnour, L. A., Green, J. E., Ambs, S., et al. (2012). S-nitrosylation of EGFR and Src activates an oncogenic signaling network in human basal-like breast cancer. *Molecular Cancer Research, 10*(9), 1203–1215. http://dx.doi.org/10.1158/1541-7786.MCR-12-0124.

Tanese, K., Grimm, E. A., & Ekmekcioglu, S. (2012). The role of melanoma tumor-derived nitric oxide in the tumor inflammatory microenvironment: Its impact on the chemokine expression profile, including suppression of CXCL10. *International Journal of Cancer, 131*(4), 891–901. http://dx.doi.org/10.1002/ijc.26451.

Tanriover, N., Ulu, M. O., Isler, C., Durak, H., Oz, B., Uzan, M., et al. (2008). Neuronal nitric oxide synthase expression in glial tumors: Correlation with malignancy and tumor proliferation. *Neurological Research, 30*(9), 940–944. http://dx.doi.org/10.1179/174313208X319099.

Thiery, J. P. (2002). Epithelial-mesenchymal transitions in tumour progression. *Nature Reviews. Cancer, 2*(6), 442–454. http://dx.doi.org/10.1038/nrc822.

Thiery, J. P. (2003). Epithelial-mesenchymal transitions in development and pathologies. *Current Opinion in Cell Biology, 15*(6), 740–746. http://dx.doi.org/10.1016/j.ceb.2003.10.006.

Thomas, D. D., Espey, M. G., Ridnour, L. A., Hofseth, L. J., Mancardi, D., Harris, C. C., et al. (2004). Hypoxic inducible factor 1alpha, extracellular signal-regulated kinase, and p53 are regulated by distinct threshold concentrations of nitric oxide. *Proceedings of the National Academy of Sciences of the United States of America, 101*(24), 8894–8899. http://dx.doi.org/10.1073/pnas.0400453101.

Thomas, D. D., Espey, M. G., Vitek, M. P., Miranda, K. M., & Wink, D. A. (2002). Protein nitration is mediated by heme and free metals through Fenton-type chemistry: An alternative to the NO/O2- reaction. *Proceedings of the National Academy of Sciences of the United States of America, 99*(20), 12691–12696. http://dx.doi.org/10.1073/pnas.202312699.

Thomas, D. D., Liu, X., Kantrow, S. P., & Lancaster, J. R., Jr. (2001). The biological lifetime of nitric oxide: Implications for the perivascular dynamics of NO and O2. *Proceedings of the National Academy of Sciences of the United States of America, 98*(1), 355–360. http://dx.doi.org/10.1073/pnas.011379598.

Thomas, D. D., Ridnour, L. A., Espey, M. G., Donzelli, S., Ambs, S., Hussain, S. P., et al. (2006). Superoxide fluxes limit nitric oxide-induced signaling. *Journal of Biological Chemistry, 281*(36), 25984–25993. http://dx.doi.org/10.1074/jbc.M602242200.

Thomas, D. D., Ridnour, L. A., Isenberg, J. S., Flores-Santana, W., Switzer, C. H., Donzelli, S., et al. (2008). The chemical biology of nitric oxide: Implications in cellular signaling. *Free Radical Biology and Medicine, 45*(1), 18–31. http://dx.doi.org/10.1016/j.freeradbiomed.2008.03.020.

Toledo, J. C., Jr., Bosworth, C. A., Hennon, S. W., Mahtani, H. A., Bergonia, H. A., & Lancaster, J. R., Jr. (2008). Nitric oxide-induced conversion of cellular chelatable iron into macromolecule-bound paramagnetic dinitrosyliron complexes. *Journal of Biological Chemistry, 283*(43), 28926–28933. http://dx.doi.org/10.1074/jbc.M707862200.

Torti, S. V., & Torti, F. M. (2011). Ironing out cancer. *Cancer Research, 71*(5), 1511–1514. http://dx.doi.org/10.1158/0008-5472.CAN-10-3614.

Toyokuni, S. (2009). Role of iron in carcinogenesis: Cancer as a ferrotoxic disease. *Cancer Science, 100*(1), 9–16. http://dx.doi.org/10.1111/j.1349-7006.2008.01001.x.

Tsurumi, Y., Murohara, T., Krasinski, K., Chen, D., Witzenbichler, B., Kearney, M., et al. (1997). Reciprocal relation between VEGF and NO in the regulation of endothelial integrity. *Nature Medicine, 3*(8), 879–886.

Urbich, C., Reissner, A., Chavakis, E., Dernbach, E., Haendeler, J., Fleming, I., et al. (2002). Dephosphorylation of endothelial nitric oxide synthase contributes to the anti-angiogenic effects of endostatin. *FASEB Journal, 16*(7), 706–708. http://dx.doi.org/10.1096/fj.01-0637fje.

Vakkala, M., Kahlos, K., Lakari, E., Paakko, P., Kinnula, V., & Soini, Y. (2000). Inducible nitric oxide synthase expression, apoptosis, and angiogenesis in in situ and invasive breast carcinomas. *Clinical Cancer Research, 6*(6), 2408–2416.

Vanin, A. F., Malenkova, I. V., & Serezhenkov, V. A. (1997). Iron catalyzes both decomposition and synthesis of *S*-nitrosothiols: Optical and electron paramagnetic resonance studies. *Nitric Oxide, 1*(3), 191–203. http://dx.doi.org/10.1006/niox.1997.0122.

Vithayathil, A. J., Ternberg, J. L., & Commoner, B. (1965). Changes in electron spin resonance signals of rat liver during chemical carcinogenesis. *Nature, 207*(5003), 1246–1249.

Vyas-Read, S., Shaul, P. W., Yuhanna, I. S., & Willis, B. C. (2007). Nitric oxide attenuates epithelial-mesenchymal transition in alveolar epithelial cells. *American Journal of Physiology Lung Cellular and Molecular Physiology, 293*(1), L212–L221. http://dx.doi.org/10.1152/ajplung.00475.2006.

Wei, D., Richardson, E. L., Zhu, K., Wang, L., Le, X., He, Y., et al. (2003). Direct demonstration of negative regulation of tumor growth and metastasis by host-inducible nitric oxide synthase. *Cancer Research, 63*(14), 3855–3859.

Wink, D. A., Hanbauer, I., Grisham, M. B., Laval, F., Nims, R. W., Laval, J., et al. (1996). Chemical biology of nitric oxide: Regulation and protective and toxic mechanisms. *Current Topics in Cellular Regulation, 34*, 159–187.

Wink, D. A., Kasprzak, K. S., Maragos, C. M., Elespuru, R. K., Misra, M., Dunams, T. M., et al. (1991). DNA deaminating ability and genotoxicity of nitric oxide and its progenitors. *Science, 254*(5034), 1001–1003.

Wink, D. A., & Laval, J. (1994). The Fpg protein, a DNA repair enzyme, is inhibited by the biomediator nitric oxide in vitro and in vivo. *Carcinogenesis, 15*(10), 2125–2129.

Wink, D. A., Ridnour, L. A., Hussain, S. P., & Harris, C. C. (2008). The reemergence of nitric oxide and cancer. *Nitric Oxide, 19*(2), 65–67. http://dx.doi.org/10.1016/j.niox.2008.05.003.

Wink, D. A., Vodovotz, Y., Laval, J., Laval, F., Dewhirst, M. W., & Mitchell, J. B. (1998). The multifaceted roles of nitric oxide in cancer. *Carcinogenesis, 19*(5), 711–721.

Xie, K., Huang, S., Dong, Z., Juang, S. H., Gutman, M., Xie, Q. W., et al. (1995). Transfection with the inducible nitric oxide synthase gene suppresses tumorigenicity and abrogates metastasis by K-1735 murine melanoma cells. *Journal of Experimental Medicine, 181*(4), 1333–1343.

Yermilov, V., Rubio, J., Becchi, M., Friesen, M. D., Pignatelli, B., & Ohshima, H. (1995). Formation of 8-nitroguanine by the reaction of guanine with peroxynitrite in vitro. *Carcinogenesis, 16*(9), 2045–2050.

Yermilov, V., Rubio, J., & Ohshima, H. (1995). Formation of 8-nitroguanine in DNA treated with peroxynitrite in vitro and its rapid removal from DNA by depurination. *FEBS Letters*, *376*(3), 207–210. http://dx.doi.org/10.1016/0014-5793(95)01281-6.

Ying, L., & Hofseth, L. J. (2007). An emerging role for endothelial nitric oxide synthase in chronic inflammation and cancer. *Cancer Research*, *67*(4), 1407–1410. http://dx.doi.org/10.1158/0008-5472.CAN-06-2149.

Yu, Y., Gutierrez, E., Kovacevic, Z., Saletta, F., Obeidy, P., Suryo Rahmanto, Y., et al. (2012). Iron chelators for the treatment of cancer. *Current Medicinal Chemistry*, *19*(17), 2689–2702. http://dx.doi.org/10.2174/092986712800609706.

Zhou, J., Schmid, T., & Brune, B. (2004). HIF-1alpha and p53 as targets of NO in affecting cell proliferation, death and adaptation. *Current Molecular Medicine*, *4*(7), 741–751.

Ziche, M., Morbidelli, L., Masini, E., Amerini, S., Granger, H. J., Maggi, C. A., et al. (1994). Nitric oxide mediates angiogenesis in vivo and endothelial cell growth and migration in vitro promoted by substance P. *Journal of Clinical Investigation*, *94*(5), 2036–2044. http://dx.doi.org/10.1172/JCI117557.

CHAPTER TWELVE

Dual Effect of Interferon (IFNγ)-Induced Nitric Oxide on Tumorigenesis and Intracellular Bacteria

Arnold H. Zea[1], Ashok Aiyar, David Tate
Stanley S. Scott Cancer Center, Louisiana State University, Health Sciences Center, New Orleans, Louisiana, USA
[1]Corresponding author: e-mail address: azea@lsuhsc.edu

Contents

Abstract

Nitric oxide (NO) is a key messenger involved in numerous physiological functions including inflammatory and immune responses. The functions of NO and their underlying mechanisms have been elucidated by extensive studies over the past 10 years. However, the complexity of the interactions between different levels of NO and multiple aspects of tumor development/progression as well as bacterial pathogenesis has led to apparently conflicting findings. The precise role of NO in bacterial and tumor pathogenesis involves a multitude of inter- and intracellular signaling pathways in which interferon gamma signaling and L-arginine metabolism are the major pathways involved in NO synthesis and regulation. The availability of the amino acid L-Arg can be a key factor to control the expression of inducible nitric oxide synthase (NOS2) and cellular NO levels. The role played by the NOS2/NO system both in bacterial pathogenesis and in tumor development is complex due to the dual role these molecules can play

Vitamins and Hormones, Volume 96
ISSN 0083-6729
http://dx.doi.org/10.1016/B978-0-12-800254-4.00012-X

promoting or inhibiting infections and cancer. This duality brings to the table a double challenge to determine the net impact of NO on cancer or bacterial behavior and to define the therapeutic role of NO-centered anticancer or antibacterial strategies. We believe that a comprehensive and dynamic understanding of the cascade of molecular and cellular events underlying tumor biology and bacterial pathogenesis that are affected by NO will allow researchers to exploit the potential antitumor and antibacterial properties of drugs interfering with NO metabolism. The contrasting roles of NO/NOS2 in these processes are clarified in this chapter.

1. INTRODUCTION

Nitric oxide (NO) is a free radical acting as a gaseous messenger, which affects various biological functions, either at low concentrations as a signal transducer in many physiological processes or at high concentrations as a cytotoxic defensive mechanism against pathogens and, perhaps, tumors by mechanisms that are not well understood (Buga, Wei, Bauer, Fukuto, & Ignarro, 1998; Ignarro, 2000). Moreover, accumulating evidence suggests that chronically elevated levels of NO are involved in the development of some human pathological conditions, such as inflammatory bowel disease, neurodegenerative disorders, and cancer (Ischiropoulos & Beckman, 2003; Lala & Chakraborty, 2001; Shah, Lyford, Gores, & Farrugia, 2004; Xu, Liu, Loizidou, Ahmed, & Charles, 2002). In mammals, NO is generated by enzymes termed nitric oxide synthase (NOS). There are three different types of NOS: neuronal NOS (nNOS or NOS1), endothelial NOS (eNOS or NOS3), and inducible NOS (iNOS or NOS2). All types of NOS catabolize L-arginine (L-Arg) to produce NO and citrulline. Pertinent to this review, NOS2-expressing cells produce NO as long as its enzymatic activity is intact and its substrate L-Arg is available (Heneka & Feinstein, 2001; Kroncke, Fehsel, Suschek, & Kolb-Bachofen, 2001). In addition to NOS2, L-Arg is also metabolized by the enzyme arginase (ARG) to produce L-ornithine, a substrate for ornithine decarboxylase (ODC) for the synthesis of polyamines that are essential for cell growth. Because ARG and NOS2 compete for L-Arg, this amino acid is a crucial component whose metabolism can modulate cell growth and cell death. Interestingly, the balance between NOS2 and ARG is competitively regulated by Th1- and Th2-type cytokines; whereas Th2 cytokines can induce ARG, NOS2 can be induced by Th1 cytokines (Munder, Eichmann, & Modolell, 1998). NOS2 can be induced by inflammatory mediators such as interferon gamma (IFNγ) and bacterial components such as lipopolysaccharide (LPS) and as such found to be expressed in macrophages, epithelial, and tumor cells (Xie et al., 1992). The induction of NOS2 is consistent with the biological activity of IFNγ that has long been associated with cytostatic/cytotoxic and antitumor mechanisms via the cell-mediated adaptive immune response. IFNγ has been used clinically to treat a variety of malignancies and other diseases (Ahlin, Larfars, Elinder, Palmblad, & Gyllenhammar, 1999; Garcia-Tunon et al., 2007; Willenborg, Staykova, Fordham, O'Brien, & Linares, 2007; Windbichler et al., 2000) *albeit* with mixed results and side effects that can be severe. Despite ample evidence implicating a role for IFNγ in tumor immune surveillance, a steady flow of reports has suggested that it may also have protumorigenic effects under certain circumstances. However, a major unanswered dichotomy remains: Is IFNγ per se responsible for the anti- or protumoral activities previously observed via alternative mechanisms? Alternatively, is the effect of IFNγ on the induction of NOS2 and NO production the primary mechanism that contributes to this activity? It has been proposed that, in fact, IFNγ treatment is a double-edged sword whose anti- and protumorigenic activities are dependent on the cellular, microenvironmental, and/or molecular context. Further, the high activity of NOS2 can lead to toxic levels of NO production, which have been implicated in a large number of several diseases (Pacher, Beckman, & Liaudet, 2007). The relationship between NO and the biology of cancer is still debated due to the possible dual role NO can play in cancer growth and inhibition. This chapter focuses in the role that IFNγ plays in NOS2 induction and NO production as an effector mechanism to control infection and tumor growth. The reversed consequences will be also discussed. It is anticipated that elucidating and understanding these contrasting effects of IFNγ will permit the development of new combined therapies to control cancer and infectious diseases.

2. INTERFERON γ

Interferons (IFNs) are a group of pleiotropic cytokines that play important roles in intercellular communication during innate and acquired immune responses as a host defensive response against viral and bacterial infections; in this context, IFNs also play a key role in cell growth and tumor surveillance (Pestka, Krause, & Walter, 2004). IFNs are divided into two main categories in mammals: type I (IFNα, IFNβ, IFNε, and IFNκ) and type II (IFNγ), both of which differ substantially in the relative potencies of their immunomodulatory and cell-surface molecular modification properties.

IFNγ is produced predominantly by natural killer (NK) and natural killer T (NKT) cells as part of the innate immune response and by CD4 Th1 and CD8 cytotoxic-T-lymphocyte (CTL) effector T cells once antigen-specific immunity develops (Trinchieri, 2010). The transcription of IFNγ-dependent genes is triggered after engagement of its receptors and activation of JAK/STAT and binding to GAS elements have been reviewed in detail elsewhere (O'Shea, Gadina, & Kanno, 2011; Schroder, Hertzog, Ravasi, & Hume, 2004). IFNγ signaling can promote a diversity of biological functions that include antiviral and antibacterial defense, cell cycle, apoptosis, inflammation, and innate and acquired immunity. IFNγ regulates the differentiation and function of many types of immune cells. It is intimately involved in all aspects of Th1-mediated immune responses, inhibits Th2 cell development, promotes the development of regulatory T (Treg) cells, activates macrophages, NK cells, induces production of chemokines, and has long been associated with cytostatic/cytotoxic and antitumor functions (Agnello et al., 2003; Ahlin et al., 1999; Brown, Lioubin, & Marquardt, 1987; Larkin, Ahmed, Wilson, & Johnson, 2013; Miller, Maher, & Young, 2009). Interestingly, to date, some controversy exists in the effect that IFNγ produces in humans. Many IFNγ functions have been well established and demonstrated in mouse models of infection and cancer. Despite the good benefits IFNγ promotes in those models, the results in human models have been ambiguous and sometimes discouraging. Results in patients differ from the outcomes of IFNγ exposure observed in *in vitro* models of human cancers and microbial infections. We believe that the lack of understanding of the mechanisms by which IFNγ functions has created a large gap in the potential therapeutic effect of this cytokine. Indeed, IFNγ can also produce many negative side effects. Some clinical trials using IFNs for cancer therapy and tuberculosis have failed (Bernabei et al., 2001; Condos, Rom, & Schluger, 1997; Lollini et al., 1993; Park et al., 2007), and it has also been reported that IFNγ can be toxic, protumorigenic, and antiapoptotic. The percentage of responses to IFNγ has shown to be as low as 5–6% when compared to treatment with IFNα where between 10% and 15% clinical responses have been observed. This decrease efficacy may result because viruses, bacteria, fungi, parasites, and even tumors have developed mechanisms to neutralize the functions of this cytokine in ways that are not well understood and need to be investigated. It is also possible that bioproducts generated by IFNγ stimulation could be modulated intracellularly as a mechanism of escape by pathogens or tumors in order to survive and persist. In a normal immune response, there is a balance between immune cells and the products they secrete to sustain an adequate immune surveillance. When that balance is altered, many adverse reactions will occur to the detriment of human health.

3. NOS2/NO

In eukaryotes, NO is generated by NOS. Three types of NOS exists in mammals, i.e., NOS1 also known as neuronal NOS (nNOS), NOS2 termed also as inducible NOS (iNOS), and NOS3 known as endothelial NOS (eNOS). While NOS1 and NOS3 have been classified as constitutive enzymes, NOS2 is inducible, found in the cytosol, Ca^{2+} independent and synthesized as a response to inflammatory stimuli (IFNγ) and therefore upregulated during inflammation and infection. Although NOS2 was originally identified and characterized in macrophages (MacMicking, Xie, & Nathan, 1997), it is present in numerous cell types including endothelial cells, fibroblasts, vascular smooth muscle, cardiac myocytes, and tumor cells (Willenborg et al., 2007). Pertinently, the NOS2 enzyme remains for extended period of hours to days and is able to generate micromolar concentrations of NO in the absence of changes in calcium levels. The expression of NOS2 can be transcriptionally upregulated by IFNγ, IFNα, IL-1, IL-2, TNFα, LPS, and hypoxia (Dachs & Tozer, 2000; Kleinert, Schwarz, & Forstermann, 2003). NOS2 gene regulation starts when IFNγ interacts with it receptors, which activates kinases of the JAK/STAT pathways leading to synthesis of the transcription factor IRF1 and stimulation of NOS2 mRNA transcription. NOS2 utilizes oxygen and electrons from NADPH to oxidize the substrate L-Arg into the intermediate N^{ω}-hydroxy-L-arginine (NOHA), which is then oxidized into NO and L-citrulline (Fig. 12.1). NO is a radical effector of the innate immune system that can directly inhibit pathogen replication and tumor growth (Bogdan, 2001). NOS2 mRNA can be measured between 30 min and 2 h following induction with IFNγ, whereas increases in NO can be detected only after 3–4 h following the stimuli, and its production can be sustained for several days because L-Arg can be *de novo* replenished within cells via citrulline (Bryk, Ochoa, Correia, Munera-Seeley, & Popovic, 2008; Zea et al., 2004). For reference purposes, it is worthy to note that (1) NOS2 gene expression can also depend upon activation of other transcription factors such as NF-κB or JNK-kinases (Au-Yeung, Mandhana, & Horvath, 2013; Hatano et al., 2001); (2) NOS2 transcription can be downregulated by steroids, TGFβ, IL-10, p53, and NO itself; and (3) *albeit* to a significantly

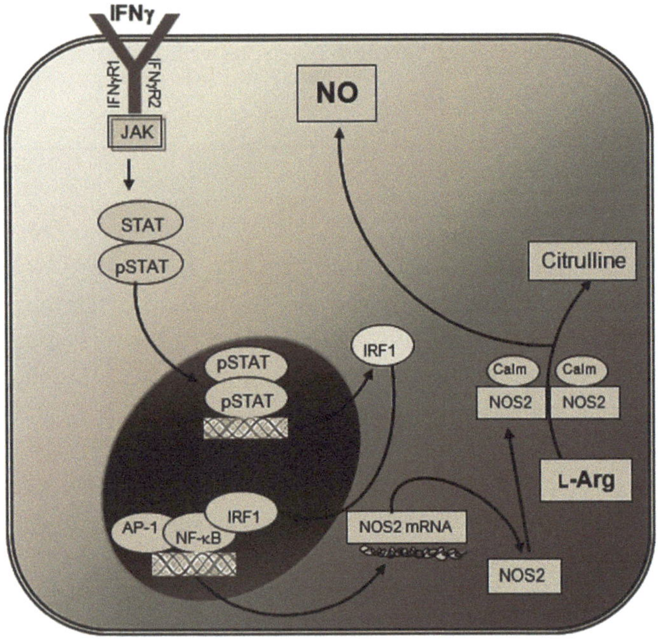

Figure 12.1 L-Arginine metabolism. L-Arg is the common substrate for ARG and NOS2. Effector functions generated by polyamines and proline synthesis (from ARG) result in cell proliferation. Effector functions by NO production from NOS2 possess antimicrobial and antitumoral effects. L-Arg can be *de novo* synthesized from citrulline through the enzymes argininosuccinate lyase and argininosuccinate synthase. (See the color plate.)

a lower extent, NOS1 and NOS3 can be also regulated at transcriptional level (Ambs et al., 1998; Forstermann, Boissel, & Kleinert, 1998; Hinz, Brune, & Pahl, 2000; Mocellin et al., 2004; Shinoda et al., 2003; Wink et al., 1998).

NO is an omnipresent intercellular messenger in all vertebrates, which modulates vasoregulation, platelet aggregation, neurotransmission, and nonspecific host defense. The biology, biochemistry, regulation, and functions of NO have been extensively reviewed by others over the years (Heinrich, Tedesco, Fukuto, & da Silva, 2014; Mocellin, Bronte, & Nitti, 2007; Wink et al., 2011). We wish to emphasize a few important points that will permit a better understanding of the role NO plays in cancer and bacterial infections: (1) The reaction of NO with superoxide ($O_2^{\bullet -}$) to form the much more powerful oxidant peroxynitrite ($ONOO^-$) is a key element in resolving the contrasting roles of NO in physiology and pathology. (2) Neither superoxide nor NO is particularly toxic *in vivo* because there are efficient means to minimize their accumulation. (3) Superoxide is rapidly removed by high concentrations of scavenging enzymes called superoxide dismutases, while NO itself is rapidly removed by its rapid diffusion through tissues into red blood cells (Joshi et al., 2002), where it is quickly converted to nitrate by reaction with oxyhemoglobin. This limits the biological half-life of NO *in vivo* to less than a second, whereas NO concentrations relevant for cellular signaling can persist for about an hour. (4) When both superoxide and NO are synthesized within a few cell diameters of each other, they will combine spontaneously to form peroxynitrite by a diffusion-limited reaction, which may play a very important role in disease (Huie & Padmaja, 1993). (5) Although NO is reported to have many potentially toxic effects, many of them are more likely mediated by its oxidation products rather than NO itself. Well-known effects of NO on DNA, mitochondrial activation/inhibition, apoptosis, autophagy, enzyme inhibition, and peroxynitrite formation have been widely discussed (Brunelli, Yermilov, & Beckman, 2001; Harris et al., 2007; Wink et al., 1991). All compiled data are showing that NO is a key messenger involved in the regulation of many physiological functions.

4. L-ARG METABOLISM

4.1. Arginase

L-Arg is a semiessential amino acid that is the *de novo* synthesized in metabolic pathways from proline, glutamine, or glutamate. Endogenous L-Arg synthesis relies on the small intestine release of citrulline into the blood, which is first metabolized into L-Arg in the proximal kidney tubules and then exported to the systemic circulation. In mammalian cells, L-Arg can be catabolized in a tissue-specific manner by NOS and ARG, which are two major catabolic enzymes capable on regulating immune responses. ARG catalyzes the hydrolysis of L-Arg to form L-ornithine and urea. L-Ornithine is further processed to polyamines via the ODC pathway (Fig. 12.2). In mammals, two genetically distinct ARG isotypes exist: cytosolic ARG1 is found predominantly in hepatocytes and ARG2 is present in numerous tissues, which is associated with mitochondrial matrix within the cell and does not intervene in urea cycle (Ash, 2004). ARG1 is present in mammalian cells of the innate immune system. In the murine system, macrophages constitute the major but not a sole source of ARG during immune responses. In resting murine macrophages, ARG1 mRNA, protein, and enzyme activities are undetectable but can be upregulated by IL-4 and IL-13 (Munder et al., 1998), whereas IL-4 and IL-10 upregulate its expression in

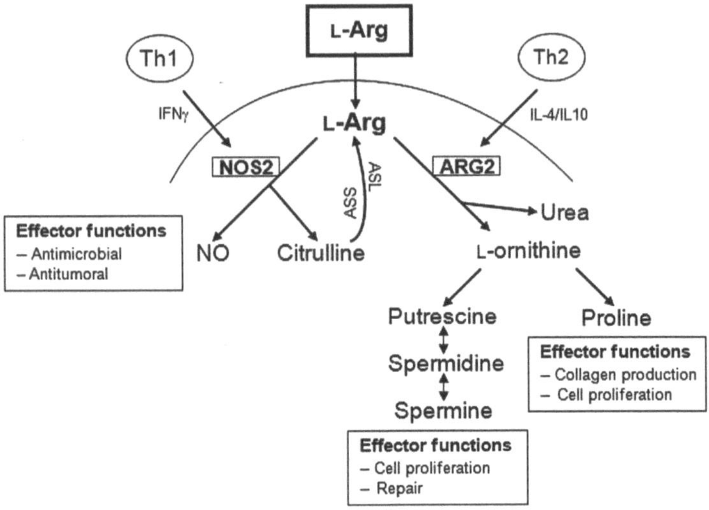

Figure 12.2 IFNγ signaling pathway toward NO production. Members of the Janus kinase 2 (JAK2)/STAT1α signaling pathways have been implicated in NOS2 gene expression and NO generation in response to different stimuli. Furthermore, it has been suggested that JAK2 is involved in IFNγ-dependent NOS2 induction. In the absence of IFNγ, following LPS stimulation, there is activation of NF-κB with subsequent NOS2 expression and NO release. L-Arg metabolism is also a key element in the synthesis of NO. In spite of the studies mentioned above, little is known about the complete mechanism underlying NO regulation in cells stimulated with IFNγ per se.

humans (Zea et al., 2005). TGFβ and GM-CSF can also induce ARG1 activity directly in different cell types of the innate immune system (Jost et al., 2003). In humans, ARG1 is constitutively expressed in granulocytes such that when death occurs at inflammation sites, it can be released into the blood to induce a profound suppression in T-cell proliferation and cytokine synthesis (Munder et al., 2006). It is not clear whether ARG1 activation within cells or its release into the extracellular milieu is the dominant event, but numerous findings have clearly demonstrated that the sustained elevation of ARG1 activity can deplete extracellular L-Arg levels and a drop in L-Arg concentrations below 50–100 μM causes loss of T-cell function and can induce T-cell anergy (Rodriguez & Ochoa, 2008; Zea et al., 2004).

4.2. NOS/NO

All three types of NOS isoenzymes metabolize L-Arg to produce NO and L-citrulline, but even though they utilize L-Arg as substrate, there is no formal proof that their activity can lead to a decrease in the availability of this amino acid. It has been demonstrated that low levels of L-Arg can control the transcription of NOS mRNA. For example, at normal concentrations of L-Arg, NOS3 should be saturated. However, increased NOS3 activity occurs with L-Arg supplementation known as the "arginine paradox" (Wu et al., 2009). L-Arg availability can be modulated by its dietary intake and endogenous production (*de novo* synthesis from citrulline), as well as by ARG. ARG reciprocally regulates NO levels in endothelial cells and macrophages by competing with NOS for the substrate L-Arg (Li et al., 2001). It has been also demonstrated in mice that NO production was reduced when citrulline availability and *de novo* arginine production were diminished. Therefore, supplementation of citrulline is believed to sustain NO production (Luiking et al., 2008). The *de novo* production of L-Arg from citrulline depends on the colocalization of NOS with the enzymes argininosuccinate synthase and argininosuccinate lyase (Wu & Brosnan, 1992). Therefore, the NOS2 enzyme can produce NO as long as it is enzymatically active and its substrate L-Arg is available. The production of NO has a handful of implications that can make the difference between health and disease. Different factors such as diet and/or L-Arg metabolism can regulate NO production. Whereas the high activity of NOS2 and NO production makes them well suited as effector molecules of cellular defense against invading microorganisms, persistent activation of NOS2 can lead to toxic levels of NO, which are deleterious for the host and can lead to dysfunctional immune responses. This issue is discussed in depth in the following sections.

5. NOS2/NO IN DISEASE

5.1. Bacterial pathogens

NOS2 has important functions in innate immunity and regulation of immune functions. Here, we briefly discuss the role of NOS2 in the pathogenesis of bacterial infection. The killing effect of NO against bacteria has been well described and appears to be very efficient. However, bacteria have also developed a series of strategies to suppress and inhibit functions of NOS2 and consequently NO. Although NO has a simple structure, it is capable of regulating various complex reactions to form products that are toxic to microbes called reactive nitrogen intermediates (Nathan, 2002; Nathan & Shiloh, 2000). Resident phagocytes in tissues represent an early sensor of infection. Macrophages may rapidly kill Gram-negative bacteria through innate recognition of LPS and subsequent NF-κB-dependent NOS2 activation (Kim & Stadtman, 1997) to activate a rapid production of reactive oxygen and nitrogen intermediates necessary to control bacterial fate. As mentioned earlier, bacteria have developed survival strategies, which involve NOS2, NO, and L-Arg metabolism, to persist by counteracting the killing ability of macrophages. Bacteria are capable to modulate NO production through L-Arg metabolism by switching macrophage phenotype from cytotoxic activated cells to apoptosis-susceptible cells. Bacteria can direct the L-Arg metabolic pathway toward the production of polyamines, then inhibiting NO production inside the host cells or by shifting cell-death program from apoptosis to necrosis as an advantage for the bacteria to survive. Therefore, why NO should be considered as a key molecule in controlling microbial infections? Enhanced levels of NOS2 and NO have been found in animal models of leishmania (Kropf et al., 2005), tuberculosis (El Kasmi et al., 2008), and *Helicobacter pylori* (Zabaleta et al., 2004). The use of NOS2 knockout mice and NO synthase inhibitors (L-NMMA) provided evidence that NO was a major factor in controlling the fate of such pathogens. These results have been further supported by the experimental use of NO donors that have shown to directly inhibit the growth of various microbes *in vitro* (Schairer, Chouake, Nosanchuk, & Friedman, 2012).

The prime role of NOS2 and the importance of NO for the control of mycobacterial growth have been well established (Herbst, Schaible, & Schneider, 2011; Mishra et al., 2013; Sharma & Banga, 2004). In the murine model of *Mycobacterium tuberculosis* infection, the importance of NO for the control of mycobacterial growth is well established. NOS2-deficient mice are more susceptible to *M. tuberculosis* infection than wild type ones. In human macrophages however, the role of NO in the control of mycobacterial infections is less clear and very controversial. Direct susceptibility of mycobacteria to reactive nitrogen intermediates *in vitro* is strain-, dose-, and time dependent (O'Brien, Carmichael, Lowrie, & Andrew, 1994). However, the mechanism by which NO mediates killing of mycobacteria by activated macrophages is unclear. It may be caused by different mechanisms such as direct toxicity, indirect interference with mycobacterial virulence, apoptosis, cytokine induction, function as a second messenger, or by manipulation of the L-Arg metabolic pathway. These mechanisms need to be further investigated so that the effect of NOS2 and NO in human tuberculosis can be understood better.

5.2. Cancer

As we discussed above, the importance of NO production (mainly by macrophage overexpression of NOS) in antimicrobial immunity is well established. The role of NOS2 in controlling cancer remains controversial. Although tumor-derived NO appears to be involved in cancer immune rejection (Hu, Dyke, Moore, Thomsen, & Brindle, 2004), the major source of NO mediating tumor killing is macrophage-derived NOS2 (Klimp, de Vries, Scherphof, & Daemen, 2002; MacMicking et al., 1997). Several reports indicate that macrophages can participate in the antitumor immune response and can selectively destroy different tumor types both *in vitro* and *in vivo*, mainly through the production of NO. Cancers potentially susceptible to the effect of activated macrophages expressing NOS2 include pancreatic, ovarian, liver, and melanoma tumors. However, on the other side of the coin, several investigators have reported the expression of NOS2 by malignant cells or within the tumor microenvironment, both at the mRNA and at the protein level. A marked expression of NOS2 has been reported in human breast, colon, head and neck, prostate, pancreas, bladder, ovarian, and lung carcinomas in addition to Kaposi's sarcoma and leukemias (Mocellin et al., 2004). Most of the NOS2 activity is found to be concentrated only in the tumor cells but not in the surrounding normal tissue. The elevated expression of NOS2 appears to correlate with poor clinical outcome in patients affected with different types of cancers. Specially, presence of NOS2 in melanoma, breast, ovarian, head and neck, and colorectal carcinomas leads researches to consider NO as a potential mediator of tumor development/progression. Different mechanisms have been proposed to understand the antitumor effects of NO that include direct damage of DNA, apoptosis, inhibition of DNA synthesis, and disruption of mitochondrial physiology (Fig. 12.3). These mechanisms lead us to believe that NO can play an important role in the development of an immune response against tumors. However, as discussed earlier, the expression of ARG, which is known to promote cancer progression metabolizes L-Arg for the synthesis of polyamines, could negate the effective action of NO in tumor immunology (Tate et al., 2008).

There is a fine line between the positive and negative actions of NO in tumor immunology. The production of NO within the tumor microenvironment is believed to promote tumor growth, mainly by stimulating angiogenesis accompanied with the release of several growth factors (VEGFs), production of prostaglandins, and enhancing tissue invasion. These stimulatory effects were confirmed by the observation that inhibiting NOS2/NO synthesis leads to the reversal of these effects with a concomitant reestablishment of antitumor responses (Serafini et al., 2004). Therefore, both direct and indirect effects appear to be involved in both tumor-promoting and tumor-suppressing roles of NOS2/NO. In the context of

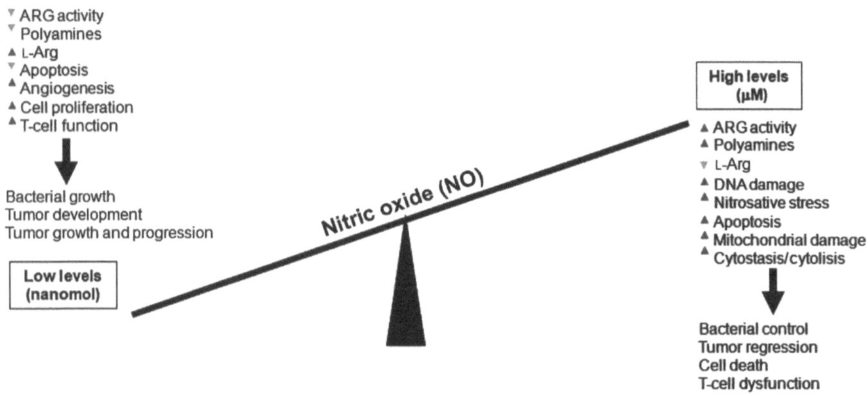

Figure 12.3 Mechanisms of action of NO on bacterial infections and cancer. NO is a very versatile molecule that can induce tumor development depending of many factors. It appears that the regulated concentration and its spatial and temporal distribution are important for tumor development and tumor progression. The effects of NO at low concentrations are considered to be protumor, whereas at high concentrations, NO can act as a potent anticancer and antibacterial agent. (See the color plate.)

immune surveillance, direct interactions of NO with tyrosyl radicals and iron proteins can result in reversible cytostasis. In the context of tumor physiology, NO prevents binding of tumor cells to the endothelium probably through low fluxes of NO. Additionally, other tumor-promoting effects such as the prevention of leukocyte infiltration, suppression of T-cell proliferation, increased vascular permeability, and angiogenesis may also be mediated through direct effects of NO. While the role of NO in cancer is multidimensional, its effects can be put into perspective based on timing, location, and concentration. As the tumor begins to grow, macrophages and NK cells use NO derived from NOS2 to kill tumor cells. However, as the tumor progresses, NO can mediate capillary leakiness, support angiogenesis, and limit leukocyte infiltration. Therefore, while some studies suggest that expression of NOS2 correlates with reduced metastases, other studies suggested that some tumors expressing NOS2 are more aggressive *in vivo* than their counterparts. The proper use of systemic NOS inhibitors and NO donors requires a comprehensive knowledge of the chemistry, biochemistry, toxicological mechanisms, and physiological mechanisms of this molecule.

6. IFNγ-INDUCED NO AS A MECHANISM TO INHIBIT RENAL CANCER GROWTH

Renal (kidney) cell cancer (RCC) accounts for 5% of all cancers. It remains one of the most resistant tumors to immunotherapy, chemotherapy, and radiotherapy. The overall survival for all patients with kidney cancer has not appreciably improved during the last 25 years, indicating the need for more effective ways to treat this disease. Immunotherapy with interferon alpha (IFNα and/or interleukin-2 (IL-2)) has been used as a first-line therapy to treat kidney cancer, but the documented clinical response rate in patients is only 10–20%, with a median survival of approximately 12 months. New targeted therapies for RCC treatment have been introduced, but they do not produce complete responses and have several adverse effects, and more importantly, their mechanisms of action are poorly understood. There is a complex relationship between immunity, tumorigenesis, and the metabolism of L-Arg (Choi et al., 2009) that can be highly relevant to RCC. Recent studies have demonstrated convincingly that ARG1, ARG2, and NOS2 play a crucial role in the tumor biology of RCC (Bronte & Zanovello, 2005; Peranzoni et al., 2007; Viola & Bronte, 2007). The ARG-dependent synthesis of polyamines is directed to enhance cell growth. In contrast, the production of NO through NOS2 induction is believed to kill tumor cells. Since ARG and NOS2 compete for L-Arg, this amino acid is a crucial component whose metabolism modulates cell growth/inhibition and suppresses T-cell function (Zea et al., 2004). One of the possible mechanisms that could explain the impairment of antitumor responses in patients with RCC is the well-documented shift from Th1 to Th2 cytokine response (Finke et al., 2008). This shift toward Th2 responses may prompt the tumor microenvironment to produce high levels of ARG and polyamines instead of initiating a NOS2-dependent production of NO that is accompanied by the lack in IFNγ production. In our laboratory, we have recently demonstrated that murine RCC cell lines present high levels of ARG2 expression that are associated with L-Arg depletion induction of T-cell dysfunction and increased tumor growth (Tate et al., 2008). For many years, treatment of RCC with IFNγ has shown minimal or no activity in the treatment of patients with RCC. Since the combination of IFNγ and IFNα treatment reflected a response rate similar to IFNα alone, the latter has been the treatment of choice for those patients for many years (Dutcher, 2003). This observation is striking since IFNγ has the most potent immunomodulatory activity among the IFNs, indicating that the mechanisms of action of IFNγ are complex, not well understood, and deserve to be revisited as a potent antitumor agent alone or in combination with other therapies with research focused on understanding the dichotomy between its effects *in vitro* and *in vivo*.

Next, we want to present data generated from a research in our laboratory to determine the mechanisms by which IFNγ can modulate L-Arg metabolism by shifting ARG-dependent cell proliferation to NOS2/NO production to produce growth inhibition of cancer cells. We have found the following: (1) a new mechanism to define the cytostatic role of NO in the inhibition of RCC cell proliferation; (2) the effect of IFNγ in modulating L-Arg metabolism and cytokine production; (3) the resistance of certain RCC cell lines to IFNγ treatment; and (4) new insights into the mechanisms by which RCC tumors may locally and systemically shut down an important effector arm (NO) of the immune response. IFNs have been widely used in preclinical models and in immunotherapy trials for kidney cancer (Haas & Hillman, 1996; Hillman et al., 1997). The activity of monotherapy with IFNα in metastatic kidney cancer has been extensively evaluated (Motzer, Bacik, Murphy, Russo, & Mazumdar, 2002) with disappointing and controversial results. However, trials using IL-2 and IFNα combination showed a clinically useful antitumor activity with a 15% overall response (Rosenberg et al., 1989; Spencer et al., 1992). In contrast, therapy with the single-agent IFNγ has shown discouraging results in patients with kidney cancer. Although preliminary studies suggested the benefits of IFNγ treatment (Ellerhorst et al., 1994), only a 4.4% overall response has been seen in phase III studies indicating that IFNγ alone has no major impact on kidney cancer treatment (Ernstoff et al., 1990). These results are striking since many IFNγ-mediated properties (e.g., antiproliferative, apoptotic, and antiangiogenic activities) have been attributed to this cytokine (Angiolillo et al., 1995). Our study was aimed to determine the mechanism(s) by which IFNγ is capable of inducing growth inhibition of murine kidney cancer cell lines. Previously, we reported that murine CL-2, CL-19, and Renca cell lines have different levels of ARG2 expression that appear to be associated with the enhancement of cell proliferation (Tate et al., 2008). The results indicate a need for the tumor cells to use ARG2 to metabolize L-Arg for the synthesis of polyamines to promote cell growth. Conversely, NO is a free radical that, at low concentrations, acts as single signal transducer, whereas at higher concentrations can act as a cytotoxic/cytostatic defensive mechanism against pathogens and perhaps tumors (Ignarro, 2000). In our study, we found that RCC cell lines in the logarithmic phase of proliferation expressed low NOS2 protein levels and low concentrations of NO (15–20 μM), whereas ARG2 activity was highly expressed in the three cell lines. These results indicate that low NOS2 and high ARG2 activity favor the growth of the cell lines and are supported by previous studies showing that high expression of ARG can inhibit NOS2 activity and NO production (Gotoh & Mori, 1999; Li et al., 2001). We believe this could be a mechanism used by tumor cells to survive by directly avoiding the cytostatic effects of NO.

Since IFNγ is a potent inducer of NOS2, we wanted to determine the *in vitro* mechanism(s) by which this cytokine is capable of inhibiting kidney cancer cell proliferation. We successfully demonstrated that NOS2 induction within the tumor cells in response to IFNγ caused a robust and sustained accumulation of endogenous NO that resulted in an 80–85% growth inhibition of CL-2 and CL-19 cell lines (Fig. 12.3). This effect was abrogated by the addition of the NOS2 inhibitor, L-NMMA, demonstrating that the reduction in cell growth was mediated by NO. It is possible that the NO-mediated cytostatic effects observed in our *in vitro* model could produce a different outcome in an *in vivo* model due to the presence of NO-producing cells infiltrating the tumor microenvironment. It is possible that the NO-mediated cytostatic effects observed in our *in vitro* model could produce different outcomes in an *in vivo* model due to the presence of NO-producing cells infiltrating the tumor microenvironment. Here, we also found that the cytostatic effect of NO in RCC cell lines was not due to apoptosis as demonstrated by others (Hussain et al., 2004), but rather by a cell cycle arrest at the G2/M phase. This possibly could be mediated by the transcriptional modulation of cyclins and/or cyclin-dependent kinases as seen in breast cancer. Since the growth-inhibitory activity of IFNγ appears to be accompanied by the inhibition of ARG activity in CL-2 and CL-19 cells, we are suggesting that this inhibition could be partially responsible for the growth-inhibitory effects of IFNγ in addition to the major cytostatic effects of NO. The importance of ARG and/or NO in the growth inhibition of RCC cells needs to be explored further by possibly using ARG and NOS2-KO mice. Altogether, our data suggest that ARG2 and NOS2 can be modulated in a coordinated fashion by IFNγ to counterbalance tumor growth and tumor inhibition. Lately, NO has emerged as a potential antiproliferative agent capable of overcoming tumor cell resistance to conventional therapeutic agents. Despite the relative beneficial effect of IFNγ-NO induction that inhibits RCC proliferation, a substantial percentage (15%) of the cells failed to respond to IFNγ treatment. For the first time, our results show that a certain subpopulation of RCC cells develop a mechanism of resistance to specific cytostatic stimuli that could largely explain the commonly observed lack of response to IFNγ in clinical trials. Mechanisms of resistance have been associated with IFNα defective JAK/STAT signaling pathways (Shang et al., 2007) that could be similar with IFNγ. Interestingly, in our study, we showed that Renca cells which, upon stimulation with

IFNγ express NOS2 mRNA, did not express NOS2 protein or NO production. Different mechanisms have been postulated to explain the lack of NOS2 expression, but a novel finding has shown that loss of NOS2 protein expression is exerted by translational inhibition specifically mediated by miR-146a (Perske et al., 2010). The goal of this study was not to understand the control of NOS2 expression at the translational level, but to highlight that different kidney cancer cell types can vary in their sensitivity to IFNγ and, therefore, to NO. These findings will have a high translational relevance because permitting us to characterize RCC cells better and to determine their sensitivity to IFN treatment. It is possible that human RCC could present with similar characteristics to those seen in Renca cells. Therefore, these cells (or patients) will not benefit from IFNγ treatment not because IFNγ does not work, rather, because these cells found a way to become resistant to this cytokine as a mechanism of survival. Therefore, in patients, the identification of resistance to IFN treatment could be extremely important for a more personalized and effective treatment of RCC.

Production of either stimulatory or inhibitory cytokines by infiltrating or neoplastic cells may have more of an impact on the development of an effective antitumor response than in the promotion of tumor growth. When we tested cytokine production before IFNγ treatment, IL-1β, IL-6, TGFβ, and VEGF were the most predominant cytokines in all three cell lines. Conversely, IFNγ did not have an effect on the production of VEGF and TGFβ indicating that the inhibition of these cytokines can be mediated by other signaling pathways and that their presence can contribute to the maintenance of tumor growth and angiogenesis. A combination therapy of IFNγ with other drugs (e.g., sunitinib) that targets VEGF receptors could advance the treatment of kidney cancer.

Altogether, these data suggest that ARG2 and NOS2 can be modulated in a coordinated fashion by IFNγ to counterbalance tumor growth and/or tumor inhibition. These findings represent a step forward in the treatment of RCC because the presence or absence of ARG, NOS2, and NO as a diagnostic marker will permit clinicians to identify different biological and functional signatures than can be determined by the sensitivity of IFNγ treatment. It is also possible that patients with RCC might present similar characteristics to those seen in Renca cells and patients could have minimal or none clinical responses. In addition, patients with RCC would benefit tremendously from IFNγ treatment if the mechanisms associated with resistance to this cytokine are identified.

In summary, the improved understanding of IFNγ biology in the inhibition of tumor cell proliferation in a NOS2/NO-dependent pathway will provide new insights in promoting the use of this cytokine, not only for the treatment of kidney cancer but also for the treatment of other tumors where the use of IFNs is indicated. IFNγ can play a key role in modulating L-Arg metabolism by inhibiting ARG activity and by enhancing the endogenous production of NO to inhibit the proliferation of kidney cancer cells. Due to the numerous conflicting results about the role of NOS2 and NO in cancer, a full understanding of the multifaceted role of NO in cancer is imperative.

7. CONCLUSIONS AND FUTURE DIRECTIONS

The primary function of the immune system is to eliminate pathogens and tumors (Miller et al., 2009). IFNγ plays an important role in the capacity of the immune system to eliminate pathogens but more importantly to mediate in the destruction of cancerous cells. The importance of the antibacterial functions of NO is highlighted by the array of different defense mechanisms that intracellular pathogens have evolved to survive or avoid damage by NO. NO also has important function in modulating immune responses to intracellular pathogens. The understanding of the function of NOS2 in immunity against intracellular pathogens will require a dissection of the dual functions of NO. Both direct and indirect effects appear to be involved in both the tumor-promoting and tumor-suppressing roles of NO. It appears that these effects of NO such as L-Arg metabolism, IFNγ, NOS2, PGE2, T-cell dysfunction, and immune suppression, which are dictated by both amount and flux, may be an important determinant whether NO promotes or inhibit tumor growth. We have learned from results derived from our research that heterogeneity of tumors can generate a variety of responses following NO production. Therefore, the biological activity of NO depends on the source of NO generation (mainly in our case from IFNγ), concentration, and tumor microenvironment that can direct the function of NO. The effects of NO at low concentrations could be characterized as protumor growth, while at high concentrations NO can acts as a potent anticancer agent promoting cell cycle arrest and inhibiting angiogenesis. Having a better understanding of the roles NO play in cancer is very important to determine the proper use of systemic of IFNγ, NOS inhibitors, and NO donors that can lead to important strategies for chemoprevention of cancer. However, very importantly, a comprehensive knowledge of the chemistry and physiological properties of NO is required.

ACKNOWLEDGMENTS

This work was supported by Stanley S. Scott Cancer Center Funds and the NIH-NIGMS COBRE Grant P20 GM103501.

REFERENCES

Agnello, D., Lankford, C. S., Bream, J., Morinobu, A., Gadina, M., O'Shea, J. J., et al. (2003). Cytokines and transcription factors that regulate T helper cell differentiation: New players and new insights. *Journal of Clinical Immunology, 23*, 147–161.

Ahlin, A., Larfars, G., Elinder, G., Palmblad, J., & Gyllenhammar, H. (1999). Gamma interferon treatment of patients with chronic granulomatous disease is associated with augmented production of nitric oxide by polymorphonuclear neutrophils. *Clinical and Diagnostic Laboratory Immunology, 6*, 420–424.

Ambs, S., Ogunfusika, M. O., Merriam, W. G., Bennett, W. P., Billiar, T. R., & Harris, C. C. (1998). Up-regulation of inducible nitric oxide synthase expression in cancer-prone p53 knockout mice. *Proceedings of the National Academy of Sciences of the United States of America, 95*, 8823–8828.

Angiolillo, A. L., Sgadari, C., Taub, D. D., Liao, F., Farber, J. M., Maheshwari, S., et al. (1995). Human interferon-inducible protein 10 is a potent inhibitor of angiogenesis in vivo. *The Journal of Experimental Medicine, 182*, 155–162.

Ash, D. E. (2004). The structure and function of arginases. *The Journal of Nutrition, 134*(10 suppl), 2760S–2764S.

Au-Yeung, N., Mandhana, R., & Horvath, C. M. (2013). Transcriptional regulation by STAT1 and STAT2 in the interferon JAK-STAT pathway. *JAK-STAT, 2*, e23931.

Bernabei, P., Allione, A., Rigamonti, L., Bosticardo, M., Losana, G., Borghi, I., et al. (2001). Regulation of interferon-gamma receptor (INF-gammaR) chains: A peculiar way to rule the life and death of human lymphocytes. *European Cytokine Network, 12*, 6–14.

Bogdan, C. (2001). Nitric oxide and the immune response. *Nature Immunology, 2*, 907–916.

Bronte, V., & Zanovello, P. (2005). Regulation of immune responses by L-arginine metabolism. *Nature Reviews Immunology, 5*, 641–654.

Brown, T. J., Lioubin, M. N., & Marquardt, H. (1987). Purification and characterization of cytostatic lymphokines produced by activated human T lymphocytes. Synergistic antiproliferative activity of transforming growth factor beta 1, interferon-gamma, and oncostatin M for human melanoma cells. *Journal of Immunology, 139*, 2977–2983.

Brunelli, L., Yermilov, V., & Beckman, J. S. (2001). Modulation of catalase peroxidatic and catalatic activity by nitric oxide. *Free Radical Biology and Medicine, 30*, 709–714.

Bryk, J., Ochoa, J. B., Correia, M. I., Munera-Seeley, V., & Popovic, P. J. (2008). Effect of citrulline and glutamine on nitric oxide production in RAW 264.7 cells in an arginine-depleted environment. *Journal of Parenteral and Enteral Nutrition, 32*, 377–383.

Buga, G. M., Wei, L. H., Bauer, P. M., Fukuto, J. M., & Ignarro, L. J. (1998). NG-hydroxy-L-arginine and nitric oxide inhibit Caco-2 tumor cell proliferation by distinct mechanisms. *The American Journal of Physiology, 275*, R1256–R1264.

Choi, B. S., Martinez-Falero, I. C., Corset, C., Munder, M., Modolell, M., Muller, I., et al. (2009). Differential impact of L-arginine deprivation on the activation and effector functions of T cells and macrophages. *Journal of Leukocyte Biology, 85*, 268–277.

Condos, R., Rom, W. N., & Schluger, N. W. (1997). Treatment of multidrug-resistant pulmonary tuberculosis with interferon-gamma via aerosol. *Lancet, 349*, 1513–1515.

Dachs, G. U., & Tozer, G. M. (2000). Hypoxia modulated gene expression: Angiogenesis, metastasis and therapeutic exploitation. *European Journal of Cancer, 36*, 1649–1660.

Dutcher, J. P. (2003). Interleukin-2 based therapy for kidney cancer. *Cancer Treatment and Research, 116*, 155–172.

El Kasmi, K. C., Qualls, J. E., Pesce, J. T., Smith, A. M., Thompson, R. W., Henao-Tamayo, M., et al. (2008). Toll-like receptor-induced arginase 1 in macrophages thwarts effective immunity against intracellular pathogens. *Nature Immunology*, *9*, 1399–1406.

Ellerhorst, J. A., Kilbourn, R. G., Amato, R. J., Zukiwski, A. A., Jones, E., & Logothetis, C. J. (1994). Phase II trial of low dose gamma-interferon in metastatic renal cell carcinoma. *The Journal of Urology*, *152*, 841–845.

Ernstoff, M. S., Nair, S., Bahnson, R. R., Miketic, L. M., Banner, B., Gooding, W., et al. (1990). A phase IA trial of sequential administration recombinant DNA-produced interferons: Combination recombinant interferon gamma and recombinant interferon alfa in patients with metastatic renal cell carcinoma. *Journal of Clinical Oncology*, *8*, 1637–1649.

Finke, J. H., Rini, B., Ireland, J., Rayman, P., Richmond, A., Golshayan, A., et al. (2008). Sunitinib reverses type-1 immune suppression and decreases T-regulatory cells in renal cell carcinoma patients. *Clinical Cancer Research*, *14*, 6674–6682.

Forstermann, U., Boissel, J. P., & Kleinert, H. (1998). Expressional control of the 'constitutive' isoforms of nitric oxide synthase (NOS I and NOS III). *The FASEB Journal*, *12*, 773–790.

Garcia-Tunon, I., Ricote, M., Ruiz, A. A., Fraile, B., Paniagua, R., & Royuela, M. (2007). Influence of IFN-gamma and its receptors in human breast cancer. *BMC Cancer*, *7*, 158.

Gotoh, T., & Mori, M. (1999). Arginase II downregulates nitric oxide (NO) production and prevents NO-mediated apoptosis in murine macrophage-derived RAW 264.7 cells. *The Journal of Cell Biology*, *144*, 427–434.

Haas, G. P., & Hillman, G. G. (1996). Update on the role of immunotherapy in the management of kidney cancer. *Cancer Control*, *3*, 536–541.

Harris, J., De Haro, S. A., Master, S. S., Keane, J., Roberts, E. A., Delgado, M., et al. (2007). T helper 2 cytokines inhibit autophagic control of intracellular *Mycobacterium tuberculosis*. *Immunity*, *27*, 505–517.

Hatano, E., Bennett, B. L., Manning, A. M., Qian, T., Lemasters, J. J., & Brenner, D. A. (2001). NF-kappaB stimulates inducible nitric oxide synthase to protect mouse hepatocytes from TNF-alpha- and Fas-mediated apoptosis. *Gastroenterology*, *120*, 1251–1262.

Heinrich, T. A., Tedesco, A. C., Fukuto, J. M., & da Silva, R. S. (2014). Production of reactive oxygen and nitrogen species by light irradiation of a nitrosyl phthalocyanine ruthenium complex as a strategy for cancer treatment. *Dalton Transactions*, *43*, 4021–4025.

Heneka, M. T., & Feinstein, D. L. (2001). Expression and function of inducible nitric oxide synthase in neurons. *Journal of Neuroimmunology*, *114*, 8–18.

Herbst, S., Schaible, U. E., & Schneider, B. E. (2011). Interferon gamma activated macrophages kill mycobacteria by nitric oxide induced apoptosis. *PLoS One*, *6*, e19105.

Hillman, G. G., Younes, E., Visscher, D., Hamzavi, F., Kim, S., Lam, J. S., et al. (1997). Inhibition of murine renal carcinoma pulmonary metastases by systemic administration of interferon gamma: Mechanism of action and potential for combination with interleukin 4. *Clinical Cancer Research*, *3*, 1799–1806.

Hinz, B., Brune, K., & Pahl, A. (2000). Nitric oxide inhibits inducible nitric oxide synthase mRNA expression in RAW 264.7 macrophages. *Biochemical and Biophysical Research Communications*, *271*, 353–357.

Hu, D. E., Dyke, S. O., Moore, A. M., Thomsen, L. L., & Brindle, K. M. (2004). Tumor cell-derived nitric oxide is involved in the immune-rejection of an immunogenic murine lymphoma. *Cancer Research*, *64*, 152–161.

Huie, R. E., & Padmaja, S. (1993). The reaction of no with superoxide. *Free Radical Research Communications*, *18*, 195–199.

Hussain, S. P., Trivers, G. E., Hofseth, L. J., He, P., Shaikh, I., Mechanic, L. E., et al. (2004). Nitric oxide, a mediator of inflammation, suppresses tumorigenesis. *Cancer Research*, *64*, 6849–6853.

Ignarro, L. J. (2000). The unique role of nitric oxide as a signaling molecule in the cardiovascular system. *Italian Heart Journal, 1*(Suppl. 3), S28–S29.

Ischiropoulos, H., & Beckman, J. S. (2003). Oxidative stress and nitration in neurodegeneration: Cause, effect, or association? *The Journal of Clinical Investigation, 111*, 163–169.

Joshi, M. S., Ferguson, T. B., Jr., Han, T. H., Hyduke, D. R., Liao, J. C., Rassaf, T., et al. (2002). Nitric oxide is consumed, rather than conserved, by reaction with oxyhemoglobin under physiological conditions. *Proceedings of the National Academy of Sciences of the United States of America, 99*, 10341–10346.

Jost, M. M., Ninci, E., Meder, B., Kempf, C., Van, R. N., Hua, J., et al. (2003). Divergent effects of GM-CSF and TGFbeta1 on bone marrow-derived macrophage arginase-1 activity, MCP-1 expression, and matrix metalloproteinase-12: A potential role during arteriogenesis. *The FASEB Journal, 17*, 2281–2283.

Kim, I. Y., & Stadtman, T. C. (1997). Inhibition of NF-kappaB DNA binding and nitric oxide induction in human T cells and lung adenocarcinoma cells by selenite treatment. *Proceedings of the National Academy of Sciences of the United States of America, 94*, 12904–12907.

Kleinert, H., Schwarz, P. M., & Forstermann, U. (2003). Regulation of the expression of inducible nitric oxide synthase. *Biological Chemistry, 384*, 1343–1364.

Klimp, A. H., de Vries, E. G., Scherphof, G. L., & Daemen, T. (2002). A potential role of macrophage activation in the treatment of cancer. *Critical Reviews in Oncology/Hematology, 44*, 143–161.

Kroncke, K. D., Fehsel, K., Suschek, C., & Kolb-Bachofen, V. (2001). Inducible nitric oxide synthase-derived nitric oxide in gene regulation, cell death and cell survival. *International Immunopharmacology, 1*, 1407–1420.

Kropf, P., Fuentes, J. M., Fahnrich, E., Arpa, L., Herath, S., Weber, V., et al. (2005). Arginase and polyamine synthesis are key factors in the regulation of experimental leishmaniasis in vivo. *The FASEB Journal, 19*, 1000–1002.

Lala, P. K., & Chakraborty, C. (2001). Role of nitric oxide in carcinogenesis and tumour progression. *The Lancet Oncology, 2*, 149–156.

Larkin, J., III, Ahmed, C. M., Wilson, T. D., & Johnson, H. M. (2013). Regulation of interferon gamma signaling by suppressors of cytokine signaling and regulatory T cells. *Frontiers in Immunology, 4*, 469.

Li, H., Meininger, C. J., Hawker, J. R., Jr., Haynes, T. E., Kepka-Lenhart, D., Mistry, S. K., et al. (2001). Regulatory role of arginase I and II in nitric oxide, polyamine, and proline syntheses in endothelial cells. *American Journal of Physiology, Endocrinology and Metabolism, 280*, E75–E82.

Lollini, P. L., Bosco, M. C., Cavallo, F., De, G. C., Giovarelli, M., Landuzzi, L., et al. (1993). Inhibition of tumor growth and enhancement of metastasis after transfection of the gamma-interferon gene. *International Journal of Cancer, 55*, 320–329.

Luiking, Y. C., Hallemeesch, M. M., van de Poll, M. C., Dejong, C. H., de Jonge, W. J., Lamers, W. H., et al. (2008). Reduced citrulline availability by OTC deficiency in mice is related to reduced nitric oxide production. *American Journal of Physiology, Endocrinology and Metabolism, 295*, E1315–E1322.

MacMicking, J., Xie, Q. W., & Nathan, C. (1997). Nitric oxide and macrophage function. *Annual Review of Immunology, 15*, 323–350.

Miller, C. H., Maher, S. G., & Young, H. A. (2009). Clinical use of interferon-gamma. *The Annals of the New York Academy of Sciences, 1182*, 69–79.

Mishra, B. B., Rathinam, V. A., Martens, G. W., Martinot, A. J., Kornfeld, H., Fitzgerald, K. A., et al. (2013). Nitric oxide controls the immunopathology of tuberculosis by inhibiting NLRP3 inflammasome-dependent processing of IL-1beta. *Nature Immunology, 14*, 52–60.

Mocellin, S., Bronte, V., & Nitti, D. (2007). Nitric oxide, a double edged sword in cancer biology: Searching for therapeutic opportunities. *Medicinal Research Reviews*, *27*, 317–352.

Mocellin, S., Provenzano, M., Rossi, C. R., Pilati, P., Scalerta, R., Lise, M., et al. (2004). Induction of endothelial nitric oxide synthase expression by melanoma sensitizes endothelial cells to tumor necrosis factor-driven cytotoxicity. *Clinical Cancer Research*, *10*, 6879–6886.

Motzer, R. J., Bacik, J., Murphy, B. A., Russo, P., & Mazumdar, M. (2002). Interferon-alfa as a comparative treatment for clinical trials of new therapies against advanced renal cell carcinoma. *Journal of Clinical Oncology*, *20*, 289–296.

Munder, M., Eichmann, K., & Modolell, M. (1998). Alternative metabolic states in murine macrophages reflected by the nitric oxide synthase/arginase balance: Competitive regulation by CD4 + T cells correlates with Th1/Th2 phenotype. *The Journal of Immunology*, *160*, 5347–5354.

Munder, M., Schneider, H., Luckner, C., Giese, T., Langhans, C. D., Fuentes, J. M., et al. (2006). Suppression of T-cell functions by human granulocyte arginase. *Blood*, *108*, 1627–1634.

Nathan, C. (2002). Inducible nitric oxide synthase in the tuberculous human lung. *American Journal of Respiratory and Critical Care Medicine*, *166*, 130–131.

Nathan, C., & Shiloh, M. U. (2000). Reactive oxygen and nitrogen intermediates in the relationship between mammalian hosts and microbial pathogens. *Proceedings of the National Academy of Sciences of the United States of America*, *97*, 8841–8848.

O'Brien, L., Carmichael, J., Lowrie, D. B., & Andrew, P. W. (1994). Strains of *Mycobacterium tuberculosis* differ in susceptibility to reactive nitrogen intermediates in vitro. *Infection and Immunity*, *62*, 5187–5190.

O'Shea, J. J., Gadina, M., & Kanno, Y. (2011). Cytokine signaling: Birth of a pathway. *The Journal of Immunology*, *187*, 5475–5478.

Pacher, P., Beckman, J. S., & Liaudet, L. (2007). Nitric oxide and peroxynitrite in health and disease. *Physiological Reviews*, *87*, 315–424.

Park, S. K., Cho, S., Lee, I. H., Jeon, D. S., Hong, S. H., Smego, R. A., Jr., et al. (2007). Subcutaneously administered interferon-gamma for the treatment of multidrug-resistant pulmonary tuberculosis. *International Journal of Infectious Diseases*, *11*, 434–440.

Peranzoni, E., Marigo, I., Dolcetti, L., Ugel, S., Sonda, N., Taschin, E., et al. (2007). Role of arginine metabolism in immunity and immunopathology. *Immunobiology*, *212*, 795–812.

Perske, C., Lahat, N., Sheffy, L. S., Bitterman, H., Hemmerlein, B., & Rahat, M. A. (2010). Loss of inducible nitric oxide synthase expression in the mouse renal cell carcinoma cell line RENCA is mediated by microRNA miR-146a. *The American Journal of Pathology*, *177*, 2046–2054.

Pestka, S., Krause, C. D., & Walter, M. R. (2004). Interferons, interferon-like cytokines, and their receptors. *Immunological Reviews*, *202*, 8–32.

Rodriguez, P. C., & Ochoa, A. C. (2008). Arginine regulation by myeloid derived suppressor cells and tolerance in cancer: Mechanisms and therapeutic perspectives. *Immunological Reviews*, *222*, 180–191.

Rosenberg, S. A., Lotze, M. T., Yang, J. C., Linehan, W. M., Seipp, C., Calabro, S., et al. (1989). Combination therapy with interleukin-2 and alpha-interferon for the treatment of patients with advanced cancer. *Journal of Clinical Oncology*, *7*, 1863–1874.

Schairer, D. O., Chouake, J. S., Nosanchuk, J. D., & Friedman, A. J. (2012). The potential of nitric oxide releasing therapies as antimicrobial agents. *Virulence*, *3*, 271–279.

Schroder, K., Hertzog, P. J., Ravasi, T., & Hume, D. A. (2004). Interferon-gamma: An overview of signals, mechanisms and functions. *Journal of Leukocyte Biology*, *75*, 163–189.

Serafini, P., Carbley, R., Noonan, K. A., Tan, G., Bronte, V., & Borrello, I. (2004). High-dose granulocyte-macrophage colony-stimulating factor-producing vaccines impair the immune response through the recruitment of myeloid suppressor cells. *Cancer Research*, *64*, 6337–6343.

Shah, V., Lyford, G., Gores, G., & Farrugia, G. (2004). Nitric oxide in gastrointestinal health and disease. *Gastroenterology*, *126*, 903–913.

Shang, D., Qi, Q. H., Wang, B. Z., Chen, H. L., Bi, W., & Guan, F. L. (2007). Role of polymorphonuclear neutrophil apoptosis and expression of Fas and caspase-3 in the systemic inflammatory response syndrome. *Zhongguo Wei Zhong Bing Ji Jiu Yi Xue*, *19*, 11–13.

Sharma, S. K., & Banga, A. (2004). Diagnostic utility of pleural fluid IFN-gamma in tuberculosis pleural effusion. *Journal of Interferon and Cytokine Research*, *24*, 213–217.

Shinoda, J., McLaughlin, K. E., Bell, H. S., Swaroop, G. R., Yamaguchi, S., Holmes, M. C., et al. (2003). Molecular mechanisms underlying dexamethasone inhibition of iNOS expression and activity in C6 glioma cells. *Glia*, *42*, 68–76.

Spencer, W. F., Linehan, W. M., Walther, M. M., Haas, G. P., Lotze, M. T., Topalian, S. L., et al. (1992). Immunotherapy with interleukin-2 and alpha-interferon in patients with metastatic renal cell cancer with in situ primary cancers: A pilot study. *The Journal of Urology*, *147*, 24–30.

Tate, D. J., Jr., Vonderhaar, D. J., Caldas, Y. A., Metoyer, T., Patterson, J. R., Aviles, D. H., et al. (2008). Effect of arginase II on L-arginine depletion and cell growth in murine cell lines of renal cell carcinoma. *Journal of Hematology and Oncology*, *1*, 14.

Trinchieri, G. (2010). Type I interferon: Friend or foe? *The Journal of Experimental Medicine*, *207*, 2053–2063.

Viola, A., & Bronte, V. (2007). Metabolic mechanisms of cancer-induced inhibition of immune responses. *Seminars in Cancer Biology*, *17*, 309–316.

Willenborg, D. O., Staykova, M., Fordham, S., O'Brien, N., & Linares, D. (2007). The contribution of nitric oxide and interferon gamma to the regulation of the neuro-inflammation in experimental autoimmune encephalomyelitis. *Journal of Neuroimmunology*, *191*, 16–25.

Windbichler, G. H., Hausmaninger, H., Stummvoll, W., Graf, A. H., Kainz, C., Lahodny, J., et al. (2000). Interferon-gamma in the first-line therapy of ovarian cancer: A randomized phase III trial. *British Journal of Cancer*, *82*, 1138–1144.

Wink, D. A., Hines, H. B., Cheng, R. Y., Switzer, C. H., Flores-Santana, W., Vitek, M. P., et al. (2011). Nitric oxide and redox mechanisms in the immune response. *Journal of Leukocyte Biology*, *89*, 873–891.

Wink, D. A., Kasprzak, K. S., Maragos, C. M., Elespuru, R. K., Misra, M., Dunams, T. M., et al. (1991). DNA deaminating ability and genotoxicity of nitric oxide and its progenitors. *Science*, *254*, 1001–1003.

Wink, D. A., Vodovotz, Y., Laval, J., Laval, F., Dewhirst, M. W., & Mitchell, J. B. (1998). The multifaceted roles of nitric oxide in cancer. *Carcinogenesis*, *19*, 711–721.

Wu, G., Bazer, F. W., Davis, T. A., Kim, S. W., Li, P., Marc, R. J., et al. (2009). Arginine metabolism and nutrition in growth, health and disease. *Amino Acids*, *37*, 153–168.

Wu, G. Y., & Brosnan, J. T. (1992). Macrophages can convert citrulline into arginine. *Biochemical Journal*, *281*(Pt 1), 45–48.

Xie, Q. W., Cho, H. J., Calaycay, J., Mumford, R. A., Swiderek, K. M., Lee, T. D., et al. (1992). Cloning and characterization of inducible nitric oxide synthase from mouse macrophages. *Science*, *256*, 225–228.

Xu, W., Liu, L. Z., Loizidou, M., Ahmed, M., & Charles, I. G. (2002). The role of nitric oxide in cancer. *Cell Research*, *12*, 311–320.

Zabaleta, J., McGee, D. J., Zea, A. H., Hernandez, C. P., Rodriguez, P. C., Sierra, R. A., et al. (2004). Helicobacter pylori arginase inhibits T cell proliferation and reduces the expression of the TCR zeta-chain (CD3zeta). *The Journal of Immunology*, *173*, 586–593.

Zea, A. H., Rodriguez, P. C., Atkins, M. B., Hernandez, C., Signoretti, S., Zabaleta, J., et al. (2005). Arginase-producing myeloid suppressor cells in renal cell carcinoma patients: A mechanism of tumor evasion. *Cancer Research*, *65*, 3044–3048.

Zea, A. H., Rodriguez, P. C., Culotta, K. S., Hernandez, C. P., DeSalvo, J., Ochoa, J. B., et al. (2004). L-Arginine modulates CD3zeta expression and T cell function in activated human T lymphocytes. *Cellular Immunology*, *232*, 21–31.

CHAPTER THIRTEEN

Antiobesogenic Role of Endothelial Nitric Oxide Synthase

Brian E. Sansbury[*,†], **Bradford G. Hill**[*,†,‡,1]
[*]Diabetes and Obesity Center, Institute of Molecular Cardiology, Louisville, Kentucky, USA
[†]Department of Physiology and Biophysics, Louisville, Kentucky, USA
[‡]Department of Biochemistry and Molecular Biology, University of Louisville School of Medicine, Louisville, Kentucky, USA
[1]Corresponding author: e-mail address: bradford.hill@louisville.edu

Contents

Abstract

The prevalence of obesity has increased remarkably in the past four decades. Because obesity can promote the development of type 2 diabetes and cardiovascular disease, understanding the mechanisms that engender weight gain and discovering safe antiobesity therapies are of critical importance. In particular, the gaseous signaling molecule, nitric oxide (NO), appears to be a central factor regulating adiposity and systemic metabolism. Obese and diabetic states are characterized by a deficit in bioavailable NO, with such decreases commonly attributed to downregulation of endothelial NO synthase (eNOS), loss of eNOS activity, or quenching of NO by its reaction with oxygen radicals. Gain-of-function studies, in which vascular-derived NO has been increased pharmacologically or genetically, reveal remarkable actions of NO on body composition and systemic metabolism. This review addresses the metabolic actions of eNOS and the potential therapeutic utility of harnessing its antiobesogenic effects.

1. INTRODUCTION

The rising prevalence of obesity is a principal health challenge in the United States and abroad. As of 2008, 10% of adults were obese, and

Vitamins and Hormones, Volume 96
ISSN 0083-6729
http://dx.doi.org/10.1016/B978-0-12-800254-4.00013-1

approximately 1.5 billion were overweight (Ahima, 2011). In the United States, recent estimates indicate that greater than one-third of adults and 17% of children are obese (Ogden, Carroll, Kit, & Flegal, 2012). This is associated with an increase in prediabetic states, with >30% of the U.S. population meeting the criteria for prediabetes (Ervin, 2009; Roger et al., 2011). Furthermore, obesity is associated with multiple other comorbidities including cardiovascular disease and cancer (Calle, Thun, Petrelli, Rodriguez, & Heath, 1999). The current high prevalence of obesity has also equated to a substantial economic burden of nearly $150 billion per year in healthcare costs (Zamosky, 2013).

While lifestyle and lack of exercise are undeniable risk factors for weight gain (Hu, Li, Colditz, Willett, & Manson, 2003; Robinson, 1999; Roger et al., 2011; Smith et al., 1994; Williamson et al., 1993), an excess consumption of foods appears to be one of the key factors in the epidemic of obesity. In the United States, the average human consumption of calories has increased by at least 200 kcal/day per person in the past three decades, and this is partly attributable to an increase in the intake of high-energy-density foods (Briefel & Johnson, 2004; Duffey & Popkin, 2007; Kant & Graubard, 2004; Nielsen & Popkin, 2003; Popkin et al., 2006; Wang, Bleich, & Gortmaker, 2008). Such dietary habits negatively affect a broad range of cardiovascular functions and promote the onset of type 2 diabetes (T2D) (Roger et al., 2011).

Insulin resistance is a cardinal feature of T2D and has been identified in multiple prospective studies as the initial defect promoting the development of disease (Reaven & Chen, 1996). It is typically defined as a decrease in sensitivity to the metabolic actions of insulin. Insulin maintains glucose homeostasis by promoting glucose uptake in skeletal muscle and by suppressing hepatic glucose production (Muniyappa, Montagnani, Koh, & Quon, 2007). Loss of insulin signaling therefore leads to hyperinsulinemia, hyperglycemia, and T2D. Hence, any treatment strategy to prevent diabetes must necessarily target insulin resistance.

Nitric oxide (NO) has emerged as a critical regulator of both adiposity and insulin sensitivity. In obese and diabetic states, the bioavailability of NO is decreased in both animal models (Bender, Herrick, Lott, & Klabunde, 2007; Kim et al., 2008) and adult and adolescent humans (Gruber et al., 2008; Higashi et al., 2001). Because the availability of NO is dependent upon its generation and degradation, lower levels observed in obese states may be due to downregulation of NOS, diminished NOS activity, or by reaction of NO with reactive oxygen species such as superoxide.

In particular, eNOS abundance and activity is reported to decrease remarkably in obese and diabetic states and, as discussed below, is likely a central feature regulating body composition.

2. eNOS IS IMPORTANT FOR REGULATING VASCULAR AND METABOLIC FUNCTION

The nitric oxide synthase (NOS) family of enzymes catalyze NADPH- and O_2-dependent oxidation of L-arginine to L-citrulline, producing NO in the process (Alderton, Cooper, & Knowles, 2001; Hill, Dranka, Bailey, Lancaster, & Darley-Usmar, 2010). NO synthesis depends also on the availability of several cofactors, including flavin adenine dinucleotide, flavin mononucleotide, tetrahydrobiopterin (BH_4), as well as the prosthetic group, heme (Li & Poulos, 2005). Endothelial NOS (eNOS), the primary subject of this review, is primarily expressed in the vascular endothelium, but has also been identified in neurons, epithelial cells, and cardiomyocytes (Dudzinski & Michel, 2007). Its activity is controlled by Ca^{2+} and calmodulin, posttranslational modifications (Oess, Icking, Fulton, Govers, & Muller-Esterl, 2006; Sessa, 2004), and shear stress (Balligand, Feron, & Dessy, 2009; Kone, Kuncewicz, Zhang, & Yu, 2003). Other isoforms of NOS, i.e., neuronal NOS (nNOS) and inducible NOS (iNOS), are commonly expressed in different tissues and cells, and, in general produce higher quantities of NO (Hill et al., 2010).

While NO is well known to have diverse biological actions (including regulation of learning and memory, platelet aggregation, leukocyte–endothelial interactions, immune function, and angiogenesis/arteriogenesis, Forstermann & Sessa, 2012), it is most renowned for its vascular actions. The discovery that endothelial cells control acetylcholine-induced relaxation of smooth muscle (Furchgott & Zawadzki, 1980) was one of several impetuses responsible for the designation of NO as endothelial-derived relaxing factor (EDRF). Following Furchgott's findings, a series of studies showed that NO synthesized by eNOS in endothelial cells diffuses into the tunica media where it activates soluble guanylate cyclase (sGC), generating cyclic GMP (cGMP) and eliciting vessel relaxation (Forstermann, Mulsch, Bohme, & Busse, 1986; Gryglewski, Moncada, & Palmer, 1986; Moncada, Palmer, & Gryglewski, 1986; Palmer, Ferrige, & Moncada, 1987; Rajfer, Aronson, Bush, Dorey, & Ignarro, 1992; Rapoport, Draznin, & Murad, 1983; Rapoport & Murad, 1983).

Although NO's most-celebrated role is a result of its reaction with the metalloprotein sGC, NO and its oxidation species have the ability to react with other biomolecules as well. NO primarily reacts with ferrous iron and other radical species, with the highest affinity interactions being with the iron-containing proteins sGC, cytochrome *c* oxidase, and hemoglobin. The presence of other radical species such as superoxide ($O_2^{\bullet -}$) can result in the formation of products such as peroxynitrite (Beckman, 2009; Pacher, Beckman, & Liaudet, 2007; Szabo, Ischiropoulos, & Radi, 2007; Trujillo, Ferrer-Sueta, & Radi, 2008), which has different biomolecular targets and causes nitration of tyrosine residues (Hill et al., 2010). NO can also react with O_2 to form oxidized species such as N_2O_3, which can S-nitrosate or promote the S-oxidization of protein side chains (Hill & Bhatnagar, 2007; West, Hill, Xuan, & Bhatnagar, 2006). Also, NO reacts with thiyl radicals to form S-nitrosated proteins. Cysteinyl thiols of glutathione and proteins are commonly targets of NO and its oxidized species and become not only S-nitrosated or S-oxidized ($SO_{2/3}$), but S-glutathiolated as well (Hill & Bhatnagar, 2012; West et al., 2006). These modifications frequently modulate enzyme activity (Hill & Bhatnagar, 2007).

In addition to its vasodilatory actions, NO modulates oxygen delivery to cells and tissues by regulating oxygen binding and release from hemoglobin. It regulates oxygen consumption as well by binding and inhibiting cytochrome *c* oxidase, with such binding dependent on both mitochondrial activity and the O_2 level (Cooper & Giulivi, 2007; Shiva et al., 2005). Hence, it can extend O_2 gradients in tissues by regulating hemoglobin action and by inhibiting O_2 consumption in mitochondria (Thomas, Liu, Kantrow, & Lancaster, 2001). Exposure of cells to relatively high concentrations of NO promotes mitochondrial biogenesis (Kelly & Scarpulla, 2004; Nisoli et al., 2003, 2004), thereby increasing overall respiratory capacity.

3. NO BIOAVAILABILITY IS DECREASED IN OBESE AND DIABETIC STATES

Several studies link a decrease in eNOS-derived NO to diabetes. A T (−786)C variant of the eNOS gene is associated with insulin resistance (Ohtoshi et al., 2002; Vecoli et al., 2012; Yoshimura et al., 2003), along with several other genetic variants in the eNOS locus, which are associated with T2D (Monti et al., 2003). eNOS variants also appear to increase susceptibility for insulin resistance, hypertriglyceridemia, and low HDL (Gonzalez-Sanchez et al., 2007) and worsen endothelial function in individuals prone to T2D (Rittig et al., 2008).

Beyond eNOS polymorphisms, a primary mechanism by which NO is decreased in obesity is through diminished expression of eNOS. A decrease in eNOS abundance occurs in both adipose tissue and skeletal muscle of obese humans and rodents (Georgescu et al., 2011; Kraus et al., 2012; Perez-Matute, Neville, Tan, Frayn, & Karpe, 2009; Sansbury et al., 2012 Valerio et al., 2006). In particular, it appears that the cytokine tumor necrosis factor-α (TNFα), which is implicated in the initiation of insulin resistance (Hotamisligil, Shargill, & Spiegelman, 1993), downregulates eNOS abundance (Anderson, Rahmutula, & Gardner, 2004; Lai, Mohamed, Monge, & Stewart, 2003; Michel & Lamas, 1992; Neumann, Gertzberg, & Johnson, 2004; Valerio et al., 2006) by decreasing the stability of eNOS mRNA (Alonso, Sanchez de Miguel, Monton, Casado, & Lopez-Farre, 1997; Sanchez de Miguel et al., 1999), thereby shortening its half-life (Yoshizumi, Perrella, Burnett, & Lee, 1993). Destabilization of the eNOS message may be due, at least in part, to upregulation of elongation factor 1-α1 (Yan, You, Chen, Liao, & Sun, 2008).

The NO-producing activity of eNOS is also diminished in metabolic disease. Conditions related with nutrient excess were shown to upregulate caveolin-1, a negative regulator of eNOS (Ju, Zou, Venema, & Venema, 1997; Michel, Feron, Sacks, & Michel, 1997), in the aorta of obese rats (Yang et al., 2007). Furthermore, ceramide (which increases in abundance in obese states, Bikman & Summers, 2011) decreases eNOS activity by disrupting the eNOS–Akt complex from HSP90 (Zhang et al., 2012).

Critical changes in eNOS phosphorylation occur in obesity as well. The eNOS enzyme can be phosphorylated at several sites, including: tyrosine (Y) residues—Y81 and Y567; serine (S) residues—S114, S615, S633, and S1177; and threonine (T) residues—T495 (for review, see Kolluru, Siamwala, & Chatterjee, 2010; Rafikov et al., 2011). In particular, phosphorylation of eNOS at serine 1177 (Ser1176 in mice) increases NO output from the enzyme (McCabe, Fulton, Roman, & Sessa, 2000) and is diminished by nutrient excess (Elrod et al., 2006; Li et al., 2013; Taguchi, Kobayashi, Matsumoto, & Kamata, 2011; Zhong et al., 2007) or high-fat feeding (Kim et al., 2007, 2008; Sansbury et al., 2012; Symons et al., 2009) in mice; it is similarly decreased in obese rats (Naruse et al., 2006; Park et al., 2013; Zecchin et al., 2007) and pigs (Low Wang et al., 2013). This eNOS phosphorylation site is regulated by Akt (Dimmeler et al., 1999), which is stimulated by insulin (Hermann, Assmus, Urbich, Zeiher, & Dimmeler, 2000). Insulin receptor signaling stimulates the Akt–eNOS pathway, which is known to regulate postprandial blood flow and nutrient delivery to peripheral tissues. Consequently, endothelial insulin resistance is sufficient to decrease NO bioavailability and promote endothelial dysfunction (Duncan et al., 2008), and diminished eNOS phosphorylation due to insulin resistance appears to be responsible for diminished glucose uptake in skeletal muscle of high-fat-fed mice (Kubota et al., 2011).

Loss of eNOS phosphorylation under conditions of nutrient excess may be due to several factors, one of which is fatty acids, which can promote insulin resistance (Kim et al., 2008). Elevated free fatty acids (FFAs; e.g., palmitic acid) decrease NO production or availability in humans (Steinberg et al., 1997, 2000), animal models (Du et al., 2006), isolated arteries, and cultured cells (Edirisinghe, McCormick Hallam, & Kappagoda, 2006; Kim et al., 2005). Insulin resistance due to FFAs is likely caused, at least in part, by activation of Toll-like receptor 2 (Jang, Kim, Hwang, Quon, & Kim, 2013) or Toll-like receptor 4 and NF-κB (Kim et al., 2005, 2007). In addition, hyperglycemia was suggested to increase O-linked *N*-acetylglucosamine (O-GlcNAc) modification of eNOS, which diminishes its activity (Du et al., 2001). A PKCβII-mediated diminishment in Akt and eNOS responsiveness to insulin has also been reported (Naruse et al., 2006; Park et al., 2013), and an apparently Akt-independent impairment of eNOS phosphorylation may occur (Symons et al., 2009). It is unclear whether or how each of these signaling pathways integrates to modulate NO production in obese and diabetic states.

eNOS activity and NO generation are dependent on proper enzyme coupling, which is regulated by cofactors, dimerization (Rodriguez-Crespo, Gerber, & Ortiz de Montellano, 1996; Rodriguez-Crespo & Ortiz de Montellano, 1996), and posttranslational modifications (Alp & Channon, 2004; Forstermann & Sessa, 2012; Li & Forstermann, 2013; Zweier, Chen, & Druhan, 2011). The cofactor BH_4 is critical for optimal eNOS activity, and it is depleted by excessive levels of reactive oxygen or nitrogen species (Channon, 2012). Obese and diabetic states are associated with decreased BH_4 and increased levels of its oxidized form, BH_2 (Cai, Khoo, & Channon, 2005; Chander et al., 2004; Ding & Triggle, 2010; Pannirselvam, Verma, Anderson, & Triggle, 2002; Shinozaki et al., 1999). This is important because deficiency in BH_4 or elevations in BH_2 uncouple NOS, resulting in superoxide production and peroxynitrite generation (Alp & Channon, 2004). Indeed, decreases in the BH_4 to BH_2 ratio are responsible for glucose-induced eNOS uncoupling (Crabtree, Smith, Lam, Goligorsky, & Gross, 2008), and replenishment of BH_4 pools is an effective treatment in multiple pathologies (e.g., Alp & Channon, 2004; Crabtree & Channon, 2011; Forstermann & Li, 2011; Kietadisorn, Juni, & Moens, 2012; Li & Forstermann, 2013).

Commonly, 3-nitrotyrosine (3-NT) modifications are found at sites of eNOS uncoupling, and 3-NT-modified proteins are observed in abundance in tissues from obese and diabetic animals (Brodsky et al., 2004; Chander et al., 2004; Molnar et al., 2005; Sansbury et al., 2012). In line with a potential role of peroxynitrite in regulating eNOS function, diabetic patients showed elevated levels 3-NT protein adducts, which localized with caveolae; these patients demonstrated diminished-flow-mediated dilation of coronary arterioles (Cassuto et al., 2014), which was rescued by the BH_4 supplement, sepiapterin (Cassuto et al., 2014). Numerous additional studies also demonstrate a role for reactive species such as peroxynitrite and superoxide to promote eNOS uncoupling (Bitar et al., 2005; Dikalova et al., 2010; Landmesser et al., 2003; Satoh et al., 2005; Xu, Xie, Reece, Pimental, & Zou, 2006). Nonetheless, the contribution of reactive species-induced damage to endothelial function remains unclear. One study suggests that, rather than uncoupling eNOS, superoxide activates the enzyme (Zhang et al., 2008), leaving open the possibility that loss of NO bioavailability could be due to quenching of NO and not to uncoupling of the enzyme. However, multiple other factors, such as asymmetric dimethyl arginine, insufficient L-arginine levels, or glutathio(ny)lation of the eNOS enzyme, can promote eNOS uncoupling and endothelial dysfunction as well (Chen et al., 2010; Forstermann & Sessa, 2012; Lei, Luo, Qin, & Xia, 2013; Risbano & Gladwin, 2013; Toutouzas, Riga, Stefanadi, & Stefanadis, 2008), suggesting that uncoupling of the enzyme contributes to decreases in NO production.

So, does obesity itself decrease NO availability? Obesity in humans is associated with decreased blood flow in response to shear stress (Arcaro et al., 1999), bradykinin (Laine et al., 1998; Van Guilder, Stauffer, Greiner, & Desouza, 2008), methacholine (Steinberg et al., 1996), substance P and acetylcholine (Van Guilder et al., 2008), and insulin (Tack, Ong, Lutterman, & Smits, 1998; Westerbacka et al., 1999), which would appear to suggest that obesity is causally linked with decreased vascular NO bioavailability. Other studies support this hypothesis as well (Andersson et al., 2010; Bhattacharjee, Alotaibi, Kheirandish-Gozal, Capdevila, &

Gozal, 2010; Bhattacharjee et al., 2012; Georgescu et al., 2011; Grassi et al., 2010; Gupta et al., 2010; Han, Patel, Lteif, Chisholm, & Mather, 2011; Lambert et al., 2010; Mahmud, Hill, Cuerden, & Clarson, 2009; Miadi-Messaoud et al., 2010; Parikh et al., 2009; Sturm et al., 2009; Weil et al., 2011). Nevertheless, the question remains: Does increased adiposity somehow decrease eNOS-derived NO and availability, or are losses in vascular NO due only to conditions associated with obesity? Several studies suggest that the state of being corpulent is *not* causative in decreasing vascular NO bioavailability. For example, morbidly obese humans appear to have endothelial dysfunction only when insulin resistance is present (El Assar et al., 2013), and, severely obese humans, in the absence of insulin resistance, have better flow-mediated dilation compared with both normal and obese insulin-sensitive subjects (Biasucci et al., 2010). Moreover, in overweight (insulin-sensitive) individuals, capillary recruitment may actually be higher compared with lean controls (Czernichow et al., 2010). Hence, it appears that either insulin resistance or conditions directly linked with the insulin-resistant phenotype are to blame for loss of NO bioavailability in obesity.

4. REGULATION OF OBESITY AND INSULIN RESISTANCE BY eNOS

Does eNOS-derived NO affect insulin resistance and obesity? This question has been addressed by multiple pharmacological and genetic studies, which, collectively, have helped clarify critical roles for eNOS-derived NO in regulating obesity and insulin resistance. Human studies show that L-arginine supplementation has favorable effects on adiposity and insulin sensitivity (Alizadeh et al., 2012; Bogdanski et al., 2012, 2013; Lucotti et al., 2006; Monti et al., 2013; Suliburska, Bogdanski, Szulinska, Pupek-Musialik, & Jablecka, 2014; Wascher et al., 1997). Results from animal studies also show that L-arginine decreases fat mass, increases muscle mass, and improves insulin sensitivity (Clemmensen, Madsen, Smajilovic, Holst, & Brauner-Osborne, 2012). Dietary L-arginine supplementation in rats increases brown fat and skeletal muscle mass and reduces serum concentrations of triglycerides, glucose, homocysteine, FFAs, dimethylarginines, and leptin (Fu et al., 2005; Jobgen et al., 2009). L-arginine has a similar effect on pigs (Tan et al., 2009).

Interestingly, sildenafil—which prevents the degradation of cGMP and is used to treat erectile dysfunction in humans—increases insulin sensitivity and prevents obesity in high-fat-fed mice (Ayala et al., 2007), potentially by promoting "browning" of white adipose tissue (Mitschke et al., 2013). Sildenafil increases mitochondrial biogenesis in human adipose tissue *ex vivo* as well (De Toni et al., 2011). Other compounds that activate the NO pathway also support a role for NO in improving insulin sensitivity. Beraprost (a stable prostaglandin analog), when given to endothelial-specific insulin receptor substrate 2 knockout mice, restored eNOS phosphorylation, capillary recruitment, and insulin and glucose delivery to skeletal muscle (Kubota et al., 2011). Additionally, S-nitrosation in response to L-arginine, insulin, or sodium nitroprusside was shown to be important for regulating vascular endothelial insulin uptake and transendothelial transport (Wang, Wang, Aylor, & Barrett, 2013). Thus, it appears that NO may regulate obesity and insulin resistance by both cGMP-dependent and -independent pathways.

Although chronic treatment with NOS inhibitors causes weight loss and promotes insulin sensitivity in animals (Morley & Flood, 1994; Stricker-Krongrad, Beck, & Burlet, 1996; Tsuchiya et al., 2007), their acute application causes systemic insulin resistance (Baron et al., 1995), in part by promoting metabolic changes in the liver (Meshkani & Adeli, 2009). Furthermore, BH_4, which is oxidized to BH_2 in the diabetic state (Meininger et al., 2000, 2004; Xu et al., 2007), administered to STZ-treated mice lowered blood glucose levels in an eNOS-dependent manner. Increasing BH_4 was shown to also improve glucose tolerance and insulin sensitivity in *ob/ob* mice (Abudukadier et al., 2013). This was suggested to be due to eNOS-mediated activation of AMPK in the liver (Abudukadier et al., 2013), which suppresses hepatic glucose production (Viollet et al., 2009). Therefore, eNOS uncoupling in liver appears to negatively regulate systemic glucose metabolism in obese, diabetic states.

Genetic models in which eNOS has been deleted or overexpressed have helped to further elucidate the mechanisms by which NO regulates obesity and insulin resistance. Deletion of eNOS causes insulin resistance, hyperlipidemia, and hypertension (Duplain et al., 2001), and partial deletion of the gene can exaggerate insulin resistance, glucose intolerance, and hypertension under conditions of nutrient excess (Cook et al., 2003, 2004). Mice in which both eNOS and nNOS are absent show similar results, with deletion of eNOS appearing to be responsible for insulin resistance in both skeletal muscle and liver (Shankar, Wu, Shen, Zhu, & Baron, 2000). Similarly, mice lacking eNOS, nNOS, and iNOS (i.e., triple knockout mice) show increased visceral obesity, hypertension, hypertriglyceridemia, and impaired glucose tolerance (Nakata et al., 2008).

The metabolic phenotype caused by eNOS deletion or otherwise low-endothelial-derived NO appears to relate directly to changes in substrate metabolism in liver, skeletal muscle, and adipose tissue. In skeletal muscle, eNOS KO mice have lower mitochondrial content and fatty acid oxidation than WT mice, and they demonstrate markedly lower energy expenditure (Le Gouill et al., 2007). Supplementation of eNOS KO mice with nitrate, which can be reduced to nitrite and NO in the body, decreased not only blood pressure, but visceral adipose tissue and triglycerides as well (Carlstrom et al., 2010).

Our studies in mice overexpressing eNOS suggest a remarkable ability of eNOS to regulate metabolism and body composition. Mice overexpressing eNOS in the vasculature show an antiobesogenic phenotype characterized by resistance to accumulation of white adipose tissue in response to a high-fat diet, a higher metabolic rate, resistance to diet-induced hyperinsulinemia, and remarkably lower plasma levels of FFAs and triglycerides (Sansbury et al., 2012). As shown in Fig. 13.1, overexpression of eNOS resulted in decreased weight gain on a high-fat diet, which was due to diminished expansion of the adipose tissue.

An eNOS phosphomimetic point mutant mouse model (Atochin & Huang, 2010; Kashiwagi et al., 2013) showed a very similar phenotype: mutation of eNOS ser1176 to an aspartic acid increased endothelial NO production as well as promoted resistance to diet-induced weight gain and hyperinsulinemia, whereas mutation of this residue to an alanine promoted insulin resistance and permitted the development of an obese state (Huang, 2009; Kashiwagi et al., 2013).

That eNOS KO mice have elevated plasma levels of triglycerides and FFAs compared with WT mice (Cook et al., 2003; Duplain et al., 2001), while eNOS transgenic mice show diminished abundance of the lipids (Sansbury et al., 2012) suggests that eNOS regulates lipid oxidation or synthesis. Indeed, eNOS KO mice show diminished fat oxidation capacity in skeletal muscle (Le Gouill et al., 2007). Furthermore, administration of a NOS inhibitor to rats increases serum triglycerides and diminishes fatty acid oxidation in the liver (Khedara, Kawai, Kayashita, & Kato, 1996), potentially by decreasing carnitine palmitoyl transferase activity (Khedara, Goto, Morishima, Kayashita, & Kato, 1999). NOS inhibitor-dependent decreases in fatty acid oxidation occur in heart as well (Recchia et al., 2002). In hepatocytes, NO donors increase β-oxidation in a cGMP-dependent manner by inhibiting acetyl CoA carboxylase, thereby stimulating the activity of carnitine palmitoyl transferase (Garcia-Villafranca, Guillen, & Castro, 2003). NO also diminishes fatty acid

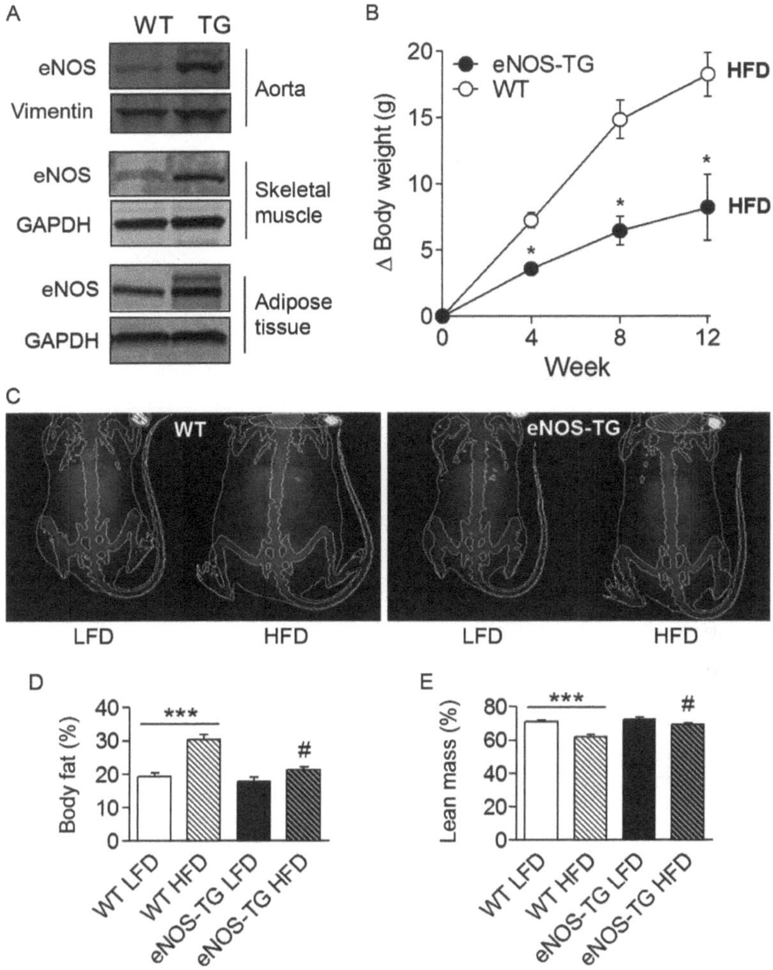

Figure 13.1 Overexpression of eNOS prevents obesity. (A) Immunoblot analysis: eNOS expression in aorta, skeletal muscle, and adipose tissue of wild-type (WT) and eNOS transgenic (eNOS-TG) mice. (B) Change in body weight during high-fat feeding: mice were fed a high-fat diet (HFD) for up to 12 weeks, and the change in body weight was measured. *$p<0.05$ vs. WT. (C) Representative Dexascan images of WT and eNOS-TG mice fed a low-fat diet (LFD) or HFD for 6 weeks. (D and E) Dexascan analysis of body fat and lean mass percentage. ***$p<0.001$ vs. indicated group; #$p<0.05$ vs. WT HFD. *Adapted with permission from Sansbury et al. (2012).* (See the color plate.)

synthesis in hepatocytes (Garcia-Villafranca et al., 2003), which is consistent with studies showing that inhibitors of NOS (Goto et al., 1999) or deletion of eNOS increases lipid synthesis in liver (Schild et al., 2008). Similarly, in skeletal muscle, loss of eNOS increases neolipogenic gene expression while decreasing those genes that promote fatty acid oxidation

(Le Gouill et al., 2007). These data suggest that eNOS may regulate peroxisome proliferator activated receptor (PPAR)-α, which is well known to regulate lipid metabolism (Lefebvre, Chinetti, Fruchart, & Staels, 2006). Indeed, our studies show that overexpression of eNOS increases PPARα expression in adipose tissue (Sansbury et al., 2012), suggesting that endothelial-derived NO increases the molecular machinery required to program cells to burn fat. However, it is possible that NO primes fat oxidation in other ways as well. For example, recent studies demonstrate S-nitrosation of multiple enzymes involved in metabolism. In particular, very long chain acyl-coA dehydrogenase, a liver enzyme important in β-oxidation, was shown to be nitrosated at Cys238, which increases the catalytic efficiency of the enzyme. This modification was dependent on eNOS activity, as nitrosation of the enzyme was absent in eNOS KO mice (Gould, Doulias, Tenopoulou, Raju, & Ischiropoulos, 2013). Lastly, it is possible that NO-induced increases in mitochondrial mass (Nisoli et al., 2003, 2004; Piantadosi & Suliman, 2012) could be sufficient to increase metabolic rate and prevent obesity. This would be consistent with studies showing that cGMP-dependent increases in mitochondrial biogenesis prevent obesity (Miyashita et al., 2009) as well as several other studies demonstrating a link between augmented mitochondrial mass and resistance to diet-induced weight gain (e.g., Fang et al., 2011; Hwang et al., 2009; Yadav et al., 2011; Yamamoto et al., 2011).

5. SYNOPSIS

Collectively, these studies suggest that eNOS-derived NO has powerful antiobesity and insulin-sensitizing effects. It is likely that the enzyme increases fat oxidation and diminishes lipid synthesis in tissues such as liver, skeletal muscle, and fat. The relatively low levels of adiposity and plasma FFAs and triglycerides in models in which eNOS is overexpressed or permanently activated is consistent with this mechanism (Kashiwagi et al., 2013; Sansbury et al., 2012). The favorable effects of eNOS on glucose metabolism and insulin sensitivity appear to be due to its ability to stimulate the transport of insulin and glucose to key peripheral tissues such as skeletal muscle and to regulate gluconeogenesis. In addition, eNOS overexpression or activation prevents diet-induced hyperinsulinemia (Kashiwagi et al., 2013; Sansbury et al., 2012), suggesting that it could impact glucose metabolism by regulating insulin secretion. Exploiting the beneficial metabolic actions of eNOS is a promising prospect for antiobesity therapies.

ACKNOWLEDGMENTS

The authors acknowledge support from NIH Grants GM103492 and HL078825.

REFERENCES

Abudukadier, A., Fujita, Y., Obara, A., Ohashi, A., Fukushima, T., Sato, Y., et al. (2013). Tetrahydrobiopterin has a glucose-lowering effect by suppressing hepatic gluconeogenesis in an endothelial nitric oxide synthase-dependent manner in diabetic mice. *Diabetes, 62*(9), 3033–3043.

Ahima, R. S. (2011). Digging deeper into obesity. *Journal of Clinical Investigation, 121*(6), 2076–2079.

Alderton, W. K., Cooper, C. E., & Knowles, R. G. (2001). Nitric oxide synthases: Structure, function and inhibition. *Biochemical Journal, 357*(Pt 3), 593–615.

Alizadeh, M., Safaeiyan, A., Ostadrahimi, A., Estakhri, R., Daneghian, S., Ghaffari, A., et al. (2012). Effect of L-arginine and selenium added to a hypocaloric diet enriched with legumes on cardiovascular disease risk factors in women with central obesity: A randomized, double-blind, placebo-controlled trial. *Annals of Nutrition & Metabolism, 60*(2), 157–168.

Alonso, J., Sanchez de Miguel, L., Monton, M., Casado, S., & Lopez-Farre, A. (1997). Endothelial cytosolic proteins bind to the 3' untranslated region of endothelial nitric oxide synthase mRNA: Regulation by tumor necrosis factor alpha. *Molecular and Cellular Biology, 17*(10), 5719–5726.

Alp, N. J., & Channon, K. M. (2004). Regulation of endothelial nitric oxide synthase by tetrahydrobiopterin in vascular disease. *Arteriosclerosis, Thrombosis, and Vascular Biology, 24*(3), 413–420.

Anderson, H. D., Rahmutula, D., & Gardner, D. G. (2004). Tumor necrosis factor-alpha inhibits endothelial nitric-oxide synthase gene promoter activity in bovine aortic endothelial cells. *Journal of Biological Chemistry, 279*(2), 963–969.

Andersson, J., Sjostrom, L. G., Karlsson, M., Wiklund, U., Hultin, M., Karpe, F., et al. (2010). Dysregulation of subcutaneous adipose tissue blood flow in overweight postmenopausal women. *Menopause, 17*(2), 365–371.

Arcaro, G., Zamboni, M., Rossi, L., Turcato, E., Covi, G., Armellini, F., et al. (1999). Body fat distribution predicts the degree of endothelial dysfunction in uncomplicated obesity. *International Journal of Obesity and Related Metabolic Disorders, 23*(9), 936–942.

Atochin, D. N., & Huang, P. L. (2010). Endothelial nitric oxide synthase transgenic models of endothelial dysfunction. *Pflügers Archiv, 460*(6), 965–974.

Ayala, J. E., Bracy, D. P., Julien, B. M., Rottman, J. N., Fueger, P. T., & Wasserman, D. H. (2007). Chronic treatment with sildenafil improves energy balance and insulin action in high fat-fed conscious mice. *Diabetes, 56*(4), 1025–1033.

Balligand, J. L., Feron, O., & Dessy, C. (2009). eNOS activation by physical forces: From short-term regulation of contraction to chronic remodeling of cardiovascular tissues. *Physiological Reviews, 89*(2), 481–534.

Baron, A. D., Zhu, J. S., Marshall, S., Irsula, O., Brechtel, G., & Keech, C. (1995). Insulin resistance after hypertension induced by the nitric oxide synthesis inhibitor L-NMMA in rats. *American Journal of Physiology, 269*(4 Pt 1), E709–715.

Beckman, J. S. (2009). Understanding peroxynitrite biochemistry and its potential for treating human diseases. *Archives of Biochemistry and Biophysics, 484*(2), 114–116.

Bender, S. B., Herrick, E. K., Lott, N. D., & Klabunde, R. E. (2007). Diet-induced obesity and diabetes reduce coronary responses to nitric oxide due to reduced bioavailability in isolated mouse hearts. *Diabetes, Obesity & Metabolism, 9*(5), 688–696.

Bhattacharjee, R., Alotaibi, W. H., Kheirandish-Gozal, L., Capdevila, O. S., & Gozal, D. (2010). Endothelial dysfunction in obese non-hypertensive children without evidence of sleep disordered breathing. *BMC Pediatrics*, *10*, 8.

Bhattacharjee, R., Kim, J., Alotaibi, W. H., Kheirandish-Gozal, L., Capdevila, O. S., & Gozal, D. (2012). Endothelial dysfunction in children without hypertension: Potential contributions of obesity and obstructive sleep apnea. *Chest*, *141*(3), 682–691.

Biasucci, L. M., Graziani, F., Rizzello, V., Liuzzo, G., Guidone, C., De Caterina, A. R., et al. (2010). Paradoxical preservation of vascular function in severe obesity. *American Journal of Medicine*, *123*(8), 727–734.

Bikman, B. T., & Summers, S. A. (2011). Ceramides as modulators of cellular and whole-body metabolism. *Journal of Clinical Investigation*, *121*(11), 4222–4230.

Bitar, M. S., Wahid, S., Mustafa, S., Al-Saleh, E., Dhaunsi, G. S., & Al-Mulla, F. (2005). Nitric oxide dynamics and endothelial dysfunction in type II model of genetic diabetes. *European Journal of Pharmacology*, *511*(1), 53–64.

Bogdanski, P., Suliburska, J., Grabanska, K., Musialik, K., Cieslewicz, A., Skoluda, A., et al. (2012). Effect of 3-month L-arginine supplementation on insulin resistance and tumor necrosis factor activity in patients with visceral obesity. *European Review for Medical and Pharmacological Sciences*, *16*(6), 816–823.

Bogdanski, P., Szulinska, M., Suliburska, J., Pupek-Musialik, D., Jablecka, A., & Witmanowski, H. (2013). Supplementation with L-arginine favorably influences plasminogen activator inhibitor type 1 concentration in obese patients. A randomized, double blind trial. *Journal of Endocrinological Investigation*, *36*(4), 221–226.

Briefel, R. R., & Johnson, C. L. (2004). Secular trends in dietary intake in the United States. *Annual Review of Nutrition*, *24*, 401–431.

Brodsky, S. V., Gealekman, O., Chen, J., Zhang, F., Togashi, N., Crabtree, M., et al. (2004). Prevention and reversal of premature endothelial cell senescence and vasculopathy in obesity-induced diabetes by ebselen. *Circulation Research*, *94*(3), 377–384.

Cai, S., Khoo, J., & Channon, K. M. (2005). Augmented BH4 by gene transfer restores nitric oxide synthase function in hyperglycemic human endothelial cells. *Cardiovascular Research*, *65*(4), 823–831.

Calle, E. E., Thun, M. J., Petrelli, J. M., Rodriguez, C., & Heath, C. W., Jr. (1999). Body-mass index and mortality in a prospective cohort of U.S. adults. *New England Journal of Medicine*, *341*(15), 1097–1105.

Carlstrom, M., Larsen, F. J., Nystrom, T., Hezel, M., Borniquel, S., Weitzberg, E., et al. (2010). Dietary inorganic nitrate reverses features of metabolic syndrome in endothelial nitric oxide synthase-deficient mice. *Proceedings of the National Academy of Sciences of the United States of America*, *107*(41), 17716–17720.

Cassuto, J., Dou, H., Czikora, I., Szabo, A., Patel, V. S., Kamath, V., et al. (2014). Peroxynitrite disrupts endothelial caveolae leading to eNOS uncoupling and diminished flow-mediated dilation in coronary arterioles of diabetic patients. *Diabetes*, *63*(4), 1381–1393.

Chander, P. N., Gealekman, O., Brodsky, S. V., Elitok, S., Tojo, A., Crabtree, M., et al. (2004). Nephropathy in Zucker diabetic fat rat is associated with oxidative and nitrosative stress: Prevention by chronic therapy with a peroxynitrite scavenger ebselen. *Journal of the American Society of Nephrology*, *15*(9), 2391–2403.

Channon, K. M. (2012). Tetrahydrobiopterin: A vascular redox target to improve endothelial function. *Current Vascular Pharmacology*, *10*(6), 705–708.

Chen, C. A., Wang, T. Y., Varadharaj, S., Reyes, L. A., Hemann, C., Talukder, M. A., et al. (2010). S-glutathionylation uncouples eNOS and regulates its cellular and vascular function. *Nature*, *468*(7327), 1115–1118.

Clemmensen, C., Madsen, A. N., Smajilovic, S., Holst, B., & Brauner-Osborne, H. (2012). L-arginine improves multiple physiological parameters in mice exposed to diet-induced metabolic disturbances. *Amino Acids*, *43*(3), 1265–1275.

Cook, S., Hugli, O., Egli, M., Menard, B., Thalmann, S., Sartori, C., et al. (2004). Partial gene deletion of endothelial nitric oxide synthase predisposes to exaggerated high-fat diet-induced insulin resistance and arterial hypertension. *Diabetes*, *53*(8), 2067–2072.

Cook, S., Hugli, O., Egli, M., Vollenweider, P., Burcelin, R., Nicod, P., et al. (2003). Clustering of cardiovascular risk factors mimicking the human metabolic syndrome X in eNOS null mice. *Swiss Medical Weekly*, *133*(25–26), 360–363.

Cooper, C. E., & Giulivi, C. (2007). Nitric oxide regulation of mitochondrial oxygen consumption II: Molecular mechanism and tissue physiology. *American Journal of Physiology. Cell Physiology*, *292*(6), C1993–C2003.

Crabtree, M. J., & Channon, K. M. (2011). Synthesis and recycling of tetrahydrobiopterin in endothelial function and vascular disease. *Nitric Oxide*, *25*(2), 81–88.

Crabtree, M. J., Smith, C. L., Lam, G., Goligorsky, M. S., & Gross, S. S. (2008). Ratio of 5,6,7,8-tetrahydrobiopterin to 7,8-dihydrobiopterin in endothelial cells determines glucose-elicited changes in NO vs. superoxide production by eNOS. *American Journal of Physiology. Heart and Circulatory Physiology*, *294*(4), H1530–H1540.

Czernichow, S., Greenfield, J. R., Galan, P., Bastard, J. P., Charnaux, N., Samaras, K., et al. (2010). Microvascular dysfunction in healthy insulin-sensitive overweight individuals. *Journal of Hypertension*, *28*(2), 325–332.

De Toni, L., Strapazzon, G., Gianesello, L., Caretta, N., Pilon, C., Bruttocao, A., et al. (2011). Effects of type 5-phosphodiesterase inhibition on energy metabolism and mitochondrial biogenesis in human adipose tissue ex vivo. *Journal of Endocrinological Investigation*, *34*(10), 738–741.

Dikalova, A. E., Gongora, M. C., Harrison, D. G., Lambeth, J. D., Dikalov, S., & Griendling, K. K. (2010). Upregulation of Nox1 in vascular smooth muscle leads to impaired endothelium-dependent relaxation via eNOS uncoupling. *American Journal of Physiology. Heart and Circulatory Physiology*, *299*(3), H673–H679.

Dimmeler, S., Fleming, I., Fisslthaler, B., Hermann, C., Busse, R., & Zeiher, A. M. (1999). Activation of nitric oxide synthase in endothelial cells by Akt-dependent phosphorylation. *Nature*, *399*(6736), 601–605.

Ding, H., & Triggle, C. R. (2010). Endothelial dysfunction in diabetes: Multiple targets for treatment. *Pflügers Archiv*, *459*(6), 977–994.

Du, X. L., Edelstein, D., Dimmeler, S., Ju, Q., Sui, C., & Brownlee, M. (2001). Hyperglycemia inhibits endothelial nitric oxide synthase activity by posttranslational modification at the Akt site. *Journal of Clinical Investigation*, *108*(9), 1341–1348.

Du, X., Edelstein, D., Obici, S., Higham, N., Zou, M. H., & Brownlee, M. (2006). Insulin resistance reduces arterial prostacyclin synthase and eNOS activities by increasing endothelial fatty acid oxidation. *Journal of Clinical Investigation*, *116*(4), 1071–1080.

Dudzinski, D. M., & Michel, T. (2007). Life history of eNOS: Partners and pathways. *Cardiovascular Research*, *75*(2), 247–260.

Duffey, K. J., & Popkin, B. M. (2007). Shifts in patterns and consumption of beverages between 1965 and 2002. *Obesity (Silver Spring)*, *15*(11), 2739–2747.

Duncan, E. R., Crossey, P. A., Walker, S., Anilkumar, N., Poston, L., Douglas, G., et al. (2008). Effect of endothelium-specific insulin resistance on endothelial function in vivo. *Diabetes*, *57*(12), 3307–3314.

Duplain, H., Burcelin, R., Sartori, C., Cook, S., Egli, M., Lepori, M., et al. (2001). Insulin resistance, hyperlipidemia, and hypertension in mice lacking endothelial nitric oxide synthase. *Circulation*, *104*(3), 342–345.

Edirisinghe, I., McCormick Hallam, K., & Kappagoda, C. T. (2006). Effect of fatty acids on endothelium-dependent relaxation in the rabbit aorta. *Clinical Science (London)*, *111*(2), 145–151.

El Assar, M., Ruiz de Adana, J. C., Angulo, J., Pindado Martinez, M. L., Hernandez Matias, A., & Rodriguez-Manas, L. (2013). Preserved endothelial function in human obesity in the absence of insulin resistance. *Journal of Translational Medicine*, *11*, 263.

Elrod, J. W., Duranski, M. R., Langston, W., Greer, J. J., Tao, L., Dugas, T. R., et al. (2006). eNOS gene therapy exacerbates hepatic ischemia-reperfusion injury in diabetes: A role for eNOS uncoupling. *Circulation Research*, *99*(1), 78–85.

Ervin, R. B. (2009). Prevalence of metabolic syndrome among adults 20 years of age and over, by sex, age, race and ethnicity, and body mass index: United States, 2003–2006. *National Health Statistics Report*, (13), 1–7.

Fang, S., Suh, J. M., Atkins, A. R., Hong, S. H., Leblanc, M., Nofsinger, R. R., et al. (2011). Corepressor SMRT promotes oxidative phosphorylation in adipose tissue and protects against diet-induced obesity and insulin resistance. *Proceedings of the National Academy of Sciences of the United States of America*, *108*(8), 3412–3417.

Forstermann, U., & Li, H. (2011). Therapeutic effect of enhancing endothelial nitric oxide synthase (eNOS) expression and preventing eNOS uncoupling. *British Journal of Pharmacology*, *164*(2), 213–223.

Forstermann, U., Mulsch, A., Bohme, E., & Busse, R. (1986). Stimulation of soluble guanylate cyclase by an acetylcholine-induced endothelium-derived factor from rabbit and canine arteries. *Circulation Research*, *58*(4), 531–538.

Forstermann, U., & Sessa, W. C. (2012). Nitric oxide synthases: Regulation and function. *European Heart Journal*, *33*(7), 829–837, 837a-837d.

Fu, W. J., Haynes, T. E., Kohli, R., Hu, J., Shi, W., Spencer, T. E., et al. (2005). Dietary L-arginine supplementation reduces fat mass in Zucker diabetic fatty rats. *Journal of Nutrition*, *135*(4), 714–721.

Furchgott, R. F., & Zawadzki, J. V. (1980). The obligatory role of endothelial cells in the relaxation of arterial smooth muscle by acetylcholine. *Nature*, *288*(5789), 373–376.

Garcia-Villafranca, J., Guillen, A., & Castro, J. (2003). Involvement of nitric oxide/cyclic GMP signaling pathway in the regulation of fatty acid metabolism in rat hepatocytes. *Biochemical Pharmacology*, *65*(5), 807–812.

Georgescu, A., Popov, D., Constantin, A., Nemecz, M., Alexandru, N., Cochior, D., et al. (2011). Dysfunction of human subcutaneous fat arterioles in obesity alone or obesity associated with type 2 diabetes. *Clinical Science (London)*, *120*(10), 463–472.

Gonzalez-Sanchez, J. L., Martinez-Larrad, M. T., Saez, M. E., Zabena, C., Martinez-Calatrava, M. J., & Serrano-Rios, M. (2007). Endothelial nitric oxide synthase haplotypes are associated with features of metabolic syndrome. *Clinical Chemistry*, *53*(1), 91–97.

Goto, T., Ohnomi, S., Khedara, A., Kato, N., Ogawa, H., & Yanagita, T. (1999). Feeding the nitric oxide synthase inhibitor L-N(omega)nitroarginine elevates serum very low density lipoprotein and hepatic triglyceride synthesis in rats. *Journal of Nutrition and Biochemistry*, *10*(5), 274–278.

Gould, N., Doulias, P. T., Tenopoulou, M., Raju, K., & Ischiropoulos, H. (2013). Regulation of protein function and signaling by reversible cysteine S-nitrosylation. *Journal of Biological Chemistry*, *288*(37), 26473–26479.

Grassi, G., Seravalle, G., Scopelliti, F., Dell'Oro, R., Fattori, L., Quarti-Trevano, F., et al. (2010). Structural and functional alterations of subcutaneous small resistance arteries in severe human obesity. *Obesity (Silver Spring)*, *18*(1), 92–98.

Gruber, H. J., Mayer, C., Mangge, H., Fauler, G., Grandits, N., & Wilders-Truschnig, M. (2008). Obesity reduces the bioavailability of nitric oxide in juveniles. *International Journal of Obesity*, *32*(5), 826–831.

Gryglewski, R. J., Moncada, S., & Palmer, R. M. (1986). Bioassay of prostacyclin and endothelium-derived relaxing factor (EDRF) from porcine aortic endothelial cells. *British Journal of Pharmacology*, *87*(4), 685–694.

Gupta, A. K., Cornelissen, G., Greenway, F. L., Dhoopati, V., Halberg, F., & Johnson, W. D. (2010). Abnormalities in circadian blood pressure variability and endothelial function: Pragmatic markers for adverse cardiometabolic profiles in asymptomatic obese adults. *Cardiovascular Diabetology, 9*, 58.

Han, K. A., Patel, Y., Lteif, A. A., Chisholm, R., & Mather, K. J. (2011). Contributions of dysglycaemia, obesity, and insulin resistance to impaired endothelium-dependent vasodilation in humans. *Diabetes/Metabolism Research and Reviews, 27*(4), 354–361.

Hermann, C., Assmus, B., Urbich, C., Zeiher, A. M., & Dimmeler, S. (2000). Insulin-mediated stimulation of protein kinase Akt: A potent survival signaling cascade for endothelial cells. *Arteriosclerosis, Thrombosis, and Vascular Biology, 20*(2), 402–409.

Higashi, Y., Sasaki, S., Nakagawa, K., Matsuura, H., Chayama, K., & Oshima, T. (2001). Effect of obesity on endothelium-dependent, nitric oxide-mediated vasodilation in normotensive individuals and patients with essential hypertension. *American Journal of Hypertension, 14*(10), 1038–1045.

Hill, B. G., & Bhatnagar, A. (2007). Role of glutathiolation in preservation, restoration and regulation of protein function. *IUBMB Life, 59*(1), 21–26.

Hill, B. G., & Bhatnagar, A. (2012). Protein S-glutathiolation: Redox-sensitive regulation of protein function. *Journal of Molecular and Cellular Cardiology, 52*(3), 559–567.

Hill, B. G., Dranka, B. P., Bailey, S. M., Lancaster, J. R., Jr., & Darley-Usmar, V. M. (2010). What part of NO don't you understand? Some answers to the cardinal questions in nitric oxide biology. *Journal of Biological Chemistry, 285*(26), 19699–19704.

Hotamisligil, G. S., Shargill, N. S., & Spiegelman, B. M. (1993). Adipose expression of tumor necrosis factor-alpha: Direct role in obesity-linked insulin resistance. *Science, 259*(5091), 87–91.

Hu, F. B., Li, T. Y., Colditz, G. A., Willett, W. C., & Manson, J. E. (2003). Television watching and other sedentary behaviors in relation to risk of obesity and type 2 diabetes mellitus in women. *JAMA, 289*(14), 1785–1791.

Huang, P. L. (2009). eNOS, metabolic syndrome and cardiovascular disease. *Trends in Endocrinology and Metabolism, 20*(6), 295–302.

Hwang, J. H., Kim, D. W., Jo, E. J., Kim, Y. K., Jo, Y. S., Park, J. H., et al. (2009). Pharmacological stimulation of NADH oxidation ameliorates obesity and related phenotypes in mice. *Diabetes, 58*(4), 965–974.

Jang, H. J., Kim, H. S., Hwang, D. H., Quon, M. J., & Kim, J. A. (2013). Toll-like receptor 2 mediates high-fat diet-induced impairment of vasodilator actions of insulin. *American Journal of Physiology, Endocrinology and Metabolism, 304*(10), E1077–E1088.

Jobgen, W., Meininger, C. J., Jobgen, S. C., Li, P., Lee, M. J., Smith, S. B., et al. (2009). Dietary L-arginine supplementation reduces white fat gain and enhances skeletal muscle and brown fat masses in diet-induced obese rats. *Journal of Nutrition, 139*(2), 230–237.

Ju, H., Zou, R., Venema, V. J., & Venema, R. C. (1997). Direct interaction of endothelial nitric-oxide synthase and caveolin-1 inhibits synthase activity. *Journal of Biological Chemistry, 272*(30), 18522–18525.

Kant, A. K., & Graubard, B. I. (2004). Eating out in America, 1987–2000: Trends and nutritional correlates. *Preventive Medicine, 38*(2), 243–249.

Kashiwagi, S., Atochin, D. N., Li, Q., Schleicher, M., Pong, T., Sessa, W. C., et al. (2013). eNOS phosphorylation on serine 1176 affects insulin sensitivity and adiposity. *Biochemical and Biophysical Research Communications, 431*(2), 284–290.

Kelly, D. P., & Scarpulla, R. C. (2004). Transcriptional regulatory circuits controlling mitochondrial biogenesis and function. *Genes and Development, 18*(4), 357–368.

Khedara, A., Goto, T., Morishima, M., Kayashita, J., & Kato, N. (1999). Elevated body fat in rats by the dietary nitric oxide synthase inhibitor, L-N omega nitroarginine. *Bioscience, Biotechnology, and Biochemistry, 63*(4), 698–702.

Khedara, A., Kawai, Y., Kayashita, J., & Kato, N. (1996). Feeding rats the nitric oxide synthase inhibitor, L-N(omega)nitroarginine, elevates serum triglyceride and cholesterol and lowers hepatic fatty acid oxidation. *Journal of Nutrition, 126*(10), 2563–2567.

Kietadisorn, R., Juni, R. P., & Moens, A. L. (2012). Tackling endothelial dysfunction by modulating NOS uncoupling: New insights into its pathogenesis and therapeutic possibilities. *American Journal of Physiology, Endocrinology and Metabolism, 302*(5), E481–E495.

Kim, F., Pham, M., Luttrell, I., Bannerman, D. D., Tupper, J., Thayer, J., et al. (2007). Toll-like receptor-4 mediates vascular inflammation and insulin resistance in diet-induced obesity. *Circulation Research, 100*(11), 1589–1596.

Kim, F., Pham, M., Maloney, E., Rizzo, N. O., Morton, G. J., Wises, B. E., et al. (2008). Vascular inflammation, insulin resistance, and reduced nitric oxide production precede the onset of peripheral insulin resistance. *Arteriosclerosis, Thrombosis, and Vascular Biology, 28*(11), 1982–1988.

Kim, F., Tinseling, K. A., Rice, J., Pham, M., Hajji, L., Gallis, B. M., et al. (2005). Free fatty acid impairment of nitric oxide production in endothelial cells is mediated by IKKbeta. *Arteriosclerosis, Thrombosis, and Vascular Biology, 25*(5), 989–994.

Kolluru, G. K., Siamwala, J. H., & Chatterjee, S. (2010). eNOS phosphorylation in health and disease. *Biochimie, 92*(9), 1186–1198.

Kone, B. C., Kuncewicz, T., Zhang, W., & Yu, Z. Y. (2003). Protein interactions with nitric oxide synthases: Controlling the right time, the right place, and the right amount of nitric oxide. *American Journal of Physiology. Renal Physiology, 285*(2), F178–F190.

Kraus, R. M., Houmard, J. A., Kraus, W. E., Tanner, C. J., Pierce, J. R., Choi, M. D., et al. (2012). Obesity, insulin resistance, and skeletal muscle nitric oxide synthase. *Journal of Applied Physiology (Bethesda, Md. 1985), 113*(5), 758–765.

Kubota, T., Kubota, N., Kumagai, H., Yamaguchi, S., Kozono, H., Takahashi, T., et al. (2011). Impaired insulin signaling in endothelial cells reduces insulin-induced glucose uptake by skeletal muscle. *Cell Metabolism, 13*(3), 294–307.

Lai, P. F., Mohamed, F., Monge, J. C., & Stewart, D. J. (2003). Downregulation of eNOS mRNA expression by TNFalpha: Identification and functional characterization of RNA-protein interactions in the 3'UTR. *Cardiovascular Research, 59*(1), 160–168.

Laine, H., Yki-Jarvinen, H., Kirvela, O., Tolvanen, T., Raitakari, M., Solin, O., et al. (1998). Insulin resistance of glucose uptake in skeletal muscle cannot be ameliorated by enhancing endothelium-dependent blood flow in obesity. *Journal of Clinical Investigation, 101*(5), 1156–1162.

Lambert, E., Sari, C. I., Dawood, T., Nguyen, J., McGrane, M., Eikelis, N., et al. (2010). Sympathetic nervous system activity is associated with obesity-induced subclinical organ damage in young adults. *Hypertension, 56*(3), 351–358.

Landmesser, U., Dikalov, S., Price, S. R., McCann, L., Fukai, T., Holland, S. M., et al. (2003). Oxidation of tetrahydrobiopterin leads to uncoupling of endothelial cell nitric oxide synthase in hypertension. *Journal of Clinical Investigation, 111*(8), 1201–1209.

Le Gouill, E., Jimenez, M., Binnert, C., Jayet, P. Y., Thalmann, S., Nicod, P., et al. (2007). Endothelial nitric oxide synthase (eNOS) knockout mice have defective mitochondrial beta-oxidation. *Diabetes, 56*(11), 2690–2696.

Lefebvre, P., Chinetti, G., Fruchart, J. C., & Staels, B. (2006). Sorting out the roles of PPAR alpha in energy metabolism and vascular homeostasis. *Journal of Clinical Investigation, 116*(3), 571–580.

Lei, H., Luo, S., Qin, H., & Xia, Y. (2013). Molecular mechanisms of endothelial NO synthase uncoupling. *Current Pharmaceutical Design.*

Li, Q., Atochin, D., Kashiwagi, S., Earle, J., Wang, A., Mandeville, E., et al. (2013). Deficient eNOS phosphorylation is a mechanism for diabetic vascular dysfunction contributing to increased stroke size. *Stroke, 44*(11), 3183–3188.

Li, H., & Forstermann, U. (2013). Uncoupling of endothelial NO synthase in atherosclerosis and vascular disease. *Current Opinion in Pharmacology*, *13*(2), 161–167.
Li, H., & Poulos, T. L. (2005). Structure-function studies on nitric oxide synthases. *Journal of Inorganic Biochemistry*, *99*(1), 293–305.
Low Wang, C. C., Lu, L., Leitner, J. W., Sarraf, M., Gianani, R., Draznin, B., et al. (2013). Arterial insulin resistance in Yucatan micropigs with diet-induced obesity and metabolic syndrome. *Journal of Diabetes and its Complications*, *27*(4), 307–315.
Lucotti, P., Setola, E., Monti, L. D., Galluccio, E., Costa, S., Sandoli, E. P., et al. (2006). Beneficial effects of a long-term oral L-arginine treatment added to a hypocaloric diet and exercise training program in obese, insulin-resistant type 2 diabetic patients. *American Journal of Physiology, Endocrinology and Metabolism*, *291*(5), E906–E912.
Mahmud, F. H., Hill, D. J., Cuerden, M. S., & Clarson, C. L. (2009). Impaired vascular function in obese adolescents with insulin resistance. *Journal of Pediatrics*, *155*(5), 678–682.
McCabe, T. J., Fulton, D., Roman, L. J., & Sessa, W. C. (2000). Enhanced electron flux and reduced calmodulin dissociation may explain "calcium-independent" eNOS activation by phosphorylation. *Journal of Biological Chemistry*, *275*(9), 6123–6128.
Meininger, C. J., Cai, S., Parker, J. L., Channon, K. M., Kelly, K. A., Becker, E. J., et al. (2004). GTP cyclohydrolase I gene transfer reverses tetrahydrobiopterin deficiency and increases nitric oxide synthesis in endothelial cells and isolated vessels from diabetic rats. *FASEB Journal*, *18*(15), 1900–1902.
Meininger, C. J., Marinos, R. S., Hatakeyama, K., Martinez-Zaguilan, R., Rojas, J. D., Kelly, K. A., et al. (2000). Impaired nitric oxide production in coronary endothelial cells of the spontaneously diabetic BB rat is due to tetrahydrobiopterin deficiency. *Biochemical Journal*, *349*(Pt. 1), 353–356.
Meshkani, R., & Adeli, K. (2009). Hepatic insulin resistance, metabolic syndrome and cardiovascular disease. *Clinical Biochemistry*, *42*(13–14), 1331–1346.
Miadi-Messaoud, H., Chouchane, A., Abderrazek, E., Debbabi, H., Zaouali-Ajina, M., Tabka, Z., et al. (2010). Obesity-induced impairment of endothelium-dependent vasodilation in Tunisian women. *International Journal of Obesity*, *34*(2), 273–279.
Michel, J. B., Feron, O., Sacks, D., & Michel, T. (1997). Reciprocal regulation of endothelial nitric-oxide synthase by Ca^{2+}–calmodulin and caveolin. *Journal of Biological Chemistry*, *272*(25), 15583–15586.
Michel, T., & Lamas, S. (1992). Molecular cloning of constitutive endothelial nitric oxide synthase: Evidence for a family of related genes. *Journal of Cardiovascular Pharmacology*, *20*(Suppl. 12), S45–S49.
Mitschke, M. M., Hoffmann, L. S., Gnad, T., Scholz, D., Kruithoff, K., Mayer, P., et al. (2013). Increased cGMP promotes healthy expansion and browning of white adipose tissue. *FASEB Journal*, *27*(4), 1621–1630.
Miyashita, K., Itoh, H., Tsujimoto, H., Tamura, N., Fukunaga, Y., Sone, M., et al. (2009). Natriuretic peptides/cGMP/cGMP-dependent protein kinase cascades promote muscle mitochondrial biogenesis and prevent obesity. *Diabetes*, *58*(12), 2880–2892.
Molnar, J., Yu, S., Mzhavia, N., Pau, C., Chereshnev, I., & Dansky, H. M. (2005). Diabetes induces endothelial dysfunction but does not increase neointimal formation in high-fat diet fed C57BL/6J mice. *Circulation Research*, *96*(11), 1178–1184.
Moncada, S., Palmer, R. M., & Gryglewski, R. J. (1986). Mechanism of action of some inhibitors of endothelium-derived relaxing factor. *Proceedings of the National Academy of Sciences of the United States of America*, *83*(23), 9164–9168.
Monti, L. D., Barlassina, C., Citterio, L., Galluccio, E., Berzuini, C., Setola, E., et al. (2003). Endothelial nitric oxide synthase polymorphisms are associated with type 2 diabetes and the insulin resistance syndrome. *Diabetes*, *52*(5), 1270–1275.
Monti, L. D., Casiraghi, M. C., Setola, E., Galluccio, E., Pagani, M. A., Quaglia, L., et al. (2013). L-arginine enriched biscuits improve endothelial function and glucose

metabolism: A pilot study in healthy subjects and a cross-over study in subjects with impaired glucose tolerance and metabolic syndrome. *Metabolism, 62*(2), 255–264.

Morley, J. E., & Flood, J. F. (1994). Effect of competitive antagonism of NO synthetase on weight and food intake in obese and diabetic mice. *American Journal of Physiology, 266*(1 Pt 2), R164–R168.

Muniyappa, R., Montagnani, M., Koh, K. K., & Quon, M. J. (2007). Cardiovascular actions of insulin. *Endocrine Reviews, 28*(5), 463–491.

Nakata, S., Tsutsui, M., Shimokawa, H., Suda, O., Morishita, T., Shibata, K., et al. (2008). Spontaneous myocardial infarction in mice lacking all nitric oxide synthase isoforms. *Circulation, 117*(17), 2211–2223.

Naruse, K., Rask-Madsen, C., Takahara, N., Ha, S. W., Suzuma, K., Way, K. J., et al. (2006). Activation of vascular protein kinase C-beta inhibits Akt-dependent endothelial nitric oxide synthase function in obesity-associated insulin resistance. *Diabetes, 55*(3), 691–698.

Neumann, P., Gertzberg, N., & Johnson, A. (2004). TNF-alpha induces a decrease in eNOS promoter activity. *American Journal of Physiology. Lung Cellular and Molecular Physiology, 286*(2), L452–L459.

Nielsen, S. J., & Popkin, B. M. (2003). Patterns and trends in food portion sizes, 1977–1998. *JAMA, 289*(4), 450–453.

Nisoli, E., Clementi, E., Paolucci, C., Cozzi, V., Tonello, C., Sciorati, C., et al. (2003). Mitochondrial biogenesis in mammals: The role of endogenous nitric oxide. *Science, 299*(5608), 896–899.

Nisoli, E., Falcone, S., Tonello, C., Cozzi, V., Palomba, L., Fiorani, M., et al. (2004). Mitochondrial biogenesis by NO yields functionally active mitochondria in mammals. *Proceedings of the National Academy of Sciences of the United States of America, 101*(47), 16507–16512.

Oess, S., Icking, A., Fulton, D., Govers, R., & Muller-Esterl, W. (2006). Subcellular targeting and trafficking of nitric oxide synthases. *Biochemical Journal, 396*(3), 401–409.

Ogden, C. L., Carroll, M. D., Kit, B. K., & Flegal, K. M. (2012). Prevalence of obesity in the United States, 2009–2010. *NCHS Data Brief, 00*(82), 1–8.

Ohtoshi, K., Yamasaki, Y., Gorogawa, S., Hayaishi-Okano, R., Node, K., Matsuhisa, M., et al. (2002). Association of (−)786T-C mutation of endothelial nitric oxide synthase gene with insulin resistance. *Diabetologia, 45*(11), 1594–1601.

Pacher, P., Beckman, J. S., & Liaudet, L. (2007). Nitric oxide and peroxynitrite in health and disease. *Physiological Reviews, 87*(1), 315–424.

Palmer, R. M., Ferrige, A. G., & Moncada, S. (1987). Nitric oxide release accounts for the biological activity of endothelium-derived relaxing factor. *Nature, 327*(6122), 524–526.

Pannirselvam, M., Verma, S., Anderson, T. J., & Triggle, C. R. (2002). Cellular basis of endothelial dysfunction in small mesenteric arteries from spontaneously diabetic (db/db −/−) mice: Role of decreased tetrahydrobiopterin bioavailability. *British Journal of Pharmacology, 136*(2), 255–263.

Parikh, N. I., Keyes, M. J., Larson, M. G., Pou, K. M., Hamburg, N. M., Vita, J. A., et al. (2009). Visceral and subcutaneous adiposity and brachial artery vasodilator function. *Obesity (Silver Spring), 17*(11), 2054–2059.

Park, K., Li, Q., Rask-Madsen, C., Mima, A., Mizutani, K., Winnay, J., et al. (2013). Serine phosphorylation sites on IRS2 activated by angiotensin II and protein kinase C to induce selective insulin resistance in endothelial cells. *Molecular and Cellular Biology, 33*(16), 3227–3241.

Perez-Matute, P., Neville, M. J., Tan, G. D., Frayn, K. N., & Karpe, F. (2009). Transcriptional control of human adipose tissue blood flow. *Obesity (Silver Spring), 17*(4), 681–688.

Piantadosi, C. A., & Suliman, H. B. (2012). Redox regulation of mitochondrial biogenesis. *Free Radical Biology & Medicine, 53*(11), 2043–2053.

Popkin, B. M., Armstrong, L. E., Bray, G. M., Caballero, B., Frei, B., & Willett, W. C. (2006). A new proposed guidance system for beverage consumption in the United States. *American Journal of Clinical Nutrition, 83*(3), 529–542.

Rafikov, R., Fonseca, F. V., Kumar, S., Pardo, D., Darragh, C., Elms, S., et al. (2011). ENOS activation and NO function: Structural motifs responsible for the posttranslational control of endothelial nitric oxide synthase activity. *Journal of Endocrinology, 210*(3), 271–284.

Rajfer, J., Aronson, W. J., Bush, P. A., Dorey, F. J., & Ignarro, L. J. (1992). Nitric oxide as a mediator of relaxation of the corpus cavernosum in response to nonadrenergic, noncholinergic neurotransmission. *New England Journal of Medicine, 326*(2), 90–94.

Rapoport, R. M., Draznin, M. B., & Murad, F. (1983). Endothelium-dependent relaxation in rat aorta may be mediated through cyclic GMP-dependent protein phosphorylation. *Nature, 306*(5939), 174–176.

Rapoport, R. M., & Murad, F. (1983). Agonist-induced endothelium-dependent relaxation in rat thoracic aorta may be mediated through cGMP. *Circulation Research, 52*(3), 352–357.

Reaven, G. M., & Chen, Y. D. (1996). Insulin resistance, its consequences, and coronary heart disease. Must we choose one culprit? *Circulation, 93*(10), 1780–1783.

Recchia, F. A., Osorio, J. C., Chandler, M. P., Xu, X., Panchal, A. R., Lopaschuk, G. D., et al. (2002). Reduced synthesis of NO causes marked alterations in myocardial substrate metabolism in conscious dogs. *American Journal of Physiology, Endocrinology and Metabolism, 282*(1), E197–E206.

Risbano, M. G., & Gladwin, M. T. (2013). Therapeutics targeting of dysregulated redox equilibrium and endothelial dysfunction. *Handbook of Experimental Pharmacology, 218*, 315–349.

Rittig, K., Holder, K., Stock, J., Tschritter, O., Peter, A., Stefan, N., et al. (2008). Endothelial NO-synthase intron 4 polymorphism is associated with disturbed in vivo nitric oxide production in individuals prone to type 2 diabetes. *Hormone and Metabolic Research, 40*(1), 13–17.

Robinson, T. N. (1999). Reducing children's television viewing to prevent obesity: A randomized controlled trial. *JAMA, 282*(16), 1561–1567.

Rodriguez-Crespo, I., Gerber, N. C., & Ortiz de Montellano, P. R. (1996). Endothelial nitric-oxide synthase. Expression in Escherichia coli, spectroscopic characterization, and role of tetrahydrobiopterin in dimer formation. *Journal of Biological Chemistry, 271*(19), 11462–11467.

Rodriguez-Crespo, I., & Ortiz de Montellano, P. R. (1996). Human endothelial nitric oxide synthase: Expression in Escherichia coli, coexpression with calmodulin, and characterization. *Archives of Biochemistry and Biophysics, 336*(1), 151–156.

Roger, V. L., Go, A. S., Lloyd-Jones, D. M., Benjamin, E. J., Berry, J. D., Borden, W. B., et al. (2011). Heart disease and stroke statistics–2012 update: A report from the American Heart Association. *Circulation.*

Sanchez de Miguel, L., Alonso, J., Gonzalez-Fernandez, F., de la Osada, J., Monton, M., Rodriguez-Feo, J. A., et al. (1999). Evidence that an endothelial cytosolic protein binds to the 3'-untranslated region of endothelial nitric oxide synthase mRNA. *Journal of Vascular Research, 36*(3), 201–208.

Sansbury, B. E., Cummins, T. D., Tang, Y., Hellmann, J., Holden, C. R., Harbeson, M. A., et al. (2012). Overexpression of endothelial nitric oxide synthase prevents diet-induced obesity and regulates adipocyte phenotype. *Circulation Research, 111*(9), 1176–1189.

Satoh, M., Fujimoto, S., Haruna, Y., Arakawa, S., Horike, H., Komai, N., et al. (2005). NAD(P)H oxidase and uncoupled nitric oxide synthase are major sources of glomerular superoxide in rats with experimental diabetic nephropathy. *American Journal of Physiology. Renal Physiology, 288*(6), F1144–F1152.

Schild, L., Dombrowski, F., Lendeckel, U., Schulz, C., Gardemann, A., & Keilhoff, G. (2008). Impairment of endothelial nitric oxide synthase causes abnormal fat and glycogen deposition in liver. *Biochimica et Biophysica Acta*, *1782*(3), 180–187.

Sessa, W. C. (2004). eNOS at a glance. *Journal of Cell Science*, *117*(Pt 12), 2427–2429.

Shankar, R. R., Wu, Y., Shen, H. Q., Zhu, J. S., & Baron, A. D. (2000). Mice with gene disruption of both endothelial and neuronal nitric oxide synthase exhibit insulin resistance. *Diabetes*, *49*(5), 684–687.

Shinozaki, K., Kashiwagi, A., Nishio, Y., Okamura, T., Yoshida, Y., Masada, M., et al. (1999). Abnormal biopterin metabolism is a major cause of impaired endothelium-dependent relaxation through nitric oxide/O2- imbalance in insulin-resistant rat aorta. *Diabetes*, *48*(12), 2437–2445.

Shiva, S., Oh, J. Y., Landar, A. L., Ulasova, E., Venkatraman, A., Bailey, S. M., et al. (2005). Nitroxia: The pathological consequence of dysfunction in the nitric oxide-cytochrome c oxidase signaling pathway. *Free Radical Biology and Medicine*, *38*(3), 297–306.

Smith, D. E., Lewis, C. E., Caveny, J. L., Perkins, L. L., Burke, G. L., & Bild, D. E. (1994). Longitudinal changes in adiposity associated with pregnancy. The CARDIA study. Coronary artery risk development in young adults study. *JAMA*, *271*(22), 1747–1751.

Steinberg, H. O., Chaker, H., Leaming, R., Johnson, A., Brechtel, G., & Baron, A. D. (1996). Obesity/insulin resistance is associated with endothelial dysfunction. Implications for the syndrome of insulin resistance. *Journal of Clinical Investigation*, *97*(11), 2601–2610.

Steinberg, H. O., Paradisi, G., Hook, G., Crowder, K., Cronin, J., & Baron, A. D. (2000). Free fatty acid elevation impairs insulin-mediated vasodilation and nitric oxide production. *Diabetes*, *49*(7), 1231–1238.

Steinberg, H. O., Tarshoby, M., Monestel, R., Hook, G., Cronin, J., Johnson, A., et al. (1997). Elevated circulating free fatty acid levels impair endothelium-dependent vasodilation. *Journal of Clinical Investigation*, *100*(5), 1230–1239.

Stricker-Krongrad, A., Beck, B., & Burlet, C. (1996). Nitric oxide mediates hyperphagia of obese Zucker rats: Relation to specific changes in the microstructure of feeding behavior. *Life Sciences*, *58*(1), L9–L15.

Sturm, W., Sandhofer, A., Engl, J., Laimer, M., Molnar, C., Kaser, S., et al. (2009). Influence of visceral obesity and liver fat on vascular structure and function in obese subjects. *Obesity (Silver Spring)*, *17*(9), 1783–1788.

Suliburska, J., Bogdanski, P., Szulinska, M., Pupek-Musialik, D., & Jablecka, A. (2014). Changes in mineral status are associated with improvements in insulin sensitivity in obese patients following L-arginine supplementation. *European Journal of Nutrition*, *53*(2), 387–393.

Symons, J. D., McMillin, S. L., Riehle, C., Tanner, J., Palionyte, M., Hillas, E., et al. (2009). Contribution of insulin and Akt1 signaling to endothelial nitric oxide synthase in the regulation of endothelial function and blood pressure. *Circulation Research*, *104*(9), 1085–1094.

Szabo, C., Ischiropoulos, H., & Radi, R. (2007). Peroxynitrite: Biochemistry, pathophysiology and development of therapeutics. *Nature Reviews. Drug Discovery*, *6*(8), 662–680.

Tack, C. J., Ong, M. K., Lutterman, J. A., & Smits, P. (1998). Insulin-induced vasodilatation and endothelial function in obesity/insulin resistance. Effects of troglitazone. *Diabetologia*, *41*(5), 569–576.

Taguchi, K., Kobayashi, T., Matsumoto, T., & Kamata, K. (2011). Dysfunction of endothelium-dependent relaxation to insulin via PKC-mediated GRK2/Akt activation in aortas of ob/ob mice. *American Journal of Physiology. Heart and Circulatory Physiology*, *301*(2), H571–H583.

Tan, B., Yin, Y., Liu, Z., Li, X., Xu, H., Kong, X., et al. (2009). Dietary L-arginine supplementation increases muscle gain and reduces body fat mass in growing-finishing pigs. *Amino Acids*, *37*(1), 169–175.

Thomas, D. D., Liu, X., Kantrow, S. P., & Lancaster, J. R., Jr. (2001). The biological lifetime of nitric oxide: Implications for the perivascular dynamics of NO and O_2. *Proceedings of the National Academy of Sciences of the United States of America*, *98*(1), 355–360.

Toutouzas, K., Riga, M., Stefanadi, E., & Stefanadis, C. (2008). Asymmetric dimethylarginine (ADMA) and other endogenous nitric oxide synthase (NOS) inhibitors as an important cause of vascular insulin resistance. *Hormone and Metabolic Research*, *40*(9), 655–659.

Trujillo, M., Ferrer-Sueta, G., & Radi, R. (2008). Peroxynitrite detoxification and its biologic implications. *Antioxidants and Redox Signaling*, *10*(9), 1607–1620.

Tsuchiya, K., Sakai, H., Suzuki, N., Iwashima, F., Yoshimoto, T., Shichiri, M., et al. (2007). Chronic blockade of nitric oxide synthesis reduces adiposity and improves insulin resistance in high fat-induced obese mice. *Endocrinology*, *148*(10), 4548–4556.

Valerio, A., Cardile, A., Cozzi, V., Bracale, R., Tedesco, L., Pisconti, A., et al. (2006). TNF-alpha downregulates eNOS expression and mitochondrial biogenesis in fat and muscle of obese rodents. *Journal of Clinical Investigation*, *116*(10), 2791–2798.

Van Guilder, G. P., Stauffer, B. L., Greiner, J. J., & Desouza, C. A. (2008). Impaired endothelium-dependent vasodilation in overweight and obese adult humans is not limited to muscarinic receptor agonists. *American Journal of Physiology. Heart and Circulatory Physiology*, *294*(4), H1685–H1692.

Vecoli, C., Andreassi, M. G., Liga, R., Colombo, M. G., Coceani, M., Carpeggiani, C., et al. (2012). T(−786)−>C polymorphism of the endothelial nitric oxide synthase gene is associated with insulin resistance in patients with ischemic or non ischemic cardiomyopathy. *BMC Medical Genomics*, *13*, 92.

Viollet, B., Guigas, B., Leclerc, J., Hebrard, S., Lantier, L., Mounier, R., et al. (2009). AMP-activated protein kinase in the regulation of hepatic energy metabolism: From physiology to therapeutic perspectives. *Acta Physiologica (Oxford, England)*, *196*(1), 81–98.

Wang, Y. C., Bleich, S. N., & Gortmaker, S. L. (2008). Increasing caloric contribution from sugar-sweetened beverages and 100% fruit juices among US children and adolescents, 1988–2004. *Pediatrics*, *121*(6), e1604–e1614.

Wang, H., Wang, A. X., Aylor, K., & Barrett, E. J. (2013). Nitric oxide directly promotes vascular endothelial insulin transport. *Diabetes*, *62*(12), 4030–4042.

Wascher, T. C., Graier, W. F., Dittrich, P., Hussain, M. A., Bahadori, B., Wallner, S., et al. (1997). Effects of low-dose L-arginine on insulin-mediated vasodilatation and insulin sensitivity. *European Journal of Clinical Investigation*, *27*(8), 690–695.

Weil, B. R., Westby, C. M., Van Guilder, G. P., Greiner, J. J., Stauffer, B. L., & DeSouza, C. A. (2011). Enhanced endothelin-1 system activity with overweight and obesity. *American Journal of Physiology. Heart and Circulatory Physiology*, *301*(3), H689–H695.

West, M. B., Hill, B. G., Xuan, Y. T., & Bhatnagar, A. (2006). Protein glutathiolation by nitric oxide: An intracellular mechanism regulating redox protein modification. *FASEB Journal*, *20*(10), 1715–1717.

Westerbacka, J., Vehkavaara, S., Bergholm, R., Wilkinson, I., Cockcroft, J., & Yki-Jarvinen, H. (1999). Marked resistance of the ability of insulin to decrease arterial stiffness characterizes human obesity. *Diabetes*, *48*(4), 821–827.

Williamson, D. F., Madans, J., Anda, R. F., Kleinman, J. C., Kahn, H. S., & Byers, T. (1993). Recreational physical activity and ten-year weight change in a US national cohort. *International Journal of Obesity and Related Metabolic Disorders*, *17*(5), 279–286.

Xu, J., Wu, Y., Song, P., Zhang, M., Wang, S., & Zou, M. H. (2007). Proteasome-dependent degradation of guanosine 5'-triphosphate cyclohydrolase I causes tetrahydrobiopterin deficiency in diabetes mellitus. *Circulation*, *116*(8), 944–953.

Xu, J., Xie, Z., Reece, R., Pimental, D., & Zou, M. H. (2006). Uncoupling of endothelial nitric oxidase synthase by hypochlorous acid: Role of NAD(P)H oxidase-derived superoxide and peroxynitrite. *Arteriosclerosis, Thrombosis, and Vascular Biology, 26*(12), 2688–2695.

Yadav, H., Quijano, C., Kamaraju, A. K., Gavrilova, O., Malek, R., Chen, W., et al. (2011). Protection from obesity and diabetes by blockade of TGF-beta/Smad3 signaling. *Cell Metabolism, 14*(1), 67–79.

Yamamoto, T., Habata, Y., Matsumoto, Y., Yasuhara, Y., Hashimoto, T., Hamajyo, H., et al. (2011). Apelin-transgenic mice exhibit a resistance against diet-induced obesity by increasing vascular mass and mitochondrial biogenesis in skeletal muscle. *Biochimica et Biophysica Acta, 1810*(9), 853–862.

Yan, G., You, B., Chen, S. P., Liao, J. K., & Sun, J. (2008). Tumor necrosis factor-alpha downregulates endothelial nitric oxide synthase mRNA stability via translation elongation factor 1-alpha 1. *Circulation Research, 103*(6), 591–597.

Yang, N., Ying, C., Xu, M., Zuo, X., Ye, X., Liu, L., et al. (2007). High-fat diet up-regulates caveolin-1 expression in aorta of diet-induced obese but not in diet-resistant rats. *Cardiovascular Research, 76*(1), 167–174.

Yoshimura, T., Hisatomi, A., Kajihara, S., Yasutake, T., Ogawa, Y., Mizuta, T., et al. (2003). The relationship between insulin resistance and polymorphisms of the endothelial nitric oxide synthase gene in patients with coronary artery disease. *Journal of Atherosclerosis and Thrombosis, 10*(1), 43–47.

Yoshizumi, M., Perrella, M. A., Burnett, J. C., Jr., & Lee, M. E. (1993). Tumor necrosis factor downregulates an endothelial nitric oxide synthase mRNA by shortening its half-life. *Circulation Research, 73*(1), 205–209.

Zamosky, L. (2013). The obesity epidemic. While America swallows $147 billion in obesity-related healthcare costs, physicians called on to confront the crisis. *Medical Economics, 90*(4), 14–17.

Zecchin, H. G., Priviero, F. B., Souza, C. T., Zecchin, K. G., Prada, P. O., Carvalheira, J. B., et al. (2007). Defective insulin and acetylcholine induction of endothelial cell-nitric oxide synthase through insulin receptor substrate/Akt signaling pathway in aorta of obese rats. *Diabetes, 56*(4), 1014–1024.

Zhang, Q. J., Holland, W. L., Wilson, L., Tanner, J. M., Kearns, D., Cahoon, J. M., et al. (2012). Ceramide mediates vascular dysfunction in diet-induced obesity by PP2A-mediated dephosphorylation of the eNOS-Akt complex. *Diabetes, 61*(7), 1848–1859.

Zhang, Q., Malik, P., Pandey, D., Gupta, S., Jagnandan, D., Belin de Chantemele, E., et al. (2008). Paradoxical activation of endothelial nitric oxide synthase by NADPH oxidase. *Arteriosclerosis, Thrombosis, and Vascular Biology, 28*(9), 1627–1633.

Zhong, J. C., Yu, X. Y., Huang, Y., Yung, L. M., Lau, C. W., & Lin, S. G. (2007). Apelin modulates aortic vascular tone via endothelial nitric oxide synthase phosphorylation pathway in diabetic mice. *Cardiovascular Research, 74*(3), 388–395.

Zweier, J. L., Chen, C. A., & Druhan, L. J. (2011). S-glutathionylation reshapes our understanding of endothelial nitric oxide synthase uncoupling and nitric oxide/reactive oxygen species-mediated signaling. *Antioxidants & Redox Signaling, 14*(10), 1769–1775.

CHAPTER FOURTEEN

Nitric Oxide and Cerebrovascular Regulation

Seyed Nasrollah Tabatabaei[‡], Hélène Girouard[*,†,1]
[*]Department of pharmacology, Faculty of medicine, Université de Montréal, Montreal, Quebec, Canada
[†]Research Center of the Institut Universitaire de Gériatrie de Montréal, Montreal, Quebec, Canada
[‡]Department of Computer and Software Engineering and Institute of Biomedical Engineering, École Polytechnique de Montréal
[1]Corresponding author: e-mail address: helene.girouard@umontreal.ca

Contents

Abstract

The cerebrovascular regulation involves highly complex mechanisms to assure that the brain is perfused at all times. These mechanisms depend on all components of the neurovascular units: neurons, glia, and vascular cells. All these cell types can produce nitric oxide (NO), a powerful vasodilator through different NO synthases. Many studies underlined the key role of NO in the maintenance of resting cerebral blood flow (CBF) as well as in the mechanisms that control cerebrovascular tone: autoregulation and neurovascular coupling. However, although the role of NO in the control of CBF has been largely investigated, the complexity of the NO system and the lack of specific NO synthase inhibitors led to still unresolved questions such as the origin of NO and the

Vitamins and Hormones, Volume 96
ISSN 0083-6729
http://dx.doi.org/10.1016/B978-0-12-800254-4.00014-3

pathways by which it controls the vascular tone. In this chapter, the role of NO in the regulation of CBF is critically reviewed and discussed in the context of the neurovascular unit and the general principles of cerebrovascular regulation.

1. INTRODUCTION

The brain is a complex and heterogenous organ extremely dependent on oxygen and glucose from circulating blood. Hence, to optimize neuronal function and survival, the brain is equipped with fine mechanisms for a precise spatial and temporal control of cerebral blood flow (CBF). The mechanisms underlying CBF regulation have been the subject of inquisition for more than 100 years and involve many vasoactive factors. These factors include ions, metabolic byproducts, vasoactive neurotransmitters and vasoactive mediators derived from glia, blood vessels, or blood itself. One vasoactive mediator was first discovered by Bredt and colleagues (Girouard & Iadecola, 2006) and identified as nitric oxide (NO). NO is a powerful vasodilator originally identified in the endothelium and also produced in the brain. This later led to yet another discovery which identified NO as a neurotransmitter and as a CBF modulator.

CBF is regulated by two main mechanisms: autoregulation and neurovascular coupling (NVC). Autoregulation is an autoprotective mechanism ensuring a constant CBF despite arterial pressure fluctuations between 50 and 160 mm Hg. This mechanism thus in one hand compensates for decreases in arterial pressure in order to maintain a constant oxygen and substrate supply to the brain, and on the other hand, it protects small vessel walls from high arterial pressure. The second main mechanism that regulates CBF is NVC, which is the dynamic link between hemodynamic modulations and energy needs of active neurons. This mechanism depends on the interaction between all components of the neurovascular unit (neurons, astrocytes, and microvessels). Nevertheless, NO plays a central role in CBF regulation by modulating both autoregulation and NVC. In the present chapter, this role within the two mentioned regulatory mechanisms will be critically discussed. Note that, cerebral circulation may be modulated by NO itself, other primary or secondary products of its chemical reaction, or by various substances such as nitrosothiols (S-NO), peroxynitrite, and metal-nitrosyl products (Bredt & Snyder, 1989; Stamler, 1994; Wei, Kontos, & Beckman, 1996). Moreover, age and many pathological conditions may alter the effect of NO on regulating CBF. Therefore, for simplicity reasons, the scope of this review will remain focused exclusively on the effect of NO and its physiological role in regulating CBF.

2. SOURCES OF NO IN THE NEUROVASCULAR UNIT

Neurons, glial cells, and vascular cells work in concert to maintain brain homeostasis. This integrated system is called the neurovascular unit. All cell types of the neurovascular unit can produce NO from one or more of the four NO synthase (NOS) isoforms: calcium-independent inducible NOS (iNOS) (Kukreja, Wei, Kontos, & Bates, 1993), calcium-dependent and constitutively expressed neuronal NOS (nNOS) (Cho et al., 1992), endothelial NOS (eNOS) (Bredt, Hwang, & Snyder, 1990), and red blood cell NOS (RBC-NOS) (Jubelin & Gierman, 1996; Palmer, Ferrige, & Moncada, 1987). Although the name of an isoform is an accurate indication for the cell type in which it was originally discovered, it does not remain exclusive to that cell type. For instance, nNOS was observed in interneurons, some pyramidal cells, and nitrergic nerves originating from pterygopalatine ganglia, but it has also been reported in vascular smooth muscle and endothelium of rat cerebral parenchymal vessels (Kleinbongard et al., 2006) as well as in the smooth muscle of bovine and rat common carotid artery (Hama-Tomioka et al., 2012). In astrocytes, the presence of either nNOS or eNOS had been observed in mice, rats, and hamsters (Boulanger et al., 1998; Brophy, Knoepp, Xin, & Pollock, 2000; Caillol, Devinoy, Lacroix, & Schirar, 2000; Gabbott & Bacon, 1996; Lin, Taktakishvili, & Talman, 2007; Wiencken & Casagrande, 1999). However, electronic microscopy observations by Wang, Lessard, Trudeau, and Girouard (2013) did not reveal any trace of nNOS in vascular smooth muscle cells, endothelium, or glial processes of rat somatosensory cortex. In addition to the endothelium and glia, the endothelial isoform had been found in rat vascular smooth muscle cells (Wang, Hitron, Iadecola, & Pickel, 2005), platelets (Comini et al., 1996; Wallerath et al., 1997), and neurons of the rat hippocampus (Sase & Michel, 1995). The RBC-NOS isoform has common regulatory mechanisms with eNOS because its activity depends on the intracellular Ca^{2+} level and phosphorylation at serine 1177 regulated by phosphatidylinositol-3 kinase (PI3K) (Dinerman, Dawson, Schell, Snowman, & Snyder, 1994). However, due to the lack of normal cell organelles, RBC-NOS seems to be devoid of various eNOS regulatory mechanisms associated with microdomains.

Another important source of NO in the brain is nitrite. Nitrite is a permanent constituent of blood in all animal species at concentrations that vary considerably with diet. Human blood contains 150–300 nM of nitrite (Dejam et al., 2005; Ulker et al., 2011), which is similar to what is found in rats (Kleinbongard et al., 2003). In general, concentrations of nitrite vary between 0.5 and 20 μM among different mammalian tissues (Dejam et al., 2005; Samouilov, Woldman, Zweier, & Khramtsov, 2007). Nitrite can be converted into NO through the nitrite reductase activity of deoxyhemoglobin (Feelisch et al., 2008), xanthine oxidase (Cosby et al., 2003), aldehyde oxidase (Li, Cui, Kundu, Alzawahra, & Zweier, 2008), carbonic anhydrase (Li, Kundu, & Zweier, 2009), and all isoforms of NOS (Aamand et al., 2009).

3. RESTING CBF

Under physiological conditions and in response to intraluminal pressure, arterioles display a myogenic tone, which sets the stage for autoregulation as well as NVC and it determines the level of the resting CBF. There are a number of reports involving NO in the regulation of resting CBF and vascular tone. The role of NO in the control of resting CBF was determined by NOS inhibitors, guanylate cyclase (GC) inhibitors, and endothelial disruption (Mikula, Durocher, Martasek, Mutus, & Slama-Schwok, 2009; Rosenblum, 1988; Sobey & Faraci, 1997b). Systemic or topical administration of selective or nonselective NOS inhibitors decreases resting CBF in many mammals such as mice, rats, rabbits, dogs, pigs, fetal sheep, goats, monkeys, and humans in a range between 10% and 60% (Dreier et al., 1998; Iadecola, 1993; Toda & Okamura, 2003). Endogenous NOS inhibition constricts cerebral arteries in both *in vitro* and *in vivo* models (Duchemin, Boily, Sadekova, & Girouard, 2012; Faraci, 1991; Geary, Krause, & Duckles, 2000; Prado, Watson, Kuluz, & Dietrich, 1992; You, Johnson, Marrelli, Mombouli, & Bryan, 1999). Interestingly, the involvement of NOS in the maintenance of resting CBF in a conscious animal is identical to that of an anesthetized one (Bonvento, Seylaz, & Lacombe, 1994; You, Johnson, Marrelli, & Bryan, 1999).

NOS inhibitors are Nω-substituted L-arginine analogues that inhibit NOS competitively (Iadecola, Zhang, & Xu, 1994). Most commonly used NOS inhibitors are nitro-L-arginine methyl ester (L-NAME), nitro-L-arginine (L-NA), and *N*-monomethyl L-arginine (L-NMMA). These compounds block the conversion of L-arginine to L-citrulline without affecting cyclic guanosine monophosphate (cGMP) production induced by NO donors such as nitroprusside (Moncada, 1992b; Palmer, Rees, Ashton, &

Moncada, 1988). These inhibitors are nonselectives and block the inducible and the constitutively expressed NOS. The specific effects of these inhibitors are stereospecific and, hence, the nonspecific effects can be verified by testing both L- and D-enantiomers.

The effect of NO on the tone of cerebral blood vessels is assessed with the use of exogenous NO (acidified $NaNO_2$) (Archer, 1993; Moncada, 1992a), the NO synthesis precursor, L-arginine, and NO donors, including nitroglycerin and sodium nitroprusside (SNP). These drugs induce a dose-dependent dilation of cerebral arteries that are not affected by NOS inhibitors, but are abolished by the soluble GC inhibitor, methylene blue, and the NO scavenger, oxyhemoglobin. Systemic administration of L-arginine (500 mg/kg over 30 min) increases CBF velocity in humans (Furchgott & Vanhoutte, 1989) and local superfusion over a cranial window dilates cortical pial arteries in a dose-dependent manner in mice and piglets (Okamoto, Etani, Yagita, Kinoshita, & Nukada, 2001).

Primates such as Japanese monkeys respond to NOS inhibitors with constriction of the anterior and middle cerebral arteries. This effect can be reversed by L-arginine (Busija, Leffler, & Wagerle, 1990; Rosenblum, Nishimura, & Nelson, 1990). However, Joshi et al. (Toda, Tanaka, Ayajiki, & Okamura, 2000) noted that, during isoflurane anesthesia, intracarotid SNP does not increase the CBF in healthy baboons, but increases the cGMP concentrations in an *in vitro*-isolated vessel. This suggests that SNP is ineffective when delivered by the intra-arterial route. In humans, on the other hand, intravenous injection of L-NMMA (1.5 mg/kg) does not affect the resting CBF in young (average age of 25 years) individuals but decreases CBF significantly in older subjects (average age of 78 years) (Joshi et al., 2003). Despite this, infusion of L-arginine significantly increases CBF in young human subjects attesting the presence of functional NOS in their tissues (Kamper et al., 2004). Therefore, for humans, age is a contributing factor for the involvement of the NOS pathway in CBF regulation. This difference between rodents, monkeys, and humans concerning the involvement of NO in the control of the resting CBF cannot be explained by a difference in the mode of injection of the NOS inhibitor. Indeed, systemic L-NMMA (30 mg/kg) administration decreases resting CBF by 20–30% (Okamoto et al., 2001) and topical application of L-NMMA (5–10 μM) constricts the basilar artery in rats (Tanaka et al., 1991). It can be argued that the absence of an effect from L-NMMA on CBF in the young subjects is caused by a lack of effectiveness of L-NMMA at young age. However, this seems less likely as systemic administration of L-NMMA (Faraci, 1990) normalizes the

CBF increase induced by hypoxia, while the resting CBF remained constant in young individuals.

Thus, contribution of NO in maintaining the resting CBF varies between species and brain regions. For example, in cat, NO does not contribute to the resting diameter or the control of pial arteries diameter in the cerebral cortex (Van Mil et al., 2002), whereas, in regions rich in nNOS such as the cerebellum and the pituitary, NOS inhibition attenuates CBF (Kovach et al., 1992; Wei, Kukreja, & Kontos, 1992).

While the acute NOS blockade that was described above generally attenuates the resting CBF, chronic effects seem absent. In fact, chronic (1 week) oral administration of L-NAME in the hippocampus does not alter CBF. However, reversal of chronic NOS blockade by L-arginine infusion evokes hypothalamic hyperemia. This is evidence for a compensatory vasodilator mechanism during chronic L-NAME administration (Bredt et al., 1990; Hortobagyi et al., 2007; Vincent & Hope, 1992). It also suggests that this compensatory mechanism is independent of vasodilator prostanoids but that a reduced release of vasoconstrictor prostanoids may contribute to the normalization of hypothalamic blood flow after chronic loss of NO.

The mechanisms underlying the effects of NOS inhibition are not entirely demystified. It is now well accepted that the effects of endogenous NOS inhibition are not secondary to changes in systemic blood pressure glucose or oxygen consumption. Systemic administration of nonselective NOS inhibitors change arterial pressure before CBF and at lower concentrations than what is needed to change CBF (Hortobagyi et al., 2007). The lack of change in oxygen or glucose consumption after NOS blockade suggests that the effects of NOS inhibition hemodynamic are not secondary to an alteration of cerebral energetic metabolism (Iadecola & Zhang, 1994; Iadecola, Zhang, & Xu, 1993).

To date, studies suggest that vascular signaling by NO is primarily mediated by smooth muscle cGMP, which is converted by NO-activated GC from guanosine triphosphate. Effects of cGMP occur through cGMP-dependent protein kinases (PKGs) or cGMP-gated cation channels (Kelly, Ritchie, & Arbuthnott, 1995; Macrae, Dawson, Norrie, & McCulloch, 1993; McPherson, Kirsch, Ghaly, & Traystman, 1995; Saito, Wilson, Hanley, & Traystman, 1994). In pressurized and isolated mice cerebral arteries, under NOS inhibition, the resting diameter is reestablished by the cGMP analogue, 8-Br-cGMP, while it is decreased in the presence of the soluble GC blocker, ODQ (*i*H [1,2,4]oxadiazole [4,3-a]quinoxalin-1-one) (Gao, Dhanakoti, Tolsa, & Raj, 1999; Piggott et al., 2006). However, Doughty and Langton

(Sobey & Faraci, 1997a) did not observe any effect of ODQ on the tone of rat isolated and pressurized cerebral arteries.

In vivo, 3-morpholinosydnonimine (SIN-1) or *S*-nitroso-*N*-acetylpenicillamine (SNAP) reestablish resting CBF after NOS inhibition with L-NNA, and 8-Br-cGMP restores the resting CBF level (Doughty & Langton, 1998). NO also stimulates the release of prostaglandins (PGs) (Lindauer, Megow, Matsuda, & Dirnagl, 1999). However, the inhibition of prostanoid synthesis by indomethacin does not mimic the contractile effect of NOS blockade. L-Arginine analogue NOS inhibitors have been shown to increase the release of thromboxane A_2 under *in vivo* and *in vitro* experimental conditions, suggesting an inhibitory effect of endogenous NOS on thromboxane production (Salvemini, 1997). In the rats' middle cerebral and basilar arteries, L-NA-induced contractions are attenuated after administration of the cyclooxygenase (COX) inhibitor, indomethacin, the thromboxane A_2 synthase inhibitor, ridogrel, or the thromboxane receptor antagonist, ICI 192605 (Hellyer, Johnson, & Olson, 1997; Stadler et al., 1993). These results suggest that endogenous NO maintain the resting cerebrovascular tone by regulation (suppression) of the synthesis and/or release of thromboxane A_2. Other studies have shown constitutive thromboxane release from cerebrovascular endothelial cells (Benyo, Gorlach, & Wahl, 1998a, 1998b).

The NOS isoform responsible for maintaining CBF tone is not yet identified. The main underlying problem is the lack of specific inhibitors. Clearly, the iNOS isoform is not involved because its specific inhibitor aminoguanidine does not modulate the resting CBF (Spatz et al., 1993). The relatively specific nNOS inhibitor, 7-nitro-indazole (7-NI), had been mostly used but special attention has to be paid to the mode of administration of 7-NI since it becomes specific exclusively *in vivo*. Intraperitoneal injections of 7-NI lower the CBF baseline in awake rats (laser Doppler on the somatosensory cortex) (Iadecola, Zhang, & Xu, 1995) (autoradiography in the barrel cortex) (Montecot, Borredon, Seylaz, & Pinard, 1997) as well as in anesthetized mice (laser Doppler on the somatosensory cortex) (Gotoh et al., 2001). It also lowers cerebral capillary flow in anesthetized rats (intravital microscopy on the right parietal cortex) (Girouard, Park, Anrather, Zhou, & Iadecola, 2007; Girouard et al., 2009), global CBF in anesthetized rats (autoradiography) (Hudetz, Shen, & Kampine, 1998), and global CBF in cats (positron emission tomography-PET scan) (Cholet, Seylaz, Lacombe, & Bonvento, 1997). The decreased cortical CBF ranges from about 10% to 60% depending on the 7-NI concentration and the cortical areas (Hayashi, Osawa, & Tohyama, 2002). The effects of 7-NI on CBF reach its maximal intensity within 30 min and the autoradiography technique shows that this effect is present in most of the brain structures. *In vivo* administration of 7-NI does not change body temperature, blood gas tensions, pH, glycemia, hematocrit level, mean arterial blood pressure, or heart rate (Cholet et al., 1997; Hayashi, Katsumi, et al., 2002; Kelly et al., 1995). Also, 7-NI and other nonspecific NOS inhibitors do not alter resting cerebral glucose or global oxygen utilization (Cholet et al., 1997). These results collectively support the idea that 7-NI induces global and appreciable decrease of resting CBF without altering brain metabolism. Resetting of the flow–metabolism relationship observed after 7-NI administration indicates that NO directly participates in the maintenance of the resting cerebrovascular tone.

The activity of eNOS is regulated at multiple steps, including transcription, substrate availability, interactions with calcium, calmodulin, and enzymatic cofactors such as FAD, FMN, nicotinamide adenine dinucleotide phosphate (NADPH), and tetrahydrobiopterin (BH4), protein–protein interactions with hsp90 and caveolins, and posttranslational modifications. eNOS is also regulated by dimerization and precise subcellular localization to caveolae (Cholet et al., 1997; Iadecola & Xu, 1994; Takahashi, Cook, Jehle, Kennedy, & Sokoloff, 1995; Wang et al., 1993). Cavtratin, a protein that has been used both *in vivo* and *in vitro*, complexes with active eNOS to block the generation of NO (Feron et al., 1996; Shaul, 2002). This peptide contains portions of the sequence of caveolin-1, the eNOS scaffolding protein. Cortical superfusion with this peptide does not modify the arteriolar diameter in the rat cortex (Gratton et al., 2003; Hatakeyama et al., 2006). L-N5-(1-iminoethyl)ornithine (L-NIO), a relatively selective eNOS inhibitor, reduces resting CBF in rats (Bauser-Heaton, Song, & Bohlen, 2008). However, L-NIO may also inhibit nNOS and, compared to 7-NI, the later study did not show any superior effect of L-NIO on CBF. These results suggest a preponderant role for nNOS on the maintenance of vascular tone in physiological situations. Nonetheless, it does not exclude a participation of eNOS in situations involving changes in the blood pressure or in the occurrence of pathological incidents.

3.1. Studies in transgenic mice

Resting CBF evaluated with the hydrogen clearance did not significantly differ among wild-type, nNOS-, and eNOS-null mice (Oyama et al., 2010).

Glucose consumption is not altered by NOS gene deletions (Browne SE, Ayata C et al., JCBFM, 1999). Interestingly, the preserved resting CBF in eNOS-null mice occurs despite the resulting hypertension caused by the gene deletion. CBF is probably preserved by autoregulatory mechanisms and the compensatory action of perivascular nNOS. However, intracisternal injections with replication-defective adenovirus containing the bovine eNOS gene developed transient increase in CBF, suggesting that NO from eNOS can modulate resting CBF in rats (Atochin et al., 2003).

There are at least three nNOS alternative-splice variants: nNOSα, nNOSβ, and nNOSγ (Luders et al., 2000). The mitochondrial NOS corresponds to the nNOSα variant and could be inhibited by 7-NI. In a culture of rat hippocampal neurons, it has been demonstrated that nNOS is completely localized into mitochondria and activated by *N*-methyl-D-aspartic (NMDA) (Putzke, Seidel, Huang, & Wolf, 2000; Saur, Paehge, Schusdziarra, & Allescher, 2000). That is why, identifying the source of NO using 7-NI is challenging. In addition, the nNOSα variant is specifically knocked out in the nNOS-null mice (Marks, Boriboun, & Wang, 2005), which means that these mice lack both neuronal and mitochondrial NOS.

Therefore, the use of transgenic and "knockout" mice to study the physiological role of specific proteins is not devoid of pitfalls (Putzke et al., 2000). For example, the lack of a protein from the early stages of development may induce compensatory changes in related systems aimed at restoring the function exerted by the missing protein. Accordingly, one should not assume that genetically engineered mice with a deficiency in selected protein(s) are identical to wild-type mice in all other aspects. Thus, pharmacological inhibitors remain an important tool that, when complemented with studies using null mice, provide strong evidence in favor of or against the involvement of a specific protein.

3.2. Vasomotion

At resting state, arterioles exhibit slow spontaneous rhythmic diameter and blood flow oscillation of about 0.1 Hz defined as vasomotion. In the brain, vasomotion has been described in awake and anesthetized animals, in *in vitro* preparations as well as in humans (Williams & Wagner, 2000) and may be influenced by a variety of physiological factors (Brown, Key, & Lovick, 2002; Dirnagl, Kaplan, Jacewicz, & Pulsinelli, 1989;

Filosa, Bonev, & Nelson, 2004; Fujii, Heistad, & Faraci, 1990; Hundley, Renaldo, Levasseur, & Kontos, 1988; Rayshubskiy et al., 2013; Vetri, Menicucci, Lapi, Gemignani, & Colantuoni, 2007). One of these factors is NO that can alter the temporal pattern of the flow. Following administration of NOS inhibitors, synchronized high amplitude oscillations within a common feeding vessel have been reported (Fujii et al., 1990). This vasomotion is significantly attenuated during topical application of NO donors such as SIN-1 or SNAP. NOS inhibitors induced oscillations propagate to downstream branches (Adachi, Inanami, & Sato, 1992; Dirnagl, Lindauer, & Villringer, 1993a; Lindauer, Megow, Matsuda, et al., 1999). However, during superfusion with the cGMP analog, 8-bromo-cGMP, no change in NOS inhibition-induced rhythmic vasomotion is observed. GC blockade by ODQ also induces rhythmic oscillations of resting CBF, with the amplitudes being much smaller than after NOS inhibition with L-NNA (Behzadi & Liu, 2005; Griffith & Kilbourn, 1996). The physiological importance and the underlying mechanisms of vasomotion are not well understood. Overall, it seems that vasomotion is primarily myogenic and depends on the smooth muscle cells cytosolic oscillator, Ca^{2+}. This oscillator interacts with the membrane to establish membrane-potential changes that will ultimately determine the synchronization. The mechanism by which NO alters Ca^{2+} oscillation is difficult to predict due to the multiple targets of the NO/cGMP pathway and their differential effects on Ca^{2+} dynamics. The effect of NOS inhibition is different from blockade of the soluble GC/cGMP. This suggests that NO may reduce oscillations by acting on the smooth muscle cell membrane potential through direct activation of the calcium-dependent potassium channels (BK_{Ca}). A major suggestion for the role of vasomotion is that it regulates blood flow at branch points (Lindauer, Megow, Matsuda, et al., 1999) limiting the amount of blood to areas of the vascular network, but maintaining flow at low perfusion.

3.3. Shear stress

Fluid shear stress from blood flow acting on the endothelium is crucial in the regulation of vascular tone. Shear stress is a hemodynamic force acting parallel to the vessel walls thereby creating a frictional shear force on the surface of the endothelium. This causes arteries to undergo adaptation responses to both chronic and acute changes in blood flow. The adaptation is mediated in part by the production of endothelial-derived vasoactive mediators (Ursino, Cavalcanti, Bertuglia, & Colantuoni, 1996). These vasodilators relax the larger arteries and amplify the increase of flow, but there is limited evidence supporting this possibility in the neocortical microcirculation. NO produced by eNOS and additional endothelial-derived relaxing factors are key elements in the control of the flow response (Busse & Fleming, 2003). Hydrogen peroxide, produced after eNOS (Drouin & Thorin, 2009; You, Golding, & Bryan, 2005) and NADPH-oxidase (Drouin & Thorin, 2009) activation, was identified as a key contributor to shear-dependent dilation of rodent cerebral arteries, but not by everyone (Paravicini, Miller, Drummond, & Sobey, 2006). However, in the study by You et al. (2005) a cell permeant hydrogen peroxide scavenger was not used. It would also be interesting to conduct experiments as in the study by Drouin and Thorin (You et al., 2005) on isolated and pressurized arteries in the presence of BH4, a cofactor which promotes the production of NO over reactive oxygen species (Drouin & Thorin, 2009).

Increased shear stress can distort and activate mitochondrial-driven events possibly through filamentous connections with the glycocalyx, which results in production of dilator agents (Wever, van Dam, van Rijn, de Groot, & Rabelink, 1997). In fact, mitochondrial-depolarizing agents modulate vascular smooth muscle tone in cerebral arteries by inducing the release of vasoactive mediators such as NO (Liu, Li, Bubolz, Zhang, & Gutterman, 2008). The NO release results from activation of eNOS via increased endothelial $[Ca^{2+}]_i$ and activation of PI3K leading to Akt phosphorylation (Katakam et al., 2009).

Shear stress could also activate NOS in red blood cells (Katakam et al., 2013). Although blood cells can produce a considerable amount of NO, this aspect is rarely considered in neurovascular studies. However, the possible involvement of NO from blood cells in the control of vascular tone is inevitable. In this regard, both endothelial and red blood cells would play a role in CBF autoregulation. Interestingly, red blood cells contain functional NMDA receptors, which upon activation increase NOS-dependent NO production (Ulker, Sati, Celik-Ozenci, Meiselman, & Baskurt, 2009). This may explain the drop in resting CBF in the presence of the NMDA inhibitor MK-801 although the effect of 7-NI on the RBC-NOS remains unknown.

Although a flow-mediated dilation in cerebral arteries may occur (Makhro et al., 2010), some authors have questioned the presence of this phenomenon in these arteries where instead of dilations, constrictions were observed (Fujii, Heistad, & Faraci, 1991). However, all these studies were performed in isolated vessels where the rheologic environment is not necessarily well reproduced.

4. AUTOREGULATION

Cerebrovascular resistance varies in response to changes in transmural pressure and blood flow so that the flow remains constant. The changes in resistance are the result of vasodilation and vasoconstriction of the pial and parenchymal vessels (Bryan, Marrelli, Steenberg, Schildmeyer, & Johnson, 2001; Bryan, Steenberg, & Marrelli, 2001; Madden & Christman, 1999). Several studies have investigated the role of NO in the mechanisms involved in autoregulation.

4.1. Lower limit

To examine the role of NO in the lower limit of autoregulation (i.e., low systemic blood pressure of about 60 mm Hg), arterial pressure is decreased in a stepwise manner or with one level of hypotension using different methods such as hemorrhage, ganglionic blockade, or administration of a potassium channel activator (Shapiro, Stromberg, Lee, & Wiederhielm, 1971). Majority of reports have failed to observe any inhibition effect in the lower end of CBF autoregulation after intravenous infusion of a NOS inhibitor (Dieguez et al., 1999). However, topical application of an NOS inhibitor over the cortex has demonstrated a role of NO in the lower limit of CBF autoregulation (Jones et al., 2003; Jones, Radinsky, Furlan, Chyatte, & Perez-Trepichio, 1999). Indeed, L-NAME superfusion over a cranial window both raised the lower limit of CBF autoregulation and depressed the height of the CBF autoregulatory curve (Buchanan & Phillis, 1993; Chapman, Cipolla, & Chan, 2013; Dieguez et al., 1999; Saito et al., 1994; Thompson, Pluta, Girton, & Oldfield, 1996; Wang, Pelligrino, Koenig, & Albrecht, 1994). This suggests that the increase in blood pressure induced by systemic administration of NOS inhibitors may conceal the NO effect on the lower limit of autoregulation.

In healthy subjects, the rate of rise in the flow velocity of the middle cerebral artery was compared with that of arterial blood pressure after a stepwise fall in arterial blood pressure using bilateral thigh cuffs that were inflated suprasystolically for 3 min, and suddenly deflated afterward (Jones et al., 2003, 1999). This protocol was repeated after a bolus injection of L-NMMA or noradrenaline as a control for the blood pressure increase. Interestingly, the mean change in autoregulation index after noradrenaline administration at a similar pressor dose was greater than the change after the bolus injection of L-NMMA. This suggests that NO mediates, at least partly, the lower limit of the cerebral autoregulation.

The source of NO that plays a role in autoregulation is not clear. A study showing nonspecific NOS-induced autoregulation dysfunction failed to demonstrate similar effects with the nNOS inhibitor 7-NI (White, Vallance, & Markus, 2000). However, in this study, the 7-NI was applied topically over the cortex whereas, in reality, it is most effective and best specific when administered systemically. Another study suggested a contribution of nNOS to CBF at the lower limit of autoregulation using an NMDA receptor blocker (Toyoda et al., 1997). This effect, however, does not exclude the participation of other vasodilators such as carbon monoxide that is produced upon NMDA activation (Bauser-Heaton & Bohlen, 2007). Nonetheless, some of the findings in this regard seem divergent; for instance, an increase in the lower limit of autoregulation has been reported in eNOS knockout mice (Leffler, Parfenova, Fedinec, Basuroy, & Tcheranova, 2006) but remained normal in nNOS-knockout mice (Huang et al., 1996). This indicates that eNOS is involved in the vasodilation in response to decreases in arterial pressure. Conversely, Bauser-Heaton and Bohlen (Huang et al., 1994) demonstrated that most of the NO produced during mild to moderate reductions in brain oxygen availability associated with systemic hypotension originate from nNOS.

4.2. Upper limit

To examine the upper limit of autoregulation, arterial pressure is gradually elevated. When arterial pressure exceeds an upper limit, breakthrough of autoregulation occurs, cerebral blood vessels dilate, and CBF dramatically increases (Bauser-Heaton & Bohlen, 2007). This breakthrough may be delayed to a much higher blood pressure if the arterial baroreflex is peripherally or centrally interrupted (MacKenzie et al., 1976; Skinhoj & Strandgaard, 1973). In addition, this breakthrough point shifts to higher arterial pressure when the nitroxidergic innervation to the cerebral vessels is interrupted (Talman, Dragon, & Ohta, 1994; Talman & Nitschke Dragon, 2002) or when NO synthesis is inhibited by systemic administration of a NOS inhibitor (Talman & Dragon, 2000). Essentially all of the nitroxidergic innervation of forebrain vessels derives from neurons of the pterygopalatine ganglia (Talman & Dragon, 1995), which are bilateral and lie near the floor of the orbit where they send ganglionic projections into the cranium through the mesial orbital wall (Toda, Ayajiki, Yoshida, Kimura, & Okamura, 1993;

Yoshida et al., 1993). Stimulation of these ganglia leads to cerebral arterial dilatation (Hara, Zhang, Kuroyanagi, & Kobayashi, 1993). Some reports indicate that inhibition of the ganglia or their innervation leads to cerebral vasoconstriction (Seylaz et al., 1988; Toda, Ayajiki, Tanaka, & Okamura, 2000; Toda, Tanaka, et al., 2000). Therefore, this pathway exerts a tonic vasodilator influence. Thus, NO generated by nNOS in parasympathetic nerves from the pterygopalatine ganglia is involved in cerebral vasodilatation during acute hypertension (Toda, Ayajiki, et al., 2000; Toda, Tanaka, et al., 2000).

4.3. Dynamic autoregulation

Constant subtle oscillations in blood pressure are accompanied by spontaneous CBF fluctuations (Talman & Nitschke Dragon, 2007) that has been referred to as "dynamic cerebral autoregulation." This phenomenon is in contrast to static cerebral autoregulation which refers to CBF control in response to large changes in arterial pressure (Hughson, Edwards, O'Leary, & Shoemaker, 2001; Panerai et al., 2002; Zhang, Zuckerman, Giller, & Levine, 1998). In humans, beat-to-beat analysis shows that inhibition of tonic production of NO does not alter dynamic cerebral autoregulation.

Overall studies using local administration of NOS inhibitors suggest that NO play a role in CBF autoregulation both in the lower and upper limits but not in the dynamic autoregulation. Furthermore, it seems that eNOS is the main source of NO at the lower limit of autoregulation, while nNOS from nitroxidergic innervation is potentially the source at the upper limit. However, at low blood pressure levels where oxygen is scarce, the effects of eNOS-derived NO compared to that of nNOS-derived NO are negligible. In fact, studies that showed no effect of NOS inhibition in CBF autoregulation used systemic mode of administration of NOS inhibitors, which may directly exert various physiological effects that may mask the involvement of NO in CBF autoregulation.

5. NEUROVASCULAR COUPLING (FIG. 14.1)

It has been known for more than a century that local neuronal activity is tightly coupled to the local regulation of CBF through a complex mechanism called NVC. The finding by Bredt and colleague (Paulson, Strandgaard, & Edvinsson, 1990) that NO, a strong vasodilator, is produced in the brain upon neuronal activation led to the presumption that it could

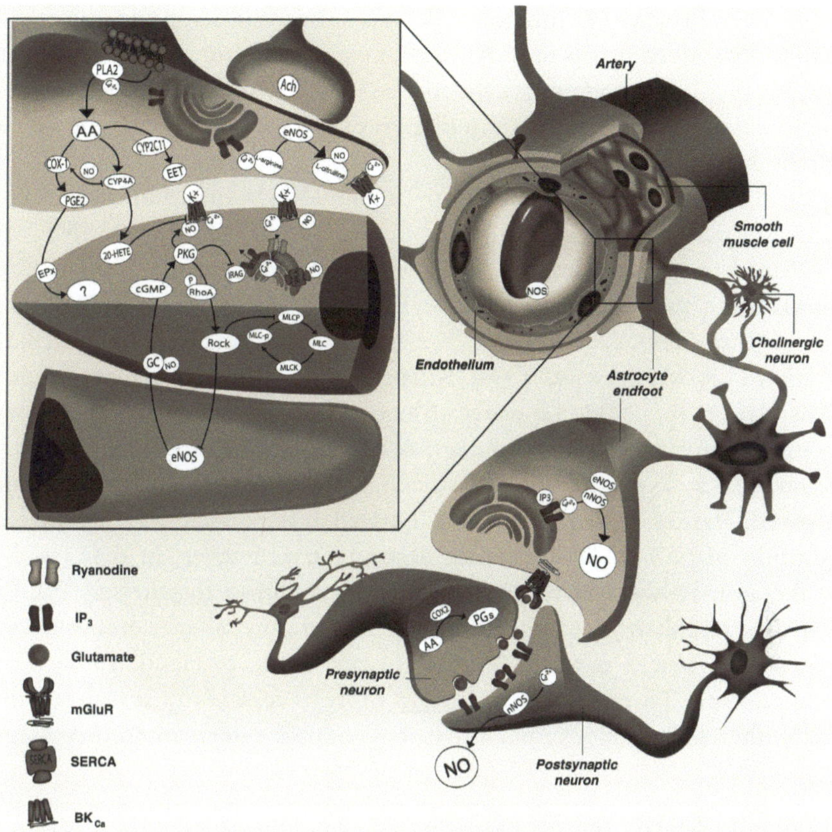

Figure 14.1 Nitric oxide (NO) in neurovascular coupling (NVC). NO may be released from erythrocytes, endothelial cells, smooth muscle cells, astrocyte, interneurons, and some pyramidal cells. This gas may be involved in the initiation of NVC through a cyclic guanosine monophosphate (cGMP)-dependent pathway or modulate NVC by interacting with different cGMP-independent pathways. NO could be released from neuronal NO synthase (nNOS) in nitrergic interneurons and act on astrocytic guanylate cyclase (GC), which increases the Ca^{2+}-activated K^+ channel (BK_{Ca}) opening probability and the subsequent release of K^+ in the perivascular space. Astrocytic NOS can be activated following activation of cholinergic or glutamatergic receptors by acetylcholine (Ach) and glutamate released from cholinergic neurons and pyramidal cells, respectively. Binding to either receptors lead to inositol triphosphate (IP_3) formation and the subsequent release of Ca^{2+} from the sarcoplasmic reticulum or the entry of Ca^{2+} through membrane channels. The increase in intracellular Ca^{2+} activates BK_{Ca} channel and catalyzes the phosphorylation of arachidonic acid (AA) by phospholipase A_2 (PLA_2). AA metabolites (prostaglandin (PG)E_2, 20-hydroxyeicosatetraenoic acid (20-HETE), and epoxyeicosatrienoic acids (EETs)) will, in turn, modulate cerebral blood flow (CBF). NO may interact with different astrocytic pathways such as Ca^{2+} release, Ca^{2+}-activated K^+

(Continued)

control cerebrovascular functions. Several other studies provided evidence that NO is indeed involved in NVC. For example, in anesthetized mice and rats, NOS inhibitors attenuate the increase in CBF induced by activation of the somatosensory cortex by stimulation of the sciatic nerve or the vibrissa (Bredt & Snyder, 1989).

In contrast to studies in anesthetized animals, systemic administration or intracisternal delivery of NOS inhibitors such as L-NNA, L-NAME, and L-NMMA in awake animals does not affect the CBF response to neocortical stimulation (Bonvento, Cholet, & Seylaz, 2000; Cholet et al., 1997; Dirnagl, Niwa, Lindauer, & Villringer, 1994; Girouard et al., 2007; Irikura, Maynard, & Moskowitz, 1994; Kitaura et al., 2007; Lindauer, Megow, Matsuda, et al., 1999; Lindauer, Megow, Schultze, Weber, & Dirnagl, 1996; Ma, Ayata, Huang, Fishman, & Moskowitz, 1996; Ngai, Meno, & Winn, 1995; Northington, Matherne, & Berne, 1992; Peng, Zhang, Alkayed, Harder, & Koehler, 2004). This led to the conclusion that the contribution of NO to NVC is state dependent. However, in these studies, awake animals were restrained, a stressful condition that lowers PaCO2, increases circulating catecholamines and therefore alters cerebrovascular regulation (Adachi et al., 1994; Greenberg, Sohn, & Hand, 1999; Wang et al., 1993). To address this issue, a more recent study in unrestrained unanesthetized animals showed that they exhibit a decrease in the vascular

Figure 14.1—Cont'd channel (BK_{Ca}) opening and cyclooxygenase (COX)-1 activity. NO can also decrease 20-HETE formation through the inhibition of CYP4A. Vascular smooth muscle cells are the final integrators of NO-induced signals coming from the endothelial NOS (eNOS), the blood, neurons, and astrocytes. In the smooth muscle cells, NO directly mediate vasodilation through the increases in cGMP-protein kinase G (PKG) activity and BK_{Ca} channel opening probability. Another PKG-dependent pathway is the IP_3 receptor-associated cGMP-kinase substrate (IRAG), an endoplasmic reticulum-anchored membrane protein. IRAG interacts specifically in a trimeric complex with the PKG1β isoform and the IP_3 receptor I and, upon phosphorylation, it reduces the intracellular Ca^{2+} release from the intracellular stores. In addition, NO can control intracellular Ca^{2+} concentrations by increasing Ca^{2+} reuptake into intracellular stores through activation of the sarcoplasmic reticulum ATPase (SERCA). These NO-dependent mechanisms will lead to the subsequent decrease in the cytosolic Ca^{2+} concentration as well as the inhibition of RhoA phosphorylation. Both phenomenona will converge toward a decrease in the contractile state of vascular smooth muscle cells, which is determined by the phosphorylation level of myosin light chains (MLC). MLC phosphorylation is initiated by the MLC kinase (MLCK) activation by Ca^{2+}–calmodulin interaction, while RhoA/Rho kinase signaling pathway inactivates the MLC phosphatase (MLCP) by phosphorylation resulting in increased unphosphorylated MLCP and sustained smooth muscle contraction. NOS may be inhibited, in turn, by RhoA kinase (ROCK). (See the color plate.)

response to neuronal activation following unspecific NOS and specific nNOS inhibition (Lacombe & Seylaz, 1984).

5.1. Sources of NO involved in NVC

Several studies support the idea that the NO involved in NVC is neurogenic. Systemic administration of the nNOS inhibitor, 7-NI, reduces the amplitude of NVC in different experimental paradigms of neuronal activation such as whisker stimulation and electric stimulation of mice forepaw (Gotoh et al., 2001). In eNOS- and nNOS-knockout mice, activity-dependent vascular changes are similar to those in wild-type mice. In eNOS-null mice, NVC is suppressed by NOS inhibitors (Bonvento et al., 2000; Cholet, Bonvento, & Seylaz, 1996; Girouard et al., 2007; Liu, Li, Falck, et al., 2008), while in nNOS-knockout mice, it remains intact (Ayata, Ma, Meng, Huang, & Moskowitz, 1996). This is an indication of an exclusive involvement of NO derived from nNOS. A more recent study, however, showed an implication of both eNOS and nNOS in NVC depending on the level of stimulation, eNOS plays a prominent role at low intensity stimulation, while nNOS is more involved at high intensity stimulation (Ma et al., 1996).

The effects of nNOS inhibition on NVC may be the consequence of a damped neuronal activity (de Labra et al., 2009). Most studies demonstrated that NOS inhibition has no significant effect or a very slight effect on evoked potentials in the somatosensory cortex (Jayakumar et al., 1999; Lonart, Wang, & Johnson, 1992; Manzoni & Bockaert, 1993; Schuman & Madison, 1994) or cerebral glucose activation (Burke & Buhrle, 2006; Hoffmeyer, Enager, Thomsen, & Lauritzen, 2007; Lindauer, Megow, Schultze, Weber, & Dirnagl, 1996). However, in other studies, NOS inhibition decreases the amplitude of evoked potentials induced by electrical stimulation of the forepaw or the sciatic nerve (Cholet et al., 1997). Notwithstanding the observed effects on activated potentials, NOS inhibition is always accompanied by CBF reduction in response to neuronal stimulation which supports a role of NO in the modulation of NVC.

NO is one of the vasodilators that diffuse the most rapidly (Ngai et al., 1995; Stefanovic, Schwindt, Hoehn, & Silva, 2007) and may thus account for the initiation of the vascular response to neuronal activation. Such a role of NO is plausible since its release from neurons depends on intracellular Ca^{2+} elevation (Wood & Garthwaite, 1994), which occurs following fast neuronal depolarization (10–12 ms) (Lauritzen, 2005) and/or activation of ionotropic receptors. In addition, the CBF response to 1 s stimulation of the mouse hindpaw is reduced by about 50% in nNOS-knockout mice or after topical application of L-NA (Petersen, Grinvald, & Sakmann, 2003). NO may also be involved in the maintenance (up to 15 min) of NVC. In this case, activation of metabotropic receptors leading to Ca^{2+} release from intracellular stores would ensure a slower release of NO (Kitaura et al., 2007). In support of this idea, it is known that NO concentrations increase at the beginning of the stimulation and remain elevated during at least 2 min of neuronal stimulation (Perea & Araque, 2005). In addition, topical application of L-NNA and systemic administration of 7-NI damper the entire NVC during long stimulus of more than 60 s (de Labra et al., 2009).

nNOS is present in nitrergic nerves originating from pterygopalatine ganglia, interneurons, and some pyramidal cells (Fig. 14.2). Immunohistochemistry for the nicotinamide adenine dinucleotide phosphate-diaphorase (NADPHd), a specific histochemical marker for nNOS (Dirnagl, Lindauer, & Villringer, 1993b; Dirnagl et al., 1994; Lindauer et al., 1999; Ngai et al., 1995; Peng et al., 2004), shows that nitrergic nerves are strategically positioned in proximity to cerebral arteries. NADPHd positive fibers have been found around pial arteries as well as parenchymal vessels. Pial arteries are innervated by perivascular nitrergic nerves that originate from sphenopalatine, otic, and trigeminal ganglia (Bredt et al., 1991; Huang, Dawson, Bredt, Snyder, & Fishman, 1993; Matsumoto, Nakane, Pollock, Kuk, & Forstermann, 1993; Fig. 14.2). Fibers close to parenchymal arteries have been identified as γ-aminobutyric acid (GABA)ergic interneurons. NOS interneurons have access to the parenchymal arterioles but not to the arterioles and arteries proximal to the Virchow–Robin space (Abadia-Fenoll, 1969; Suzuki et al., 1994; Fig. 14.2). Around 80% of NADPHd positive cells in the rat cortex are GABAergic interneurons and they account for 2% of all interneurons (Abadia-Fenoll, 1969; Busija, 1993) and represent about 15% of cortical neurons in rodents (Valtschanoff et al., 1993). Using patch-clamp recordings, biocytin labeling, and single-cell reverse transcriptase-PCR, nNOS expression was demonstrated in 9% of fast spiking parvalbumin (PV) interneurons, 6% of adapting somatosatin (SOM) interneurons, 2% of adapting vasoactive intestinal peptide (VIP) interneurons, 0% of bursting VIP interneurons, and 26% of adapting neuropeptide (NP)Y interneurons (Gabbott, Dickie, Vaid, Headlam, & Bacon, 1997). The distribution of these interneurons subtypes in association with rat cortical microvessels in layer I–III indicates a privileged redistribution of NPY

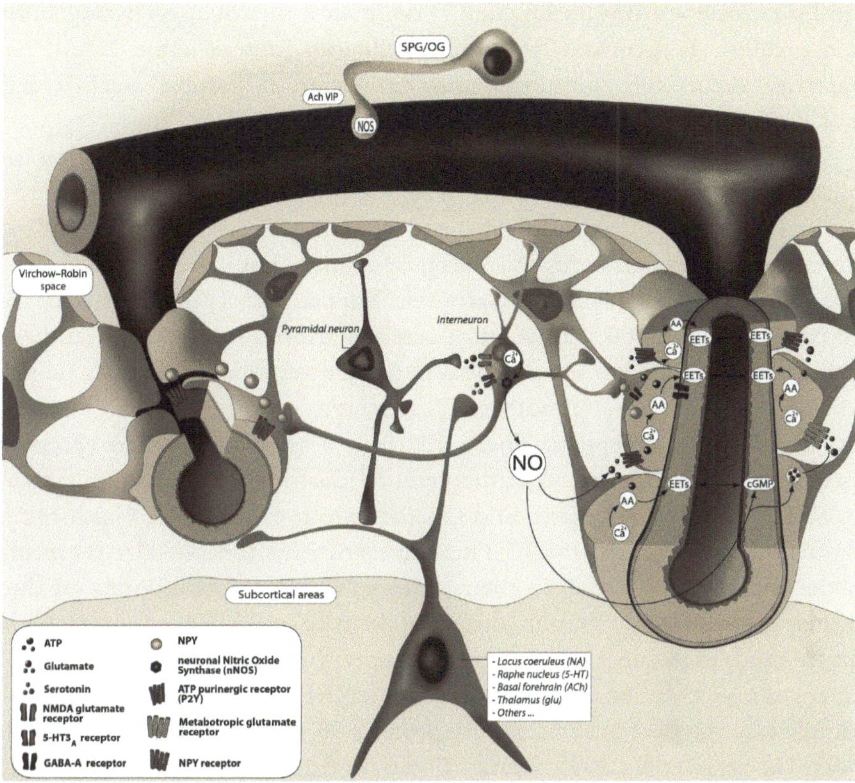

Figure 14.2 Proposed mechanisms by which NO interneurons may control cerebral blood flow. A pial artery gives rise to two parenchymal descending arteries (left and right of the figure). The pial artery is lying over a layer of astrocytes called glia limitans, while parenchymal arteries are covered by astrocytes specialized projections called astrocytic endfeet excepted for the portion separated by the Virchow–Robin space. Pial arteries are innervated by "extrinsic nerves" coming from the peripheral nervous system. Nitrergic nerves originate from the sphenopalatine or the otic ganglion and can corelease acetylcholine (Ach) and the vasoactive intestinal peptide (VIP) in addition to NO. Following local neuronal activity, pyramidal cells (black cells) stimulate nitrergic interneurons (yellow cell), through their glutamate acid *N*-methyl-D-aspartic (NMDA) receptors. The influx of calcium (Ca^{2+}) into the cell allows activation of the neuronal NO synthase (nNOS) and the production of NO. NO diffuses freely through cell membranes and reaches the astrocytic endfeet where it could modulate the release of various vasoactive metabolites. NO may be involved in retrograde vasodilation either through astrocyte signaling modulation or by acting directly on smooth muscle cells. Indeed, NO can increase the release of adenosine triphosphate (ATP) and glutamate from astrocytes. ATP released from vesicles goes from one astrocyte to the next 2PY purinergic receptors, creating a Ca^{2+} wave. These Ca^{2+} waves enhance the production of arachidonic acid (AA) derived vasodilation mediators like epoxyeicosatrienoic acids (EETs) or activate large conductance Ca^{2+}-dependent K^+ channels to induce arterial

(Continued)

and NOS interneurons in the vicinity of cortical microvessels (Karagiannis et al., 2009). In primates, pyramidal cells also present some NADPHd reactivity or nNOS immunostaining in different cortical areas (Cauli et al., 2004).

The NO release from neurons may result from the stimulation of different glutamatergic receptors. For example, when applied to the cerebral cortex, NMDA, kainate, and α-amino-3-hydroxy-5-methyl-4-isoxazolepropionic acid (AMPA) receptor agonists result in NO production (Barone & Kennedy, 2000; Garbossa, Fontanella, Tomasi, Ducati, & Vercelli, 2005) as well as dilation of cerebral arteries in all species studied. Although glutamate has the potential to induce vasodilation though activation of any of the three ionotropic glutamate receptors, this effect mostly involves NMDA receptors. Thus, MK-801, a selective NMDA receptor antagonist, blocks almost all of the cerebral dilator responses to glutamate or NMDA in rabbits, piglets, and rats (Bhardwaj et al., 1997; Garthwaite, Southam, & Anderton, 1989). This preference for the NMDA receptors in the cortex may be due to their location in the cerebral cortex or due to their greater affinity for glutamate than kainate or AMPA receptors. Glutamate binds to its NMDA ionotropic receptors to produce NO by the activation of nNOS, an enzyme tethered to the NMDA NR1 subunit receptor complex by the postsynaptic density protein-95 (Faraci, Breese, & Heistad, 1993; Girouard et al., 2009; Meng, Tobin, & Busija, 1995; Pelligrino, Gay, Baughman, & Wang, 1996; Pelligrino, Wang, Koenig, & Albrecht, 1995). Ionotropic NMDA receptor stimulation results in Ca^{2+} influx, membrane depolarization, activation of nNOS, and the subsequent production of NO, which diffuses out of the neurons and acts on smooth muscle cells or pericytes in local capillaries, arterioles, and pial arteries to cause vasodilatation (Christopherson, Hillier, Lim, & Bredt, 1999). Hemodynamic response induced by NMDA is mediated by NO-dependent activation of sGC through the cGMP effector kinase PKG (Faraci & Brian, 1995;

Figure 14.2—Cont'd vasodilation. NO also directly increases astrocytic Ca^{2+} and might therefore facilitate the production of vasodilators. Finally, NO binds to smooth muscle guanylate cyclase and increase the cyclic guanosine monophosphate (cGMP) production. Stimulated nitrergic interneurons also release acid γ-aminobutyrique (GABA), which could act either on astrocytes or smooth muscles GABA-A receptors to cause vasodilation. However, GABA seems to restrain the astrocytic signal propagation. Simultaneously, Neuropeptide Y (NPY) is released most probably to a remote vascular bed (on the left of the illustration) to cause a vasoconstriction and concentrate the blood flow close to the neuronal activation. (See the color plate.)

Garthwaite, Garthwaite, Palmer, & Moncada, 1989; Girouard et al., 2009; Hamilton, Vayro, Wigley, & Butt, 2010; Meng et al., 1995; Pelligrino et al., 1996).

Dilation to kainate is also attenuated by inhibition of NOS (Girouard et al., 2009) as well as heme oxygenase (HO) and COX (Faraci, Breese, & Heistad, 1994). Arterial dilation to AMPA is not only restricted by superfusion of NOS inhibitors but it is also attenuated by adenosine A(2A) and A(2B) receptor antagonists and the HO inhibitor chromium mesoporphyrin (Bari, Louis, & Busija, 1997; Ohata, Cao, & Koehler, 2006).

Interestingly, depending on the afferent pathway and intensity of the stimulation, distinct population of neurons may be recruited. Electrical stimulation of the corticocortical (transcallosal) or thalamocortical (infraorbital) pathways results in increased cortical CBF associated with the recruitment of distinct populations of interneurons and COX2 pyramidal cells (Ohata et al., 2006). The frequency of the stimulation will determine which of these cells will be solicited. For example, analysis of c-Fos expression shows that transcallosal stimulation at 30 Hz is associated to the activation of a large fraction of SOM–NOS interneurons together with a subset of VIP interneurons, but there was a silencing of those containing PV. At 4 Hz, transcallosal stimulation recruits more COX2 pyramidal cells and less SOM interneurons (Enager et al., 2009). Thalamocortical stimulation at 4 Hz is associated with increased c-Fos expression in COX2 expressing pyramidal cells and specific subtypes of inhibitory GABA interneurons that contain PV, SOM, and NOS (Enager et al., 2009). The amplitude of the blood flow responses to these stimulations is strongly correlated with the percentage of the SOM eNOS subset of GABAergic interneurons recruited (Kubota, Hattori, & Yui, 1994).

In addition to the frequency, the solicited brain area may determine the involvement of NOS interneurons in NVC. NOS interneuron displays an anteroposterior gradient of density, with the highest concentration in the medial prefrontal, frontal, and cingulate cortices (Enager et al., 2009). This distribution is in accordance with the effects of NOS inhibitors on NVC in function of the sollicitated neuronal pathway. In rats under perioral somatosensory stimulation, NO plays a role in NVC in the somatosensory area and the thalamus but less so in the trigeminal primary nucleus (Enager et al., 2009). During vibrissal stimulation in unanesthetized rats, NOS inhibition with L-NAME attenuates the increase in CBF in the ventroposteromedial thalamic nucleus and 7-NI reduces the CBF in both the ventroposteromedial thalamic nucleus and the barrel cortex. NOS inhibitors do not significantly affect CBF in the spinal trigeminal and the principal sensory trigeminal nuclei (Cholet et al., 1997).

Cortical NOS interneurons receive projection from the basalocortical acetylcholine (Ach), the brainstem 5-hydroxytryptamine (5-HT) and the locus coeruleus (noradrenaline) afferents in both rats and humans (Gotoh et al., 2001). Most of these nNOS-containing neurons contact local blood vessels. Nitrergic interneurons thus appear exceptionally well positioned to relay Ach, 5-HT, and noradrenergic afferent information to blood vessels (Tong & Hamel, 1999; Toussay, Basu, Lacoste, & Hamel, 2013; Vaucher, Linville, & Hamel, 1997; Fig. 14.2). However, CBF increase elicited by stimulation of the basal forebrain failed to be attenuate by the nNOS inhibitor, 7-NI (Estrada & DeFelipe, 1998). On the contrary, non-selective NOS inhibitors such as L-N^G-nitroarginine (L-NNA) attenuate the CBF increase induced by stimulation of the basal forebrain (Iadecola & Zhang, 1996; Zhang, Xu, & Iadecola, 1995). Cholinergic basalocortical afferents contact cortical microvessels or the surrounding astrocytes in rat and human (Raszkiewicz, Linville, Kerwin, Wagenaar, & Arneric, 1992) and could as a result directly activate eNOS. The lack of an additive CBF effect of the coapplication of muscarinic receptor antagonist, atropine, and nonselective NOS inhibitor, L-NNA, on the cerebral cortex compared to L-NNA alone, suggest that Ach increases CBF possibly through eNOS activation (Mesulam, Hersh, Mash, & Geula, 1992; Tong & Hamel, 1999; Vaucher & Hamel, 1995). More recently, Lecrux et al. (Zhang et al., 1995) demonstrated that the hyperemic response to basal forebrain stimulation is largely mediated by glutamate released from pyramidal cells and by epoxyeicosatrienoic acids (EETs). They also showed that the metabolic gliotoxins fluorocitrate and fluoroacetate, blocked the evoked CBF response by about 50% suggesting an important contribution of astrocytes. Thus, the role of GABA interneurons in this pathway remains to be elucidated. On the contrary, following stimulation of the locus coeruleus, a significant proportion (>30%) of NO-producing interneurons is recruited and inhibition of the NO-sensitive GC with ODQ significantly reduces the CBF response (Lecrux, Kocharyan, Sandoe, Tong, & Hamel, 2012). Activation of 5-HT3A-expressing interneurons which receives serotoninergic innervation from raphe nuclei mostly induces NO-mediated vasodilation and, less frequently, NPY-mediated vasoconstriction (Toussay et al., 2013). These effects are absent in Pet-1 knockout mice which display a drastic depletion of cortical serotoninergic fibers (Perrenoud et al., 2012) supporting a role of the interaction between serotoninergic fibers and NPY/NOS interneuron in the vasomotor control (Kiyasova et al., 2011). Overall, these results suggest that cortical NOS interneurons mediate the CBF increase following activation of the raphe nuclei and the locus coeruleus, while it plays a minor role if any in the CBF response to basal forebrain stimulation.

5.2. NO and NVC in the cerebellum

The cerebellum has provided a simple and well-characterized model to study the mechanisms underlying NVC. The cerebellar cortex is endowed with two major excitatory synaptic inputs converging on cerebellar Purkinje cells, the only output neurons of the cerebellum: the parallel and the climbing fibers. The parallel fibers are axons of cerebellar granule cells that make synaptic contacts with Purkinje cell dendrites and molecular layer interneurons (Perrenoud et al., 2012). The climbing fibers originate from the contralateral inferior olive and innervate Purkinje cell dendrites and interneurons (Palay, 1974). The main transmitter of parallel fibers is glutamate (Palay, 1974) while climbing fibers release glutamate, aspartate, or *N*-acetyl-aspartyl-glutamate (Ito, 1991). In the cerebellar molecular layer, NOS is present in interneurons, mainly basket and stellate cells but not in Purkinje cells (Renno, Lee, & Beitz, 1997; Wiklund, Toggenburger, & Cuenod, 1982). It is therefore likely that NO is produced by molecular layer interneurons and relays neuronal activity coming from the parallel and the climbing fibers. Indeed, NOS or GC inhibitors attenuate the CBF increase induced by stimulation of parallel or climbing fibers. Single-cell-evoked firing by patch-clamp stimulations of single stellate cells in acute slices of rat cerebellum clearly demonstrates that stellate interneurons mediate NVC via NO release (Bredt et al., 1990; Southam, Morris, & Garthwaite, 1992; Vincent & Kimura, 1992). In agreement with this study, in cyclin D_2-null mice which lack the cell cycle protein cyclin D_2 and present a marked reduction in the number of stellate neurons (Rancillac et al., 2006), the hemodynamic response to neuronal stimulation is markedly reduced whereas the field potentials of the activated neurons remain unchanged (Huard, Forster, Carter, Sicinski, & Ross, 1999). In addition, the magnitude of the attenuation of NVC is similar between cyclin D_2- and nNOS-null mice and cyclin D_2-deficient mice treated with 7-NI, does not show further attenuation of the response (Yang, Huard, Beitz, Ross, & Iadecola, 2000).

5.3. Permissive or obligatory role of NO

NO can act through different pathways and may thus serve as the ultimate mediator of smooth muscle relaxation as well as a permissive factor by facilitating the action of other vasodilators. This is done by either acting directly on smooth muscle cells through the GC/cGMP system or indirectly by having a permissive role. For example, in the cerebral cortex, there is evidence that NO is a permissive factor of NVC. This conclusion is based on the observation that the reduction of the magnitude of the NVC reduced by NOS inhibition could be restored by the normalization of basal NO levels with the infusion of the NO precursor, L-arginine NO (Yang et al., 2000). In the cerebellum, NOS inhibition attenuates the CBF increase to parallel fiber activation but NO donors are unable to reverse this effect (Lindauer et al., 1999; Northington et al., 1992).

In addition, in the cerebral cortex, the increase in CBF produced by activation of the facial whiskers is not attenuated in nNOS-deficient mice (Yang & Iadecola, 1997). In contrast, the increase in CBF produced by activation of the cerebellum is attenuated in these mice. This finding suggests that, whereas in the cerebral cortex other vasoactive factors are able to compensate completely for the absence of neuronal NO, in the cerebellum such full compensation does not occur and the flow response is attenuated. These results support the concept that NO plays an obligatory role in the cerebellum, whereas it is rather permissive in other brain regions.

6. CONCLUSION

NO is involved in the maintenance of resting CBF, the autoregulation as well as in the control of the vascular responses to neuronal activation. However, there are still disagreements about the sources of NO and the underlying mechanisms for each physiological phenomenon. The heterogeneity of the brain and neuronal stimulations in addition to the multiple sources of NO add to the complexity of this subject. As pharmacological approaches and existing transgenic mice present evident limitations when it comes to cell-type specificity, new techniques to better characterize the role of NO in cerebrovascular regulation need to be developed. Actually, the most promising tool is conditional transgenic optogenetic silencing Cre driver lines in different type of cells in combination with pharmacological approaches.

ACKNOWLEDGMENTS

Hélène Girouard was supported by funds from the Canadian Institutes in Health Research (CIHR), the Natural Sciences and Engineering Research Council of Canada (NSERC), the "Fonds de Recherche en Santé du Québec" (FRSQ), Heart and Stroke Foundation of Canada (HSFC), and the Canada Found for Innovation (CFI). She was also the holder of new investigator awards from the FRSQ and the HSFC.

REFERENCES

Aamand, R., Dalsgaard, T., Jensen, F. B., Simonsen, U., Roepstorff, A., & Fago, A. (2009). Generation of nitric oxide from nitrite by carbonic anhydrase: A possible link between metabolic activity and vasodilation. *American Journal of Physiology. Heart and Circulatory Physiology, 297*(6), H2068–H2074. http://dx.doi.org/10.1152/ajpheart.00525.2009.

Abadia-Fenoll, F. (1969). Structure of the leptomeninx and cerebral vessels of the cat. II. Constituent elements of the intracerebral vessel wall (analogy with the meninges). *Angiology, 20*(9), 535–562.

Adachi, T., Inanami, O., & Sato, A. (1992). Nitric oxide (NO) is involved in increased cerebral cortical blood flow following stimulation of the nucleus basalis of Meynert in anesthetized rats. *Neuroscience Letters, 139*(2), 201–204.

Adachi, K., Takahashi, S., Melzer, P., Campos, K. L., Nelson, T., Kennedy, C., et al. (1994). Increases in local cerebral blood flow associated with somatosensory activation are not mediated by NO. *The American Journal of Physiology, 267*(6 Pt. 2), H2155–H2162.

Archer, S. (1993). Measurement of nitric oxide in biological models. *The FASEB Journal,* 7(2), 349–360.

Atochin, D. N., Demchenko, I. T., Astern, J., Boso, A. E., Piantadosi, C. A., & Huang, P. L. (2003). Contributions of endothelial and neuronal nitric oxide synthases to cerebrovascular responses to hyperoxia. *Journal of Cerebral Blood Flow and Metabolism, 23*(10), 1219–1226. http://dx.doi.org/10.1097/01.WCB.0000089601.87125.E4.

Ayata, C., Ma, J., Meng, W., Huang, P., & Moskowitz, M. A. (1996). L-NA-sensitive rCBF augmentation during vibrissal stimulation in type III nitric oxide synthase mutant mice. *Journal of Cerebral Blood Flow and Metabolism, 16*(4), 539–541. http://dx.doi.org/10.1097/00004647-199607000-00002.

Bari, F., Louis, T. M., & Busija, D. W. (1997). Kainate-induced cerebrovascular dilation is resistant to ischemia in piglets. *Stroke, 28*(6), 1272–1276, discussion 1277.

Barone, P., & Kennedy, H. (2000). Non-uniformity of neocortex: Areal heterogeneity of NADPH-diaphorase reactive neurons in adult macaque monkeys. *Cerebral Cortex, 10*(2), 160–174.

Bauser-Heaton, H. D., & Bohlen, H. G. (2007). Cerebral microvascular dilation during hypotension and decreased oxygen tension: A role for nNOS. *American Journal of Physiology. Heart and Circulatory Physiology, 293*(4), H2193–H2201. http://dx.doi.org/10.1152/ajpheart.00190.2007.

Bauser-Heaton, H. D., Song, J., & Bohlen, H. G. (2008). Cerebral microvascular nNOS responds to lowered oxygen tension through a bumetanide-sensitive cotransporter and sodium-calcium exchanger. *American Journal of Physiology. Heart and Circulatory Physiology, 294*(5), H2166–H2173. http://dx.doi.org/10.1152/ajpheart.01074.2007.

Behzadi, Y., & Liu, T. T. (2005). An arteriolar compliance model of the cerebral blood flow response to neural stimulus. *NeuroImage, 25*(4), 1100–1111. http://dx.doi.org/10.1016/j.neuroimage.2004.12.057.

Benyo, Z., Gorlach, C., & Wahl, M. (1998a). Involvement of thromboxane A2 in the mediation of the contractile effect induced by inhibition of nitric oxide synthesis in isolated rat

middle cerebral arteries. *Journal of Cerebral Blood Flow and Metabolism, 18*(6), 616–618. http://dx.doi.org/10.1097/00004647-199806000-00003.
Benyo, Z., Gorlach, C., & Wahl, M. (1998b). Role of nitric oxide and thromboxane in the maintenance of cerebrovascular tone. *Kidney International. Supplement, 67*, S218–S220.
Bhardwaj, A., Northington, F. J., Ichord, R. N., Hanley, D. F., Traystman, R. J., & Koehler, R. C. (1997). Characterization of ionotropic glutamate receptor-mediated nitric oxide production in vivo in rats. *Stroke, 28*(4), 850–856, discussion 856–857.
Bonvento, G., Cholet, N., & Seylaz, J. (2000). Sustained attenuation of the cerebrovascular response to a 10 min whisker stimulation following neuronal nitric oxide synthase inhibition. *Neuroscience Research, 37*(2), 163–166.
Bonvento, G., Seylaz, J., & Lacombe, P. (1994). Widespread attenuation of the cerebrovascular reactivity to hypercapnia following inhibition of nitric oxide synthase in the conscious rat. *Journal of Cerebral Blood Flow and Metabolism, 14*(5), 699–703. http://dx.doi.org/10.1038/jcbfm.1994.90.
Boulanger, C. M., Heymes, C., Benessiano, J., Geske, R. S., Levy, B. I., & Vanhoutte, P. M. (1998). Neuronal nitric oxide synthase is expressed in rat vascular smooth muscle cells: Activation by angiotensin II in hypertension. *Circulation Research, 83*(12), 1271–1278.
Bredt, D. S., Hwang, P. M., Glatt, C. E., Lowenstein, C., Reed, R. R., & Snyder, S. H. (1991). Cloned and expressed nitric oxide synthase structurally resembles cytochrome P-450 reductase. *Nature, 351*(6329), 714–718. http://dx.doi.org/10.1038/351714a0.
Bredt, D. S., Hwang, P. M., & Snyder, S. H. (1990). Localization of nitric oxide synthase indicating a neural role for nitric oxide. *Nature, 347*(6295), 768–770. http://dx.doi.org/10.1038/347768a0.
Bredt, D. S., & Snyder, S. H. (1989). Nitric oxide mediates glutamate-linked enhancement of cGMP levels in the cerebellum. *Proceedings of the National Academy of Sciences of the United States of America, 86*(22), 9030–9033.
Brophy, C. M., Knoepp, L., Xin, J., & Pollock, J. S. (2000). Functional expression of NOS 1 in vascular smooth muscle. *American Journal of Physiology. Heart and Circulatory Physiology, 278*(3), H991–H997.
Browne, S. E., Ayata, C., Huang, P. L., Moskowitz, M. A., & Beal, M. F. (1999). The cerebral metabolic consequences of nitric oxide synthase deficiency: Glucose utilization in endothelial and neuronal nitric oxide synthase null mice. *Journal of Cerebral Blood Flow and Metabolism, 19*(2), 144–148.
Brown, L. A., Key, B. J., & Lovick, T. A. (2002). Inhibition of vasomotion in hippocampal cerebral arterioles during increases in neuronal activity. *Autonomic Neuroscience, 95*(1–2), 137–140.
Bryan, R. M., Jr., Marrelli, S. P., Steenberg, M. L., Schildmeyer, L. A., & Johnson, T. D. (2001). Effects of luminal shear stress on cerebral arteries and arterioles. *American Journal of Physiology Heart and Circulatory Physiology, 280*(5), H2011–H2022.
Bryan, R. M., Jr., Steenberg, M. L., & Marrelli, S. P. (2001). Role of endothelium in shear stress-induced constrictions in rat middle cerebral artery. *Stroke, 32*(6), 1394–1400.
Buchanan, J. E., & Phillis, J. W. (1993). The role of nitric oxide in the regulation of cerebral blood flow. *Brain Research, 610*(2), 248–255.
Burke, M., & Buhrle, Ch. (2006). BOLD response during uncoupling of neuronal activity and CBF. *NeuroImage, 32*(1), 1–8. http://dx.doi.org/10.1016/j.neuroimage.2006.03.035.
Busija, D. W. (1993). The role of central neural pathways in the regulation of cerebral blood flow. In J. W. Phillis (Ed.), *The regulation cerebral blood flow* (pp. 65–77). New York: CRC Press.
Busija, D. W., Leffler, C. W., & Wagerle, L. C. (1990). Mono-L-arginine-containing compounds dilate piglet pial arterioles via an endothelium-derived relaxing factor-like substance. *Circulation Research, 67*(6), 1374–1380.

Busse, R., & Fleming, I. (2003). Regulation of endothelium-derived vasoactive autacoid production by hemodynamic forces. *Trends in Pharmacological Sciences*, *24*(1), 24–29.

Caillol, M., Devinoy, E., Lacroix, M. C., & Schirar, A. (2000). Endothelial and neuronal nitric oxide synthases are present in the suprachiasmatic nuclei of Syrian hamsters and rats. *The European Journal of Neuroscience*, *12*(2), 649–661.

Cauli, B., Tong, X. K., Rancillac, A., Serluca, N., Lambolez, B., Rossier, J., et al. (2004). Cortical GABA interneurons in neurovascular coupling: Relays for subcortical vasoactive pathways. *The Journal of Neuroscience*, *24*(41), 8940–8949. http://dx.doi.org/10.1523/JNEUROSCI.3065-04.2004.

Chapman, A. C., Cipolla, M. J., & Chan, S. L. (2013). Effect of pregnancy and nitric oxide on the myogenic vasodilation of posterior cerebral arteries and the lower limit of cerebral blood flow autoregulation. *Reproductive Sciences*, *20*(9), 1046–1054. http://dx.doi.org/10.1177/1933719112473661.

Cho, H. J., Xie, Q. W., Calaycay, J., Mumford, R. A., Swiderek, K. M., Lee, T. D., et al. (1992). Calmodulin is a subunit of nitric oxide synthase from macrophages. *The Journal of Experimental Medicine*, *176*(2), 599–604.

Cholet, N., Bonvento, G., & Seylaz, J. (1996). Effect of neuronal NO synthase inhibition on the cerebral vasodilatory response to somatosensory stimulation. *Brain Research*, *708*(1–2), 197–200.

Cholet, N., Seylaz, J., Lacombe, P., & Bonvento, G. (1997). Local uncoupling of the cerebrovascular and metabolic responses to somatosensory stimulation after neuronal nitric oxide synthase inhibition. *Journal of Cerebral Blood Flow and Metabolism*, *17*(11), 1191–1201. http://dx.doi.org/10.1097/00004647-199711000-00008.

Christopherson, K. S., Hillier, B. J., Lim, W. A., & Bredt, D. S. (1999). PSD-95 assembles a ternary complex with the N-methyl-D-aspartic acid receptor and a bivalent neuronal NO synthase PDZ domain. *The Journal of Biological Chemistry*, *274*(39), 27467–27473.

Comini, L., Bachetti, T., Gaia, G., Pasini, E., Agnoletti, L., Pepi, P., et al. (1996). Aorta and skeletal muscle NO synthase expression in experimental heart failure. *Journal of Molecular and Cellular Cardiology*, *28*(11), 2241–2248. http://dx.doi.org/10.1006/jmcc.1996.0216.

Cosby, K., Partovi, K. S., Crawford, J. H., Patel, R. P., Reiter, C. D., Martyr, S., et al. (2003). Nitrite reduction to nitric oxide by deoxyhemoglobin vasodilates the human circulation. *Nature Medicine*, *9*(12), 1498–1505. http://dx.doi.org/10.1038/nm954.

de Labra, C., Rivadulla, C., Espinosa, N., Dasilva, M., Cao, R., & Cudeiro, J. (2009). Different sources of nitric oxide mediate neurovascular coupling in the lateral geniculate nucleus of the cat. *Frontiers in Systems Neuroscience*, *3*, 9. http://dx.doi.org/10.3389/neuro.06.009.2009.

Dejam, A., Hunter, C. J., Pelletier, M. M., Hsu, L. L., Machado, R. F., Shiva, S., et al. (2005). Erythrocytes are the major intravascular storage sites of nitrite in human blood. *Blood*, *106*(2), 734–739. http://dx.doi.org/10.1182/blood-2005-02-0567.

Dieguez, G., Fernandez, N., Sanchez, M. A., Martinez, M. A., Garcia-Villalon, A. L., Monge, L., et al. (1999). Role of nitric oxide in the cerebral circulation during hypotension after hemorrhage, ganglionic blockade and diazoxide in awake goats. *Brain Research*, *851*(1–2), 133–140.

Dinerman, J. L., Dawson, T. M., Schell, M. J., Snowman, A., & Snyder, S. H. (1994). Endothelial nitric oxide synthase localized to hippocampal pyramidal cells: Implications for synaptic plasticity. *Proceedings of the National Academy of Sciences of the United States of America*, *91*(10), 4214–4218.

Dirnagl, U., Kaplan, B., Jacewicz, M., & Pulsinelli, W. (1989). Continuous measurement of cerebral cortical blood flow by laser-Doppler flowmetry in a rat stroke model. *Journal of Cerebral Blood Flow and Metabolism*, *9*(5), 589–596. http://dx.doi.org/10.1038/jcbfm.1989.84.

Dirnagl, U., Lindauer, U., & Villringer, A. (1993a). Nitric oxide synthase blockade enhances vasomotion in the cerebral microcirculation of anesthetized rats. *Microvascular Research*, *45*(3), 318–323. http://dx.doi.org/10.1006/mvre.1993.1028.
Dirnagl, U., Lindauer, U., & Villringer, A. (1993b). Role of nitric oxide in the coupling of cerebral blood flow to neuronal activation in rats. *Neuroscience Letters*, *149*(1), 43–46.
Dirnagl, U., Niwa, K., Lindauer, U., & Villringer, A. (1994). Coupling of cerebral blood flow to neuronal activation: Role of adenosine and nitric oxide. *The American Journal of Physiology*, *267*(1 Pt. 2), H296–H301.
Doughty, J. M., & Langton, P. D. (1998). A transient dilatation of pressurised rat cerebral arteries during rapid pressure increases is mediated by nitric oxide. *Pflügers Archiv*, *436*(2), 220–226.
Dreier, J. P., Korner, K., Ebert, N., Gorner, A., Rubin, I., Back, T., et al. (1998). Nitric oxide scavenging by hemoglobin or nitric oxide synthase inhibition by N-nitro-L-arginine induces cortical spreading ischemia when K^+ is increased in the subarachnoid space. *Journal of Cerebral Blood Flow and Metabolism*, *18*(9), 978–990. http://dx.doi.org/10.1097/00004647-199809000-00007.
Drouin, A., & Thorin, E. (2009). Flow-induced dilation is mediated by Akt-dependent activation of endothelial nitric oxide synthase-derived hydrogen peroxide in mouse cerebral arteries. *Stroke*, *40*(5), 1827–1833. http://dx.doi.org/10.1161/STROKEAHA.108.536805.
Duchemin, S., Boily, M., Sadekova, N., & Girouard, H. (2012). The complex contribution of NOS interneurons in the physiology of cerebrovascular regulation. *Frontiers in Neural Circuits*, *6*, 51. http://dx.doi.org/10.3389/fncir.2012.00051.
Enager, P., Piilgaard, H., Offenhauser, N., Kocharyan, A., Fernandes, P., Hamel, E., et al. (2009). Pathway-specific variations in neurovascular and neurometabolic coupling in rat primary somatosensory cortex. *Journal of Cerebral Blood Flow and Metabolism*, *29*(5), 976–986. http://dx.doi.org/10.1038/jcbfm.2009.23.
Estrada, C., & DeFelipe, J. (1998). Nitric oxide-producing neurons in the neocortex: Morphological and functional relationship with intraparenchymal microvasculature. *Cerebral Cortex*, *8*(3), 193–203.
Faraci, F. M. (1990). Role of nitric oxide in regulation of basilar artery tone in vivo. *The American Journal of Physiology*, *259*(4 Pt. 2), H1216–H1221.
Faraci, F. M. (1991). Role of endothelium-derived relaxing factor in cerebral circulation: Large arteries vs. microcirculation. *The American Journal of Physiology*, *261*(4 Pt. 2), H1038–H1042.
Faraci, F. M., Breese, K. R., & Heistad, D. D. (1993). Nitric oxide contributes to dilatation of cerebral arterioles during seizures. *The American Journal of Physiology*, *265*(6 Pt. 2), H2209–H2212.
Faraci, F. M., Breese, K. R., & Heistad, D. D. (1994). Responses of cerebral arterioles to kainate. *Stroke*, *25*(10), 2080–2083, discussion 2084.
Faraci, F. M., & Brian, J. E., Jr. (1995). 7-Nitroindazole inhibits brain nitric oxide synthase and cerebral vasodilatation in response to N-methyl-D-aspartate. *Stroke*, *26*(11), 2172–2175, discussion 2176.
Feelisch, M., Fernandez, B. O., Bryan, N. S., Garcia-Saura, M. F., Bauer, S., Whitlock, D. R., et al. (2008). Tissue processing of nitrite in hypoxia: An intricate interplay of nitric oxide-generating and -scavenging systems. *The Journal of Biological Chemistry*, *283*(49), 33927–33934. http://dx.doi.org/10.1074/jbc.M806654200.
Feron, O., Belhassen, L., Kobzik, L., Smith, T. W., Kelly, R. A., & Michel, T. (1996). Endothelial nitric oxide synthase targeting to caveolae. Specific interactions with caveolin isoforms in cardiac myocytes and endothelial cells. *The Journal of Biological Chemistry*, *271*(37), 22810–22814.

Filosa, J. A., Bonev, A. D., & Nelson, M. T. (2004). Calcium dynamics in cortical astrocytes and arterioles during neurovascular coupling. *Circulation Research*, *95*(10), e73–e81. http://dx.doi.org/10.1161/01.RES.0000148636.60732.2e.

Fujii, K., Heistad, D. D., & Faraci, F. M. (1990). Vasomotion of basilar arteries in vivo. *The American Journal of Physiology*, *258*(6 Pt. 2), H1829–H1834.

Fujii, K., Heistad, D. D., & Faraci, F. M. (1991). Flow-mediated dilatation of the basilar artery in vivo. *Circulation Research*, *69*(3), 697–705.

Furchgott, R. F., & Vanhoutte, P. M. (1989). Endothelium-derived relaxing and contracting factors. *The FASEB Journal*, *3*(9), 2007–2018.

Gabbott, P. L., & Bacon, S. J. (1996). Localisation of NADPH diaphorase activity and NOS immunoreactivity in astroglia in normal adult rat brain. *Brain Research*, *714*(1–2), 135–144.

Gabbott, P. L., Dickie, B. G., Vaid, R. R., Headlam, A. J., & Bacon, S. J. (1997). Local-circuit neurones in the medial prefrontal cortex (areas 25, 32 and 24b) in the rat: Morphology and quantitative distribution. *The Journal of Comparative Neurology*, *377*(4), 465–499.

Gao, Y., Dhanakoti, S., Tolsa, J. F., & Raj, J. U. (1999). Role of protein kinase G in nitric oxide- and cGMP-induced relaxation of newborn ovine pulmonary veins. *Journal of Applied Physiology (1985)*, *87*(3), 993–998.

Garbossa, D., Fontanella, M., Tomasi, S., Ducati, A., & Vercelli, A. (2005). Differential distribution of NADPH-diaphorase histochemistry in human cerebral cortex. *Brain Research*, *1034*(1–2), 1–10. http://dx.doi.org/10.1016/j.brainres.2004.10.049.

Garthwaite, J., Garthwaite, G., Palmer, R. M., & Moncada, S. (1989). NMDA receptor activation induces nitric oxide synthesis from arginine in rat brain slices. *European Journal of Pharmacology*, *172*(4–5), 413–416.

Garthwaite, J., Southam, E., & Anderton, M. (1989). A kainate receptor linked to nitric oxide synthesis from arginine. *Journal of Neurochemistry*, *53*(6), 1952–1954.

Geary, G. G., Krause, D. N., & Duckles, S. P. (2000). Estrogen reduces mouse cerebral artery tone through endothelial NOS- and cyclooxygenase-dependent mechanisms. *American Journal of Physiology Heart and Circulatory Physiology*, *279*(2), H511–H519.

Girouard, H., & Iadecola, C. (2006). Neurovascular coupling in the normal brain and in hypertension, stroke, and Alzheimer disease. *Journal of Applied Physiology (1985)*, *100*(1), 328–335. http://dx.doi.org/10.1152/japplphysiol.00966.2005.

Girouard, H., Park, L., Anrather, J., Zhou, P., & Iadecola, C. (2007). Cerebrovascular nitrosative stress mediates neurovascular and endothelial dysfunction induced by angiotensin II. *Arteriosclerosis, Thrombosis, and Vascular Biology*, *27*(2), 303–309. http://dx.doi.org/10.1161/01.ATV.0000253885.41509.25.

Girouard, H., Wang, G., Gallo, E. F., Anrather, J., Zhou, P., Pickel, V. M., et al. (2009). NMDA receptor activation increases free radical production through nitric oxide and NOX2. *The Journal of Neuroscience*, *29*(8), 2545–2552. http://dx.doi.org/10.1523/JNEUROSCI.0133-09.2009.

Gotoh, J., Kuang, T. Y., Nakao, Y., Cohen, D. M., Melzer, P., Itoh, Y., et al. (2001). Regional differences in mechanisms of cerebral circulatory response to neuronal activation. *American Journal of Physiology. Heart and Circulatory Physiology*, *280*(2), H821–H829.

Gratton, J. P., Lin, M. I., Yu, J., Weiss, E. D., Jiang, Z. L., Fairchild, T. A., et al. (2003). Selective inhibition of tumor microvascular permeability by cavtratin blocks tumor progression in mice. *Cancer Cell*, *4*(1), 31–39.

Greenberg, J. H., Sohn, N. W., & Hand, P. J. (1999). Nitric oxide and the cerebral-blood-flow response to somatosensory activation following deafferentation. *Experimental Brain Research*, *129*(4), 541–550.

Griffith, O. W., & Kilbourn, R. G. (1996). Nitric oxide synthase inhibitors: Amino acids. *Methods in Enzymology, 268*, 375–392.

Hama-Tomioka, K., Kinoshita, H., Nakahata, K., Kondo, T., Azma, T., Kawahito, S., et al. (2012). Roles of neuronal nitric oxide synthase, oxidative stress, and propofol in N-methyl-D-aspartate-induced dilatation of cerebral arterioles. *British Journal of Anaesthesia, 108*(1), 21–29. http://dx.doi.org/10.1093/bja/aer368.

Hamilton, N., Vayro, S., Wigley, R., & Butt, A. M. (2010). Axons and astrocytes release ATP and glutamate to evoke calcium signals in NG2-glia. *Glia, 58*(1), 66–79. http://dx.doi.org/10.1002/glia.20902.

Hara, H., Zhang, Q. J., Kuroyanagi, T., & Kobayashi, S. (1993). Parasympathetic cerebrovascular innervation: An anterograde tracing from the sphenopalatine ganglion in the rat. *Neurosurgery, 32*(5), 822–827, discussion 827.

Hatakeyama, T., Pappas, P. J., Hobson, R. W., 2nd., Boric, M. P., Sessa, W. C., & Duran, W. N. (2006). Endothelial nitric oxide synthase regulates microvascular hyperpermeability in vivo. *The Journal of Physiology, 574*(Pt. 1), 275–281. http://dx.doi.org/10.1113/jphysiol.2006.108175.

Hayashi, T., Katsumi, Y., Mukai, T., Inoue, M., Nagahama, Y., Oyanagi, C., et al. (2002). Neuronal nitric oxide has a role as a perfusion regulator and a synaptic modulator in cerebellum but not in neocortex during somatosensory stimulation—An animal PET study. *Neuroscience Research, 44*(2), 155–165.

Hayashi, S., Osawa, T., & Tohyama, K. (2002). Comparative observations on corneas, with special reference to Bowman's layer and Descemet's membrane in mammals and amphibians. *Journal of Morphology, 254*(3), 247–258. http://dx.doi.org/10.1002/jmor.10030.

Hellyer, P. W., Johnson, L. W., & Olson, N. C. (1997). Effect of NG-nitro-L-arginine-methyl-ester on cardiopulmonary function and biosynthesis of cyclooxygenase products during porcine endotoxemia. *Critical Care Medicine, 25*(6), 1051–1058.

Hoffmeyer, H. W., Enager, P., Thomsen, K. J., & Lauritzen, M. J. (2007). Nonlinear neurovascular coupling in rat sensory cortex by activation of transcallosal fibers. *Journal of Cerebral Blood Flow and Metabolism, 27*(3), 575–587. http://dx.doi.org/10.1038/sj.jcbfm.9600372.

Hortobagyi, L., Kis, B., Hrabak, A., Horvath, B., Huszty, G., Schweer, H., et al. (2007). Adaptation of the hypothalamic blood flow to chronic nitric oxide deficiency is independent of vasodilator prostanoids. *Brain Research, 1131*(1), 129–137. http://dx.doi.org/10.1016/j.brainres.2006.11.009.

Huang, P. L., Dawson, T. M., Bredt, D. S., Snyder, S. H., & Fishman, M. C. (1993). Targeted disruption of the neuronal nitric oxide synthase gene. *Cell, 75*(7), 1273–1286.

Huang, Z., Huang, P. L., Ma, J., Meng, W., Ayata, C., Fishman, M. C., et al. (1996). Enlarged infarcts in endothelial nitric oxide synthase knockout mice are attenuated by nitro-L-arginine. *Journal of Cerebral Blood Flow and Metabolism, 16*(5), 981–987. http://dx.doi.org/10.1097/00004647-199609000-00023.

Huang, Z., Huang, P. L., Panahian, N., Dalkara, T., Fishman, M. C., & Moskowitz, M. A. (1994). Effects of cerebral ischemia in mice deficient in neuronal nitric oxide synthase. *Science, 265*(5180), 1883–1885.

Huard, J. M., Forster, C. C., Carter, M. L., Sicinski, P., & Ross, M. E. (1999). Cerebellar histogenesis is disturbed in mice lacking cyclin D2. *Development, 126*(9), 1927–1935.

Hudetz, A. G., Shen, H., & Kampine, J. P. (1998). Nitric oxide from neuronal NOS plays critical role in cerebral capillary flow response to hypoxia. *The American Journal of Physiology, 274*(3 Pt. 2), H982–H989.

Hughson, R. L., Edwards, M. R., O'Leary, D. D., & Shoemaker, J. K. (2001). Critical analysis of cerebrovascular autoregulation during repeated head-up tilt. *Stroke, 32*(10), 2403–2408.

Hundley, W. G., Renaldo, G. J., Levasseur, J. E., & Kontos, H. A. (1988). Vasomotion in cerebral microcirculation of awake rabbits. *The American Journal of Physiology*, *254*(1 Pt. 2), H67–H71.

Iadecola, C. (1993). Regulation of the cerebral microcirculation during neural activity: Is nitric oxide the missing link? *Trends in Neurosciences*, *16*(6), 206–214.

Iadecola, C., & Xu, X. (1994). Nitro-L-arginine attenuates hypercapnic cerebrovasodilation without affecting cerebral metabolism. *The American Journal of Physiology*, *266*(2 Pt. 2), R518–R525.

Iadecola, C., & Zhang, F. (1994). Nitric oxide-dependent and -independent components of cerebrovasodilation elicited by hypercapnia. *The American Journal of Physiology*, *266*(2 Pt. 2), R546–R552.

Iadecola, C., & Zhang, F. (1996). Permissive and obligatory roles of NO in cerebrovascular responses to hypercapnia and acetylcholine. *The American Journal of Physiology*, *271*(4 Pt. 2), R990–R1001.

Iadecola, C., Zhang, F., & Xu, X. (1993). Role of nitric oxide synthase-containing vascular nerves in cerebrovasodilation elicited from cerebellum. *The American Journal of Physiology*, *264*(4 Pt. 2), R738–R746.

Iadecola, C., Zhang, F., & Xu, X. (1994). SIN-1 reverses attenuation of hypercapnic cerebrovasodilation by nitric oxide synthase inhibitors. *The American Journal of Physiology*, *267*(1 Pt. 2), R228–R235.

Iadecola, C., Zhang, F., & Xu, X. (1995). Inhibition of inducible nitric oxide synthase ameliorates cerebral ischemic damage. *The American Journal of Physiology*, *268*(1 Pt. 2), R286–R292.

Irikura, K., Maynard, K. I., & Moskowitz, M. A. (1994). Importance of nitric oxide synthase inhibition to the attenuated vascular responses induced by topical L-nitroarginine during vibrissal stimulation. *Journal of Cerebral Blood Flow and Metabolism*, *14*(1), 45–48. http://dx.doi.org/10.1038/jcbfm.1994.7.

Ito, M. (1991). The cellular basis of cerebellar plasticity. *Current Opinion in Neurobiology*, *1*(4), 616–620.

Jayakumar, A. R., Sujatha, R., Paul, V., Asokan, C., Govindasamy, S., & Jayakumar, R. (1999). Role of nitric oxide on GABA, glutamic acid, activities of GABA-T and GAD in rat brain cerebral cortex. *Brain Research*, *837*(1–2), 229–235.

Jones, S. C., Easley, K. A., Radinsky, C. R., Chyatte, D., Furlan, A. J., & Perez-Trepichio, A. D. (2003). Nitric oxide synthase inhibition depresses the height of the cerebral blood flow-pressure autoregulation curve during moderate hypotension. *Journal of Cerebral Blood Flow and Metabolism*, *23*(9), 1085–1095. http://dx.doi.org/10.1097/01.WCB.0000081202.00668.FB.

Jones, S. C., Radinsky, C. R., Furlan, A. J., Chyatte, D., & Perez-Trepichio, A. D. (1999). Cortical NOS inhibition raises the lower limit of cerebral blood flow-arterial pressure autoregulation. *The American Journal of Physiology*, *276*(4 Pt. 2), H1253–H1262.

Joshi, S., Hartl, R., Sun, L. S., Libow, A. D., Wang, M., Pile-Spellman, J., et al. (2003). Despite in vitro increase in cyclic guanosine monophosphate concentrations, intracarotid nitroprusside fails to augment cerebral blood flow of healthy baboons. *Anesthesiology*, *98*(2), 412–419.

Jubelin, B. C., & Gierman, J. L. (1996). Erythrocytes may synthesize their own nitric oxide. *American Journal of Hypertension*, *9*(12 Pt. 1), 1214–1219. http://dx.doi.org/10.1016/S0895-7061(96)00257-9.

Kamper, A. M., Spilt, A., de Craen, A. J., van Buchem, M. A., Westendorp, R. G., & Blauw, G. J. (2004). Basal cerebral blood flow is dependent on the nitric oxide pathway in elderly but not in young healthy men. *Experimental Gerontology*, *39*(8), 1245–1248. http://dx.doi.org/10.1016/j.exger.2004.04.001.

Karagiannis, A., Gallopin, T., David, C., Battaglia, D., Geoffroy, H., Rossier, J., et al. (2009). Classification of NPY-expressing neocortical interneurons. *The Journal of Neuroscience*, *29*(11), 3642–3659. http://dx.doi.org/10.1523/JNEUROSCI.0058-09.2009.

Katakam, P. V., Domoki, F., Snipes, J. A., Busija, A. R., Jarajapu, Y. P., & Busija, D. W. (2009). Impaired mitochondria-dependent vasodilation in cerebral arteries of Zucker obese rats with insulin resistance. *American Journal of Physiology. Regulatory, Integrative and Comparative Physiology, 296*(2), R289–R298. http://dx.doi.org/10.1152/ajpregu.90656.2008.

Katakam, P. V., Wappler, E. A., Katz, P. S., Rutkai, I., Institoris, A., Domoki, F., et al. (2013). Depolarization of mitochondria in endothelial cells promotes cerebral artery vasodilation by activation of nitric oxide synthase. *Arteriosclerosis, Thrombosis, and Vascular Biology, 33*(4), 752–759. http://dx.doi.org/10.1161/ATVBAHA.112.300560.

Kelly, P. A., Ritchie, I. M., & Arbuthnott, G. W. (1995). Inhibition of neuronal nitric oxide synthase by 7-nitroindazole: Effects upon local cerebral blood flow and glucose use in the rat. *Journal of Cerebral Blood Flow and Metabolism, 15*(5), 766–773. http://dx.doi.org/10.1038/jcbfm.1995.97.

Kitaura, H., Uozumi, N., Tohmi, M., Yamazaki, M., Sakimura, K., Kudoh, M., et al. (2007). Roles of nitric oxide as a vasodilator in neurovascular coupling of mouse somatosensory cortex. *Neuroscience Research, 59*(2), 160–171. http://dx.doi.org/10.1016/j.neures.2007.06.1469.

Kiyasova, V., Fernandez, S. P., Laine, J., Stankovski, L., Muzerelle, A., Doly, S., et al. (2011). A genetically defined morphologically and functionally unique subset of 5-HT neurons in the mouse raphe nuclei. *The Journal of Neuroscience, 31*(8), 2756–2768. http://dx.doi.org/10.1523/JNEUROSCI.4080-10.2011.

Kleinbongard, P., Dejam, A., Lauer, T., Rassaf, T., Schindler, A., Picker, O., et al. (2003). Plasma nitrite reflects constitutive nitric oxide synthase activity in mammals. *Free Radical Biology and Medicine, 35*(7), 790–796.

Kleinbongard, P., Schulz, R., Rassaf, T., Lauer, T., Dejam, A., Jax, T., et al. (2006). Red blood cells express a functional endothelial nitric oxide synthase. *Blood, 107*(7), 2943–2951. http://dx.doi.org/10.1182/blood-2005-10-3992.

Kovach, A. G., Szabo, C., Benyo, Z., Csaki, C., Greenberg, J. H., & Reivich, M. (1992). Effects of NG-nitro-L-arginine and L-arginine on regional cerebral blood flow in the cat. *The Journal of Physiology, 449*, 183–196.

Kubota, Y., Hattori, R., & Yui, Y. (1994). Three distinct subpopulations of GABAergic neurons in rat frontal agranular cortex. *Brain Research, 649*(1–2), 159–173.

Kukreja, R. C., Wei, E. P., Kontos, H. A., & Bates, J. N. (1993). Nitric oxide and S-nitroso-L-cysteine as endothelium-derived relaxing factors from acetylcholine in cerebral vessels in cats. *Stroke, 24*(12), 2010–2014, discussion 2014–2015.

Lacombe, P., & Seylaz, J. (1984). Significance of the cerebrovascular effects of immobilization stress in the rabbit. *Journal of Cerebral Blood Flow and Metabolism, 4*(3), 397–406. http://dx.doi.org/10.1038/jcbfm.1984.58.

Lauritzen, M. (2005). Reading vascular changes in brain imaging: Is dendritic calcium the key? *Nature Reviews Neuroscience, 6*(1), 77–85. http://dx.doi.org/10.1038/nrn1589.

Lecrux, C., Kocharyan, A., Sandoe, C. H., Tong, X. K., & Hamel, E. (2012). Pyramidal cells and cytochrome P450 epoxygenase products in the neurovascular coupling response to basal forebrain cholinergic input. *Journal of Cerebral Blood Flow and Metabolism, 32*(5), 896–906. http://dx.doi.org/10.1038/jcbfm.2012.4.

Leffler, C. W., Parfenova, H., Fedinec, A. L., Basuroy, S., & Tcheranova, D. (2006). Contributions of astrocytes and CO to pial arteriolar dilation to glutamate in newborn pigs. *American Journal of Physiology. Heart and Circulatory Physiology, 291*(6), H2897–H2904. http://dx.doi.org/10.1152/ajpheart.00722.2006.

Li, H., Cui, H., Kundu, T. K., Alzawahra, W., & Zweier, J. L. (2008). Nitric oxide production from nitrite occurs primarily in tissues not in the blood: Critical role of xanthine oxidase and aldehyde oxidase. *The Journal of Biological Chemistry*, *283*(26), 17855–17863. http://dx.doi.org/10.1074/jbc.M801785200.

Li, H., Kundu, T. K., & Zweier, J. L. (2009). Characterization of the magnitude and mechanism of aldehyde oxidase-mediated nitric oxide production from nitrite. *The Journal of Biological Chemistry*, *284*(49), 33850–33858. http://dx.doi.org/10.1074/jbc.M109.019125.

Lin, L. H., Taktakishvili, O., & Talman, W. T. (2007). Identification and localization of cell types that express endothelial and neuronal nitric oxide synthase in the rat nucleus tractus solitarii. *Brain Research*, *1171*, 42–51. http://dx.doi.org/10.1016/j.brainres.2007.07.057.

Lindauer, U., Megow, D., Matsuda, H., & Dirnagl, U. (1999). Nitric oxide: A modulator, but not a mediator, of neurovascular coupling in rat somatosensory cortex. *The American Journal of Physiology*, *277*(2 Pt. 2), H799–H811.

Lindauer, U., Megow, D., Schultze, J., Weber, J. R., & Dirnagl, U. (1996). Nitric oxide synthase inhibition does not affect somatosensory evoked potentials in the rat. *Neuroscience Letters*, *216*(3), 207–210.

Liu, Y., Li, H., Bubolz, A. H., Zhang, D. X., & Gutterman, D. D. (2008). Endothelial cytoskeletal elements are critical for flow-mediated dilation in human coronary arterioles. *Medical and Biological Engineering and Computing*, *46*(5), 469–478. http://dx.doi.org/10.1007/s11517-008-0331-1.

Liu, X., Li, C., Falck, J. R., Roman, R. J., Harder, D. R., & Koehler, R. C. (2008). Interaction of nitric oxide, 20-HETE, and EETs during functional hyperemia in whisker barrel cortex. *American Journal of Physiology. Heart and Circulatory Physiology*, *295*(2), H619–H631. http://dx.doi.org/10.1152/ajpheart.01211.2007.

Lonart, G., Wang, J., & Johnson, K. M. (1992). Nitric oxide induces neurotransmitter release from hippocampal slices. *European Journal of Pharmacology*, *220*(2–3), 271–272.

Luders, J. C., Weihl, C. C., Lin, G., Ghadge, G., Stoodley, M., Roos, R. P., et al. (2000). Adenoviral gene transfer of nitric oxide synthase increases cerebral blood flow in rats. *Neurosurgery*, *47*(5), 1206–1214, discussion 1214–1205.

Ma, J., Ayata, C., Huang, P. L., Fishman, M. C., & Moskowitz, M. A. (1996). Regional cerebral blood flow response to vibrissal stimulation in mice lacking type I NOS gene expression. *The American Journal of Physiology*, *270*(3 Pt 2), H1085–H1090.

MacKenzie, E. T., Strandgaard, S., Graham, D. I., Jones, J. V., Harper, A. M., & Farrar, J. K. (1976). Effects of acutely induced hypertension in cats on pial arteriolar caliber, local cerebral blood flow, and the blood-brain barrier. *Circulation Research*, *39*(1), 33–41.

Macrae, I. M., Dawson, D. A., Norrie, J. D., & McCulloch, J. (1993). Inhibition of nitric oxide synthesis: Effects on cerebral blood flow and glucose utilisation in the rat. *Journal of Cerebral Blood Flow and Metabolism*, *13*(6), 985–992. http://dx.doi.org/10.1038/jcbfm.1993.123.

Madden, J. A., & Christman, N. J. (1999). Integrin signaling, free radicals, and tyrosine kinase mediate flow constriction in isolated cerebral arteries. *The American Journal of Physiology*, *277*(6 Pt 2), H2264–H2271.

Makhro, A., Wang, J., Vogel, J., Boldyrev, A. A., Gassmann, M., Kaestner, L., et al. (2010). Functional NMDA receptors in rat erythrocytes. *American Journal of Physiology. Cell Physiology*, *298*(6), C1315–C1325. http://dx.doi.org/10.1152/ajpcell.00407.2009.

Manzoni, O., & Bockaert, J. (1993). Nitric oxide synthase activity endogenously modulates NMDA receptors. *Journal of Neurochemistry*, *61*(1), 368–370.

Marks, J. D., Boriboun, C., & Wang, J. (2005). Mitochondrial nitric oxide mediates decreased vulnerability of hippocampal neurons from immature animals to NMDA. *The Journal of Neuroscience*, *25*(28), 6561–6575. http://dx.doi.org/10.1523/JNEUROSCI.1450-05.2005.

Matsumoto, T., Nakane, M., Pollock, J. S., Kuk, J. E., & Forstermann, U. (1993). A correlation between soluble brain nitric oxide synthase and NADPH-diaphorase activity is only seen after exposure of the tissue to fixative. *Neuroscience Letters*, *155*(1), 61–64.

McPherson, R. W., Kirsch, J. R., Ghaly, R. F., & Traystman, R. J. (1995). Effect of nitric oxide synthase inhibition on the cerebral vascular response to hypercapnia in primates. *Stroke*, *26*(4), 682–687.

Meng, W., Tobin, J. R., & Busija, D. W. (1995). Glutamate-induced cerebral vasodilation is mediated by nitric oxide through N-methyl-D-aspartate receptors. *Stroke*, *26*(5), 857–862, discussion 863.

Mesulam, M. M., Hersh, L. B., Mash, D. C., & Geula, C. (1992). Differential cholinergic innervation within functional subdivisions of the human cerebral cortex: A choline acetyltransferase study. *The Journal of Comparative Neurology*, *318*(3), 316–328. http://dx.doi.org/10.1002/cne.903180308.

Mikula, I., Durocher, S., Martasek, P., Mutus, B., & Slama-Schwok, A. (2009). Isoform-specific differences in the nitrite reductase activity of nitric oxide synthases under hypoxia. *The Biochemical Journal*, *418*(3), 673–682. http://dx.doi.org/10.1042/BJ20080987.

Moncada, S. (1992a). The 1991 Ulf von Euler lecture. The L-arginine: Nitric oxide pathway. *Acta Physiologica Scandinavica*, *145*(3), 201–227. http://dx.doi.org/10.1111/j.1748-1716.1992.tb09359.x.

Moncada, S. (1992b). Nitric oxide gas: Mediator, modulator, and pathophysiologic entity. *The Journal of Laboratory and Clinical Medicine*, *120*(2), 187–191.

Montecot, C., Borredon, J., Seylaz, J., & Pinard, E. (1997). Nitric oxide of neuronal origin is involved in cerebral blood flow increase during seizures induced by kainate. *Journal of Cerebral Blood Flow and Metabolism*, *17*(1), 94–99. http://dx.doi.org/10.1097/00004647-199701000-00012.

Ngai, A. C., Meno, J. R., & Winn, H. R. (1995). L-NNA suppresses cerebrovascular response and evoked potentials during somatosensory stimulation in rats. *The American Journal of Physiology*, *269*(5 Pt 2), H1803–H1810.

Northington, F. J., Matherne, G. P., & Berne, R. M. (1992). Competitive inhibition of nitric oxide synthase prevents the cortical hyperemia associated with peripheral nerve stimulation. *Proceedings of the National Academy of Sciences of the United States of America*, *89*(14), 6649–6652.

Ohata, H., Cao, S., & Koehler, R. C. (2006). Contribution of adenosine A2A and A2B receptors and heme oxygenase to AMPA-induced dilation of pial arterioles in rats. *American Journal of Physiology. Regulatory, Integrative and Comparative Physiology*, *291*(3), R728–R735. http://dx.doi.org/10.1152/ajpregu.00757.2005.

Okamoto, M., Etani, H., Yagita, Y., Kinoshita, N., & Nukada, T. (2001). Diminished reserve for cerebral vasomotor response to L-arginine in the elderly: Evaluation by transcranial Doppler sonography. *Gerontology*, *47*(3), 131–135. http://dx.doi.org/10.1159/000052786.

Oyama, N., Yagita, Y., Sasaki, T., Omura-Matsuoka, E., Terasaki, Y., Sugiyama, Y., et al. (2010). An angiotensin II type 1 receptor blocker can preserve endothelial function and attenuate brain ischemic damage in spontaneously hypertensive rats. *Journal of Neuroscience Research*, *88*(13), 2889–2898. http://dx.doi.org/10.1002/jnr.22441.

Palay, S. L. (1974). *Cerebellar cortex*. New York, NY: Springer-Verlag.

Palmer, R. M., Ferrige, A. G., & Moncada, S. (1987). Nitric oxide release accounts for the biological activity of endothelium-derived relaxing factor. *Nature*, *327*(6122), 524–526. http://dx.doi.org/10.1038/327524a0.

Palmer, R. M., Rees, D. D., Ashton, D. S., & Moncada, S. (1988). L-Arginine is the physiological precursor for the formation of nitric oxide in endothelium-dependent relaxation. *Biochemical and Biophysical Research Communications*, *153*(3), 1251–1256.

Panerai, R. B., Hudson, V., Fan, L., Mahony, P., Yeoman, P. M., Hope, T., et al. (2002). Assessment of dynamic cerebral autoregulation based on spontaneous fluctuations in arterial blood pressure and intracranial pressure. *Physiological Measurement*, *23*(1), 59–72.

Paravicini, T. M., Miller, A. A., Drummond, G. R., & Sobey, C. G. (2006). Flow-induced cerebral vasodilatation in vivo involves activation of phosphatidylinositol-3 kinase, NADPH-oxidase, and nitric oxide synthase. *Journal of Cerebral Blood Flow and Metabolism*, *26*(6), 836–845. http://dx.doi.org/10.1038/sj.jcbfm.9600235.

Paulson, O. B., Strandgaard, S., & Edvinsson, L. (1990). Cerebral autoregulation. *Cerebrovascular and Brain Metabolism Reviews*, *2*(2), 161–192.

Pelligrino, D. A., Gay, R. L., 3rd., Baughman, V. L., & Wang, Q. (1996). NO synthase inhibition modulates NMDA-induced changes in cerebral blood flow and EEG activity. *The American Journal of Physiology*, *271*(3 Pt 2), H990–H995.

Pelligrino, D. A., Wang, Q., Koenig, H. M., & Albrecht, R. F. (1995). Role of nitric oxide, adenosine, N-methyl-D-aspartate receptors, and neuronal activation in hypoxia-induced pial arteriolar dilation in rats. *Brain Research*, *704*(1), 61–70.

Peng, X., Zhang, C., Alkayed, N. J., Harder, D. R., & Koehler, R. C. (2004). Dependency of cortical functional hyperemia to forepaw stimulation on epoxygenase and nitric oxide synthase activities in rats. *Journal of Cerebral Blood Flow and Metabolism*, *24*(5), 509–517. http://dx.doi.org/10.1097/00004647-200405000-00004.

Perea, G., & Araque, A. (2005). Properties of synaptically evoked astrocyte calcium signal reveal synaptic information processing by astrocytes. *The Journal of Neuroscience*, *25*(9), 2192–2203. http://dx.doi.org/10.1523/JNEUROSCI.3965-04.2005.

Perrenoud, Q., Rossier, J., Ferezou, I., Geoffroy, H., Gallopin, T., Vitalis, T., et al. (2012). Activation of cortical 5-HT(3) receptor-expressing interneurons induces NO mediated vasodilatations and NPY mediated vasoconstrictions. *Frontiers in Neural Circuits*, *6*, 50. http://dx.doi.org/10.3389/fncir.2012.00050.

Petersen, C. C., Grinvald, A., & Sakmann, B. (2003). Spatiotemporal dynamics of sensory responses in layer 2/3 of rat barrel cortex measured in vivo by voltage-sensitive dye imaging combined with whole-cell voltage recordings and neuron reconstructions. *The Journal of Neuroscience*, *23*(4), 1298–1309.

Piggott, L. A., Hassell, K. A., Berkova, Z., Morris, A. P., Silberbach, M., & Rich, T. C. (2006). Natriuretic peptides and nitric oxide stimulate cGMP synthesis in different cellular compartments. *The Journal of General Physiology*, *128*(1), 3–14. http://dx.doi.org/10.1085/jgp.200509403.

Prado, R., Watson, B. D., Kuluz, J., & Dietrich, W. D. (1992). Endothelium-derived nitric oxide synthase inhibition. Effects on cerebral blood flow, pial artery diameter, and vascular morphology in rats. *Stroke*, *23*(8), 1118–1123, discussion 1124.

Putzke, J., Seidel, B., Huang, P. L., & Wolf, G. (2000). Differential expression of alternatively spliced isoforms of neuronal nitric oxide synthase (nNOS) and N-methyl-D-aspartate receptors (NMDAR) in knockout mice deficient in nNOS alpha (nNOS alpha(delta/delta) mice). *Brain Research. Molecular Brain Research*, *85*(1–2), 13–23.

Rancillac, A., Rossier, J., Guille, M., Tong, X. K., Geoffroy, H., Amatore, C., et al. (2006). Glutamatergic control of microvascular tone by distinct GABA neurons in the cerebellum. *The Journal of Neuroscience*, *26*(26), 6997–7006. http://dx.doi.org/10.1523/JNEUROSCI.5515-05.2006.

Raszkiewicz, J. L., Linville, D. G., Kerwin, J. F., Jr., Wagenaar, F., & Arneric, S. P. (1992). Nitric oxide synthase is critical in mediating basal forebrain regulation of cortical cerebral circulation. *Journal of Neuroscience Research*, *33*(1), 129–135. http://dx.doi.org/10.1002/jnr.490330116.

Rayshubskiy, A., Wojtasiewicz, T. J., Mikell, C. B., Bouchard, M. B., Timerman, D., Youngerman, B. E., et al. (2014). Direct, intraoperative observation of ~0.1 Hz hemodynamic oscillations in awake human cortex: Implications for fMRI. *NeuroImage*, *87*, 323–331. http://dx.doi.org/10.1016/j.neuroimage.2013.10.044.

Renno, W. M., Lee, J. H., & Beitz, A. J. (1997). Light and electron microscopic immunohistochemical localization of N-acetylaspartylglutamate (NAAG) in the olivocerebellar pathway of the rat. *Synapse*, *26*(2), 140–154. http://dx.doi.org/10.1002/(SICI)1098-2396(199706)26:2<140::AID-SYN5>3.0.CO;2-8.

Rosenblum, W. I. (1988). Loss of endothelium-dependent relaxation in mouse cerebral microvessels may be rapidly reversible. *Microvascular Research*, *35*(1), 132–138.

Rosenblum, W. I., Nishimura, H., & Nelson, G. H. (1990). Endothelium-dependent L-Arg- and L-NMMA-sensitive mechanisms regulate tone of brain microvessels. *The American Journal of Physiology*, *259*(5 Pt 2), H1396–H1401.

Saito, S., Wilson, D. A., Hanley, D. F., & Traystman, R. J. (1994). Nitric oxide synthase does not contribute to cerebral autoregulatory phenomenon in anesthetized dogs. *Journal of the Autonomic Nervous System*, *49*(Suppl.), S73–S76.

Salvemini, D. (1997). Regulation of cyclooxygenase enzymes by nitric oxide. *Cellular and Molecular Life Sciences*, *53*(7), 576–582.

Samouilov, A., Woldman, Y. Y., Zweier, J. L., & Khramtsov, V. V. (2007). Magnetic resonance study of the transmembrane nitrite diffusion. *Nitric Oxide*, *16*(3), 362–370. http://dx.doi.org/10.1016/j.niox.2006.12.006.

Sase, K., & Michel, T. (1995). Expression of constitutive endothelial nitric oxide synthase in human blood platelets. *Life Sciences*, *57*(22), 2049–2055.

Saur, D., Paehge, H., Schusdziarra, V., & Allescher, H. D. (2000). Distinct expression of splice variants of neuronal nitric oxide synthase in the human gastrointestinal tract. *Gastroenterology*, *118*(5), 849–858.

Schuman, E. M., & Madison, D. V. (1994). Nitric oxide and synaptic function. *Annual Review of Neuroscience*, *17*, 153–183. http://dx.doi.org/10.1146/annurev.ne.17.030194.001101.

Seylaz, J., Hara, H., Pinard, E., Mraovitch, S., MacKenzie, E. T., & Edvinsson, L. (1988). Effect of stimulation of the sphenopalatine ganglion on cortical blood flow in the rat. *Journal of Cerebral Blood Flow and Metabolism*, *8*(6), 875–878. http://dx.doi.org/10.1038/jcbfm.1988.145.

Shapiro, H. M., Stromberg, D. D., Lee, D. R., & Wiederhielm, C. A. (1971). Dynamic pressures in the pial arterial microcirculation. *The American Journal of Physiology*, *221*(1), 279–283.

Shaul, P. W. (2002). Regulation of endothelial nitric oxide synthase: Location, location, location. *Annual Review of Physiology*, *64*, 749–774. http://dx.doi.org/10.1146/annurev.physiol.64.081501.155952.

Skinhoj, E., & Strandgaard, S. (1973). Pathogenesis of hypertensive encephalopathy. *Lancet*, *1*(7801), 461–462.

Sobey, C. G., & Faraci, F. M. (1997a). Effect of nitric oxide and potassium channel agonists and inhibitors on basilar artery diameter. *The American Journal of Physiology*, *272*(1 Pt. 2), H256–H262.

Sobey, C. G., & Faraci, F. M. (1997b). Effects of a novel inhibitor of guanylyl cyclase on dilator responses of mouse cerebral arterioles. *Stroke*, *28*(4), 837–842, discussion 842–833.

Southam, E., Morris, R., & Garthwaite, J. (1992). Sources and targets of nitric oxide in rat cerebellum. *Neuroscience Letters*, *137*(2), 241–244.

Spatz, M., Stanimirovic, D., Uematsu, S., Roberts, L. J., 2nd., Bembry, J., & McCarron, R. M. (1993). Prostaglandin D2 and endothelin-1 induce the production of prostaglandin F2 alpha, 9 alpha, 11 beta-prostaglandin F2, prostaglandin E2, and thromboxane in capillary endothelium of human brain. *Prostaglandins, Leukotrienes, and Essential Fatty Acids*, *49*(4), 789–793.

Stadler, J., Harbrecht, B. G., Di Silvio, M., Curran, R. D., Jordan, M. L., Simmons, R. L., et al. (1993). Endogenous nitric oxide inhibits the synthesis of cyclooxygenase products and interleukin-6 by rat Kupffer cells. *Journal of Leukocyte Biology*, *53*(2), 165–172.

Stamler, J. S. (1994). Redox signaling: Nitrosylation and related target interactions of nitric oxide. *Cell, 78*(6), 931–936.
Stefanovic, B., Schwindt, W., Hoehn, M., & Silva, A. C. (2007). Functional uncoupling of hemodynamic from neuronal response by inhibition of neuronal nitric oxide synthase. *Journal of Cerebral Blood Flow and Metabolism, 27*(4), 741–754. http://dx.doi.org/10.1038/sj.jcbfm.9600377.
Suzuki, N., Fukuuchi, Y., Koto, A., Naganuma, Y., Isozumi, K., Konno, S., et al. (1994). Distribution and origins of cerebrovascular NADPH-diaphorase-containing nerve fibers in the rat. *Journal of the Autonomic Nervous System, 49*(Suppl.), S51–S54.
Takahashi, S., Cook, M., Jehle, J., Kennedy, C., & Sokoloff, L. (1995). Lack of effects of inhibition of nitric oxide synthesis on local glucose utilization in the rat brain. *Journal of Neurochemistry, 65*(1), 414–419.
Talman, W. T., & Dragon, D. N. (1995). Inhibition of nitric oxide synthesis extends cerebrovascular autoregulation during hypertension. *Brain Research, 672*(1–2), 48–54.
Talman, W. T., & Dragon, D. N. (2000). Parasympathetic nerves influence cerebral blood flow during hypertension in rat. *Brain Research, 873*(1), 145–148.
Talman, W. T., Dragon, D. N., & Ohta, H. (1994). Baroreflexes influence autoregulation of cerebral blood flow during hypertension. *The American Journal of Physiology, 267*(3 Pt 2), H1183–H1189.
Talman, W. T., & Nitschke Dragon, D. (2002). Inhibiting the nucleus tractus solitarii extends cerebrovascular autoregulation during hypertension. *Brain Research, 931*(1), 92–95.
Talman, W. T., & Nitschke Dragon, D. (2007). Neuronal nitric oxide mediates cerebral vasodilatation during acute hypertension. *Brain Research, 1139*, 126–132. http://dx.doi.org/10.1016/j.brainres.2007.01.008.
Tanaka, K., Gotoh, F., Gomi, S., Takashima, S., Mihara, B., Shirai, T., et al. (1991). Inhibition of nitric oxide synthesis induces a significant reduction in local cerebral blood flow in the rat. *Neuroscience Letters, 127*(1), 129–132.
Thompson, B. G., Pluta, R. M., Girton, M. E., & Oldfield, E. H. (1996). Nitric oxide mediation of chemoregulation but not autoregulation of cerebral blood flow in primates. *Journal of Neurosurgery, 84*(1), 71–78. http://dx.doi.org/10.3171/jns.1996.84.1.0071.
Toda, N., Ayajiki, K., Tanaka, T., & Okamura, T. (2000). Preganglionic and postganglionic neurons responsible for cerebral vasodilation mediated by nitric oxide in anesthetized dogs. *Journal of Cerebral Blood Flow and Metabolism, 20*(4), 700–708. http://dx.doi.org/10.1097/00004647-200004000-00007.
Toda, N., Ayajiki, K., Yoshida, K., Kimura, H., & Okamura, T. (1993). Impairment by damage of the pterygopalatine ganglion of nitroxidergic vasodilator nerve function in canine cerebral and retinal arteries. *Circulation Research, 72*(1), 206–213.
Toda, N., & Okamura, T. (2003). The pharmacology of nitric oxide in the peripheral nervous system of blood vessels. *Pharmacological Reviews, 55*(2), 271–324. http://dx.doi.org/10.1124/pr.55.2.3.
Toda, N., Tanaka, T., Ayajiki, K., & Okamura, T. (2000). Cerebral vasodilatation induced by stimulation of the pterygopalatine ganglion and greater petrosal nerve in anesthetized monkeys. *Neuroscience, 96*(2), 393–398.
Tong, X. K., & Hamel, E. (1999). Regional cholinergic denervation of cortical microvessels and nitric oxide synthase-containing neurons in Alzheimer's disease. *Neuroscience, 92*(1), 163–175.
Toussay, X., Basu, K., Lacoste, B., & Hamel, E. (2013). Locus coeruleus stimulation recruits a broad cortical neuronal network and increases cortical perfusion. *The Journal of Neuroscience, 33*(8), 3390–3401. http://dx.doi.org/10.1523/JNEUROSCI.3346-12.2013.
Toyoda, K., Fujii, K., Ibayashi, S., Nagao, T., Kitazono, T., & Fujishima, M. (1997). Role of nitric oxide in regulation of brain stem circulation during hypotension. *Journal of Cerebral*

Blood Flow and Metabolism, *17*(10), 1089–1096. http://dx.doi.org/10.1097/00004647-199710000-00011.

Ulker, P., Sati, L., Celik-Ozenci, C., Meiselman, H. J., & Baskurt, O. K. (2009). Mechanical stimulation of nitric oxide synthesizing mechanisms in erythrocytes. *Biorheology*, *46*(2), 121–132. http://dx.doi.org/10.3233/BIR-2009-0532.

Ulker, P., Yaras, N., Yalcin, O., Celik-Ozenci, C., Johnson, P. C., Meiselman, H. J., et al. (2011). Shear stress activation of nitric oxide synthase and increased nitric oxide levels in human red blood cells. *Nitric Oxide*, *24*(4), 184–191. http://dx.doi.org/10.1016/j.niox.2011.03.003.

Ursino, M., Cavalcanti, S., Bertuglia, S., & Colantuoni, A. (1996). Theoretical analysis of complex oscillations in multibranched microvascular networks. *Microvascular Research*, *51*(2), 229–249. http://dx.doi.org/10.1006/mvre.1996.0023.

Valtschanoff, J. G., Weinberg, R. J., Kharazia, V. N., Schmidt, H. H., Nakane, M., & Rustioni, A. (1993). Neurons in rat cerebral cortex that synthesize nitric oxide: NADPH diaphorase histochemistry, NOS immunocytochemistry, and colocalization with GABA. *Neuroscience Letters*, *157*(2), 157–161.

Van Mil, A. H., Spilt, A., Van Buchem, M. A., Bollen, E. L., Teppema, L., Westendorp, R. G., et al. (2002). Nitric oxide mediates hypoxia-induced cerebral vasodilation in humans. *Journal of Applied Physiology (1985)*, *92*(3), 962–966. http://dx.doi.org/10.1152/japplphysiol.00616.2001.

Vaucher, E., & Hamel, E. (1995). Cholinergic basal forebrain neurons project to cortical microvessels in the rat: Electron microscopic study with anterogradely transported *Phaseolus vulgaris* leucoagglutinin and choline acetyltransferase immunocytochemistry. *The Journal of Neuroscience*, *15*(11), 7427–7441.

Vaucher, E., Linville, D., & Hamel, E. (1997). Cholinergic basal forebrain projections to nitric oxide synthase-containing neurons in the rat cerebral cortex. *Neuroscience*, *79*(3), 827–836.

Vetri, F., Menicucci, D., Lapi, D., Gemignani, A., & Colantuoni, A. (2007). Pial arteriolar vasomotion changes during cortical activation in rats. *NeuroImage*, *38*(1), 25–33. http://dx.doi.org/10.1016/j.neuroimage.2007.07.017.

Vincent, S. R., & Hope, B. T. (1992). Neurons that say NO. *Trends in Neurosciences*, *15*(3), 108–113.

Vincent, S. R., & Kimura, H. (1992). Histochemical mapping of nitric oxide synthase in the rat brain. *Neuroscience*, *46*(4), 755–784.

Wallerath, T., Gath, I., Aulitzky, W. E., Pollock, J. S., Kleinert, H., & Forstermann, U. (1997). Identification of the NO synthase isoforms expressed in human neutrophil granulocytes, megakaryocytes and platelets. *Thrombosis and Haemostasis*, 77(1), 163–167.

Wang, H., Hitron, I. M., Iadecola, C., & Pickel, V. M. (2005). Synaptic and vascular associations of neurons containing cyclooxygenase-2 and nitric oxide synthase in rat somatosensory cortex. *Cerebral Cortex*, *15*(8), 1250–1260. http://dx.doi.org/10.1093/cercor/bhi008.

Wang, Q., Kjaer, T., Jorgensen, M. B., Paulson, O. B., Lassen, N. A., Diemer, N. H., et al. (1993). Nitric oxide does not act as a mediator coupling cerebral blood flow to neural activity following somatosensory stimuli in rats. *Neurological Research*, *15*(1), 33–36.

Wang, X., Lessard, A., Trudeau, L.-E., & Girouard, H. (2013). Astrocytes produce nitric oxide in response to cholinergic or glutamatergic stimulation. *The FASEB Journal*, *27*, 1096.13, Meeting Abstracts.

Wang, Q., Pelligrino, D. A., Koenig, H. M., & Albrecht, R. F. (1994). The role of endothelium and nitric oxide in rat pial arteriolar dilatory responses to CO2 in vivo. *Journal of Cerebral Blood Flow and Metabolism*, *14*(6), 944–951. http://dx.doi.org/10.1038/jcbfm.1994.126.

Wei, E. P., Kontos, H. A., & Beckman, J. S. (1996). Mechanisms of cerebral vasodilation by superoxide, hydrogen peroxide, and peroxynitrite. *The American Journal of Physiology*, *271*(3 Pt. 2), H1262–H1266.

Wei, E. P., Kukreja, R., & Kontos, H. A. (1992). Effects in cats of inhibition of nitric oxide synthesis on cerebral vasodilation and endothelium-derived relaxing factor from acetylcholine. *Stroke*, *23*(11), 1623–1628 discussion, 1628–1629.

Wever, R. M., van Dam, T., van Rijn, H. J., de Groot, F., & Rabelink, T. J. (1997). Tetrahydrobiopterin regulates superoxide and nitric oxide generation by recombinant endothelial nitric oxide synthase. *Biochemical and Biophysical Research Communications*, *237*(2), 340–344. http://dx.doi.org/10.1006/bbrc.1997.7069.

White, R. P., Vallance, P., & Markus, H. S. (2000). Effect of inhibition of nitric oxide synthase on dynamic cerebral autoregulation in humans. *Clinical Science (London)*, *99*(6), 555–560.

Wiencken, A. E., & Casagrande, V. A. (1999). Endothelial nitric oxide synthetase (eNOS) in astrocytes: Another source of nitric oxide in neocortex. *Glia*, *26*(4), 280–290.

Wiklund, L., Toggenburger, G., & Cuenod, M. (1982). Aspartate: Possible neurotransmitter in cerebellar climbing fibers. *Science*, *216*(4541), 78–80.

Williams, R. S., & Wagner, P. D. (2000). Transgenic animals in integrative biology: Approaches and interpretations of outcome. *Journal of Applied Physiology (1985)*, *88*(3), 1119–1126.

Wood, J., & Garthwaite, J. (1994). Models of the diffusional spread of nitric oxide: Implications for neural nitric oxide signalling and its pharmacological properties. *Neuropharmacology*, *33*(11), 1235–1244.

Yang, G., Huard, J. M., Beitz, A. J., Ross, M. E., & Iadecola, C. (2000). Stellate neurons mediate functional hyperemia in the cerebellar molecular layer. *The Journal of Neuroscience*, *20*(18), 6968–6973.

Yang, G., & Iadecola, C. (1997). Obligatory role of NO in glutamate-dependent hyperemia evoked from cerebellar parallel fibers. *The American Journal of Physiology*, *272*(4 Pt. 2), R1155–R1161.

Yoshida, K., Okamura, T., Kimura, H., Bredt, D. S., Snyder, S. H., & Toda, N. (1993). Nitric oxide synthase-immunoreactive nerve fibers in dog cerebral and peripheral arteries. *Brain Research*, *629*(1), 67–72.

You, J., Golding, E. M., & Bryan, R. M., Jr. (2005). Arachidonic acid metabolites, hydrogen peroxide, and EDHF in cerebral arteries. *American Journal of Physiology. Heart and Circulatory Physiology*, *289*(3), H1077–H1083. http://dx.doi.org/10.1152/ajpheart.01046.2004.

You, J., Johnson, T. D., Marrelli, S. P., & Bryan, R. M., Jr. (1999). Functional heterogeneity of endothelial P2 purinoceptors in the cerebrovascular tree of the rat. *The American Journal of Physiology*, *277*(3 Pt 2), H893–H900.

You, J., Johnson, T. D., Marrelli, S. P., Mombouli, J. V., & Bryan, R. M., Jr. (1999). P2u receptor-mediated release of endothelium-derived relaxing factor/nitric oxide and endothelium-derived hyperpolarizing factor from cerebrovascular endothelium in rats. *Stroke*, *30*(5), 1125–1133.

Zhang, F., Xu, S., & Iadecola, C. (1995). Role of nitric oxide and acetylcholine in neocortical hyperemia elicited by basal forebrain stimulation: Evidence for an involvement of endothelial nitric oxide. *Neuroscience*, *69*(4), 1195–1204.

Zhang, R., Zuckerman, J. H., Giller, C. A., & Levine, B. D. (1998). Transfer function analysis of dynamic cerebral autoregulation in humans. *The American Journal of Physiology*, *274*(1 Pt. 2), H233–H241.

CHAPTER FIFTEEN

Endothelial Nitric Oxide Synthase Gene Polymorphisms in Cardiovascular Disease

Cecilia Vecoli[1]
Institute of Clinical Physiology-CNR, Pisa, Italy
[1]Corresponding author: e-mail address: vecoli@ifc.cnr.it

Contents

Abstract

Endothelial nitric oxide synthase (eNOS or NOS3), an enzyme constitutively expressed especially in endothelial cells, is largely responsible for nitric oxide (NO) bioavailability at the endothelial level. Alterations in endothelial-derived NO production have been associated with various diseases, and, in humans, can be genetically determined by the presence of different polymorphisms in the eNOS gene. To date, the most studied and functionally related polymorphisms are: Glu298Asp (rs1799983), −786T/C (rs2070744), and the intron 4 variable number tandem repeat. Evidence supports a major role of these variants in increasing susceptibility to cardiovascular disease (such as coronary artery disease, myocardial infarction, hypertension, pre-eclampsia, and stroke) as well as in the onset of other complex diseases.

Vitamins and Hormones, Volume 96
ISSN 0083-6729
http://dx.doi.org/10.1016/B978-0-12-800254-4.00015-5

1. INTRODUCTION

Since the discovery of nitric oxide (NO) as the endothelium-derived relaxation factor, new roles for NO in biology continue to emerge. This is especially the case for the heart, where NO is known to govern cardiac morphometric and functional properties. NO is synthesized by three distinct isoforms of NO synthase (NOS), namely neuronal NOS (nNOS or NOS1), inducible NOS (iNOS or NOS2), and endothelial NOS (eNOS or NOS3). Both nNOS and eNOS are constitutively expressed and elicit their biological actions in a subcellular localization-dependent manner. iNOS, on the other hand, is not expressed under normal circumstance although it may be induced under certain stress conditions and is thought to be deleterious. Thus, while low NO release is known to predispose humans to various diseases such as atherosclerosis, hypertension, thrombosis, and vasospasm, a marked increase in NO concentrations has been reported to be associated with endotoxic shock and exaggerated inflammatory reactions (Förstermann & Sessa, 2012).

eNOS is mostly expressed in endothelial cells. However, it has also been detected in cardiac myocytes, platelets, certain neurons of the brain, in syncytio-trophoblasts of the human placenta, and in LLC-PK1 kidney tubular epithelial cells (Förstermann & Sessa, 2012).

The eNOS protein synthesizes NO constitutively via a reaction including the conversion of L-arginine to L-citrulline, which involves the transfer of five electrons provided by nicotinamide adenine dinucleotide phosphate. The enzyme acts as a homodimer that can be divided functionally into two major domains: a C-terminal reductase domain and an N-terminal oxygenase domain. Catalytic activity requires the presence of heme and the cofactors tetrahydrobiopterin, flavin adenine dinucleotide, flavin mononucleotide, and calmodulin (Förstermann & Sessa, 2012).

Analyses of genetically engineered animals deficient in eNOS expression (eNOS$^{-/-}$) provide support for the importance of eNOS-derived NO, especially in cardiovascular homeostasis. For instance, eNOS$^{-/-}$ mice develop numerous vascular disease states, including systemic hypertension (Duplain et al., 2001; Huang et al., 1995), platelet aggregation and thrombosis (Ramesh et al., 2011), and atherogenesis (Ponnuswamy et al., 2012). eNOS$^{-/-}$ mice are also hyperinsulinemic, dyslipidemic and have decreased insulin-stimulated glucose uptake compared with wild-type mice (Duplain et al., 2001).

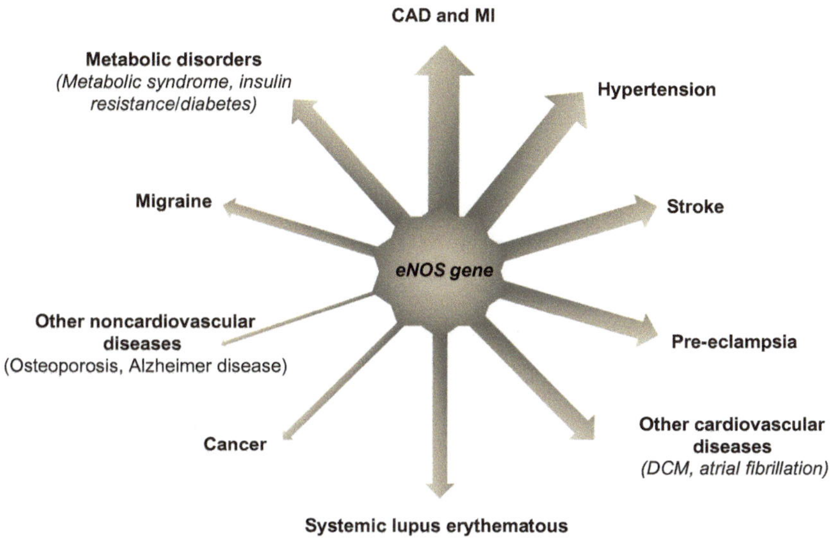

Figure 15.1 The gene–disease association network. The width of the arrow represents the strength of association in the network. (See the color plate.)

In humans, alterations in endothelial-derived NO production can be genetically determined. Several genetic polymorphisms of the gene encoding eNOS are now well known. Although a normal functional enzyme is formed, the polymorphisms can lead to altered transcription and/or processing rates, thereby interfering with normal enzyme function. The eNOS polymorphic variants have been extensively investigated to determine their relevance, mainly in cardiovascular disease (CVD), although a major role of this protein has also been recognized in the onset of other complex diseases (Fig. 15.1).

2. GENETIC VARIANTS

The *eNOS* gene was cloned in 1993 and was localized to chromosome 7q35-36 (Marsden et al., 1993). Spanning roughly 21 kb of genomic DNA, the gene comprises 26 exons that encode a 135-kDa protein containing 1203 amino acids (Fig. 15.2). Approximately 1500 bp of upstream promoter sequence have also been characterized and contain transcription factor-binding sites that mediate regulation by shear stress and estrogens, among others (Hingorani, 2001).

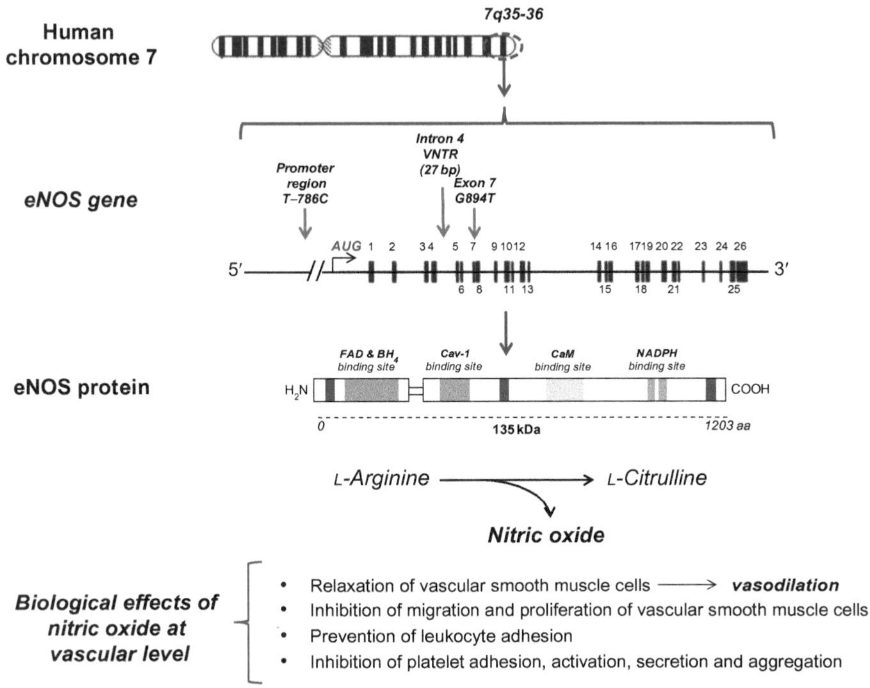

Figure 15.2 (A and B) Scheme of the gene encoding endothelial nitric oxide synthase (*eNOS* or *NOS3*). The human *NOS3* gene (located at 7q35–36) contains 26 exons that span 21 kb. Exons are described by number. AUG denotes transcription start site. Three specific polymorphisms in the *NOS3* gene are marked by arrows. (C) Scheme of the eNOS protein. Cofactors: orange, FAD = flavin adenine dinucleotide; BH4 = tetrahydrobiopterin; pink, Cav-1 = caveolin 1; yellow, CaM = calmodulin; green, NADPH = nicotinamide adenine dinucleotide phosphate. (D) Major biological effects of nitric oxide at vascular level. (See the color plate.)

The *NOS3* gene has been extensively screened for variations and many polymorphic sites have been detected. Several promoter single-nucleotide polymorphisms (SNPs) have been identified, but there is no clear evidence that any of them lies directly within the consensus sequence for a known transcription factor of NOS3. Similarly, no variations in the 3′ untranslated region able to influence RNA stability have been reported so far although more studies have identified a posttranscriptional control by miRNA (Rippe et al., 2012; Weber, Baker, Moore, & Searles, 2010).

Among the known eNOS polymorphisms, the most examined and functionally related common variants are: the Glu298Asp (894G/T, SNP id rs1799983), the −786T/C (SNP id rs2070744), and the intron 4 variable number tandem repeat (VNTR).

2.1. The Glu298Asp variant

The only SNP identified (http://www.uniprot.org/docs/humchr07) that leads to an amino acid substitution in the mature protein is the 894G/T variant, in which a guanine/thymine substitution at exon 7 leads to a glutamate/aspartate substitution at position 298. This variant alters the primary structure of the protein and has the potential to alter one or more functional properties of the enzyme directly. Some mechanistic studies have been published suggesting a functional effect of the Glu298Asp polymorphism. In particular, specific associations between the Glu298Asp polymorphism and NO synthesis have been described. Two different studies have shown the eNOS protein containing Asp at position 298 to be subject to selective proteolytic cleavage in endothelial cells and vascular tissues (Persu et al., 2002; Tesauro et al., 2000). If this observation is correct, the cleaved fragments would be expected to lack NO synthase activity. However, two other reports suggest that this observation might be an artifact (Fairchild et al., 2001; McDonald, Alp, & Channon, 2004), so further *in vitro* work is needed to resolve this issue. Finally, the Glu298Asp SNP affects NOS3 localization to caveolar membrane (Joshi, Mineo, Shaul, & Bauer, 2007).

2.2. The −786T/C variant

Given the location of −786T/C polymorphism in the promoter region of gene, studies examining the functionality of this polymorphism have focused on eNOS expression levels. This polymorphism is able to reduce the gene promoter activity by approximately 50% (Miyamoto et al., 2000; Nakayama et al., 1999). Lower eNOS mRNA and serum nitrite/nitrate levels have been found in individuals with the −786C variant (Miyamoto et al., 2000). A nuclear protein that exhibits differential binding to the promoter containing the −786T and −786C alleles has also been described (Cattaruzza et al., 2004). In such study, the functional importance of the −786C/T SNP has been investigated. The authors found that the CC genotype was associated with a deficit in NOS3 expression in human endothelial cells exposed to laminar shear stress and with a reduced NO-mediated vasomotor function. They suggested a molecular mechanism in which the insensitivity of NOS3 expression to shear stress might comprise the binding of an inhibitory DNA-binding protein to the C-type promoter of the NOS3 gene.

2.3. Intron 4 VNTR variant

The intron 4 a/b variant corresponds to a 27-bp repeat polymorphism in intron 4 of the eNOS gene. The two most common alleles contain four (allele a) and five (allele b) repeats. The intron 4 VNTR variant has been recognized for a long time as having little chance of being a functional variation. To date, conflicting associations between the intron 4 variant and NO pathway activity have been described. Some reports have indicated that carriers of this variant have lower NO plasma levels and decreased protein expression, although this finding was not supported by other studies (Casas et al., 2006). However, a functional role of this variant of eNOS gene has been found by Wang, Dudley, and Wang (2002). These authors constructed pGL3 luciferase reporter vectors by inserting an eNOS promoter fragment containing either T or C nucleotide at −786 bp at the 5′ end of the luciferase coding region and eNOS intron 4 containing either 5 × or 4 × 27-bp repeats at the 3′ end of the luciferase gene. The measurement of transcription efficiency showed a *cis*-acting role of the 27-bp repeats in eNOS promoter function. The author concluded that the 27 bp VNTR polymorphism coordinates with −786C variant and regulates the transcription levels of eNOS gene.

More recently, it is also been shown that the 27 nt repeats in eNOS intron 4 produce 27 nt small RNA, which may inhibit eNOS expression by altering histone acetylation and DNA methylation in regions adjacent to the 27 nt repeat element and core promoter (Zhang et al., 2008).

3. eNOS POLYMORPHISMS IN CVD

3.1. Coronary artery disease and myocardial infarction

Coronary artery disease (CAD) is the most complex disorder affecting public health and is the leading cause of mortality and morbidity in Western countries. It is well-accepted that endothelial dysfunction occurs in response to specific risk factors and precedes the development of coronary lesions. Several experimental studies have shown that NO inhibits many key steps of the atherosclerotic process while a defect in NO production can facilitate its progression (Cooke, 1996). As a matter of fact, the *NOS3* gene has assumed special importance in the pathogenesis of CAD as well as in myocardial infarction (MI) (Yoon, Song, Hong, & Kim, 2000). In 2006, a large meta-analysis of NOS3 polymorphisms in coronary heart disease (CHD) revealed per-allele odds ratios of 1.17 (95% confidence interval: 1.07, 1.28) for Glu298Asp, 1.17 (95% confidence interval: 1.07, 1.28) for −786T/C, and 1.12 (95% confidence interval: 1.01, 1.24) for intron 4 (Casas et al., 2006). The authors showed that small studies with more striking results could affect the associations and suggested large-scale genetic association studies using tagging polymorphisms to confirm or refute a role of the NOS3 gene in CHD. To date, large-scale genome-wide association studies have yet to confirm a role of eNOS gene (Roberts & Stewart, 2012) in the pathogenesis of CAD, while several candidate-gene studies continue to show associations between different eNOS variants and ischemic heart disease. A group in Turkey concluded that the −786T/C polymorphism of NOS3 gene might be a risk factor for CAD (Tangurek et al., 2006), similarly to a Korean study in which the association between the −786T/C polymorphism of the NOS3 gene and CAD or MI was demonstrated (Kim et al., 2007). Recently, Salimi et al. (2012) provided evidence that −786T/C polymorphism of the NOS3 gene was associated with increased risk of CAD in a population of patients with angiographically documented CAD. These authors also found that NO metabolites (nitrite and nitrate, NOx) concentrations were significantly higher in CAD patients than in controls and the plasma NOx concentrations were significantly different in −786T/C polymorphism genotypes in total population. Colombo et al. (2002, 2003) found that both the −786T/C and the Glu298Asp polymorphisms of the eNOS gene were associated with the presence and severity of CAD in the Italian population and that those individuals carrying both eNOS variants simultaneously had a fourfold higher risk of developing CAD. Other studies revealed a strong association between the Glu298Asp variants and ischemic heart disease in different populations. Clinical evidence suggested that Glu298Asp might carry a higher risk for MI in the United Kingdom and in Japan (Hibi et al., 1998; Shimasaki et al., 1998) and a strong correlation between the TT genotype of Glu298Asp variant and the occurrence of premature CAD in the Egyptian population has been shown (Abdel-Aziz & Mohamed, 2013). Antoniades et al. (2005) showed that the Glu298Asp polymorphism was associated with a higher risk for premature MI in a Greek Caucasian population. Furthermore, the presence of the T allele may be associated with a greater increase in oxidizing LDL, and proinflammatory cytokine levels during the acute phase of MI (an effect that diminishes one year after the event). These findings imply that the Glu298Asp polymorphism of the eNOS gene may modify endothelial injury, oxidative stress status, and inflammatory response during the acute phase of MI. However, data from other studies in different populations indicate that there is no association of Glu298Asp polymorphism with CAD or MI (Granath, Taylor, van Bockxmeer, & Mamotte, 2001; Poirier et al., 1999; Wang, Hsu, Ko, Ko, & Lee, 2001). Likewise, conflicting findings have been found regarding the intron 4 VNTR variant. Fatini et al. (2004) showed that the homozygosity for eNOS 4a represented an independent predisposition to acute coronary syndromes, in particular to acute MI, while Meluzín, Vasků, Kincl,

Panovský, and Srámková (2009) found no relation of intron 4 a/b polymorphisms to the presence of coronary atherosclerosis. There is a marked ethnic difference in the distribution of eNOS variants (Table 15.1) and it is likely that such ethnic differences may relate to the ethnic-specific predisposition to CAD. The haplotype analysis shows that these three polymorphisms are not randomly distributed. The Asp298 allele is negatively associated with the 4a allele, meaning that an Asp298 minor allele carrier is very unlikely to carry the VNTR 4a minor allele. The Asp298 allele is also positively associated with the −786C allele in Caucasians, with the Asp298-786C-4b haplotype found in 24% frequency in the population (Table 15.2).

3.2. Hypertension

Hypertension is a multifactorial disease involving both environmental and genetic components. Experimental studies suggest that the total disruption of eNOS gene leads *per se* to hypertension in mice (Duplain et al., 2001;

Table 15.1 Distribution of the three risk variants among the major ethnic groups

Variants	African-Americans (%)	Asians (%)	Caucasians (%)	Hispanics (%)
−786C	15	17.5	39	25
VNTR 4a	26.5[a]	11.5	16	–
Asp298	13	8.6	32	20

[a]In the African-American population, two other VNTR polymorphisms in intron 4, 4c that corresponds to six 27-bp repeats and 4y that has only two 27-bp repeats were also reported in rare cases of patients with severe cardiovascular diseases (Tanus-Santos, Desai, & Flockhart, 2001).

Table 15.2 Haplotype distribution of the three polymorphisms in major ethnic groups

Glu298Asp	−786T/C	VNTR 4b/a	African-Americans	Caucasians	Asians
Glu	T	4b	46%	46%	77%
Glu	T	4a	27%	2%	3%
Glu	C	4b	7%	4%	2%
Glu	C	4a	5	14	9%
Asp	T	4b	9%	10%	6%
Asp	T	4a	0	0	0
Asp	C	4b	6%	24%	2%
Asp	C	4a	0	0	1%

The reported data are from Tanus-Santos et al. (2001).

Huang et al., 1995), while the partial deletion predisposes to an increase in arterial hypertension under metabolic stress (Cook et al., 2004).

The inhibition of eNOS elevates blood pressure (BP) in healthy humans (Haynes, Noon, Walker, & Webb, 1993), while NO production is diminished in patients with essential hypertension, under basal conditions (Forte et al., 1997). Between the eNOS variants, the Glu298Asp is the most highlighted variant associated with hypertension. It was found to be associated with increased risk of hypertension in Caucasians (Lacolley et al., 1998) and Japaneses (Miyamoto et al., 1998), but in another two studies it was not (in neither Caucasians nor Japaneses) (Benjafield & Morris, 2000; Kato et al., 1999). Nevertheless, it is generally accepted that the polymorphic variants of eNOS gene, modifying the production of NO, may predispose to hypertension. Distinct genetic variants—not including NOS—robustly associated with (blood pressure) BP have been identified (Ehret et al., 2011; Lieb et al., 2013), but collectively these explain only a few percent of the heritability for BP phenotypes. Recently, Johnson et al. (2011) used a bespoke gene-centric array to genotype 25,118 individuals including hypertensive case–control and general population subjects, found also the NOS3 locus as associated with BP. In this study, chromatin immunoprecipitation and transcript expression data highlighted potential gene regulatory mechanisms at the NOS3 locus.

In 2011, a large meta-analysis study was conducted to investigate the association of three common eNOS polymorphisms with hypertension, from the English and Chinese-published literature (Niu & Qi, 2011). All together, the results of the study ascertained the role of eNOS 894G/T and 4b/a polymorphisms on the development of hypertension for Asian populations and −786T/C polymorphism for Caucasians. Anyway, due to the presence of heterogeneity between the selected studies the authors suggested further analyses to investigate eNOS adjacent markers in a wider context, and that future studies should focus on gene–gene and gene–environment interactions, as well as haplotype patterns.

Since hypertension is recognized worldwide as a major risk factor for CVD, its control is an important priority. Despite improved treatment options, one-third of treated hypertensive patients have BP above target thresholds. One reason for this is that individual response to pharmacologic treatments varies. Genetic background is an important determinant of this variable response. As a matter of fact, evidence increasingly suggests that genetic polymorphisms interact with antihypertensive treatment, leading to different BP responses and cardiovascular outcomes (Arnett, Claas, & Glasser, 2006; Cooke, Doshi, & Binkley, 2007). In a recent study, the pharmacogenetic associations of three NOS3 polymorphisms and three antihypertensive drugs with cardiovascular outcomes were assessed in a multicenter, double-blind clinical trial (Zhang et al., 2012). The authors observing significant associations with NOS3 variants and CHD and heart failure and significant pharmacogenetic effects for stroke and all cause mortality, proposed that NOS3 variants may potentially provide useful clinical information with respect to treatment.

3.3. Pre-eclampsia

Growing evidence indicates that reduced NO formation is associated with hypertensive disorders of pregnancy, especially pre-eclampsia (PE) (Bernardi, Constantino, Machado, Petronilho, & Dal-Pizzol, 2008; Garcia et al., 2007). PE, a condition characterized by high BP and significant amount of proteins in the urine of a pregnant woman, is not only the leading cause of maternal death in most developed countries but also the main cause of prenatal mortality and morbidity, according to the World Health Organization (Sibai, Dekker, & Kupferminc, 2005). Furthermore, PE is a significant risk factor for future CVD. In the past, many studies focused on the genetic association of the three eNOS polymorphisms with PE, but the results remain conflicting and controversial (Chen et al., 2007; Fatini, Sticchi, Gensini, et al., 2006; Kim et al., 2008; Sandrim, Palei, Sertorio, et al., 2010; Sandrim, Palei, Metzger, et al., 2010; Serrano et al., 2004). Recently, a meta-analysis performed to assess whether maternal eNOS gene polymorphisms and serum concentration of NO are related to PE risk, clearly showed that eNOS −786T/C and 4b/a polymorphisms contributed significantly to PE risk (especially in European women). This result is of utmost clinical relevance, in order to provide probable evidence of PE occurrence and new methods for its screening and prediction (Dai, Liu, Zhang, Zhang, & Wang, 2013).

3.4. Ischemic stroke

Stroke is a leading cause of disability and the second most common cause of death in adults around the world (Bonita et al., 2004). Ischemic stroke (IS), accounting for 87% of all the stroke cases, is one of the most common diseases resulting from genetic factors, environmental factors, and their interactions. To date, many case–control studies have been carried out to investigate the role of eNOS polymorphisms in the development of IS, but these have produced conflicting or inconclusive results. In 2009, Tao and Chen (2009) performed a meta-analysis in order to clarify the effect of eNOS genotype (894G/T, 4b/a) on the risk of developing IS. The authors concluded that only the G894T polymorphism might be associated with IS in overall population. After that, an expanding body of literature in both Caucasian and Asian populations provided conflict results. In the last year, two different independent meta-analyses have been performed to investigate the potential association between the three eNOS gene polymorphisms (894G/T, 4b/a, and −786T/C) and IS risk in both Asians and Caucasians (Niu et al., 2013; Yao, Chang, Jin, & He, 2013), while an additional independent study was performed only in an Asiatic population (Wang, Jiang, Wu, & Zhang, 2013). The first analysis provided strong evidence that the eNOS T-786C gene polymorphism was not associated with IS, while the 894G/T and 4b/a polymorphisms might be associated with IS, at least in Asians. Conversely, the second one established that the 894G/T, 4b/a, and −786T/C polymorphisms were associated with IS risk in Asians. A possible contrasting role of the 4b/a polymorphism in IS was indicated in Caucasian populations. The third study in Asiatic population confirmed that eNOS gene 4b/a, −786T/C, 894G/T polymorphisms were associated with IS. Anyway, these contradictory meta-analysis results suggest that more studies are required to draw solid conclusions on the relationship between eNOS polymorphisms and IS. Future studies should also consider gene–gene and gene–environment interactions, and haplotype-based effects as well as different subtypes of IS to reach a final certain conclusion.

3.5. Other CVDs

In addition to CAD, MI, hypertension, PE, and stroke, a large body of evidence has shown associations between the eNOS gene polymorphisms and other CVDs such as dilated cardiomyopathy (DCM) and atrial fibrillation (AF).

DCM is characterized by left ventricular dilatation and systolic dysfunction in the absence of associated conditions such as hypertension, valve disease, or CAD. The major causal genes implicated in the pathology of DCM are sarcomeric and cytoskeletal genes, but modifier genes and environmental factors are also considered to play a significant role in variable phenotypic expression of the disease. DCM can be also due to oxidative stress, wherein NO seems to play a key role. Recently, the possible implication of eNOS gene in the disease phenotype has been shown. NOS3 may be synergistically functioning in DCM-associated heart failure via the excessive production of NO in cardiomyocytes resulting in decreased myocardial contractility and systolic dysfunction (Matsa et al., 2013). Indeed, the haplotype C4bT carrying the recessive alleles of −786T/C, of VNRT, and of 894G/T was found to exhibit sevenfold increased risk for DCM compared to the controls. In hypertrophic cardiomyopathy, AF, which remains a major cause of morbidity and mortality worldwide, is the most commonly encountered cardiac rhythm problem among adults (Klemenz, 2011). It is associated with increased long-term risk of stroke as well as heart failure and death (Murray, Lazure, Pullen, Maltais, & Dorian, 2011). Over the last decade, a number of case–control studies have been conducted to investigate the association between eNOS gene polymorphisms and AF in humans (Bedi, McNamara, London, & Schwartzman, 2006; Fatini, Sticchi, Genuardi, et al., 2006; Giusti et al., 2007). However, the studies reported conflicting results regarding this pathology. Last year, a meta-analysis approach indicated that the −786T/C polymorphism is protective against AF risk among Caucasians though additional large studies based on diverse populations are required to validate this conclusion (Chen et al., 2012).

One paper, published in 2008, also reported that eNOS 894G/T polymorphism was associated with increased risk of congenital heart defects, providing evidence of a possible gene–environment interaction effect on congenital heart defects risk between the maternal eNOS 894G/T variant and maternal cigarette smoking during pregnancy (van Beynum et al., 2008). This observation should be interpreted with caution due to the relatively small subgroups, but highlights the importance of this gene in regulating different processes affecting the cardiovascular system.

4. eNOS POLYMORPHISMS IN OTHER DISEASES

4.1. Metabolic disorders

The major role of NO in the regulation of metabolic homeostasis is demonstrated by the insulin resistance (IR) state of eNOS$^{-/-}$ mice (Duplain et al., 2001). Similarly, the partial deletion of eNOS gene in mouse facilitates the development of IR when challenged with nutritional stress (Cook et al., 2004). In human, the T-786C variant of eNOS gene has been linked to IR both in nondiabetic subjects and type 2 diabetic patients (Monti et al., 2003; Ohtoshi et al., 2002). In a recent paper, our group confirms this association, showing that the −786T/C polymorphism was the major independent determinant of IR in a population of patients with ischemic and nonischemic cardiomyopathy (without history of diabetes) and suggested that a condition of primitive eNOS could predispose to an impairment of glucose homeostasis, which in turn was able to affect the severity of heart disease (Vecoli et al., 2012). eNOS polymorphisms are also found to be associated with many diabetic complications such as diabetic retinopathy (Zhao, Li, Zheng, & Zheng, 2012) and diabetic nephropathy (Ezzidi et al., 2008).

Evidence from epidemiological studies implied that the NOS3 gene was related to the pathogenesis of metabolic syndrome, a cluster of manifestations (abdominal obesity, increased serum level of triglycerides, decreased serum level of high-density lipoprotein cholesterol, high BP, and an increased fasting blood glucose level) that increase the risk of developing type 2 diabetes mellitus and CVD, although the findings are inconsistent and controversial (Liu, Huang, Sung, Lin, & Yeh, 2013).

4.2. Cancer

The role of NO in cancerogenesis is controversial since low concentrations of NO produced by endothelial cells can cause tumor growth by stimulating angiogenesis, while higher concentrations can cause cell death (Lancaster & Xie, 2006; Wink et al., 1998). However, a major role of eNOS in cancer development is recognized. A number of studies have reported the association between eNOS polymorphisms and breast cancer susceptibility (Yao, Fang, Zhong, & Yu, 2010), but the results are inconclusive, partially due to the small sample size in each study. Recently, a meta-analysis study showed that both eNOS Glu298Asp and −786T/C polymorphisms are associated with reduced breast cancer risk (Yao et al., 2010). This result may be biologically plausible, since both Asp298 allele and C786 allele lead to reduce basal NO production. Similarly, a significant association between the T-786C and intron 4 VNTR polymorphisms with the occurrence of prostate cancer has been found, while contrasting results have been found for the 894G/T SNP (Safarinejad, Safarinejad, Shafiei, & Safarinejad, 2013). Also in this pathology, the challenge of differences in disease, ethnicity, and sample size of the groups investigated remains open. In other types of cancer, eNOS polymorphisms have been analyzed to study the treatment response or the prognosis. For instance, it has been shown that the VNTR polymorphism in intron 4 may be associated with the progression of non-small cell lung cancer and shed light on the association between polymorphisms in the eNOS gene and survival of these patients (Fujita et al., 2010). Similarly, Demacq et al. (2010) showed that NOS3 polymorphisms were associated with high relapse risk in childhood acute lymphoblastic leukemia.

4.3. Systemic lupus erythematosus

Systemic lupus erythematosus (SLE) is a prototypic autoimmune disease that is characterized by intense inflammation and multiple organ damage, and in which immune regulation is disrupted. Significant familial aggregation studies as well as, convincing evidence of multiple genetic linkages strongly suggest that the etiology of SLE is of genetic nature. The eNOS polymorphisms have been associated with SLE by some (but by no means all) authors, which may be due to small sample sizes, low statistical power, and/or clinical heterogeneity. Recently, a meta-analysis performed in order to overcome the limitations of individual studies showed that C786T polymorphism was associated with SLE susceptibility (Lee, Lee, Choi, Ji, & Song, 2012).

4.4. Migraine

Migraine is a common, debilitating, multifactorial, neurovascular disorder predominantly affecting women. The pathogenic mechanisms involved in migraine are not completely clarified. Recently, vascular dysfunction has received special attention and migraine is now recognized as an established risk factor for cardiovascular events, particularly IS (Kurth, Chabriat, & Bousser, 2012). Candidate gene approaches focusing on genes affecting the endothelial function have been studied including the genes encoding eNOS, and these studies suggest an important involvement of this gene in migraine susceptibility (Borroni et al., 2006; Gonçalves et al., 2011).

5. CONCLUSION

A considerable amount of evidence has been accumulated evaluating the role of the NOS3 gene not only in CVD but also in other complex disorders. Unfortunately, the results are often inconclusive, partially because of the small sample size in each study, or due to differences in ethnicity, or in disease. An additional significant obstacle to evaluating the real effects of the different NOS3 polymorphisms is the lack of a reliable marker of the NOS3 gene function, partly due to the technical complexity in measuring NO production in humans. To date, the complexity of results observed for the association of each polymorphism with different pathologies, together with the absence of a definitive functional gene variant, strongly suggests that future studies evaluating the effect of the NOS3 gene should use a whole sequencing gene-based approach on larger and better-characterized populations. Indeed, it seems unlikely that studies evaluating isolated NOS3 gene variants will produce a real advance in our understanding of the NOS3 gene and the role of NO. All together, the findings collected up to now suggest a clear involvement of the eNOS gene in increasing susceptibility to the onset of different pathologies, especially in the cardiovascular field.

REFERENCES

Abdel-Aziz, T. A., & Mohamed, R. H. (2013). Association of endothelial nitric oxide synthase gene polymorphisms with classical risk factors in development of premature coronary artery disease. *Molecular Biology Reports, 40*, 3065–3071.

Antoniades, C., Tousoulis, D., Vasiliadou, C., Pitsavos, C., Chrysochoou, C., Panagiotakos, D., et al. (2005). Genetic polymorphism on endothelial nitric oxide synthase affects endothelial activation and inflammatory response during the acute phase of myocardial infarction. *Journal of the American College of Cardiology, 46*, 1101–1109.

Arnett, D. K., Claas, S. A., & Glasser, S. P. (2006). Pharmacogenetics of antihypertensive treatment. *Vascular Pharmacology, 44*, 107–118.

Bedi, M., McNamara, D., London, B., & Schwartzman, D. (2006). Genetic susceptibility to atrial fibrillation in patients with congestive heart failure. *Heart Rhythm, 3*, 808–812.

Benjafield, A. V., & Morris, B. J. (2000). Association analyses of endothelial nitric oxide synthase gene polymorphisms in essential hypertension. *American Journal of Hypertension, 13*, 994–998.

Bernardi, F., Constantino, L., Machado, R., Petronilho, F., & Dal-Pizzol, F. (2008). Plasma nitric oxide, endothelin-1, arginase and superoxide dismutase in pre-eclamptic women. *Journal of Obstetrics and Gynaecology Research, 34*, 957–963.

Bonita, R., Mendis, S., Truelsen, T., Bogousslavsky, J., Toole, J., & Yatsu, F. (2004). The global stroke initiative. *Lancet Neurology, 3*, 391–393.

Borroni, B., Rao, R., Liberini, P., Venturelli, E., Cossandi, M., Archetti, S., et al. (2006). Endothelial nitric oxide synthase (Glu298Asp) polymorphism is an independent risk factor for migraine with aura. *Headache, 46*, 1575–1579.

Casas, J. P., Cavalleri, G. L., Bautista, L. E., Smeeth, L., Humphries, S. E., & Hingorani, A. D. (2006). Endothelial nitric oxide synthase gene polymorphisms and cardiovascular disease: A HuGE review. *American Journal of Epidemiology, 164*, 921–935.

Cattaruzza, M., Guzik, T. J., Slodowski, W., Pelvan, A., Becker, J., Halle, M., et al. (2004). Shear stress insensitivity of endothelial nitric oxide synthase expression as a genetic risk factor for coronary heart disease. *Circulation Research, 95*, 841–847.

Chen, H., Chu, H., Shi, Y., Bhuyan, S. S., Li, J., Liu, S. R., et al. (2012). Association between endothelial nitric oxide synthase polymorphisms and atrial fibrillation: A meta-analysis. *Journal of Cardiovascular Translational Research, 5*, 528–534.

Chen, L. K., Huang, C. H., Yeh, H. M., Lee, C. N., Shyu, M. K., Hsieh, F. J., et al. (2007). Polymorphisms in the endothelial nitric oxide synthase gene may be protective against preeclampsia in a Chinese population. *Reproductive Sciences, 14*, 175–181.

Colombo, M. G., Andreassi, M. G., Paradossi, U., Botto, N., Manfredi, S., Masetti, S., et al. (2002). Evidence for association of a common variant of the endothelial nitric oxide synthase gene (Glu298→Asp polymorphism) to the presence, extent, and severity of coronary artery disease. *Heart, 87*, 525–528.

Colombo, M. G., Paradossi, U., Andreassi, M. G., Botto, N., Manfredi, S., Masetti, S., et al. (2003). Endothelial nitric oxide synthase gene polymorphisms and risk of coronary artery disease. *Clinical Chemistry, 49*, 389–395.

Cook, S., Hugli, O., Egli, M., Ménard, B., Thalmann, S., Sartori, C., et al. (2004). Partial gene deletion of endothelial nitric oxide synthase predisposes to exaggerated high-fat diet induced insulin resistance and arterial hypertension. *Diabetes, 53*, 2067–2072.

Cooke, J. P. (1996). Role of nitric oxide in progression and regression of atherosclerosis. *Western Journal of Medicine*, *164*, 419–424.
Cooke, G. E., Doshi, A., & Binkley, P. F. (2007). Endothelial nitric oxide synthase gene: Prospects for treatment of heart disease. *Pharmacogenomics*, *8*, 1723–1734.
Dai, B., Liu, T., Zhang, B., Zhang, X., & Wang, Z. (2013). The polymorphism for endothelial nitric oxide synthase gene, the level of nitric oxide and the risk for pre-eclampsia: A meta-analysis. *Gene*, *519*, 187–193.
Demacq, C., Vasconcellos, V. B., Izidoro-Toledo, T. C., da Silva Silveira, V., Canalle, R., Queiroz, R. G., et al. (2010). Vascular endothelial growth factor (VEGF) and endothelial nitric oxide synthase (NOS3) polymorphisms are associated with high relapse risk in childhood acute lymphoblastic leukemia (ALL). *Clinica Chimica Acta*, *411*, 1335–1340.
Duplain, H., Burcelin, R., Sartori, C., Cook, S., Egli, M., Lepori, M., et al. (2001). Insulin resistance, hyperlipidemia, and hypertension in mice lacking endothelial nitric oxide synthase. *Circulation*, *104*, 342–345.
Ehret, G. B., Munroe, P. B., Rice, K. M., Bochud, M., Johnson, A. D., Chasman, D. I., et al. (2011). Genetic variants in novel pathways influence blood pressure and cardiovascular disease risk. *Nature*, *478*, 103–109.
Ezzidi, I., Mtiraoui, N., Mohamed, M. B., Mahjoub, T., Kacem, M., & Almawi, W. Y. (2008). Association of endothelial nitric oxide synthase Glu298Asp, 4b/a, and −786T>C gene variants with diabetic nephropathy. *Journal of Diabetes and its Complications*, *22*, 331–338.
Fairchild, T. A., Fulton, D., Fontana, J. T., Gratton, J. P., McCabe, T. J., & Sessa, W. C. (2001). Acidic hydrolysis as a mechanism for the cleavage of the Glu(298)/Asp variant of human endothelial nitric-oxide synthase. *Journal of Biological Chemistry*, *276*, 26674–26679.
Fatini, C., Sofi, F., Sticchi, E., Gensini, F., Gori, A. M., Fedi, S., et al. (2004). Influence of endothelial nitric oxide synthase gene polymorphisms (G894T, 4a4b, T−786C) and hyperhomocystinemia on the predisposition to acute coronary syndromes. *American Heart Journal*, *147*, 516–521.
Fatini, C., Sticchi, E., Gensini, F., Genuardi, M., Tondi, F., Gensini, G. F., et al. (2006). Endothelial nitric oxide synthase gene influences the risk of pre-eclampsia, the recurrence of negative pregnancy events, and the maternal fetal flow. *Journal of Hypertension*, *24*, 1823–1829.
Fatini, C., Sticchi, E., Genuardi, M., Sofi, F., Gensini, F., Gori, A. M., et al. (2006). Analysis of minK and eNOS genes as candidate loci for predisposition to non-valvular atrial fibrillation. *European Heart Journal*, *27*, 1712–1718.
Förstermann, U., & Sessa, W. C. (2012). Nitric oxide synthases: Regulation and function. *European Heart Journal*, *33*, 829–837.
Forte, P., Copland, M., Smith, L. M., Milne, E., Sutherland, J., & Benjamin, N. (1997). Basal nitric oxide synthesis in essential hypertension. *Lancet*, *349*, 837–842.
Fujita, S., Masago, K., Hatachi, Y., Fukuhara, A., Hata, A., Kaji, R., et al. (2010). *BMC Medical Genetics*, *11*, 167.
Garcia, R. G., Celedón, J., Sierra-Laguado, J., Alarcón, M. A., Luengas, C., Silva, F., et al. (2007). Raised C-reactive protein and impaired flow-mediated vasodilation precede the development of preeclampsia. *American Journal of Hypertension*, *20*, 98–103.
Giusti, B., Gori, A. M., Marcucci, R., Sestini, I., Saracini, C., Sticchi, E., et al. (2007). Role of C677T and A1298C MTHFR, A2756G MTR and −786 C/T eNOS gene polymorphisms in atrial fibrillation susceptibility. *PLoS One*, *2*, e495.
Gonçalves, F. M., Martins-Oliveira, A., Speciali, J. G., Luizon, M. R., Izidoro-Toledo, T. C., Silva, P. S., et al. (2011). Endothelial nitric oxide synthase haplotypes associated with aura in patients with migraine. *DNA and Cell Biology*, *30*, 363–369.

Granath, B., Taylor, R. R., van Bockxmeer, F. M., & Mamotte, C. D. (2001). Lack of evidence for association between endothelial nitric oxide synthase gene polymorphisms and coronary artery disease in the Australian Caucasian population. *Journal of Cardiovascular Risk, 8*, 235–241.

Haynes, W. G., Noon, J. P., Walker, B. R., & Webb, D. J. (1993). Inhibition of nitric oxide synthesis increases blood pressure in healthy humans. *Journal of Hypertension, 11*, 1375–1380.

Hibi, K., Ishigami, T., Tamura, K., Mizushima, S., Nyui, N., Fujita, T., et al. (1998). Endothelial nitric oxide synthase gene polymorphism and acute myocardial infarction. *Hypertension, 32*, 521–526.

Hingorani, A. D. (2001). Polymorphisms in endothelial nitric oxide synthase and atherogenesis: John French Lecture 2000. *Atherosclerosis, 154*, 521–527.

Huang, P. L., Huang, Z., Mashimo, H., Bloch, K. D., Moskowitz, M. A., Bevan, J. A., et al. (1995). Hypertension in mice lacking the gene for endothelial nitric oxide synthase. *Nature, 377*, 239–242.

Johnson, T., Gaunt, T. R., Newhouse, S. J., Padmanabhan, S., Tomaszewski, M., Kumari, M., et al. (2011). Blood pressure loci identified with a gene-centric array. *American Journal of Human Genetics, 89*, 688–700.

Joshi, M. S., Mineo, C., Shaul, P. W., & Bauer, J. A. (2007). Biochemical consequences of the NOS3 Glu298Asp variation in human endothelium: Altered caveolar localization and impaired response to shear. *The FASEB Journal, 21*, 2655–2663.

Kato, N., Sugiyama, T., Morita, H., Nabika, T., Kurihara, H., Yamori, Y., et al. (1999). Lack of evidence for association between the endothelial nitric oxide synthase gene and hypertension. *Hypertension, 33*, 933–936.

Kim, I. J., Bae, J., Lim, S. W., Cha, D. H., Cho, H. J., Kim, S., et al. (2007). Influence of endothelial nitric oxide synthase gene polymorphisms (−786T>C, 4a4b, 894G>T) in Korean patients with coronary artery disease. *Thrombosis Research, 119*, 579–585.

Kim, Y. J., Park, B. H., Park, H., Jung, S. C., Pang, M. G., Ryu, H. M., et al. (2008). No association of the genetic polymorphisms of endothelial nitric oxide synthase, dimethylarginine dimethylaminohydrolase, and vascular endothelial growth factor with preeclampsia in Korean populations. *Twin Research and Human Genetics, 11*, 77–83.

Klemenz, B. (2011). Screening for atrial fibrillation. *British Journal of General Practice, 61*, 295.

Kurth, T., Chabriat, H., & Bousser, M. G. (2012). Migraine and stroke: A complex association with clinical implications. *Lancet Neurology, 11*, 92–100.

Lacolley, P., Gautier, S., Poirier, O., Pannier, B., Cambien, F., & Benetos, A. (1998). Nitric oxide synthase gene polymorphisms, blood pressure and aortic stiffness in normotensive and hypertensive subjects. *Journal of Hypertension, 16*, 31–35.

Lancaster, J. R., & Xie, K. (2006). Tumors face NO problems? *Cancer Research, 66*, 6459–6462.

Lee, Y. H., Lee, H. S., Choi, S. J., Ji, J. D., & Song, G. G. (2012). Associations between eNOS polymorphisms and susceptibility to systemic lupus erythematosus: A meta-analysis. *Inflammation Research, 61*, 135–141.

Lieb, W., Jansen, H., Loley, C., Pencina, M. J., Nelson, C. P., Newton-Cheh, C., et al. (2013). Genetic predisposition to higher blood pressure increases coronary artery disease risk. *Hypertension, 61*, 995–1001.

Liu, C. S., Huang, R. J., Sung, F. C., Lin, C. C., & Yeh, C. C. (2013). Association between endothelial nitric oxide synthase polymorphisms and risk of metabolic syndrome. *Disease Markers, 34*, 187–197.

Marsden, P. A., Heng, H. H., Scherer, S. W., Stewart, R. J., Hall, A. V., Shi, X. M., et al. (1993). Structure and chromosomal localization of the human constitutive endothelial nitric oxide synthase gene. *Journal of Biological Chemistry, 268*, 17478–17488.

Matsa, L. S., Rangaraju, A., Vengaldas, V., Latifi, M., Jahromi, H. M., Ananthapur, V., et al. (2013). Haplotypes of NOS3 gene polymorphisms in dilated cardiomyopathy. *PLoS One*, *8*, e70523.

McDonald, D. M., Alp, N. J., & Channon, K. M. (2004). Functional comparison of the endothelial nitric oxide synthase Glu298Asp polymorphic variants in human endothelial cells. *Pharmacogenetics*, *14*, 831–839.

Meluzín, J., Vasků, A., Kincl, V., Panovský, R., & Šrámková, T. (2009). Association of coronary artery disease, erectile dysfunction, and endothelial nitric oxide synthase polymorphisms. *Heart and Vessels*, *24*, 157–163.

Miyamoto, Y., Saito, Y., Kajiyama, N., Yoshimura, M., Shimasaki, Y., Nakayama, M., et al. (1998). Endothelial nitric oxide synthase gene is positively associated with essential hypertension. *Hypertension*, *32*, 3–8.

Miyamoto, Y., Saito, Y., Nakayama, M., Shimasaki, Y., Yoshimura, T., Yoshimura, M., et al. (2000). Replication protein A1 reduces transcription of the endothelial nitric oxide synthase gene containing a −786T>C mutation associated with coronary spastic angina. *Human Molecular Genetics*, *9*, 2629–2637.

Monti, L. D., Barlassina, C., Citterio, L., Galluccio, E., Berzuini, C., Setola, E., et al. (2003). Endothelial nitric oxide synthase polymorphisms are associated with type 2 diabetes and the insulin resistance syndrome. *Diabetes*, *52*, 1270–1275.

Murray, S., Lazure, P., Pullen, C., Maltais, P., & Dorian, P. (2011). Atrial fibrillation care: Challenges in clinical practice and educational needs assessment. *Canadian Journal of Cardiology*, *27*, 98–104.

Nakayama, M., Yasue, H., Yoshimura, M., Shimasaki, Y., Kugiyama, K., Ogawa, H., et al. (1999). −786TNC mutation in the 5′-flanking region of the endothelial nitric oxide synthase gene is associated with coronary spasm. *Circulation*, *99*, 2864–2870.

Niu, W., & Qi, Y. (2011). An updated meta-analysis of endothelial nitric oxide synthase gene: Three well-characterized polymorphisms with hypertension. *PLoS One*, *6*, e24266.

Niu, P. P., Yang, G., Zheng, B. K., Guo, Z. N., Jin, H., & Yang, Y. (2013). Relationship between endothelial nitric oxide synthase gene polymorphisms and ischemic stroke: A meta-analysis. *Acta Neurologica Scandinavica*, *128*, 202–212.

Ohtoshi, K., Yamasaki, Y., Gorogawa, S., Hayaishi-Okano, R., Node, K., Matsuhisa, M., et al. (2002). Association of (−)786T-C mutation of endothelial nitric oxide synthase gene with insulin resistance. *Diabetologia*, *45*, 1594–1601.

Persu, A., Stoenoiu, M. S., Messiaen, T., Davila, S., Robino, C., El-Khattabi, O., et al. (2002). Modifier effect of eNOS in autosomal dominant polycystic kidney disease. *Human Molecular Genetics*, *11*, 229–241.

Poirier, O., Mao, C., Mallet, C., Nicaud, V., Herrmann, S. M., Evans, A., et al. (1999). Polymorphisms of the endothelial nitric oxide synthase gene—No consistent association with myocardial infarction in the ECTIM study. *European Journal of Clinical Investigation*, *29*, 284–290.

Ponnuswamy, P., Schröttle, A., Ostermeier, E., Grüner, S., Huang, P. L., Ertl, G., et al. (2012). eNOS protects from atherosclerosis despite relevant superoxide production by the enzyme in apoE mice. *PLoS One*, *7*, e30193.

Ramesh, S., Morrell, C. N., Tarango, C., Thomas, G. D., Yuhanna, I. S., Girardi, G., et al. (2011). Antiphospholipid antibodies promote leukocyte-endothelial cell adhesion and thrombosis in mice by antagonizing eNOS via β2GPI and apoER2. *Journal of Clinical Investigation*, *121*, 120–131.

Rippe, C., Blimline, M., Magerko, K. A., Lawson, B. R., LaRocca, T. J., Donato, A. J., et al. (2012). MicroRNA changes in human arterial endothelial cells with senescence: Relation to apoptosis, eNOS and inflammation. *Experimental Gerontology*, *47*, 45–51.

Roberts, R., & Stewart, A. F. R. (2012). Genes and coronary artery disease. Where are we? *Journal of the American College of Cardiology*, *60*, 1715–1721.

Safarinejad, M. R., Safarinejad, S., Shafiei, N., & Safarinejad, S. (2013). Effects of the T−786C, G894T, and intron 4 VNTR (4a/b) polymorphisms of the endothelial nitric oxide synthase gene on the risk of prostate cancer. *Urologic Oncology*, *31*(7), 1132–1140.

Salimi, S., Naghavi, A., Firoozrai, M., Zand, H., Tavilani, H., Nakhaee, A., et al. (2012). Association of plasma nitric oxide concentration and endothelial nitric oxide synthase T−786C gene polymorphism in coronary artery disease. *Pathophysiology*, *19*, 157–162.

Sandrim, V. C., Palei, A. C., Metzger, I. F., Cavalli, R. C., Duarte, G., & Tanus-Santos, J. E. (2010). Interethnic differences in ADMA concentrations and negative association with nitric oxide formation in preeclampsia. *Clinica Chimica Acta*, *411*, 1457–1460.

Sandrim, V. C., Palei, A. C., Sertorio, J. T., Cavalli, R. C., Duarte, G., & Tanus-Santos, J. E. (2010). Effects of eNOS polymorphisms on nitric oxide formation in healthy pregnancy and in pre-eclampsia. *Molecular Human Reproduction*, *16*, 506–510.

Serrano, N. C., Casas, J. P., Díaz, L. A., Páez, C., Mesa, C. M., Cifuentes, R., et al. (2004). Endothelial NO synthase genotype and risk of preeclampsia: A multicenter case–control study. *Hypertension*, *44*, 702–707.

Shimasaki, Y., Yasue, H., Yoshimura, M., Nakayama, M., Kugiyama, K., & Ogawa, H. (1998). Association of the missense Glu298Asp variant of the endothelial nitric oxide synthase gene with myocardial infarction. *Journal of the American College of Cardiology*, *31*, 1506–1510.

Sibai, B., Dekker, G., & Kupferminc, M. (2005). Pre-eclampsia. *Lancet*, *365*, 785–799.

Tangurek, B., Ozer, N., Sayar, N., Terzi, S., Yilmaz, H., Dayi, S. U., et al. (2006). The relationship between endothelial nitric oxide synthase gene polymorphism (T−786C) and coronary artery disease in the Turkish population. *Heart and Vessels*, *21*, 285–290.

Tanus-Santos, J. E., Desai, M., & Flockhart, D. A. (2001). Effects of ethnicity on the distribution of clinically relevant endothelial nitric oxide variants. *Pharmacogenetics*, *11*, 719–725.

Tao, H. M., & Chen, G. Z. (2009). Endothelial NO synthase gene polymorphisms and risk of ischemic stroke: A meta-analysis. *Neuroscience Research*, *64*, 311–316.

Tesauro, M., Thompson, W. C., Rogliani, P., Qi, L., Chaudhary, P. P., & Moss, J. (2000). Intracellular processing of endothelial nitric oxide synthase isoforms associated with differences in severity of cardiopulmonary diseases: Cleavage of proteins with aspartate vs. glutamate at position 298. *Proceedings of the National Academy of Sciences of the United States of America*, *6*, 2832–2835.

van Beynum, I. M., Mooij, C., Kapusta, L., Heil, S., den Heijer, M., & Blom, H. J. (2008). Common 894G>T single nucleotide polymorphism in the gene coding for endothelial nitric oxide synthase (eNOS) and risk of congenital heart defects. *Clinical Chemistry and Laboratory Medicine*, *46*, 1369–1375.

Vecoli, C., Andreassi, M. G., Liga, R., Colombo, M. G., Coceani, M., Carpeggiani, C., et al. (2012). T(−786)→C polymorphism of the endothelial nitric oxide synthase gene is associated with insulin resistance in patients with ischemic or non ischemic cardiomyopathy. *BMC Medical Genetics*, *13*, 92.

Wang, J., Dudley, D., & Wang, X. L. (2002). Haplotype-specific effects on endothelial NO synthase promoter efficiency: Modifiable by cigarette smoking. *Arteriosclerosis, Thrombosis, and Vascular Biology*, *22*, e1–e4.

Wang, C. L., Hsu, L. A., Ko, Y. S., Ko, Y. L., & Lee, Y. H. (2001). Lack of association between the Glu298Asp variant of the endothelial nitric oxide synthase gene and the risk of coronary artery disease among Taiwanese. *Journal of the Formosan Medical Association*, *100*, 736–740.

Wang, M., Jiang, X., Wu, W., & Zhang, D. (2013). Endothelial NO synthase gene polymorphisms and risk of ischemic stroke in Asian population: A meta-analysis. *PLoS One*, *8*, e60472.

Weber, M., Baker, M. B., Moore, J. P., & Searles, C. D. (2010). MiR-21 is induced in endothelial cells by shear stress and modulates apoptosis and eNOS activity. *Biochemical and Biophysical Research Communications*, *393*, 643–648.

Wink, D. A., Vodovotz, Y., Laval, J., Laval, F., Dewhirst, M. W., & Mitchell, J. B. (1998). The multifaceted roles of nitric oxide in cancer. *Carcinogenesis*, *19*, 711–721.

Yao, Y. S., Chang, W. W., Jin, Y. L., & He, L. P. (2013). An updated meta-analysis of endothelial nitric oxide synthase gene: Three well-characterized polymorphisms with ischemic stroke. *Gene*, *528*(2), 84–92.

Yao, L., Fang, F., Zhong, Y., & Yu, L. (2010). The association between two polymorphisms of eNOS and breast cancer risk: A meta-analysis. *Breast Cancer Research and Treatment*, *124*, 223–227.

Yoon, Y., Song, J., Hong, S. H., & Kim, J. Q. (2000). Plasma nitric oxide concentrations and nitric oxide synthase gene polymorphisms in coronary artery disease. *Clinical Chemistry*, *46*, 1626–1630.

Zhang, X., Lynch, A. I., Davis, B. R., Ford, C. E., Boerwinkle, E., Eckfeldt, J. H., et al. (2012). Pharmacogenetic association of NOS3 variants with cardiovascular disease in patients with hypertension: The GenHAT study. *PLoS One*, 7, e34217.

Zhang, M. X., Zhang, C., Shen, Y. H., Wang, J., Li, X. N., Chen, L., et al. (2008). Effect of 27 nt small RNA on endothelial nitric-oxide synthase expression. *Molecular Biology of the Cell*, *19*, 3997–4005.

Zhao, S., Li, T., Zheng, B., & Zheng, Z. (2012). Nitric oxide synthase 3 (NOS3) 4b/a, T−786C and G894T polymorphisms in association with diabetic retinopathy susceptibility: A meta-analysis. *Ophthalmic Genetics*, *33*, 200–207.

CHAPTER SIXTEEN

Role of Nitric Oxide in Pathophysiology and Treatment of Pulmonary Hypertension

Peter Kruzliak[*,1], **Junko Maruyama**[†,‡], **Kazuo Maruyama**[†,‡]
*Department of Cardiovascular Diseases, International Clinical Research Center, St. Anne's University Hospital, Brno, Czech Republic
†Department of Anesthesiology and Critical Care Medicine, Mie University School of Medicine, Mie, Japan
‡Department of Clinical Engineering, Suzuka University of Medical Science, Mie, Japan
[1]Corresponding author: e-mail address: peter.kruzliak@savba.sk

Contents

Vitamins and Hormones, Volume 96
ISSN 0083-6729
http://dx.doi.org/10.1016/B978-0-12-800254-4.00016-7

Abstract

Pulmonary hypertension is a condition characterized by vasoconstriction, vascular cell proliferation, inflammation, microthrombosis, and vessel wall remodelation. Pulmonary endothelial cells produce vasoactive substances with vasoconstrictive as well as vasodilatative effects. The imbalance of these endothelium-derived vasoactive substances induced by endothelial dysfunction is very important in the pathogenesis of PH. One of most important substances with vasodilatative effect is nitric oxide. We provide a comprehensive insight into role of NO in the pathgenesis of PH and discuss perspectives and challenges in PH therapy based on NO administration.

1. ROLE OF NO IN VASORELAXATION AND PROLIFERATION IN PH

The pathological condition of PH is characterized by vasoconstriction, vascular cell proliferation (antiapoptosis), inflammation, and microthrombosis (Archer, Weir, & Wilkins, 2010; Farber & Loscalzo, 2004; Rabinovitch, 2008). Pulmonary endothelial cells produce vasoactive substances which function as vasorelaxants and vasoconstrictors. Decreased NO and PGI_2 levels and increased endothelin-l and TXA_2 levels have been observed in PH. The imbalance of these endothelium-derived vasoactive substances induced by endothelial dysfunction is considered the pathogenesis of PH (Archer et al., 2010; Farber & Loscalzo, 2004; Rabinovitch, 2008).

NO is a simple, low-molecular-weight free radical and a potent endothelium-derived vasodilator. NO activates soluble guanylate cyclase and stimulates cGMP production (Francis, Busch, Corbin, & Sibley, 2010). In addition to this major pathway, NO binds directly to protein kinases and phosphatases and changes their enzyme activity (Nakamura & Lipton, 2011). Thus, NO modifies the phosphorylation and dephosphorylation of the target substance. cGMP and cGMP-dependent protein kinase induced by NO promote vasorelaxation to decrease the cytosolic Ca^{2+} concentration in vascular smooth muscle cells by (1) phosphorylation of the Ca^{2+}-sensitive K^+ channel, which causes hyperpolarization and closes the voltage-dependent L-type Ca^{2+} channel, (2) direct phosphorylation of the voltage-dependent sarcolemmal Ca^{2+} channel, resulting in its closure, and (3) inhibition of IP_3-sensitive Ca^{2+} release from the sarcoplasmic reticulum (Francis et al., 2010; Hampl & Herget, 2000) (Fig. 16.1).

The histological changes in PH are specific to the medium-sized pulmonary arteries peripheral to the precapillary arteries and their circumference. Vascular structural change is associated with cellular proliferation and

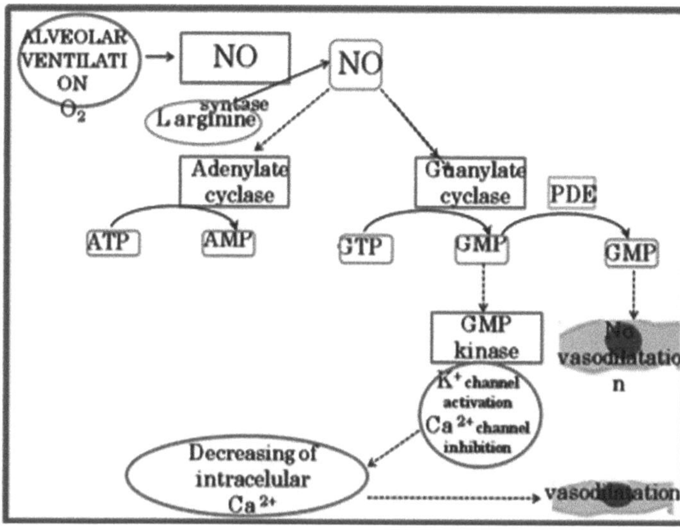

Figure 16.1 Mechanism of NO-mediated vasorelaxation. NO activates soluble guanylate cyclase and stimulates cGMP production. In addition to this major pathway, NO binds directly to protein kinases and phosphatases and changes their enzyme activity. cGMP and cGMP-dependent protein kinase induced by NO promote vasorelaxation to decrease the cytosolic Ca^{2+} concentration in vascular smooth muscle cells by (1) phosphorylation of the Ca^{2+}-sensitive K^{+} channel, which causes hyperpolarization and closes the voltage-dependent L-type Ca^{2+} channel, (2) direct phosphorylation of the voltage-dependent sarcolemmal Ca^{2+} channel, resulting in its closure, and (3) inhibition of IP_3-sensitive Ca^{2+} release from the sarcoplasmic reticulum. (See the color plate.)

migration. The principal histological feature in all forms of PH is (1) muscularization of normally nonmuscular peripheral arteries, (2) medial hypertrophy of proximal muscular arteries, and (3) an increase in intercellular connective tissue proteins, such as collagen and elastin (Archer et al., 2010; Farber & Loscalzo, 2004; Rabinovitch, 2008). The most characteristic manifestation in severe idiopathic and/or familial PH is a lumen-obliterative plexiform lesion with phenotypic abnormalities of vascular component cells (Stevens, 2005).

NO mediates antiproliferation and anticoagulation as well as vasorelaxation, and inhibits smooth muscle cell growth in a cGMP-dependent manner (Garg & Hassid, 1989). It also inhibits growth-regulating enzymes such as ribonucleotide reductase and thymidine kinase in a cGMP-independent manner (Garg, 1993; Kwon, Stuehr, & Nathan, 1991). NO inhibits the increase in ET-1 and platelet-derived growth factor-B, which are considered to be growth factors, in response to hypoxia (Kourembanas, McQuillan, Leung, & Faller, 1993).

2. ENDOTHELIAL NITRIC OXIDE SYNTHASE EXPRESSION AND ITS ACTIVITY IN RELATIONSHIP TO THE VASOREACTIVITY

Endothelial nitric oxide synthase (eNOS) is highly expressed in vascular endothelial cells, and reduced NO bioavailability may be partially associated with alteration of eNOS expression and/or activity in PH. A previous study showed that mRNA and protein expression of eNOS in rat lung and eNOS expression localized in vascular endothelial cells are upregulated in acute hypoxia (Rus, Peinado, Castro, & Del Moral, 2010). In that study, the nitrate/nitrite levels in rat lung homogenate also increased, suggesting indirectly augmented eNOS activity. In other studies using NOS inhibitors, the enhancement of eNOS activity was also observed in hypoxic pulmonary vasoconstriction (HPV) in normal rat lung (Archer, Tolins, Raij, & Weir, 1989; Fox, Paterson, & McCormack, 1994). eNOS protein expression was time-dependently increased in rats with chronic hypoxia-induced PH (Blumberg et al., 2003; Murata et al., 2005), while phosphorylated eNOS (peNOS), the active form of eNOS, was impaired (Murata et al., 2005). Monocrotaline (MCT)-induced PH rat lung has decreased expression of both eNOS and peNOS, followed by decreased NO production (Kanno, Wu, Lee, Billiar, & Ho, 2001; Mawatari et al., 2007; Pei et al., 2011). This may be due to vascular endothelial cell injury in the early phase of PH development. These results partly explain how the process of alteration in the eNOS/NO/cGMP system might modify the pathogenesis in the development of PH. Below we describe mechanisms linking the eNOS expression with the vasorelaxation. Those mechanisms are shown in Fig. 16.2.

2.1. eNOS uncoupling

Tetrahydrobiopterin (BH4) is an essential cofactor for the production of NO by eNOS (Crabtree et al., 2009). Homodimer eNOS bound to BH4 synthesizes NO by converting L-arginine to L-citrullin in the presence of O_2. In the absence of L-arginine or BH4, uncoupled eNOS in the monomer form synthesizes superoxide instead of NO. The molecular basis of eNOS uncoupling is the conversion of the NOS homodimer to the NOS monomer form (Crabtree et al., 2009). Therefore, BH4 deficiency may play a role in the pathogenesis of PH (Bevers et al., 2006; Cunnington et al., 2012; Khoo et al., 2005).

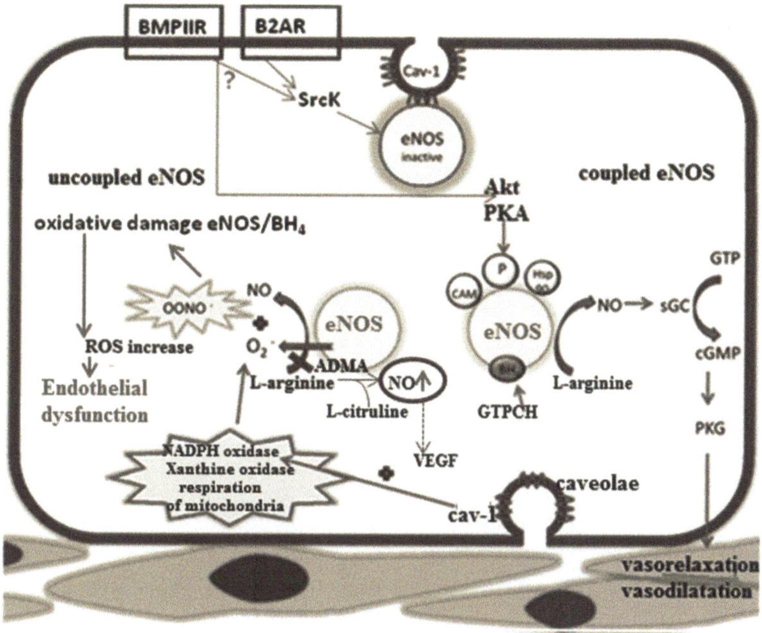

Figure 16.2 Endothelial nitric oxide synthase (eNOS) expression and its activity in relationship with the vasorelaxation. (See the color plate.)

2.2. Caveolin-1 and eNOS uncoupling

Caveolae are flask-shaped invaginations on the cell surface containing signal proteins and structural proteins called caveolin. Caveolin-1 is a major protein caveola constituent that interacts with several signaling molecules such as eNOS (Patel, Murray, & Insel, 2008). eNOS is inactivated when conjugated to caveolin-1 and activated on its dissociation from the caveolae. Loss of caveolin-1 induces chronic activation of eNOS (Zhao et al., 2009), where activated eNOS is uncoupled to produce superoxide in idiopathic PH. A progressive loss of caveolin and a subsequent reduction in eNOS expression have been observed in MCT-induced PH (Huang, Wolk, Gewitz, & Mathew, 2010), but not in chronic hypoxia-induced PH (Murata et al., 2001).

2.3. Bone morphogenetic protein receptor type II and eNOS

Mutation of bone morphogenetic protein receptor type II (BMPIIR) gene is one of the causes of familial pulmonary arterial hypertension (Archer et al., 2010; Farber & Loscalzo, 2004; Rabinovitch, 2008). BMPIIR is a member of the TGF-β superfamily of growth factor receptors. Stimulation of

BMPIIR induces eNOS phosphorylation and dissociation of eNOS from caveolin-1. It also increases the eNOS–heat-shock protein-90 (HSP90) interaction. HSP90 facilitates the dissociation of eNOS from caveolin-1 (Gangopahyay et al., 2011).

2.4. Vascular endothelial growth factor and eNOS expression

VEGF (VEGF) plays prosurvival and antiapoptotic roles and stimulates vasodilation, microvascular hyperpermeability, and angiogenesis. Plexiform lesions show striking expression of VEGF associated with endothelial proliferation (Tuder et al., 2001). VEGF stimulates NO production initially by increasing intracellular Ca^{2+} levels and subsequent Ca^{2+} -calmodulin-dependent activation of eNOS, and later by increasing intracellular eNOS message and protein levels (Hood, Meininger, Ziche, & Granger, 1998). Administration of SU5416, a VEGF receptor blocker, combined with chronic hypoxia to adult rats results in severe PH with obstructive neointimal changes (Taraseviciene-Stewart et al., 2001). However, the pathogenetic mechanism of this novel model is complex and remains to be elucidated.

3. PHYSIOLOGICAL ROLE OF NO IN PULMONARY HYPERTENSION

3.1. Role of NO in vascular basal tone

NOS inhibitors have been used to examine the physiological role of NO in pulmonary vascular basal tone. Neither L-NMMA (*N* omega-monomethyl-L-arginine) (Archer et al., 1989) nor L-NNA (*N* omega-nitro-L-arginine) (Maruyama & Maruyama, 1994; Oka et al., 1993) changed pulmonary basal tone in normal rat pulmonary arterial rings. In addition, normal isolated perfused lungs were not affected by these NOS inhibitors (Ferrario, Amin, Sugimori, Camporesi, & Hakim, 1996; Oka et al., 1993) or by L-NA (*N* omega-nitro-L-arginine) except in a few studies that showed a moderate increase with L-NAME (*N* omega-nitro-L-arginine methyl ester) (Igari et al., 1998). In chronic hypoxia, a markedly enhanced vascular tone has been observed in the presence of L-NNA (Oka et al., 1993) or L-NAME (Igari et al., 1998). Although these NOS inhibitors produced different results, the findings suggest that (1) NO is not involved in vascular basal tone in normal lung, and (2) basal NO production is potent in the pulmonary circulation in hypoxia-induced chronic PH. On exposure to acute hypoxia, NOS inhibitors augmented vascular contraction in normal (Archer et al., 1989; Bardou et al., 2001; Igari et al., 1998) and hypoxia-induced PH rats (Igari et al., 1998). These findings suggest that NO production and/or availability in HPV is increased in both normal and hypoxic PH rats.

3.2. Vasoreactivity to NO-related vasorelaxants in rat lung

Acetylcholine (Ach) and sodium nitroprusside (SNP), which are endothelium-dependent and -independent NO-related relaxing substances, respectively, are commonly used to examine functional changes in vascular endothelial and smooth muscle cells in PH.

3.2.1 Physiological changes in rats with hypoxic PH

The relaxation response to Ach is impaired in isolated conduit pulmonary arteries of rats in hypoxic PH (Elmedal, de Dam, Mulvany, & Simonsen, 2004; Mam et al., 2010; Maruyama & Maruyama, 1994; Oka, 2001; Shaul, Wells, & Horning, 1993), and the same is true for SNP (Mam et al., 2010; Maruyama & Maruyama, 1994; Oka, 2001). These results suggest (1) decreased production and/or release of NO in endothelial cells, or (2) decreased responsiveness to NO in smooth muscle cells, or (3) both. Impaired relaxation response to Ach and SNP was partially restored after cessation of chronic hypoxia. Different recovery processes in the response to Ach and SNP suggest that NO-related functional abnormalities in endothelial and smooth muscle cells occur independently. The impaired response to Ach in conduit Pas was partly restored in the presence of non-selective inhibitor of cyclooxygenase or PGH_2/TXA_2 receptor antagonist (Maruyama, Yokochi, Maruyama, & Nosaka, 1999). These results suggest (1) an imbalance between the production of vasoconstrictor and vasorelaxant prostanoids in conduit pulmonary artery endothelial cells and (2) simultaneous release of vasoconstrictor prostanoids.

The relaxation response to Ach in resistant PAs is not changed (Elmedal et al., 2004; Oka, 2001) or augmented (MacLean & McCulloch, 1998), in contrast to that in conduit PAs. It is likely that Ach-reactive NO production and/or release varies in a vascular site-specific manner. As conduit arteries produce and release more NO than peripheral arteries, they might be more susceptible to stimuli such as abnormal shear stress, circumferential wall stretch, and hypoxia.

3.2.2 Physiological changes in rats with MCT-induced PH

The relaxation response to Ach and SNP is impaired in MCT-induced rat conduit PA rings (Fullerton, Hahn, & McIntyre, 1996; Gout, Quiniou, Khandoudi, Le Dantec, & Saïag, 1999; Mam et al., 2010; Mathew,

Zeballos, Tun, & Gewitz, 1995; Ozturk & Uma, 2010). Ach-induced relaxation is impaired in the pulmonary circulation, reflecting the response of resistant PAs, whereas the SNP relaxation response has been reported to be impaired (Baber et al., 2007) or not impaired (Prié, Stewart, & Dupuis, 1998). Taken together, although vascular endothelial dysfunction is observed from proximal to peripheral PAs, no smooth muscle functional changes are apparent in peripheral PAs.

3.2.3 Experimental studies focusing on NO inhalation in PH

Studies involving animal models of PH have demonstrated that short- and long-term NO inhalation improves arterial oxygenation and reduces pulmonary arterial pressure (PAP) (Jiang et al., 2002; Katayama et al., 1995, 1994; Kobayashi et al., 2001; Maruyama et al., 1997). Low-dose NO (<5 ppm) induced a submaximal reduction in PAP in rats with chronic hypoxic PH, which does not correlate with the severity of vascular remodeling (Jiang et al., 2002). Chronic NO inhalation ameliorates the development of hypertensive pulmonary vascular changes in chronic hypoxia-induced PH in rats (Kouyoumdjian et al., 1994), but not in MCT-induced PH (Maruyama et al., 1997). Long-term NO inhalation does not facilitate the structural (Jiang et al., 2004) or functional recovery (Maruyama, Jiang, Maruyama, Takata, & Miyasaka, 2004) of the impaired NO-cGMP pathway in established hypoxic PH during the recovery process in room air after exposure to hypoxia. Clinically, the effect of inhaled NO is based on vasorelaxation. The potential role of NO as an inhibitor of vascular cell proliferation has been additionally discussed in experimental studies.

4. USE OF NO IN PATIENTS WITH PH IN CLINICAL PRACTICE

4.1. NO-related relaxation responses in patients with pulmonary hypertension

The endothelium-dependent relaxation response to acetylcholine, but not the endothelium-independent NO-related relaxation response to SNP, is impaired in isolated pulmonary arteries in patients with end-stage chronic obstructive pulmonary disease (COPD) (Dinh-Xuan et al., 1991), which might be due to the reduction of NO release from pulmonary endothelial cells. The clinically important point here is that the relaxation response to exogenous NO in PH is present even in structurally remodeled pulmonary arteries.

4.2. NO delivery in clinical use (NO donors and NO inhalation)

For many years, NO donors, pharmacologically active compounds that release NO, have been used as therapeutic agents for ischemic heart disease. These compounds are classified into organic nitrates including glyceryl trinitrate, nitrites, SNP, NONOates, and *S*-nitrosothiols (Katsumi, Nishikawa, & Hashida, 2007). However, intractable problems are known to be drug tolerance and lack of selectivity (Katsumi et al., 2007).

In 1987, NO was identified as an endothelium-derived vasorelaxing substance (Ignarro, Buga, Wood, Byrns, & Chaudhuri, 1987; Palmer, Ferrige, & Moncada, 1987). Since then, NO has been found to have clinical use in its gaseous form. The clinical effects of inhaled NO have been extensively studied and applied to the treatment of PH. Inhaled NO is a selective pulmonary vasodilator, and because of its short half-life, it does not reduce systemic blood pressure (Pepke-Zaba, Higenbottam, Dinh-Xuan, Stone, & Wallwork, 1991).

4.3. Physiological effect of NO inhalation

Selective pulmonary vasodilation decreases PAP, PVR, and right ventricular afterload. It also alters the distribution of the ventilation–perfusion (V/Q) ratio and decreases the venous admixture and, consequently, arterial oxygenation. These effects might be beneficial for treating PH (Pérez-Peñate, Julià-Serdà, Pulido-Duque, Górriz-Gómez, & Cabrera-Navarro, 2001; Shimpo et al., 1997; Vonbank et al., 2003), right ventricular failure (Argenziano et al., 1998; Rajek et al., 2000), hypoxemic respiratory failure (Adhikari et al., 2007; Dellinger et al., 1998; Maruyama, Kobayasi, Taguchi, Chikusa, & Muneyuki, 1995; Maruyama, Takeuchi, Chikusa, & Muneyuki, 1995; Neonatal Inhaled Nitric Oxide Study Group, 1997; Scherrer et al., 1996; The Australia and New Zealand Extracorporeal Membrane Oxygenation (ANZ ECMO) Influenza Investigators, 2009), and impaired pulmonary gas exchange (Barberà et al., 1996; Matsumoto et al., 1999; Yoshida, Taguchi, Gabazza, Kobayashi, et al., 1997; Yoshida, Taguchi, Gabazza, Yasui, et al., 1997).

4.4. Effects of inhaled NO on PAP

Selective pulmonary vascular relaxation was first described in patients with PH (Pepke-Zaba et al., 1991). NO inhalation decreases PAP in structurally remodeled pulmonary vasculature, but not in normal lung, perhaps because the hypertensive pulmonary artery has vascular hyperreactivity, even in the unstimulated state. In an animal study, we described how PAP was dose-dependently reduced at concentrations of 20–60 ppm and leveled off above 60 ppm (Katayama et al., 1994).

PAP reduction was seen at low doses of NO between 0.1 and 2.0 ppm in chronic hypoxia-induced PH in rats (Jiang et al., 2002). These results provided the foundation for low-dose NO inhalation in postoperative cardiac patients (Shimpo et al., 1997) and in patients with pulmonary fibrosis (Maruyama, Kobayasi, et al., 1995), respiratory failure (Maruyama, Takeuchi, et al., 1995), or COPD (Yoshida, Taguchi, Gabazza, Kobayashi, et al., 1997). In addition, these clinical experiences showed that therapeutic performance was brought about by improving arterial oxygenation rather than by reducing PAP (Gerlach, Pappert, Lewandowski, Rossaint, & Falke, 1993).

4.5. Effects of inhaled NO on arterial oxygenation

Alveolar hypoxia causes HPV in adjacent peripheral arterioles, and HPV in turn reduces the pulmonary blood flow in poorly ventilated lung units, which corrects the mismatch in the V/Q ratio, thereby preventing a further decrease in arterial oxygenation. Intrapulmonary shunt is defined as a very low V/Q ratio (<0.005), while low V/Q is defined in the range of 0.005–0.1 (normal V/Q ratio range, 0.1–10).

In some patients with COPD, inhaled NO impairs arterial oxygenation (Fig. 16.3).

5. CLINICAL APPLICATION OF NO INHALATION IN PATIENTS WITH PULMONARY HYPERTENSION

5.1. Persistent pulmonary hypertension of the neonate

Persistent pulmonary hypertension of the neonate (PPHN) is a clinical syndrome caused by failure of the pulmonary vascular transition to normal circulation after birth in full-term and near-term infants. In PPHN, increased PVR results in a right-to-left shunt across the patent ductus arteriosus and foramen ovale inducing severe hypoxemia, which is often treated with extracorporeal membrane oxygenation (ECMO). Inhaled NO reduces the right-to-left shunt and improves arterial oxygenation (Kinsella, Neish, Shaffer, & Abman, 1992; Roberts, Polaner, Lang, & Zapol, 1992). A subsequent randomized controlled trial (RCT) showed that NO inhalation reduced the need for ECMO but did not change the survival rate (Neonatal Inhaled Nitric Oxide Study Group, 1997). Inhaled NO was registered as a therapy for neonatal hypoxic respiratory failure associated with PH by the United States Food and Drug

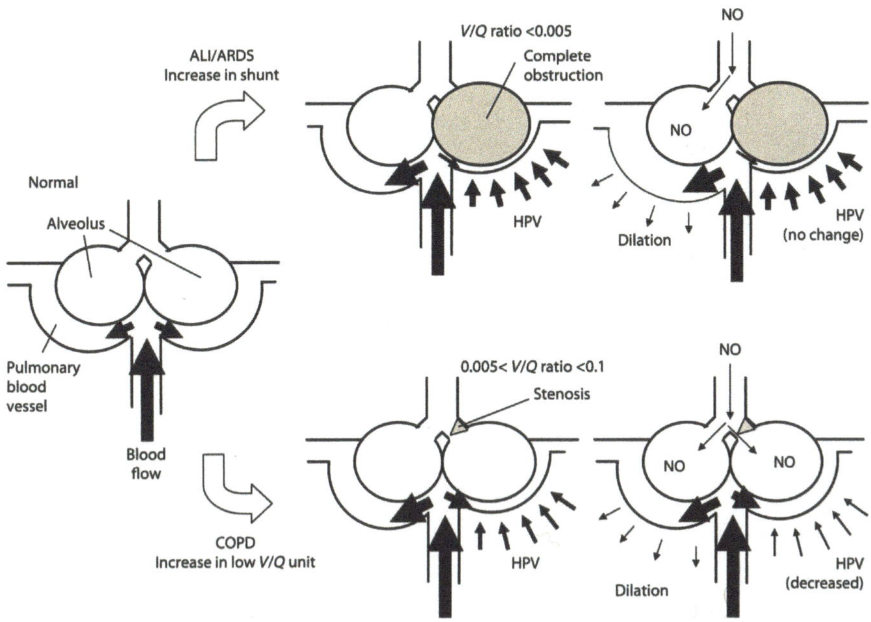

Figure 16.3 In a lung with intrapulmonary shunt, inhaled NO indirectly enhances the effect of HPV (i.e., reduces blood flow to areas with a very low V/Q ratio) by dilating the vasculature in well-ventilated units. NO then enters the ventilated unit, dilates the pulmonary vasculature in the areas permeated by inhaled NO, and increases the blood flow in the ventilated unit more than without NO inhalation. In subjects with acute lung injury and/or adult respiratory distress syndrome (ALI/ARDS), NO inhalation decreases intrapulmonary shunt. In contrast, in the areas with a low V/Q ratio (0.005 to 0.1), which are increased in patients with COPD, NO enters these poorly ventilated units and partly reverses HPV.

Administration in 1999, the European Medicine Evaluation Agency and European Commission in 2001, and the Ministry of Health, Labor and Welfare in Japan in 2009.

5.2. Bronchopulmonary dysplasia in preterm infants with respiratory distress syndrome

Bronchopulmonary dysplasia (BPD) is a chronic lung disorder commonly seen in preterm infants on prolonged mechanical ventilation. The pathogenesis of BPD is immature pulmonary vasculature and lung accompanied by inflammation, high airway pressure, and PH. Preterm infants with respiratory distress syndrome (RDS) are characterized by an absence of right-to-left shunt and different circulatory dynamics and arterial oxygenation from those of full-term infants. In infants with a birth weight <1250 g 7–21 days after birth, 20 ppm inhaled NO for 21 days was shown to shorten the period of oxygen administration and increase the survival rate (Ballard et al., 2006).

5.3. Pulmonary hypertensive crisis after cardiac surgery of congenital heart defect

Pulmonary hypertensive crisis (PHC) is a critical condition characterized by acute hypoxemia followed by right ventricular failure due to a sudden increase in pulmonary arterial resistance. According to a randomized clinical trial (RCT) constructed by Miller et al., inhaled NO at 10 ppm reduced the occurrence of severe postoperative PHC after surgical repair in infants with high preoperative pulmonary flow and pressure (Miller et al., 2000). However, more confirmatory analyses might be necessary to elucidate the effect of inhaled NO.

5.4. Acute lung injury and/or adult respiratory distress syndrome

In patients with ALI/ARDS, inhaled NO reduces intrapulmonary shunt and improves arterial oxygenation (Rossaint et al., 1993). A RCT and meta-analysis of 12 RCTs of patients with ALI/ARDS showed that inhaled NO had no significant effect on hospital mortality, duration of mechanical ventilation, or ventilation-free days, while it improved arterial oxygenation for 1–4 days, but also increased the risk of developing renal dysfunction (Adhikari et al., 2007). A subsequent systematic review revealed that inhaled NO transiently improved arterial oxygenation in the first 24 h but did not reduce mortality (Afshari, Brok, Møller, & Wetterslev, 2011). Routine use of inhaled NO in patients with ALI/ARDS is currently not recommended. However, as a rescue therapy to improve refractory hypoxemia, inhaled NO is administered. For example, NO inhalation was used in 20 of 68 patients with novel influenza A (H1N1) ARDS in a 2009 pandemic outbreak (The Australia and New Zealand Extracorporeal Membrane Oxygenation (ANZ ECMO) Influenza Investigators, 2009).

5.5. Chronic obstructive pulmonary disease

In COPD, alveolar hypoxia induces HPV and increases the area with a low V/Q ratio (0.05–0.1), while also causing PH and right ventricular failure. As described previously, NO cannot enter a nonventilated area (very low

V/*Q* units, $V/Q < 0.005$), but it can reach a low *V*/*Q* ratio area. NO reduces HPV in poorly ventilated units with lower *V*/*Q* ratios because the blood flow of poorly ventilated units increases. As a result, inhaled NO worsens the *V*/*Q* mismatch as well as arterial oxygenation in patients breathing air (Barberà et al., 1996). In contrast, combined therapy with NO and oxygen inhalation increases arterial oxygenation more than inhalation of oxygen alone (Yoshida, Taguchi, Gabazza, Kobayashi, et al., 1997). Similar findings were obtained in patients with pulmonary fibrosis (Yoshida, Taguchi, Gabazza, Yasui, et al., 1997). A subsequent RCT showed that 3-month long-term NO inhalation with oxygen decreased PAP and PVR without decreasing arterial oxygenation, and also increased cardiac output (Vonbank et al., 2003).

5.6. Other findings and future directions

Inhaled NO has been applied as a therapy for patients with congestive heart failure (Matsumoto et al., 1999), heart transplantation (Ardehali et al., 2001; Bardou et al., 2001; Stobierska-Dzierzek, Awad, & Michler, 2001), and ischemia-reperfusion injury (Mathru, Huda, Solanki, Hays, & Lang, 2007). NO inhalation is a registered therapy for PPHN; for adults however, it might be necessary to confirm the clinical applicability of NO inhalation. An animal study by Ichinose et al. showed that NO inhalation after CPR attenuates neurological and left ventricular dysfunction and improves survival (Minamishima et al., 2011). Such findings might be worthy candidates for future clinical trials.

REFERENCES

Adhikari, N. K., Burns, K. E., Friedrich, J. O., Granton, J. T., Cook, D. J., & Meade, M. O. (2007). Effect of nitric oxide on oxygenation and mortality in acute lung injury: Systematic review and meta-analysis. *British Medical Journal, 334*(7597), 779.

Afshari, A., Brok, J., Møller, A. M., & Wetterslev, J. (2011). Inhaled nitric oxide for acute respiratory distress syndrome and acute lung injury in adults and children: A systematic review with meta-analysis and trial sequential analysis. *Anesthesia and Analgesia, 112*(6), 1411–1421.

Archer, S. L., Tolins, J. P., Raij, L., & Weir, E. K. (1989). Hypoxic pulmonary vasoconstriction is enhanced by inhibition of the synthesis of an endothelium derived relaxing factor. *Biochemical and Biophysical Research Communications, 164*(3), 1198–1205.

Archer, S. L., Weir, E. K., & Wilkins, M. R. (2010). Basic science of pulmonary arterial hypertension for clinicians: New concepts and experimental therapies. *Circulation, 121*(18), 2045–2066.

Ardehali, A., Hughes, K., Sadeghi, A., Esmailian, F., Marelli, D., Moriguchi, J., et al. (2001). Inhaled nitric oxide for pulmonary hypertension after heart transplantation. *Transplantation, 72*(4), 638–641.

Argenziano, M., Choudhri, A. F., Moazami, N., Rose, E. A., Smith, C. R., Levin, H. R., et al. (1998). Randomized, double-blind trial of inhaled nitric oxide in LVAD recipients with pulmonary hypertension. *Annals of Thoracic Surgery, 65*(2), 340–345.

Baber, S. R., Deng, W., Master, R. G., Bunnell, B. A., Taylor, B. K., Murthy, S. N., et al. (2007). Intratracheal mesenchymal stem cell administration attenuates monocrotaline-induced pulmonary hypertension and endothelial dysfunction. *American Journal of Physiology. Heart and Circulatory Physiology*, *292*(2), H1120–H1128.

Ballard, R. A., Truog, W. E., Cnaan, A., Martin, R. J., Ballard, P. L., Merrill, J. D., et al. (2006). Inhaled nitric oxide in preterm infants undergoing mechanical ventilation. *New England Journal of Medicine*, *355*(4), 343–353.

Barberà, J. A., Roger, N., Roca, J., Rovira, I., Higenbottam, T. W., & Rodriguez-Roisin, R. (1996). Worsening of pulmonary gas exchange with nitric oxide inhalation in chronic obstructive pulmonary disease. *Lancet*, *347*(8999), 436–440.

Bardou, M., Goirand, F., Marchand, S., Rouget, C., Devillier, P., Dumas, J. P., et al. (2001). Hypoxic vasoconstriction of rat main pulmonary artery: Role of endogenous nitric oxide, potassium channels, and phosphodiesterase inhibition. *Journal of Cardiovascular Pharmacology*, *38*(2), 325–334.

Bevers, L. M., Braam, B., Post, J. A., van Zonneveld, A. J., Rabelink, T. J., Koomans, H. A., et al. (2006). Tetrahydrobiopterin, but not L-arginine, decreases NO synthase uncoupling in cells expressing high levels of endothelial NO synthase. *Hypertension*, *47*(1), 87–94.

Blumberg, F. C., Wolf, K., Arzt, M., Lorenz, C., Riegger, G. A., & Pfeifer, M. (2003). Effects of ET-A receptor blockade on eNOS gene expression in chronic hypoxic rat lungs. *Journal of Applied Physiology*, *94*(2), 446–452.

Crabtree, M. J., Tatham, A. L., Al-Wakeel, Y., Warrick, N., Hale, A. B., Cai, S., et al. (2009). Quantitative regulation of intracellular endothelial nitric-oxide synthase (eNOS) coupling by both tetrahydrobiopterin -eNOS stoichiometry and biopterin redox status: Insights from cells with tet-regulated GTP cyclohydrolase I expression. *Journal of Biological Chemistry*, *284*(2), 1136–1144.

Cunnington, C., Van Assche, T., Shirodaria, C., Kylintireas, I., Lindsay, A. C., Lee, J. M., et al. (2012). Systemic and vascular oxidation limits the efficacy of oral tetrahydrobiopterin treatment in patients with coronary artery disease. *Circulation*, *125*(11), 1356–1366.

Dellinger, R. P., Zimmerman, J. L., Taylor, R. W., Straube, R. C., Hauser, D. L., Criner, G. J., et al. (1998). Effects of inhaled nitric oxide in patients with acute respiratory distress syndrome: Results of a randomized phase II trial. Inhaled Nitric Oxide in ARDS Study Group. *Critical Care Medicine*, *26*(1), 15–23.

Dinh-Xuan, A. T., Higenbottam, T. W., Clelland, C. A., Pepke-Zaba, J., Cremona, G., Butt, A. Y., et al. (1991). Impairment of endothelium-dependent pulmonary-artery relaxation in chronic obstructive lung disease. *New England Journal of Medicine*, *324*(22), 1539–1547.

Elmedal, B., de Dam, M. Y., Mulvany, M. J., & Simonsen, U. (2004). The superoxide dismutase mimetic, tempol, blunts right ventricular hypertrophy in chronic hypoxic rats. *British Journal of Pharmacology*, *141*(1), 105–113.

Farber, H. W., & Loscalzo, J. (2004). Pulmonary arterial hypertension. *New England Journal of Medicine*, *351*(16), 1655–1665.

Ferrario, L., Amin, H. M., Sugimori, K., Camporesi, E. M., & Hakim, T. S. (1996). Site of action of endogenous nitric oxide on pulmonary vasculature in rats. *Pflügers Archiv*, *432*(3), 523–527.

Fox, G. A., Paterson, N. A., & McCormack, D. G. (1994). Effect of inhibition of NO synthase on vascular reactivity in a rat model of hyperdynamic sepsis. *American Journal of Physiology*, *267*(4 Pt. 2), H1377–H1382.

Francis, S. H., Busch, J. L., Corbin, J. D., & Sibley, D. (2010). cGMP-dependent protein kinases and cGMP phosphodiesterases in nitric oxide and cGMP action. *Pharmacological Reviews*, *62*(3), 525–563.

Fullerton, D. A., Hahn, A. R., & McIntyre, R. C., Jr. (1996). Mechanistic imbalance of pulmonary vasomotor control in progressive lung injury. *Surgery*, *119*(1), 98–103.

Gangopahyay, A., Oran, M., Bauer, E. M., Wertz, J. W., Comhair, S. A., Erzurum, S. C., et al. (2011). Bone morphogenetic protein receptor II is a novel mediator of endothelial nitric-oxide synthase activation. *Journal of Biological Chemistry*, *286*(38), 33134–33140.

Garg, U. C., & Hassid, A. (1989). Nitric oxide-generating vasodilators and 8-bromo-cyclic guanosine monophosphate inhibit mitogenesis and proliferation of cultured rat vascular smooth muscle cells. *Journal of Clinical Investigation*, *83*(5), 1774–1777.

Garg, U. C., & Hassid, A. (1993). Mechanisms of nitrosothiol-induced antimitogenesis in aortic smooth muscle cells. *European Journal of Pharmacology*, *237*(2–3), 243–249.

Gerlach, H., Pappert, D., Lewandowski, K., Rossaint, R., & Falke, K. J. (1993). Long-term inhalation with evaluated low doses of nitric oxide for selective improvement of oxygenation in patients with adult respiratory distress syndrome. *Intensive Care Medicine*, *19*(8), 443–449.

Gout, B., Quiniou, M. J., Khandoudi, N., Le Dantec, C., & Saïag, B. (1999). Impaired endothelium-dependent relaxation by adrenomedullin in monocrotaline-treated rat arteries. *European Journal of Pharmacology*, *380*(1), 23–30.

Hampl, V., & Herget, J. (2000). Role of nitric oxide in the pathogenesis of chronic pulmonary hypertension. *Physiological Reviews*, *80*(4), 1337–1372.

Hood, J. D., Meininger, C. J., Ziche, M., & Granger, H. J. (1998). VEGF upregulates ecNOS message, protein, and NO production in human endothelial cells. *American Journal of Physiology*, *274*(3 Pt. 2), H1054–H1058.

Huang, J., Wolk, J. H., Gewitz, M. H., & Mathew, R. (2010). Progressive endothelial cell damage in an inflammatory model of pulmonary hypertension. *Experimental Lung Research*, *36*(1), 57–66.

Igari, H., Tatsumi, K., Sugito, K., Kasahara, Y., Saito, M., Tani, T., et al. (1998). Role of EDRF in pulmonary circulation during sustained hypoxia. *Journal of Cardiovascular Pharmacology*, *31*(2), 299–305.

Ignarro, L. J., Buga, G. M., Wood, K. S., Byrns, R. E., & Chaudhuri, G. (1987). Endothelium-derived relaxing factor produced and released from artery and vein is nitric oxide. *Proceedings of the National Academy of Sciences of the United States of America*, *84*(24), 9265–9269.

Jiang, B. H., Maruyama, J., Yokochi, A., Amano, H., Mitani, Y., & Maruyama, K. (2002). Correlation of inhaled nitric-oxide induced reduction of pulmonary artery pressure and vascular changes. *European Respiratory Journal*, *20*(1), 52–58.

Jiang, B. H., Maruyama, J., Yokochi, A., Iwasaki, M., Amano, H., Mitani, Y., et al. (2004). Prolonged nitric oxide inhalation fails to regress hypoxic vascular remodeling in rat lung. *Chest*, *125*(6), 2247–2252.

Kanno, S., Wu, Y. J., Lee, P. C., Billiar, T. R., & Ho, C. (2001). Angiotensin-converting enzyme inhibitor preserves p21 and endothelial nitric oxide synthase expression in monocrotaline-induced pulmonary arterial hypertension in rats. *Circulation*, *104*(8), 945–950.

Katayama, Y., Hatanaka, K., Hayashi, T., Onoda, K., Namikawa, S., Yuasa, H., et al. (1995). Effects of inhaled nitric oxide in single lung transplantation in rats with monocrotaline-induced pulmonary hypertension. *Journal of Heart and Lung Transplantation*, *14*(3), 486–492.

Katayama, Y., Hatanaka, K., Hayashi, T., Onoda, K., Yada, I., Namikawa, S., et al. (1994). Effects of inhaled nitric oxide in rats with chemically induced pulmonary hypertension. *Respiration Physiology*, *97*(3), 301–307.

Katsumi, H., Nishikawa, M., & Hashida, M. (2007). Development of nitric oxide donors for the treatment of cardiovascular diseases. *Cardiovascular & Hematological Agents in Medicinal Chemistry*, *5*(3), 204–208.

Khoo, J. P., Zhao, L., Alp, N. J., Bendall, J. K., Nicoli, T., Rockett, K., et al. (2005). Pivotal role for endothelial tetrahydrobiopterin in pulmonary hypertension. *Circulation*, *111*(16), 2126–2133.

Kinsella, J. P., Neish, S. R., Shaffer, E., & Abman, S. H. (1992). Low-dose inhalation nitric oxide in persistent pulmonary hypertension of the newborn. *Lancet, 340*(8823), 819–820.

Kobayashi, T., Gabazza, E. C., Shimizu, S., Yasui, H., Yuda, H., Hataji, O., et al. (2001). Long-term inhalation of high-dose nitric oxide increases intraalveolar activation of coagulation system in mice. *American Journal of Respiratory and Critical Care Medicine, 163*(7), 1676–1682.

Kourembanas, S., McQuillan, L. P., Leung, G. K., & Faller, D. V. (1993). Nitric oxide regulates the expression of vasoconstrictors and growth factors by vascular endothelium under both normoxia and hypoxia. *Journal of Clinical Investigation, 92*(1), 99–104.

Kouyoumdjian, C., Adnot, S., Levame, M., Eddahibi, S., Bousbaa, H., & Raffestin, B. (1994). Continuous inhalation of nitric oxide protects against development of pulmonary hypertension in chronically hypoxic rats. *Journal of Clinical Investigation, 94*(2), 578–584.

Kwon, N. S., Stuehr, D. J., & Nathan, C. F. (1991). Inhibition of tumor cell ribonucleotide reductase by macrophage-derived nitric oxide. *Journal of Experimental Medicine, 174*(4), 761–767.

MacLean, M. R., & McCulloch, K. M. (1998). Influence of applied tension and nitric oxide on responses to endothelins in rat pulmonary resistance arteries: Effect of chronic hypoxia. *British Journal of Pharmacology, 123*(5), 991–999.

Mam, V., Tanbe, A. F., Vitali, S. H., Arons, E., Christou, H. A., & Khalil, R. A. (2010). Impaired vasoconstriction and nitric oxide-mediated relaxation in pulmonary arteries of hypoxia- and monocrotaline-induced pulmonary hypertensive rats. *Journal of Pharmacology and Experimental Therapeutics, 332*(2), 455–462.

Maruyama, J., Jiang, B. H., Maruyama, K., Takata, M., & Miyasaka, K. (2004). Prolonged nitric oxide inhalation during recovery from chronic hypoxia does not decrease nitric oxide-dependent relaxation in pulmonary arteries. *Chest, 126*(6), 1919–1925.

Maruyama, K., Kobayasi, H., Taguchi, O., Chikusa, H., & Muneyuki, M. (1995). Higher doses of inhaled nitric oxide might be less effective in improving oxygenation in a patient with interstitial pulmonary fibrosis. *Anesthesia and Analgesia, 81*(1), 210–211.

Maruyama, J., & Maruyama, K. (1994). Impaired nitric oxide-dependent responses and their recovery in hypertensive pulmonary arteries of rats. *American Journal of Physiology, 266*(6 Pt. 2), H2476–H2488.

Maruyama, J., Maruyama, K., Mitani, Y., Kitabatake, M., Yamauchi, T., & Miyasaka, K. (1997). Continuous low-dose NO inhalation does not prevent monocrotaline-induced pulmonary hypertension in rats. *American Journal of Physiology, 272*(1 Pt. 2), H517–H524.

Maruyama, K., Takeuchi, M., Chikusa, H., & Muneyuki, M. (1995). Reduction of intrapulmonary shunt by low-dose inhaled nitric oxide in a patient with late-stage respiratory distress associated with paraquat poisoning. *Intensive Care Medicine, 21*(9), 778–779.

Maruyama, J., Yokochi, A., Maruyama, K., & Nosaka, S. (1999). Acetylcholine-induced endothelium-derived contracting factor in hypoxic pulmonary hypertensive rats. *Journal of Applied Physiology, 86*(5), 1687–1695.

Mathew, R., Zeballos, G. A., Tun, H., & Gewitz, M. H. (1995). Role of nitric oxide and endothelin-1 in monocrotaline-induced pulmonary hypertension in rats. *Cardiovascular Research, 30*(5), 739–746.

Mathru, M., Huda, R., Solanki, D. R., Hays, S., & Lang, J. D. (2007). Inhaled nitric oxide attenuates reperfusion inflammatory responses in humans. *Anesthesiology, 106*(2), 275–282.

Matsumoto, A., Momomura, S., Sugiura, S., Fujita, H., Aoyagi, T., Sata, M., et al. (1999). Effect of inhaled nitric oxide on gas exchange in patients with congestive heart failure. A randomized, controlled trial. *Annals of Internal Medicine, 130*(1), 40–44.

Mawatari, E., Hongo, M., Sakai, A., Terasawa, F., Takahashi, M., Yazaki, Y., et al. (2007). Amlodipine prevents monocrotaline-induced pulmonary arterial hypertension and prolongs survival in rats independent of blood pressure lowering. *Clinical and Experimental Pharmacology & Physiology, 34*(7), 594–600.

Miller, O. I., Tang, S. F., Keech, A., Pigott, N. B., Beller, E., & Celermajer, D. S. (2000). Inhaled nitric oxide and prevention of pulmonary hypertension after congenital heart surgery: A randomised double-blind study. *Lancet, 356*(9240), 1464–1469.

Minamishima, S., Kida, K., Tokuda, K., Wang, H., Sips, P. Y., Kosugi, S., et al. (2011). Inhaled nitric oxide improves outcomes after successful cardiopulmonary resuscitation in mice. *Circulation, 124*(15), 1645–1653.

Murata, T., Kinoshita, K., Hori, M., Kuwahara, M., Tsubone, H., Karaki, H., et al. (2005). Statin protects endothelial nitric oxide synthase activity in hypoxia-induced pulmonary hypertension. *Arteriosclerosis, Thrombosis, and Vascular Biology, 25*(11), 2335–2342.

Murata, T., Yamawaki, H., Hori, M., Sato, K., Ozaki, H., & Karaki, H. (2001). Hypoxia impairs endothelium-dependent relaxation in organ cultured pulmonary artery. *European Journal of Pharmacology, 421*(1), 45–53.

Nakamura, T., & Lipton, S. A. (2011). Redox modulation by S-nitrosylation contributes to protein misfolding, mitochondrial dynamics, and neuronal synaptic damage in neurodegenerative diseases. *Cell Death and Differentiation, 18*, 1478–1486.

Neonatal Inhaled Nitric Oxide Study Group. (1997). Inhaled nitric oxide in full-term and nearly full-term infants with hypoxic respiratory failure. *New England Journal of Medicine, 336*(9), 597–604.

Oka, M. (2001). Phosphodiesterase 5 inhibition restores impaired ACh relaxation in hypertensive conduit pulmonary arteries. *American Journal of Physiology. Lung Cellular and Molecular Physiology, 280*(3), L432–L435.

Oka, M., Hasunuma, K., Webb, S. A., Stelzner, T. J., Rodman, D. M., & McMurtry, I. F. (1993). EDRF suppresses an unidentified vasoconstrictor mechanism in hypertensive rat lungs. *American Journal of Physiology, 264*, L587–L597.

Ozturk, E. I., & Uma, S. (2010). Effects of atorvastatin and L-arginine treatments on electrical field stimulation-mediated relaxations in pulmonary arterial rings of monocrotaline-induced pulmonary hypertensive rats. *Journal of Cardiovascular Pharmacology, 56*(5), 498–505.

Palmer, R. M., Ferrige, A. G., & Moncada, S. (1987). Nitric oxide release accounts for the biological activity of endothelium-derived relaxing factor. *Nature, 327*(6122), 524–526.

Patel, H. H., Murray, F., & Insel, P. A. (2008). Caveolae as organizers of pharmacologically relevant signal transduction molecules. *Annual Review of Pharmacology and Toxicology, 48*, 359–391.

Pei, Y., Ma, P., Wang, X., Zhang, W., Zhang, X., Zheng, P., et al. (2011). Rosuvastatin attenuates monocrotaline-induced pulmonary hypertension via regulation of Akt/eNOS signaling and asymmetric dimethylarginine metabolism. *European Journal of Pharmacology, 666*(1–3), 165–172.

Pepke-Zaba, J., Higenbottam, T. W., Dinh-Xuan, A. T., Stone, D., & Wallwork, J. (1991). Inhaled nitric oxide as a cause of selective pulmonary vasodilatation in pulmonary hypertension. *Lancet, 338*(8776), 1173–1174.

Pérez-Peñate, G., Julià-Serdà, G., Pulido-Duque, J. M., Górriz-Gómez, E., & Cabrera-Navarro, P. (2001). One-year continuous inhaled nitric oxide for primary pulmonary hypertension. *Chest, 119*(3), 970–973.

Prié, S., Stewart, D. J., & Dupuis, J. (1998). EndothelinA receptor blockade improves nitric oxide-mediated vasodilation in monocrotaline-induced pulmonary hypertension. *Circulation, 97*(21), 2169–2174.

Rabinovitch, M. (2008). Molecular pathogenesis of pulmonary arterial hypertension. *Journal of Clinical Investigation, 118*(7), 2372–2379.

Rajek, A., Pernerstorfer, T., Kastner, J., Mares, P., Grabenwöger, M., Sessler, D. I., et al. (2000). Inhaled nitric oxide reduces pulmonary vascular resistance more than

prostaglandin E(1) during heart transplantation. *Anesthesia and Analgesia, 90*(3), 523–530.
Roberts, J. D., Polaner, D. M., Lang, P., & Zapol, W. M. (1992). Inhaled nitric oxide in persistent pulmonary hypertension of the newborn. *Lancet, 340*(8823), 818–819.
Rossaint, R., Falke, K. J., López, F., Slama, K., Pison, U., & Zapol, W. M. (1993). Inhaled nitric oxide for the adult respiratory distress syndrome. *New England Journal of Medicine, 328*(6), 399–405.
Rus, A., Peinado, M. A., Castro, L., & Del Moral, M. L. (2010). Lung eNOS and iNOS are reoxygenation time-dependent upregulated after acute hypoxia. *Anatomical Record (Hoboken), 293*(6), 1089–1098.
Scherrer, U., Vollenweider, L., Delabays, A., Savcic, M., Eichenberger, U., Kleger, G. R., et al. (1996). Inhaled nitric oxide for high-altitude pulmonary edema. *New England Journal of Medicine, 334*(10), 624–629.
Shaul, P. W., Wells, L. B., & Horning, K. M. (1993). Acute and prolonged hypoxia attenuate endothelial nitric oxide production in rat pulmonary arteries by different mechanisms. *Journal of Cardiovascular Pharmacology, 22*(6), 819–827.
Shimpo, H., Mitani, Y., Tanaka, J., Mizumoto, T., Onoda, K., Tani, K., et al. (1997). Inhaled low-dose nitric oxide for postoperative care in patients with congenital heart defects. *Artificial Organs, 21*(1), 10–13.
Stevens, T. (2005). Molecular and cellular determinants of lung endothelial cell heterogeneity. *Chest, 128*(6 Suppl.), 558S–564S.
Stobierska-Dzierzek, B., Awad, H., & Michler, R. E. (2001). The evolving management of acute right-sided heart failure in cardiac transplant recipients. *Journal of the American College of Cardiology, 38*(4), 923–931.
Taraseviciene-Stewart, L., Kasahara, Y., Alger, L., Hirth, P., Mc Mahon, G., Waltenberger, J., et al. (2001). Inhibition of the VEGF receptor 2 combined with chronic hypoxia causes cell death-dependent pulmonary endothelial cell proliferation and severe pulmonary hypertension. *FASEB Journal, 15*(2), 427–438.
The Australia and New Zealand Extracorporeal Membrane Oxygenation (ANZ ECMO) Influenza Investigators. (2009). Extracorporeal membrane oxygenation for 2009 influenza A(H1N1) acute respiratory distress syndrome. *Journal of the American Medical Association, 302*, 1888–1895.
Tuder, R. M., Chacon, M., Alger, L., Wang, J., Taraseviciene-Stewart, L., Kasahara, Y., et al. (2001). Expression of angiogenesis-related molecules in plexiform lesions in severe pulmonary hypertension: Evidence for a process of disordered angiogenesis. *Journal of Pathology, 195*(3), 367–374.
Vonbank, K., Ziesche, R., Higenbottam, T. W., Stiebellehner, L., Petkov, V., Schenk, P., et al. (2003). Controlled prospective randomised trial on the effects on pulmonary haemodynamics of the ambulatory long term use of nitric oxide and oxygen in patients with severe COPD. *Thorax, 58*(4), 289–293.
Yoshida, M., Taguchi, O., Gabazza, E. C., Kobayashi, T., Yamakami, T., Kobayashi, H., et al. (1997). Combined inhalation of nitric oxide and oxygen in chronic obstructive pulmonary disease. *American Journal of Respiratory and Critical Care Medicine, 155*(2), 526–529.
Yoshida, M., Taguchi, O., Gabazza, E. C., Yasui, H., Kobayashi, T., Kobayashi, H., et al. (1997). The effect of low-dose inhalation of nitric oxide in patients with pulmonary fibrosis. *European Respiratory Journal, 10*(9), 2051–2054.
Zhao, Y. Y., Zhao, Y. D., Mirza, M. K., Huang, J. H., Potula, H. H., Vogel, S. M., et al. (2009). Persistent eNOS activation secondary to caveolin-1 deficiency induces pulmonary hypertension in mice and humans through PKG nitration. *Journal of Clinical Investigation, 119*(7), 2009–2018.

INDEX

Note: Page numbers followed by "*f*" indicate figures and "*t*" indicate tables.

D

E

I

N

O

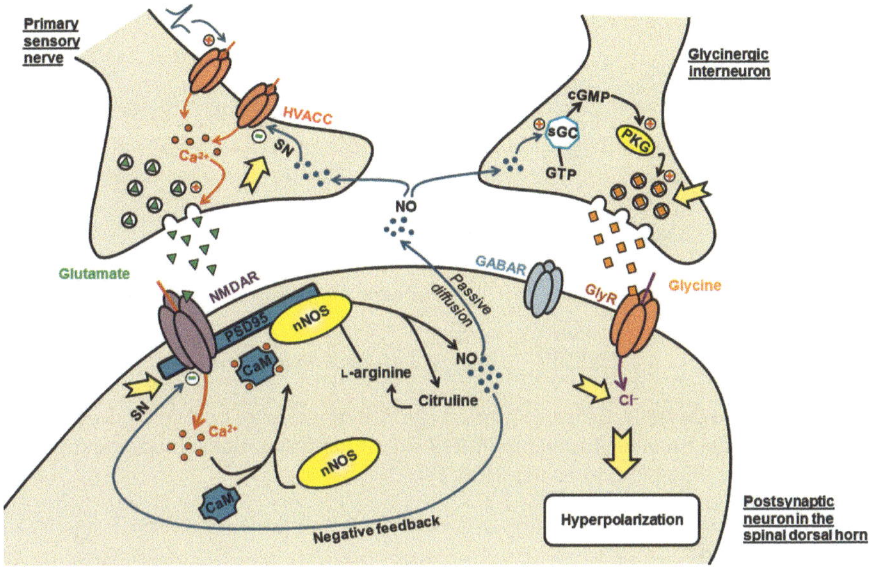

Alexis Bavencoffe *et al.*, Figure 1.1 Diagram shows that NO inhibits synaptic transmission in the dorsal horn of the spinal cord. Stimulation of primary sensory nerves by painful stimuli triggers action potentials and opening of voltage-activated calcium channels, leading to calcium influx and glutamate release from presynaptic nerve terminals. Released glutamate binds to AMPA (not shown) and NMDA receptors present on postsynaptic dorsal horn neurons to increase intracellular calcium levels and to recruit nNOS to the close proximity of NMDA receptors by the scaffolding protein PSD95 and nNOS activation via calcium/calmodulin-dependent signaling. Subsequently, nNOS converts L-arginine into NO and citruline. NO then exerts a negative feedback action by inhibiting the NMDA receptor activity via S-nitrosylation. NO also diffuses passively into the presynaptic terminal and inhibits the activity of high voltage-activated calcium channels to decrease synaptic glutamate release. Moreover, NO diffuses into the neighboring glycinergic interneurons and potentiates synaptic glycine release via stimulation of an sGC/PKG-dependent pathway. Glycine activates postsynaptic glycine receptors, leading to chloride influx. Collectively, decreased glutamate release, inhibition of NMDA receptors, and increased glycine release by NO result in hyperpolarization of the postsynaptic neuron, thereby reducing nociceptive transmission at the spinal cord level. CaM, calmodulin; cGMP, cyclic guanosine monophosphate; GABAR, γ-aminobutyric acid receptor; GlyR, glycine receptor; GTP, guanosine triphosphate; HVACC, high voltage-activated calcium channel; NMDAR, N-methyl-D-aspartate receptor; nNOS, neuronal nitric oxide synthase; NO, nitric oxide; PKG, protein kinase G; PSD95, postsynaptic density protein 95; sGC, soluble guanylyl cyclase; SN, S-nitrosylation.

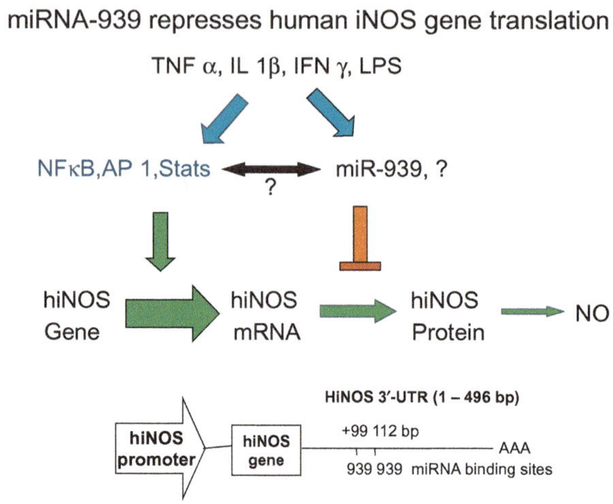

Zhong Guo and David A. Geller, Figure 2.1 Proposed model of human iNOS gene regulation in human hepatocytes with miR-939 exerting a translational blockade for iNOS protein synthesis by binding to the iNOS 3′-UTR.

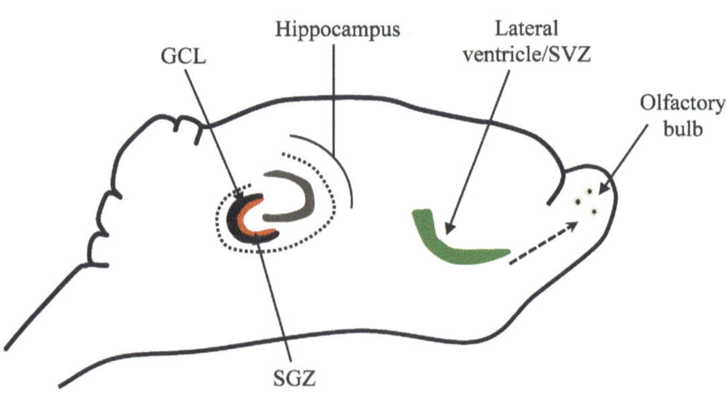

William P. Gray and Angela Cheung, Figure 4.1 Neurogenic regions of the rat brain. The subgranular zone (SGZ; red) of the dentate gyrus of the hippocampus which populates the granule cell layer (GCL; black), and the forebrain subventricular zone (SVZ) which lines the anterior lateral ventricles (green) and provides new neurons for the olfactory bulb, are areas where adult neurogenesis have been confirmed.

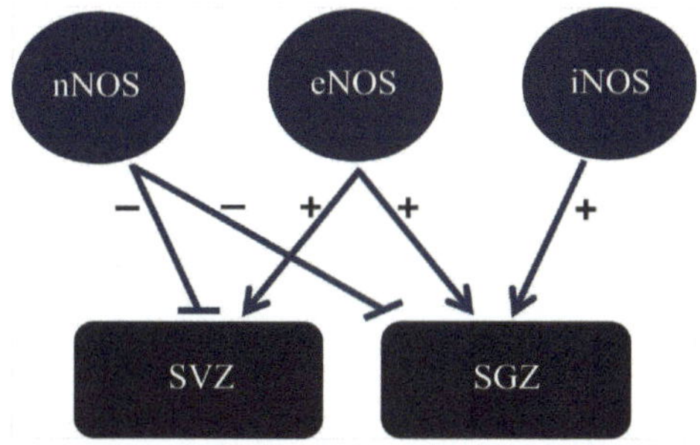

William P. Gray and Angela Cheung, Figure 4.2 The differential effects of the NOS isoforms on adult neurogenesis *in vivo*. Studies suggest that nNOS-derived NO is inhibitory, while eNOS-derived NO is proliferative, for neurogenesis in the SGZ and SVZ under both basal and ischemia-induced conditions. iNOS-derived NO, on the other hand, promotes neurogenesis in the SGZ under ischemic conditions.

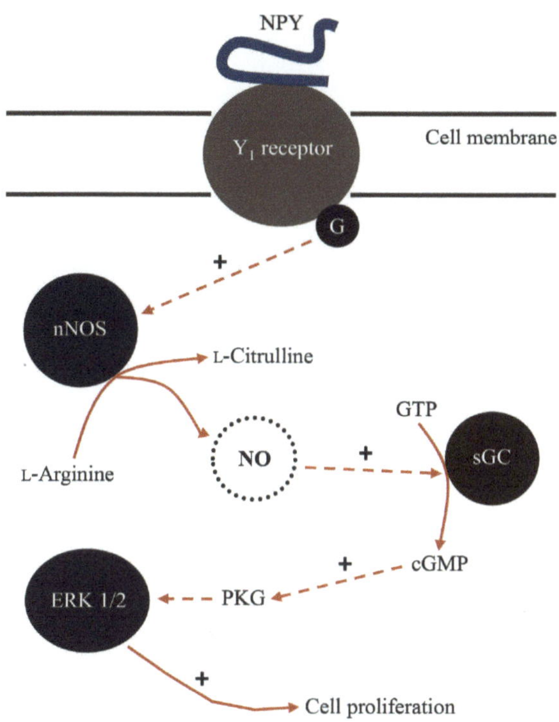

William P. Gray and Angela Cheung, Figure 4.3 Intracellular mechanisms underlying NPY-mediated neuroproliferation. NPY released by interneurons of the dentate hilus act on NPY Y1 receptors in the SGZ to initiate a neuroproliferative effect. The Y1 GPCR activates nNOS, which synthesises NO from L-arginine. The target of NO is sGC, which converts GTP into cGMP. cGMP activates PKG, which in turn leads to the activation of a range of protein targets or kinase pathways, ultimately resulting in ERK 1/2 activation. ERK 1/2 are involved in regulating the expression of genes involved in controlling cell proliferation and differentiation (Cheung et al., 2012).

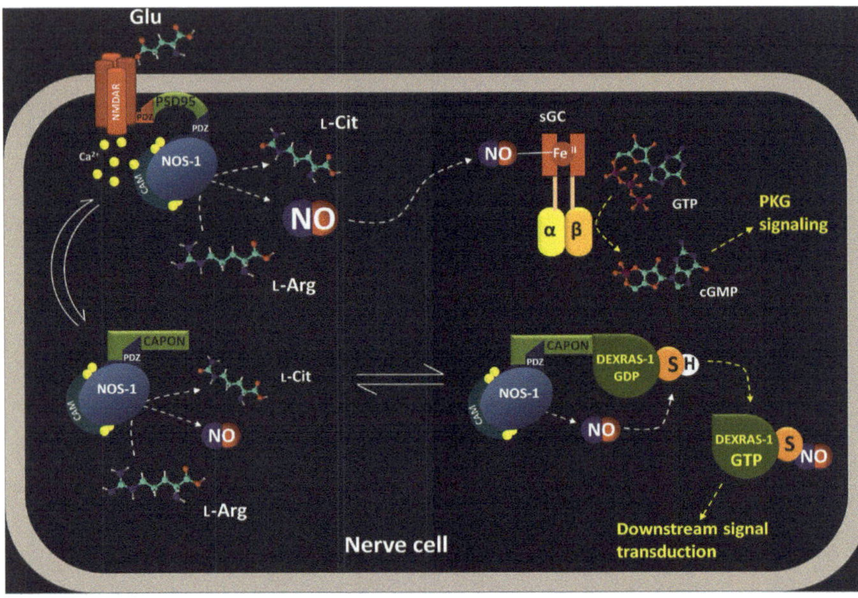

Marcelo Cossenza *et al.*, Figure 5.1 Nitric oxide synthase I-coupled platforms and NO mechanistic actions by S-nitrosylation and PKG activation in CNS. NOS I-coupled NMDA receptor through scaffold postsynaptic protein PSD-95 with PDZ domains. Opening of NMDA receptor channels promotes Ca^{2+} influx and its binding to calmodulin triggers the production of NO by L-Arg oxidation to L-citrulline. In another way, CAPON competes with PSD-95 for interaction with NOS I forcing its dissociation from the plasma membrane. DEXRAS-1 is a small GTPase protein which can interact with NOS I/CAPON complex and place them in close proximity to allow DEXRAS activation by S-nitrosylation. NO can also activate soluble guanylate cyclase (sGC) by reacting with Fe II present into its heme domain. That enzyme configuration produces rapid conversion of guanosine-5′-triphosphate (GTP) into 3′,5′-cyclic guanosine monophosphate (cGMP) to further PKG activation. This signaling mechanism is called as NO canonical pathway.

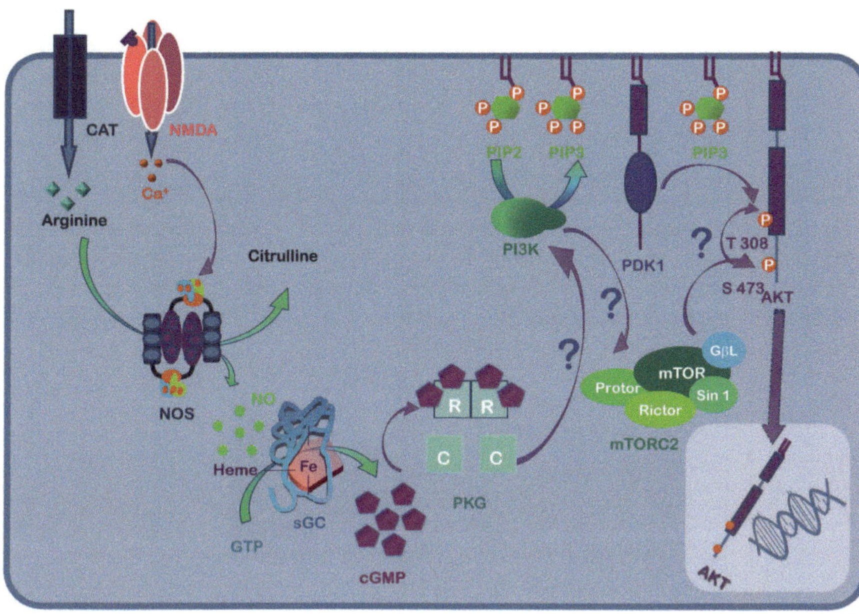

Marcelo Cossenza *et al.*, Figure 5.2 Nitric oxide increases AKT phosphorylation at residues Ser473 and Thr308. NMDA stimulation increases intracellular calcium which stimulates Nitric oxide synthase (NOS), enhancing NO production. NO interacts with soluble guanylyl cyclase (sGC) boosting cGMP production, which is capable of activating cGMP-dependent protein kinase (PKG). This kinase may enhance PI3K activity and stimulate mTORC2-dependent AKT phosphorylation at both sites, Ser473 and Thr308. Nitric oxide also enhances phospho-473 AKT localization in the nucleus.

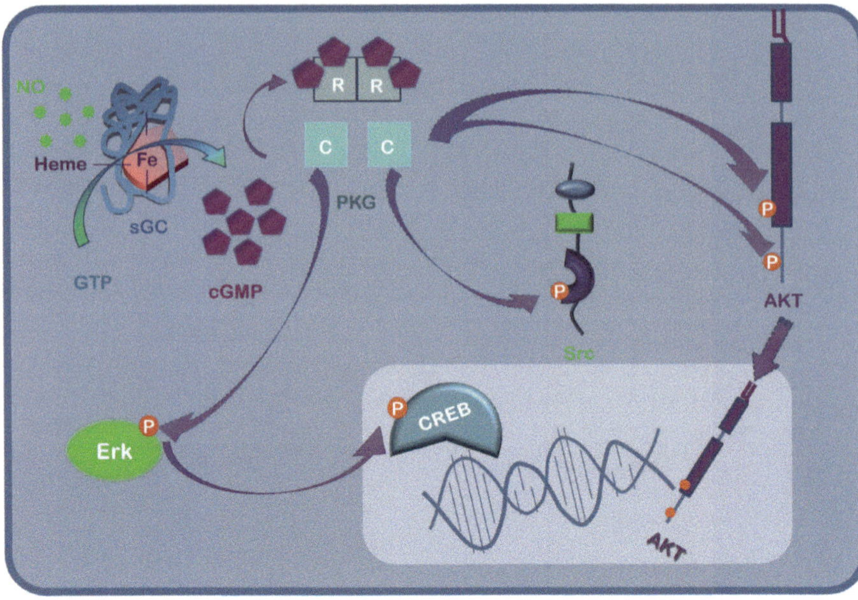

Marcelo Cossenza *et al.*, Figure 5.3 Representation of most commonly pathways activated by NO. In its canonical pathway, NO interacts with its intracellular receptor, sGC, promoting an increase of cGMP levels and PKG activation. When activated by cGMP, PKG catalytic subunits phosphorylate downstream targets, culminating in activation of Src, AKT, ERK, and CREB signaling pathways.

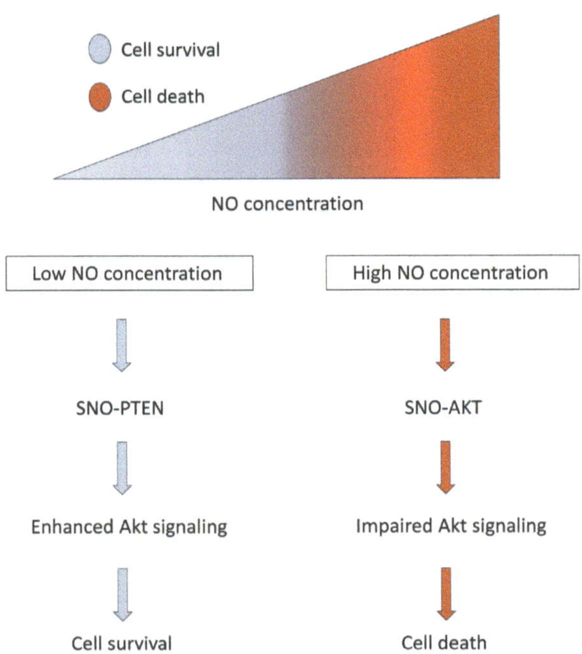

Marcelo Cossenza *et al.*, Figure 5.4 Actions of NO in neuronal signaling depend on its concentration. Low concentrations of NO in neuronal systems generally lead to cell survival by upregulating well-known protective pathways like PI3K/AKT. Higher NO concentrations are capable to promote cell death, for example, by shutting down AKT, leading cells to apoptosis. In blue, cell survival signaling; in red, cell death signaling.

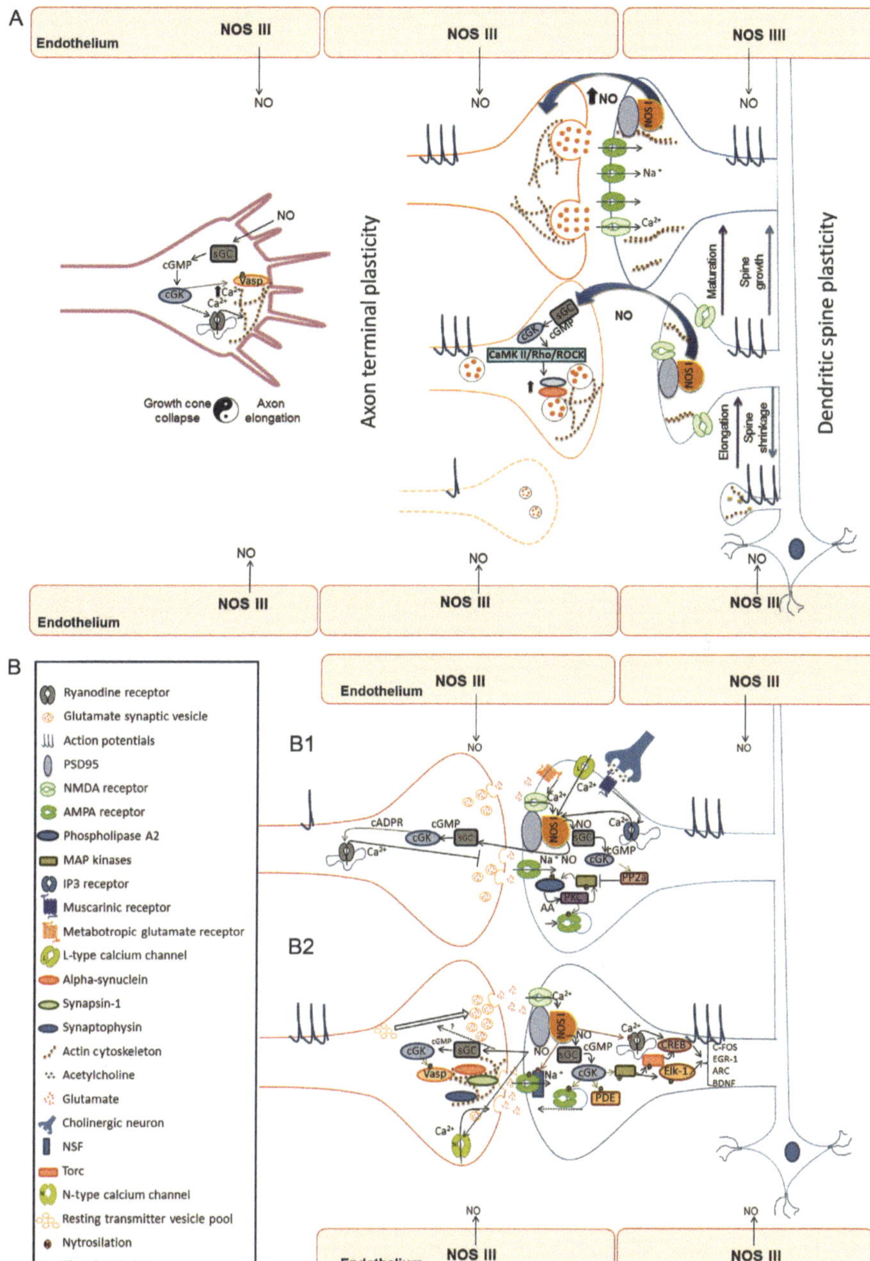

Marcelo Cossenza *et al.*, Figure 5.5 Mechanisms underlying NO-induced neuronal plasticity. (A) During early stages of development, NO modifies growth cone dynamics leading to axonal elongation or growth cone collapse (left). NO also participates in axonal and dendritic plasticity (right). Depending on spatiotemporal dynamics, both pre- and postsynaptic terminals change their morphology. Asynchronous synaptic activity

(Continued)

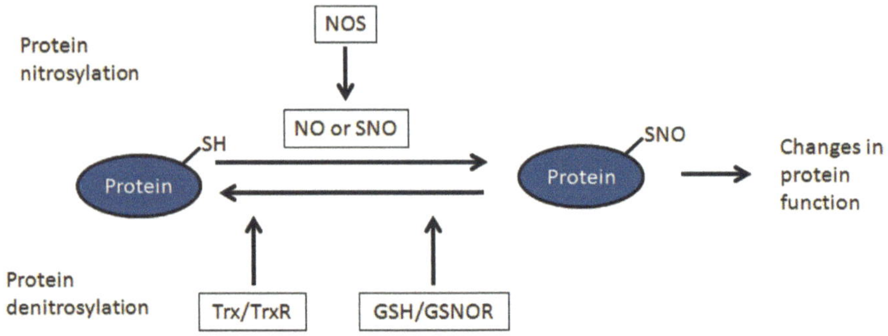

H.S. Jeffrey Man *et al.*, Figure 7.1 Equilibrium of nitrosylation and denitrosylation systems *in vivo*. Protein S-nitrosylation is dynamically regulated by "nitroso" synthases and denitrosylation systems. To induce S-nitrosylation, nitric oxide (NO) produced by nitric oxide synthases (NOSs) covalently binds to cysteine residues of target proteins, which leads to a range of effects including alterations in enzymatic activity, subcellular localization, protein–protein interaction, and stability. There are two denitrosylation systems identified *in vivo*. Both the thioredoxin reductase (TrxR) system and the S-nitrosoglutathione reductase (GSNOR) system have the ability to remove the NO group from the S-nitrolsylated cysteine residue. These pathways play important roles in physiological and pathophysiological conditions.

Figure 5.5—Cont'd leads to a retraction of axon and spine shrinkage and a synchronous activity induces NO production, neurotransmitter release, and axonal differentiation. Note that more vesicles are driven to axonal terminal. This reinforcement leads to increased NO synthesis, spine growth, and maturation, with more receptors at postsynaptic site and neurotransmitter released by presynaptic cells. (B) NO drives both LTD (B1) and LTP (B2). (B1) Noncorrelated synaptic activity leads to LTD via NO production. (B2) Correlated pre- and postsynaptic activity leads to LTP via NO production. The metabolic pathways involved are explained in the text.

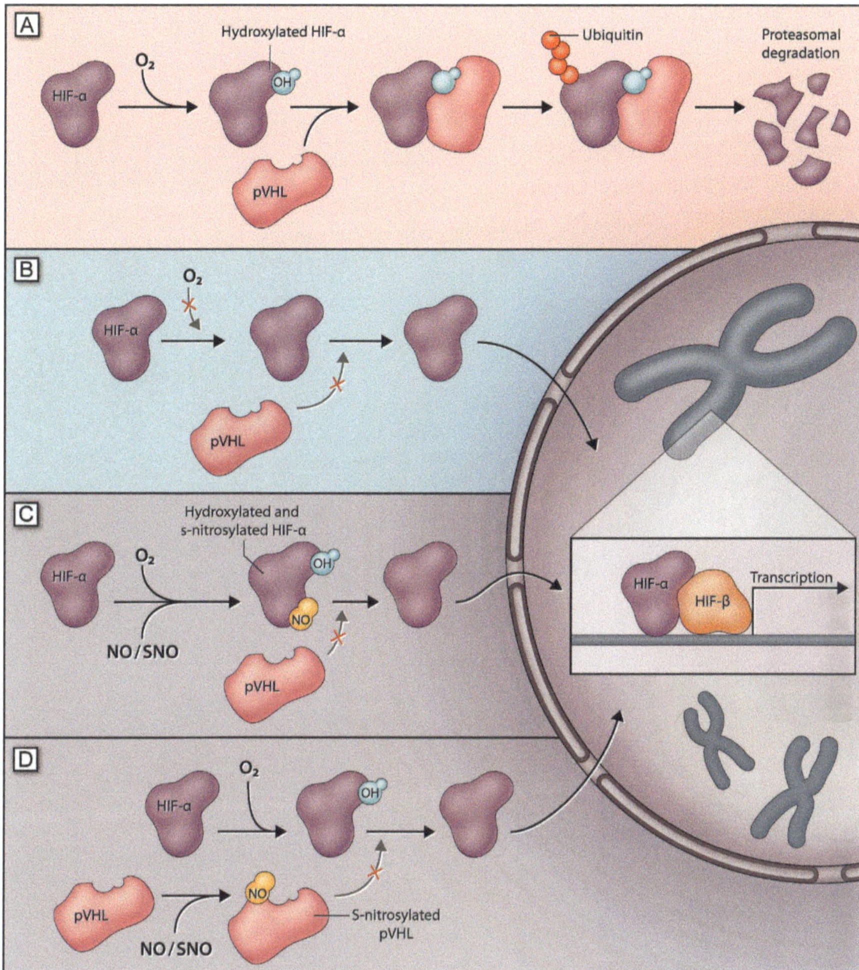

H.S. Jeffrey Man *et al.*, Figure 7.2 Regulation of hypoxia-inducible factor (HIF) by nitric oxide (NO) in the presence of oxygen. (A) In normoxia, the presence of oxygen leads to the activation of oxygen-dependent prolyl hydroxylase domains (PHDs), which hydroxylate proline residues of the HIF-α subunit. Von Hippel–Lindau protein (pVHL) recognizes and binds to the hydroxylated HIF-α, which is then subject to polyubiquitination and proteosomal degradation. Thus, HIF-α protein is normally produced and rapidly degraded in the presence of oxygen. (B) With hypoxia, oxygen-dependent PHDs do not hydroxylate HIF-α. HIF-α protein accumulates and translocates to the nucleus where it binds to the HIF-β subunit to activate target genes. (C–D) Even with enough O_2 for HIF-α hydroxylation, the presence of NO can result in HIF-α protein stabilization by S-nitrosylation of pVHL or HIF-α. S-Nitroslyation of pVHL or HIF-α prevents interaction of pVHL and HIF-α, allowing HIF-α to escape the degradation pathway. The accumulated HIF-α in the cytosol translocates to the nucleus and binds and dimerizes with HIF-1β, which can activate transcription of HIF-dependent genes.

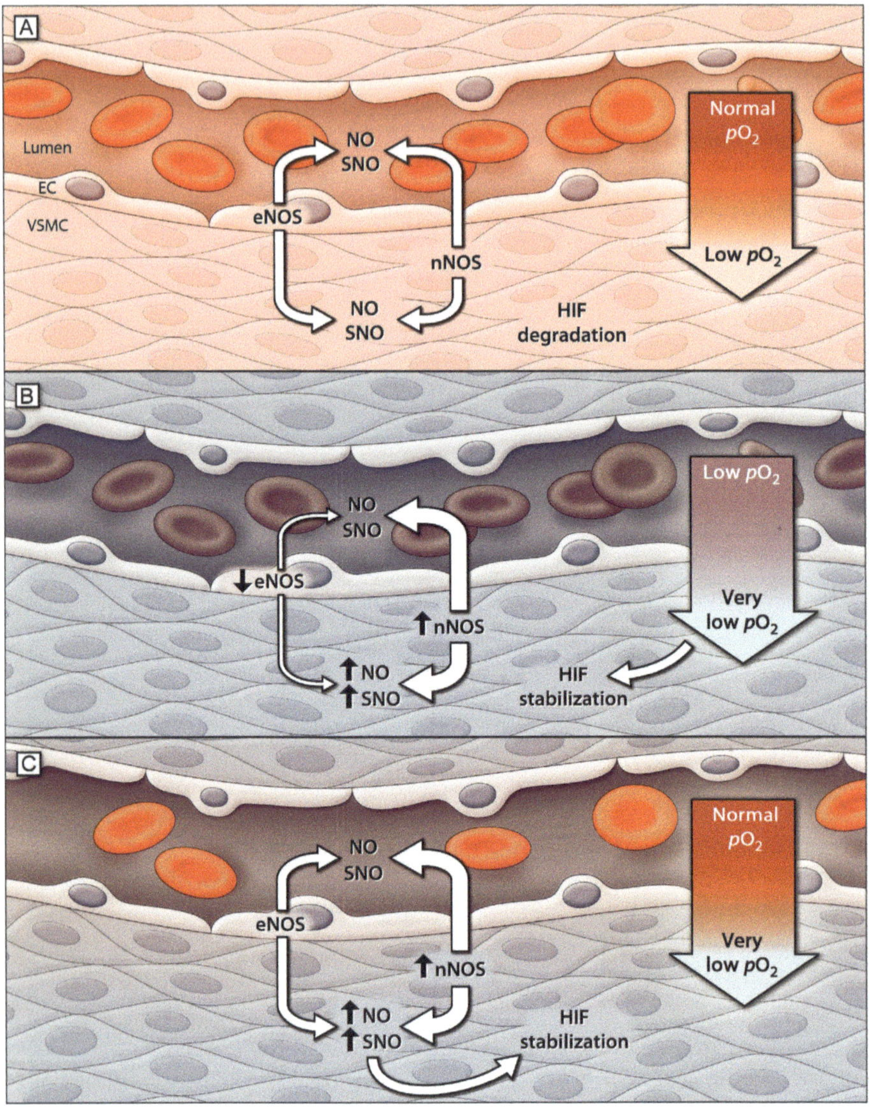

H.S. Jeffrey Man *et al.*, Figure 7.3 Mechanisms of vascular HIF stabilization in hypoxia versus anemia. (A) In normoxia (normal O_2 content, normal pO_2), constitutive expression of eNOS from the endothelium and nNOS from vascular smooth muscle contributes to the NO and SNO content in the vascular lumen and smooth muscle compartments. A normal O_2 gradient exists from blood to tissue. In anemia and hypoxia, HIF-α can be stabilized by different mechanisms despite similar reduction in blood oxygen content: (1) the classical prolyl hydroxylase pathway and (2) nNOS/NO-mediated S-nitroslyation of HIF pathway. (B) In hypoxia (low arterial pO_2, low O_2 saturation, normal hemoglobin levels), there is increased nNOS in smooth muscle but reduced eNOS expression in endothelial cells. nNOS is the main source NO/SNO in hypoxia. Due to the reduced level of free oxygen (low arterial pO_2), increased HIF-α expression is predominantly regulated via the classic prolyl hydroxylase pathways due to oxygen deprivation. Also, the O_2 gradient from blood to tissue is greatly reduced in systemic hypoxia. (C) In anemia (normal

(Continued)

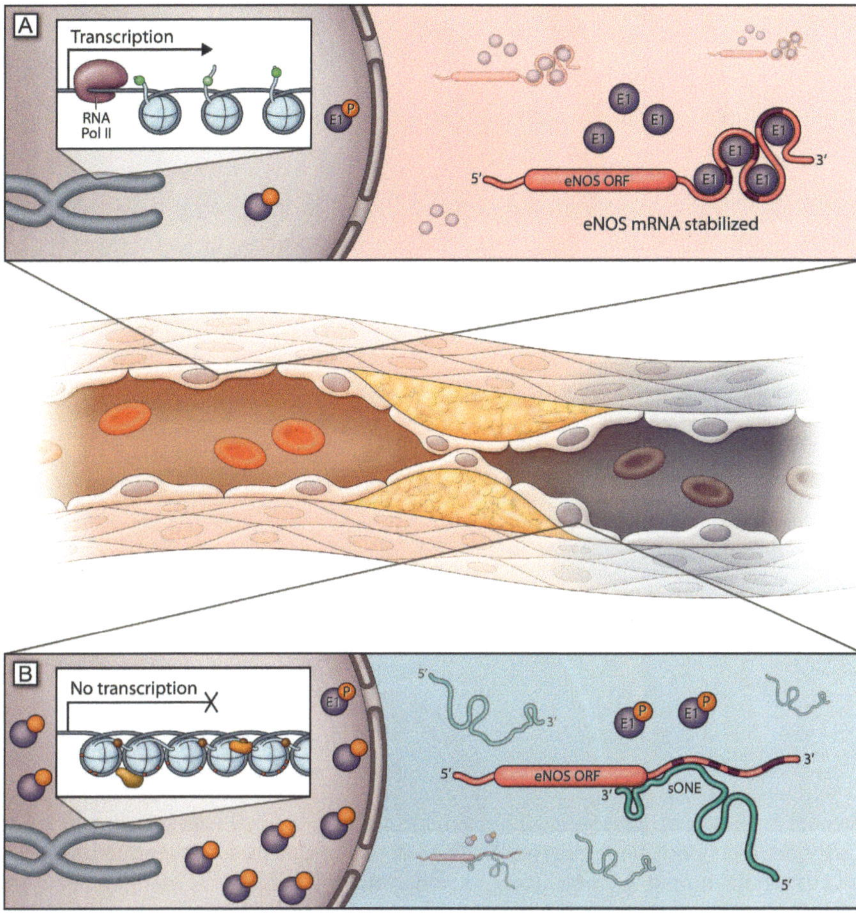

H.S. Jeffrey Man *et al.*, Figure 7.5 Effects of hypoxia on transcription and post-transcription regulation of eNOS mRNA. (A) Normoxia—with adequate oxygen availability within the cell, eNOS mRNA is actively transcribed. The chromatin adopts an open confirmation with epigenetic marks that favor transcription (green). RNA polymerase II (purple) engages at the promoter and eNOS mRNA is transcribed. In the cytoplasm, eNOS mRNA is bound and stabilized in the cytoplasm by heterogeneous ribonucleoprotein E1 (E1). (B) Hypoxia—chromatin at the eNOS promoter adopts a closed conformation with epigenetic marks that favor transcription repression (red and amber circles). RNA polymerase II cannot bind and eNOS mRNA is not transcribed. In addition, heterogeneous ribonucleoprotein E1 (E1) is phosphorylated and is more abundant in the nucleus. Phosphorylated E1 cannot bind to and stabilize eNOS mRNA. eNOS mRNA is vulnerable to destabilizing factors. At the same time, the expression of the natural *cis*-antisense transcript sONE is increased with hypoxia and can bind to eNOS mRNA. eNOS mRNA stability is reduced and eNOS mRNA and protein levels decrease with hypoxia.

Figure 7.3—Cont'd arterial pO_2, normal O_2 saturation, low hemoglobin concentration), nNOS is increased, whereas eNOS level remains unchanged. There exists a greater blood to tissue O_2 gradient in anemia compared to hypoxia. HIF-α is not stabilized via the prolyl hydroxylase pathways; rather, HIF-α stabilization occurs via nNOS-mediated S-nitrosylation of pVHL. nNOS is the major source for tissue NO/SNO levels in anemia.

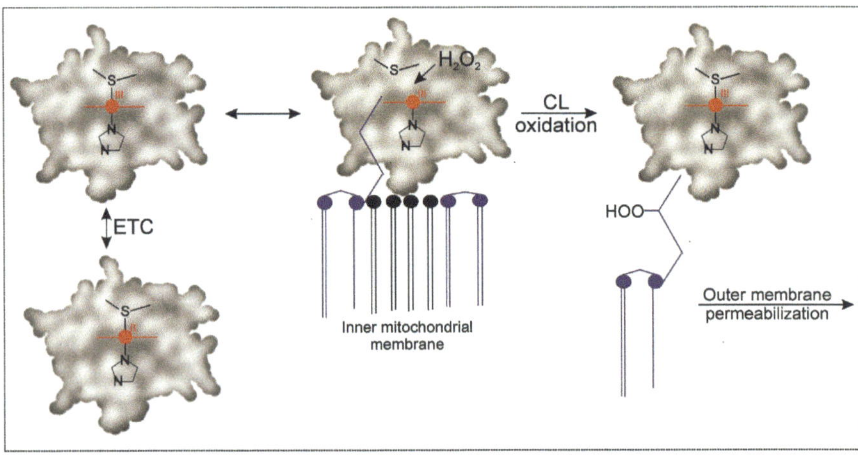

Michael A. Hough *et al.*, Figure 8.2 Schematic of the functional roles of human cyt *c* in the inner mitochondrial membrane. A small fraction of cyt *c* is considered to be bound to CL and transformed to a peroxidase by dissociation of the intrinsic methionine ligand through the insertion of one (or two) of the hydrocarbon chains. Under those oxidative stress conditions in which H_2O_2 is formed, this may oxidize the hydrocarbon catalyzed by the heme iron. The resulting lipid peroxide dissociates from the protein and eventually leads to opening of the transition pore initiating apoptosis. CL in the membrane is depicted in blue. *Adapted from figure 2B published in Journal of Biological Chemistry. Gary Silkstone, Sofia M. Kapetanaki, Ivan Husu, Marten H. Vos, Michael T. Wilson. Nitric Oxide Binds to the Proximal Heme Coordination Site of the Ferrocytochrome c/Cardiolipin Complex: FORMATION MECHANISM AND DYNAMICS. Journal of Biological Chemistry. 2010; Vol: pp.19785-19792. © the American Society for Biochemistry and Molecular Biology.*

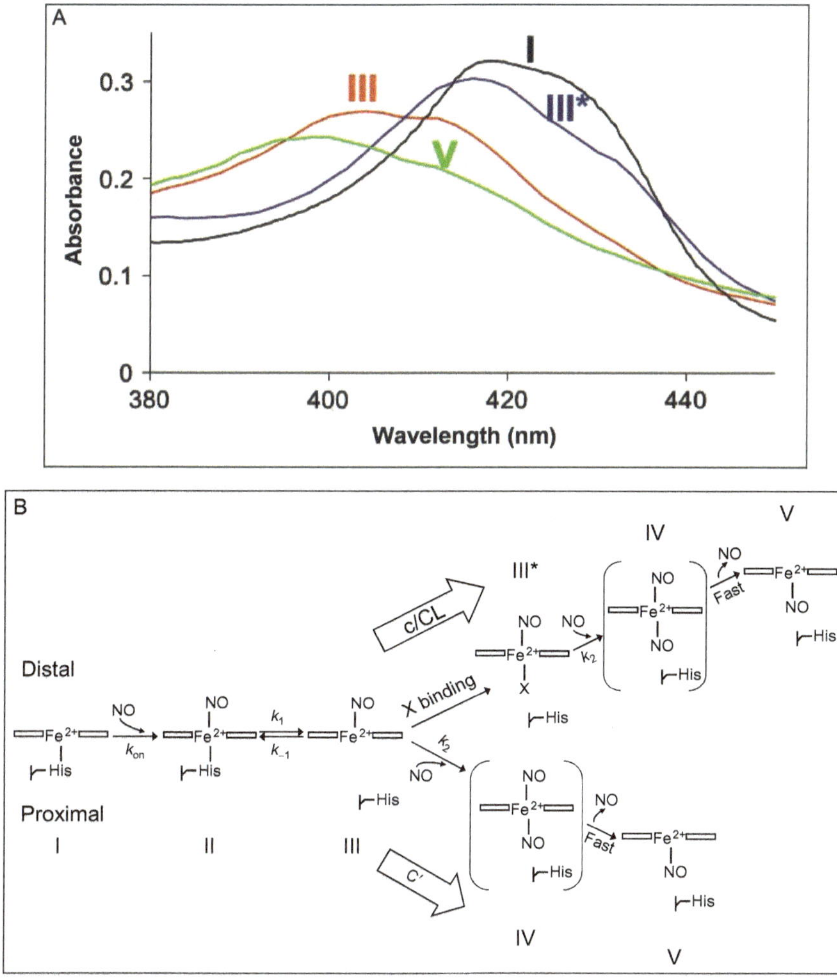

Michael A. Hough *et al.*, Figure 8.3 (A) Spectral intermediates in the reaction between the ferrous cyt *c*/CL complex (3 μM heme) and NO (60 μM) at temperature 20 °C (see also text for details and Silkstone et al., 2010). The spectra are shown in temporal sequence and color coded as follows: black, the initial spectrum of the high spin ferrous form (I); red, a penta-coordinate NO adduct populated some 20 ms after mixing the cyt *c*/CL complex with NO (III); blue, a hexa-coordinate NO species that accumulates after ~200 ms after mixing (III*); green, the final penta-coordinate NO species after 20 s (IV). The spectra are also labeled I, III, III*, and V to correspond with the species on the pathway given in Fig. 8.3B. (B) Comparison of the mechanism of NO binding to either the cyt *c*/CL complex or to Ax cyt *c*′. The steps in this pathway starting from the ferrous high spin form (I) are as follows: (a) an initial second-order association of NO with the heme at the vacant distal site to form the hexa-coordinate NO species (II). (b) The dissociation of the proximal histidine residue to form the penta-coordinate species (III). This step is slower than the initial NO-binding step in wild-type Ax cyt *c*′ allowing species (II) to be populated and observed. In the cyt *c*/CL complex, histidine dissociation is very rapid and thus species (II) is not seen. (c) In the cyt *c*/CL complex, a rearrangement of the protein occurs that leads to a second hexa-coordinate form (III*) that is not seen in Ax cyt *c*′. (d) A second NO binds to the proximal side to form the postulated intermediate (IV) that has two-coordinated NO molecules. This is very unstable and has not been directly observed. (e) NO dissociates from the distal side to yield the final penta-coordinate NO form with NO bound to the proximal side of the heme.

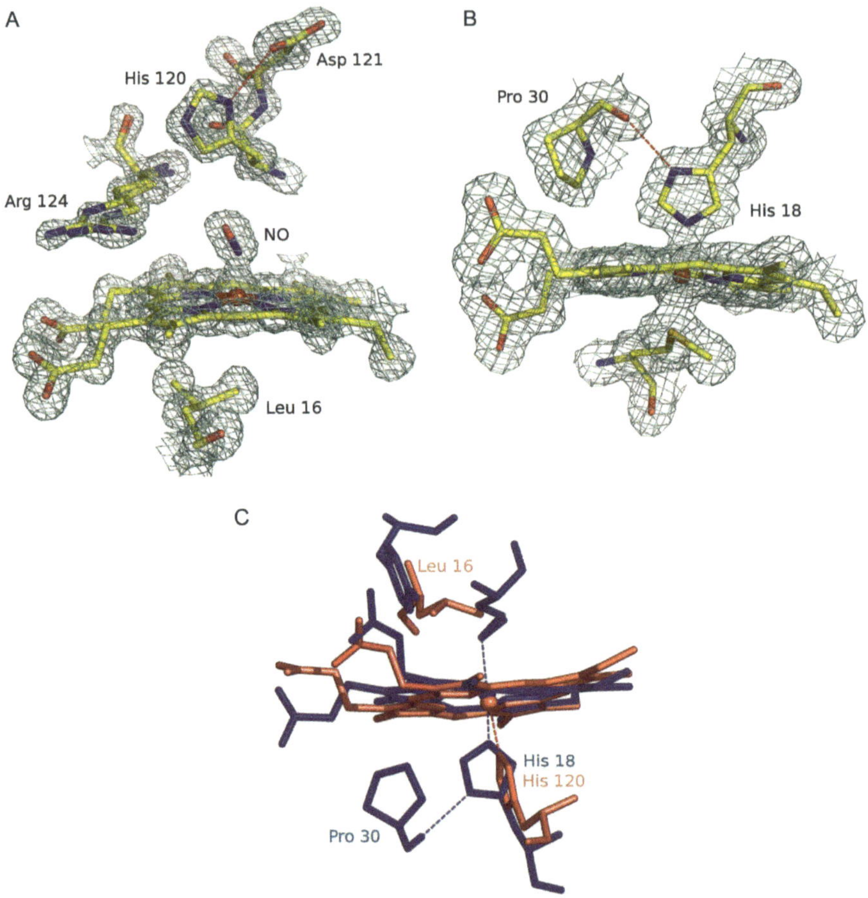

Michael A. Hough *et al.*, Figure 8.4 (A) Proximal NO binding to Ax cyt *c*′ (Hough et al., 2011; PDB accession code 2xlm). The proximal His ligand has been displaced and is stabilized in this position by a H-bond to a nearby Asp residue. Note the presence of a positively charged Arg residue in the proximal pocket and a hydrophobic Leu residue in the distal pocket. (B) Heme environment in human cyt *c* (Rajagopal et al., 2013; PDB accession code 3zcf). Note that the proximal His 18 ligand is stabilized by a hydrogen bond to the carbonyl of Pro 30. (C) Least squares superposition of the heme group in ferrous cyt *c*′ (salmon) and cyt *c* (blue). In cyt *c*, the heme is six coordinate with the Met ligand being dissociated upon formation of the cyt *c*/CL complex.

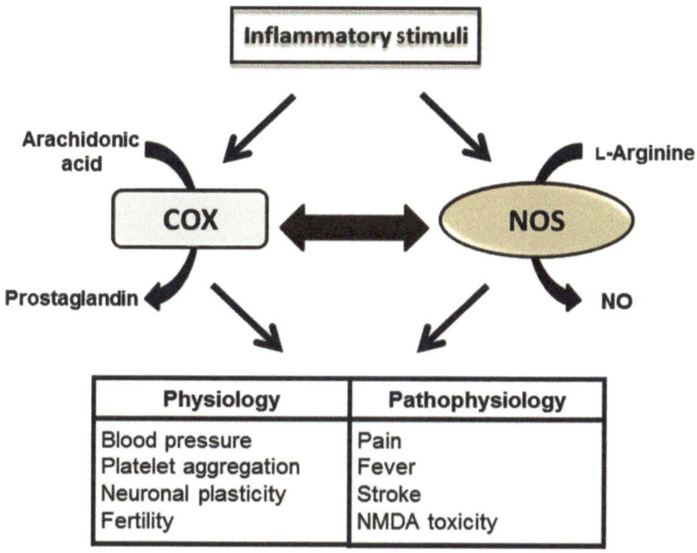

Sangwon F. Kim, Figure 9.1 The role of nitric oxide synthase (NOS) and cyclooxygenase. Both enzymes share the similar pathway for activation and their physiological as well as pathophysiological implication.

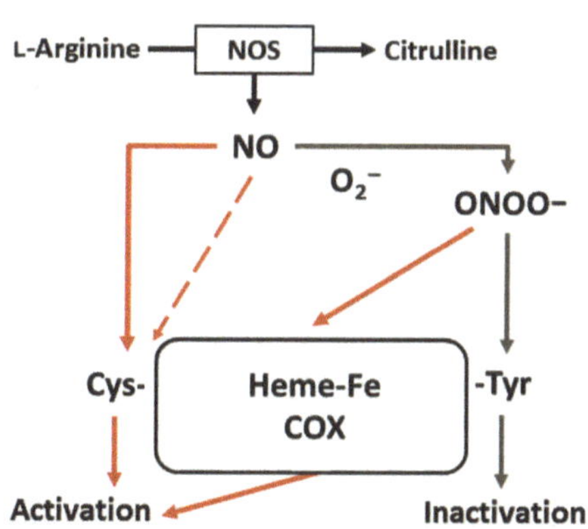

Sangwon F. Kim, Figure 9.3 Proposed mechanisms of NO-mediated regulation of prostaglandin production. Nitric oxide produced by NOS can directly modify cysteine(s) in the COXs and activate it. Moreover, superoxide can be simultaneously generated under the same environment which can enhance NO production. Superoxide rapidly interacts with NO-producing peroxynitrite which can interact with either heme or tyrosine residue(s) in COX2.

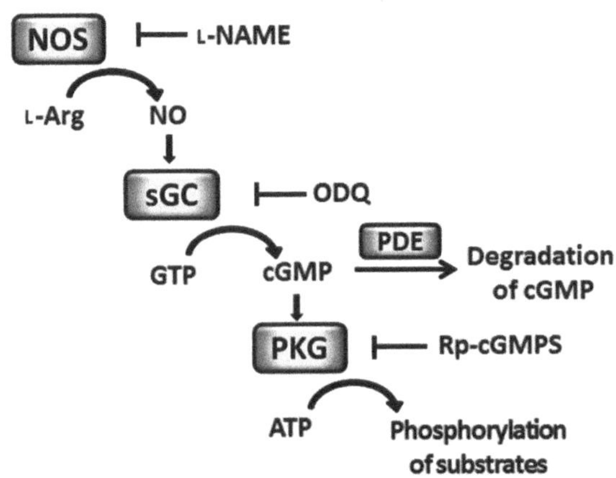

Jisha Joshua *et al.*, Figure 10.1 NO/cGMP/PKG signaling. Estrogens can activate NO synthase (NOS)-3, leading to increased NO synthesis, which stimulates soluble guanylate cyclase (sGC), leading to increased intracellular cGMP concentrations. cGMP is either degraded by phosphodiesterases (PDEs) or binds to various effectors, including cGMP-dependent protein kinases (PKGs). The following inhibitors are mentioned in the text: The NOS inhibitor L-NAME, the sGC inhibitor ODQ, and the PKG inhibitor Rp-cGMPS (abbreviated for Rp-8-pCPT-PET-cGMPS).

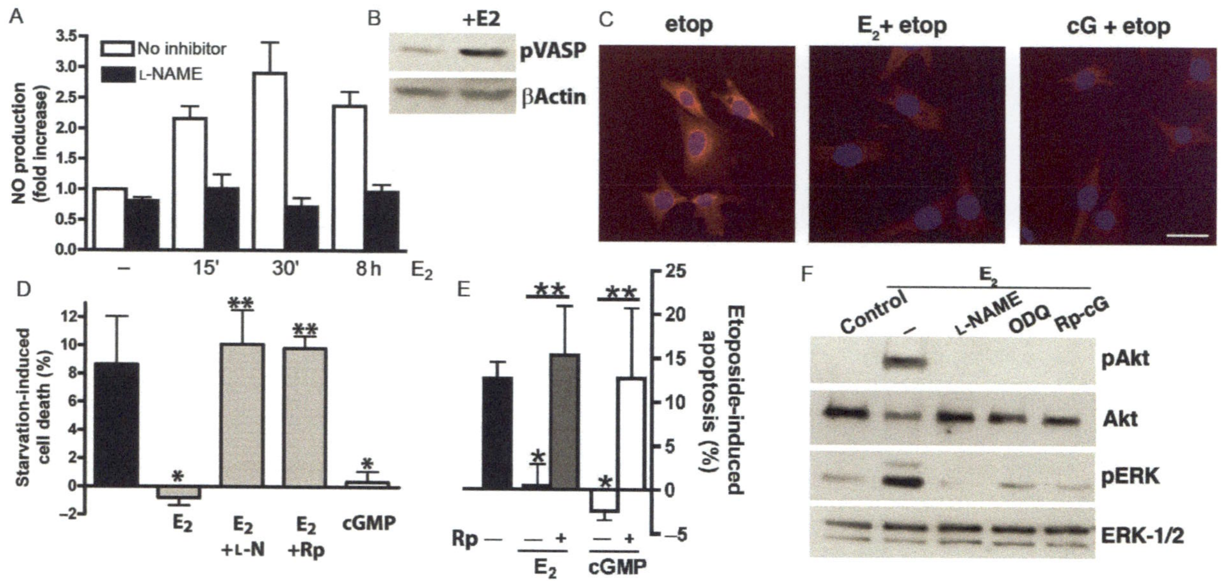

Jisha Joshua *et al.*, Figure 10.2 17β-Estradiol signals through NO/cGMP/PKG to protect osteoblasts and osteocytes from cell death. (A) MLO-Y4 osteocyte-like cells were treated with vehicle (white bars) or the NOS inhibitor L-NAME (black bars) for 1 h and received 10^{-7} M 17β-estradiol (E_2) for the indicated times. NO production was measured using the Griess reagent. (B) Cells were treated as in (A) and analyzed by Western blotting with an antibody specific for VASP phosphorylated on Ser^{259}, a preferred PKG phosphorylation site. (C) Cells were treated with etoposide (etop) to induce cell death and received 10^{-7} M E_2 or 10^{-4} M 8-pCPT-cGMP (cG) as indicated. Apoptotic cells were detected by immunofluorescence staining with an antibody specific for cleaved caspase-3 (red), and nuclei were counterstained with Hoechst 33342 (blue). (D, E) Murine primary osteoblasts (D) and MLO-Y4 osteocyte-like cells (E) received vehicle (black bar), 10^{-7} M E_2 (gray bars), or 10^{-4} M 8-pCPT-cGMP (white bars) and were serum-starved for 18 h (D) or treated with etoposide for 8 h (E) to induce apoptosis. Some cells were pretreated with L-NAME (L-N) or Rp-8-pCPT-PET-cGMPS (Rp) to inhibit NOS or PKG, respectively. Cell death was quantified by trypan blue staining (D) or TUNEL assay (E). (F) Serum-starved MLO-Y4 cells were treated with L-NAME, ODQ, or Rp-8-pCPT-PET-cGMPS to inhibit NOS, sGC, or PKG, and treated with vehicle (control) or 10^{-7} M E_2 for 5 min as indicated. Akt and Erk activation were assessed by Western blotting with phospho-specific antibodies (pAkt, pErk); loading was assessed by blotting with antibodies recognizing total Akt or Erk irrespective of their phosphorylation state. *This research was originally published in Marathe et al. (2012). ©The American Society for Biochemistry and Molecular Biology.*

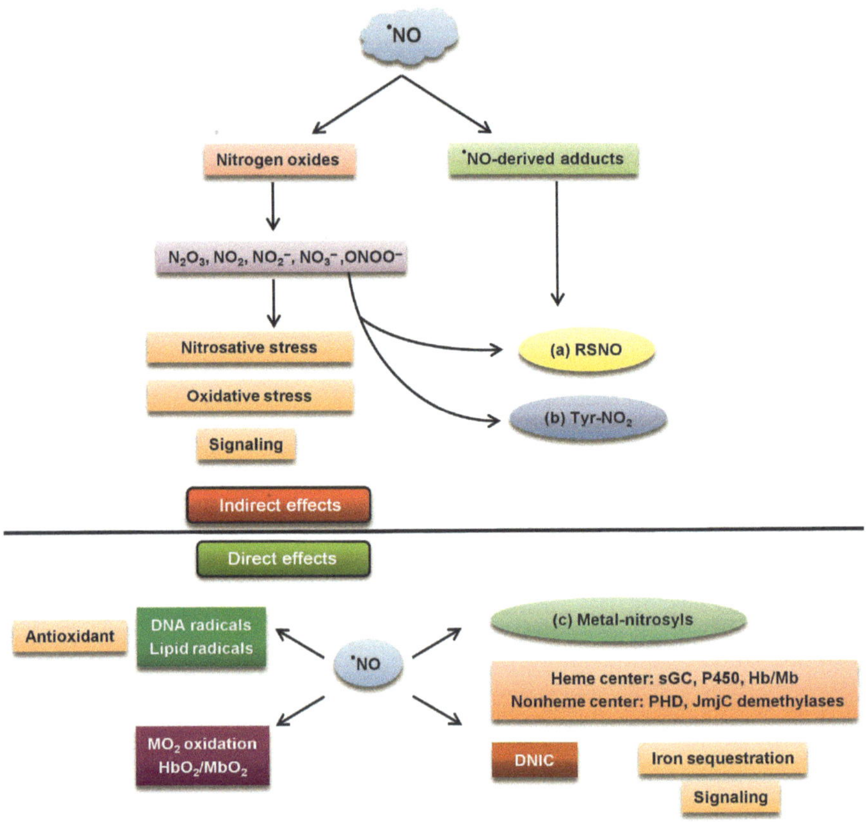

Divya Vasudevan and Douglas D. Thomas, Figure 11.1 *The chemical biology of ·NO signaling*. The biological effects of ·NO can be broken down into two categories. Direct effects are those wherein ·NO directly reacts with target molecules such as metal centers, DNA, and lipid radicals. Indirect effects are those wherein ·NO first reacts with other molecules such as O_2 or O_2^- to form nitrogen oxides that then go on to react with target molecules. Both reaction types give rise to ·NO-containing cellular adducts that can modulate signaling networks.

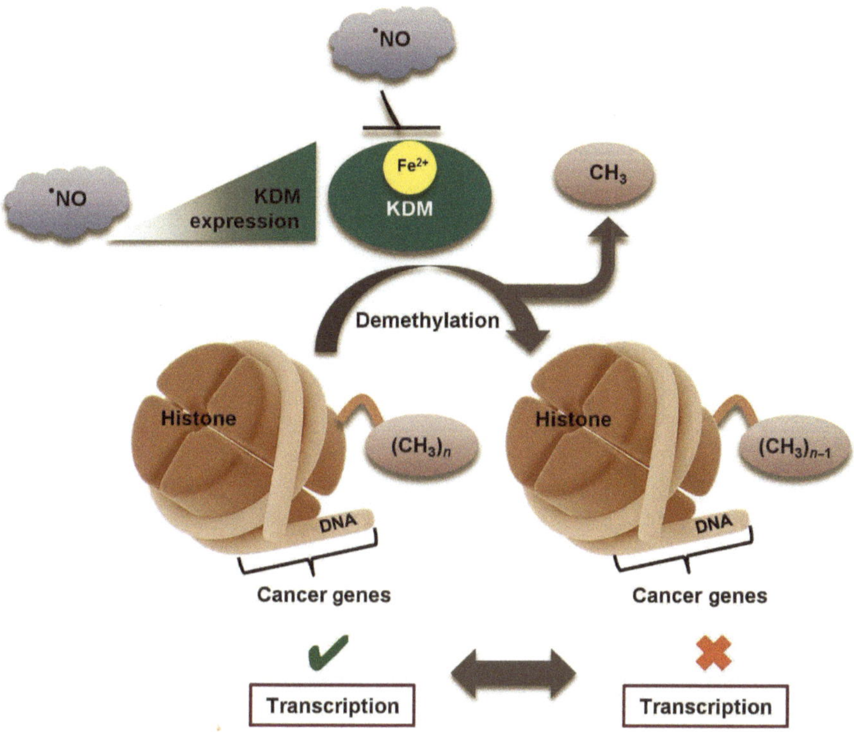

Divya Vasudevan and Douglas D. Thomas, Figure 11.2 *Mechanisms of epigenetic regulation by ·NO.* Nitric oxide can significantly alter the epigenetic landscape and regulate gene expression through a variety of mechanisms. It can directly inhibit the activity of JmjC histone demethylases (KDM) by binding to the non-heme iron atom. It can also change the expression level of a variety of histone-modifying enzymes. By changing both the degree and the location of posttranslational histone modifications, ·NO has potential to significantly alter gene transcription. This is a novel mechanism of ·NO-mediated regulation of tumor phenotype.

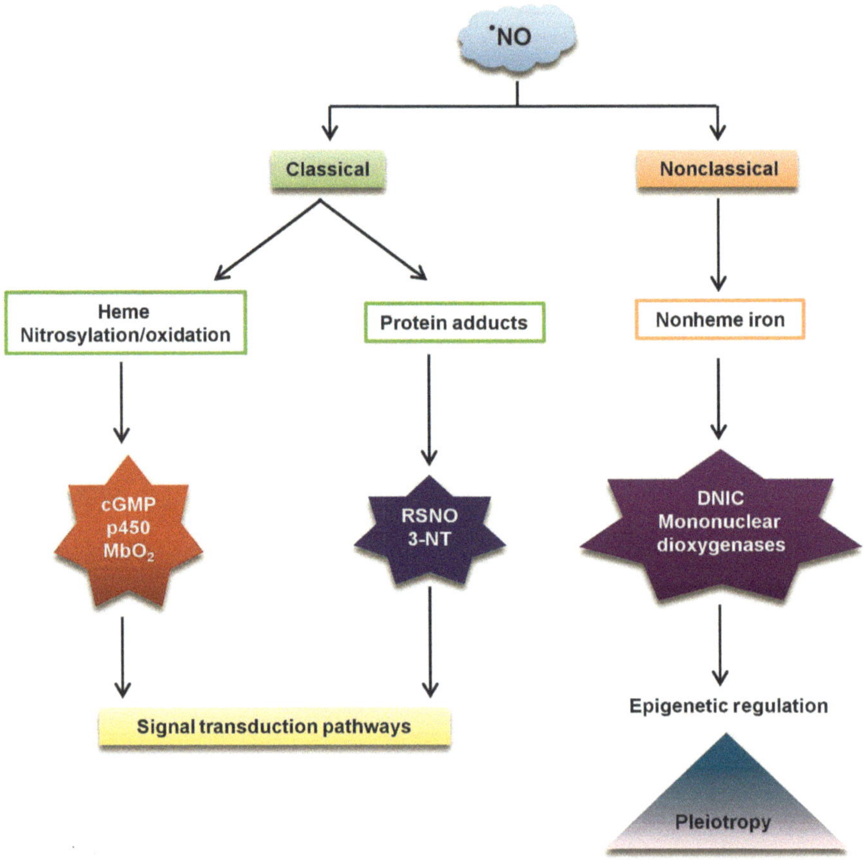

Divya Vasudevan and Douglas D. Thomas, Figure 11.3 *The pleiotropic nature ·NO signaling*. Classical ·NO signaling mechanisms such as via sGC activation, RSNO, or 3-NT formation have long been studied and account for a portion of the observed cellular responses to ·NO. Inhibition of JmjC histone demethylases and DNIC formation are consequences of the interaction of ·NO with nonheme iron and represent novel, emerging mechanisms to explain much of its pleiotropic behavior.

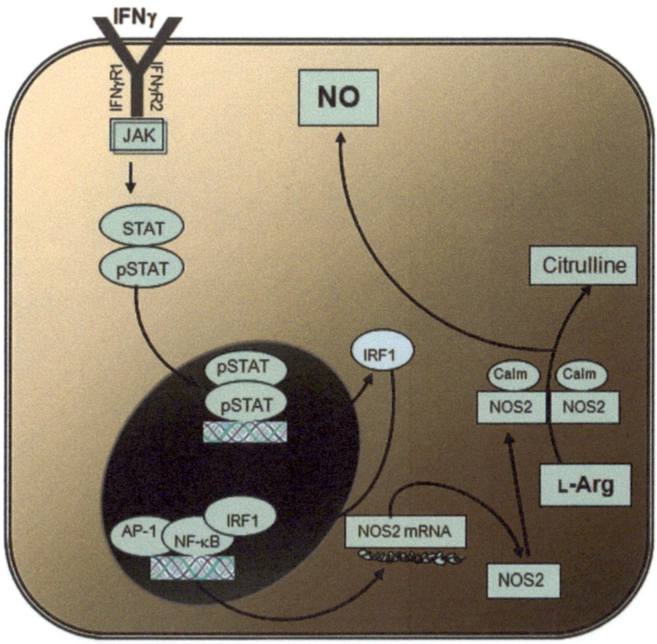

Arnold H. Zea *et al.*, Figure 12.1 L-Arginine metabolism. L-Arg is the common substrate for ARG and NOS2. Effector functions generated by polyamines and proline synthesis (from ARG) result in cell proliferation. Effector functions by NO production from NOS2 possess antimicrobial and antitumoral effects. L-Arg can be *de novo* synthesized from citrulline through the enzymes argininosuccinate lyase and argininosuccinate synthase.

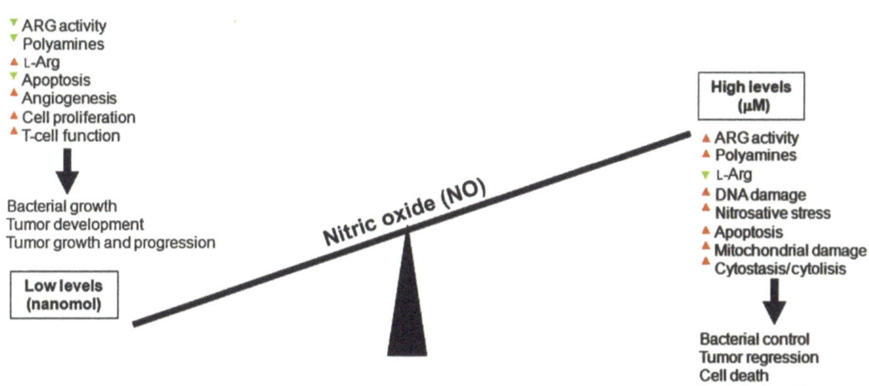

Arnold H. Zea *et al.*, Figure 12.3 Mechanisms of action of NO on bacterial infections and cancer. NO is a very versatile molecule that can induce tumor development depending of many factors. It appears that the regulated concentration and its spatial and temporal distribution are important for tumor development and tumor progression. The effects of NO at low concentrations are considered to be protumor, whereas at high concentrations, NO can act as a potent anticancer and antibacterial agent.

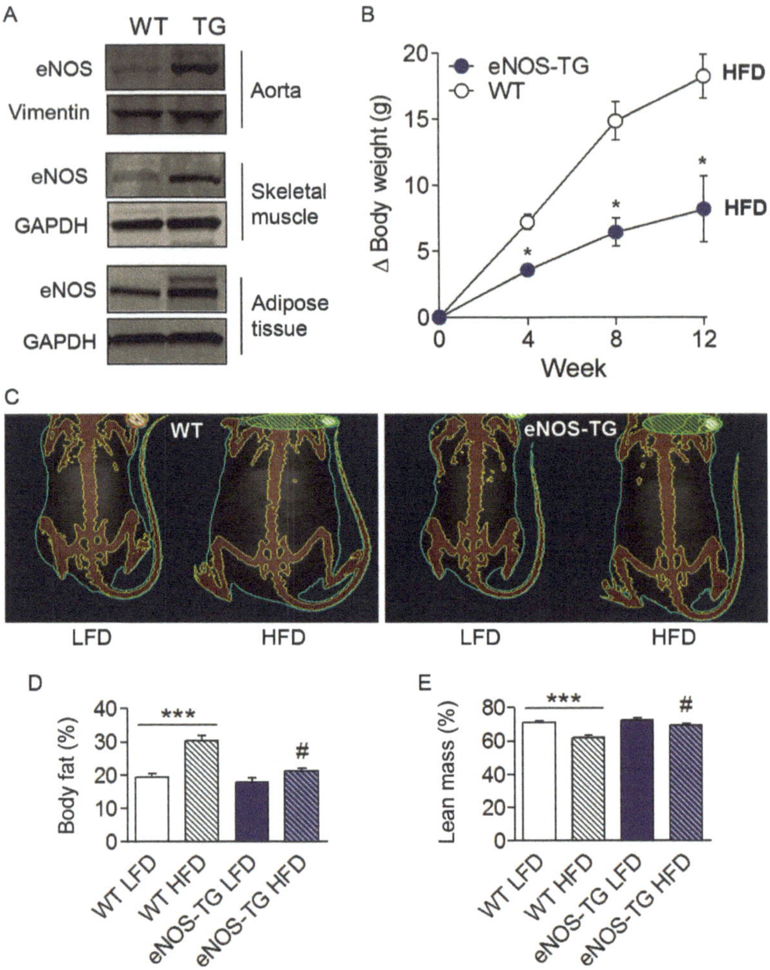

Brian E. Sansbury and Bradford G. Hill, Figure 13.1 Overexpression of eNOS prevents obesity. (A) Immunoblot analysis: eNOS expression in aorta, skeletal muscle, and adipose tissue of wild-type (WT) and eNOS transgenic (eNOS-TG) mice. (B) Change in body weight during high-fat feeding: mice were fed a high-fat diet (HFD) for up to 12 weeks, and the change in body weight was measured. $^{*}p<0.05$ vs. WT. (C) Representative Dexascan images of WT and eNOS-TG mice fed a low-fat diet (LFD) or HFD for 6 weeks. (D and E) Dexascan analysis of body fat and lean mass percentage. $^{***}p<0.001$ vs. indicated group; $\#p<0.05$ vs. WT HFD. *Adapted with permission from Sansbury et al. (2012).*

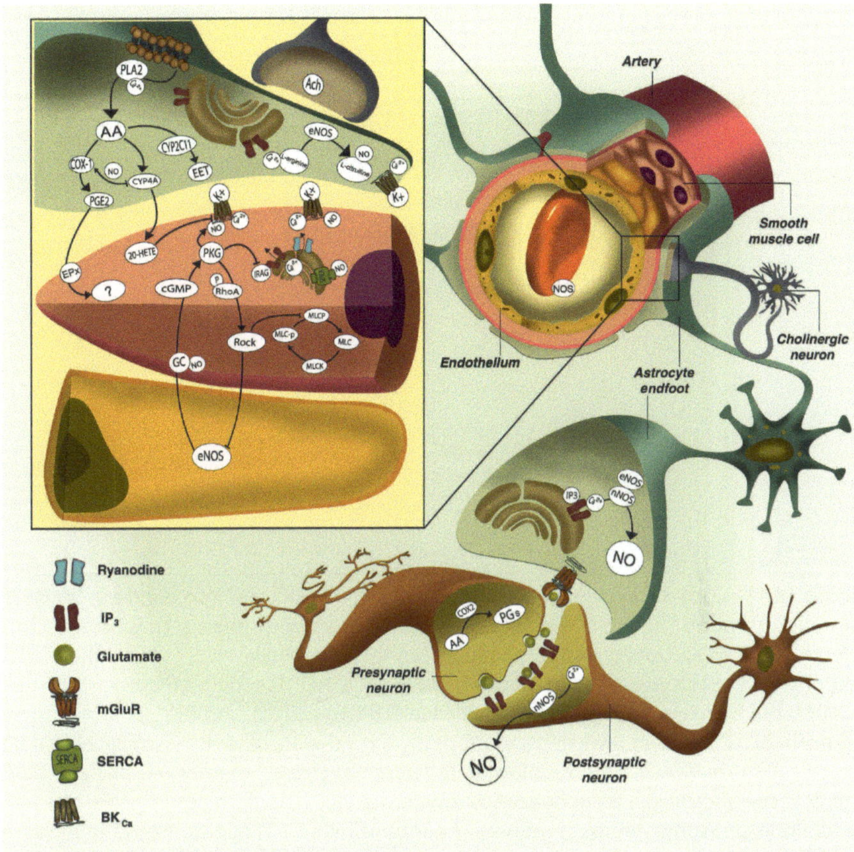

Seyed Nasrollah Tabatabaei and Hélène Girouard, Figure 14.1 Nitric oxide (NO) in neurovascular coupling (NVC). NO may be released from erythrocytes, endothelial cells, smooth muscle cells, astrocyte, interneurons, and some pyramidal cells. This gas may be involved in the initiation of NVC through a cyclic guanosine monophosphate (cGMP)-dependent pathway or modulate NVC by interacting with different cGMP-independent pathways. NO could be released from neuronal NO synthase (nNOS) in nitrergic interneurons and act on astrocytic guanylate cyclase (GC), which increases the Ca^{2+}-activated K^+ channel (BK_{Ca}) opening probability and the subsequent release of K^+ in the perivascular space. Astrocytic NOS can be activated following activation of cholinergic or glutamatergic receptors by acetylcholine (Ach) and glutamate released from cholinergic neurons and pyramidal cells, respectively. Binding to either receptors lead to inositol triphosphate (IP_3) formation and the subsequent release of Ca^{2+} from the sarcoplasmic reticulum or the entry of Ca^{2+} through membrane channels. The increase in intracellular Ca^{2+} activates BK_{Ca} channel and catalyzes the phosphorylation of arachidonic acid (AA) by phospholipase A_2 (PLA_2). AA metabolites (prostaglandin (PG)E_2, 20-hydroxyeicosatetraenoic acid (20-HETE), and epoxyeicosatrienoic acids (EETs)) will, in turn, modulate cerebral blood flow (CBF). NO may interact with different astrocytic pathways such as Ca^{2+} release, Ca^{2+}-activated K^+ channel (BK_{Ca}) opening and

(Continued)

Figure 14.1—Cont'd cyclooxygenase (COX)-1 activity. NO can also decrease 20-HETE formation through the inhibition of CYP4A. Vascular smooth muscle cells are the final integrators of NO-induced signals coming from the endothelial NOS (eNOS), the blood, neurons, and astrocytes. In the smooth muscle cells, NO directly mediate vasodilation through the increases in cGMP-protein kinase G (PKG) activity and BK_{Ca} channel opening probability. Another PKG-dependent pathway is the IP_3 receptor-associated cGMP-kinase substrate (IRAG), an endoplasmic reticulum-anchored membrane protein. IRAG interacts specifically in a trimeric complex with the PKG1β isoform and the IP_3 receptor I and, upon phosphorylation, it reduces the intracellular Ca^{2+} release from the intracellular stores. In addition, NO can control intracellular Ca^{2+} concentrations by increasing Ca^{2+} reuptake into intracellular stores through activation of the sarcoplasmic reticulum ATPase (SERCA). These NO-dependent mechanisms will lead to the subsequent decrease in the cytosolic Ca^{2+} concentration as well as the inhibition of RhoA phosphorylation. Both phenomenona will converge toward a decrease in the contractile state of vascular smooth muscle cells, which is determined by the phosphorylation level of myosin light chains (MLC). MLC phosphorylation is initiated by the MLC kinase (MLCK) activation by Ca^{2+}–calmodulin interaction, while RhoA/Rho kinase signaling pathway inactivates the MLC phosphatase (MLCP) by phosphorylation resulting in increased unphosphorylated MLCP and sustained smooth muscle contraction. NOS may be inhibited, in turn, by RhoA kinase (ROCK).

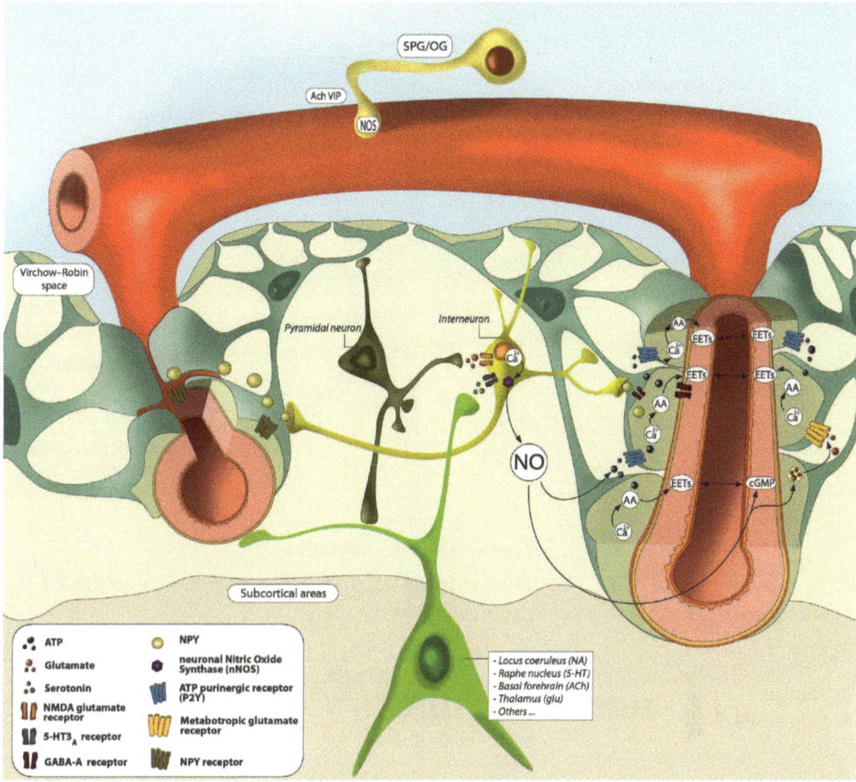

Seyed Nasrollah Tabatabaei and Hélène Girouard, Figure 14.2 Proposed mechanisms by which NO interneurons may control cerebral blood flow. A pial artery gives rise to two parenchymal descending arteries (left and right of the figure). The pial artery is lying over a layer of astrocytes called glia limitans, while parenchymal arteries are covered by astrocytes specialized projections called astrocytic endfeet excepted for the portion separated by the Virchow–Robin space. Pial arteries are innervated by "extrinsic nerves" coming from the peripheral nervous system. Nitrergic nerves originate from the sphenopalatine or the otic ganglion and can corelease acetylcholine (Ach) and the vasoactive intestinal peptide (VIP) in addition to NO. Following local neuronal activity, pyramidal cells (black cells) stimulate nitrergic interneurons (yellow cell), through their glutamate acid *N*-methyl-D-aspartic (NMDA) receptors. The influx of calcium (Ca^{2+}) into the cell allows activation of the neuronal NO synthase (nNOS) and the production of NO. NO diffuses freely through cell membranes and reaches the astrocytic endfeet where it could modulate the release of various vasoactive metabolites. NO may be involved in retrograde vasodilation either through astrocyte signaling modulation or by acting directly on smooth muscle cells. Indeed, NO can increase the release of adenosine triphosphate (ATP) and glutamate from astrocytes. ATP released from vesicles goes from one astrocyte to the next 2PY purinergic receptors, creating a Ca^{2+} wave. These Ca^{2+} waves enhance the production of arachidonic acid (AA) derived vasodilation mediators like epoxyeicosatrienoic acids (EETs) or activate large conductance Ca^{2+}-dependent K^{+} channels to induce arterial vasodilation. NO also directly increases astrocytic Ca^{2+} and might therefore facilitate the production of vasodilators. Finally, NO binds to smooth muscle guanylate cyclase and increase the cyclic guanosine monophosphate (cGMP) production. Stimulated nitrergic interneurons also release acid γ-aminobutyrique (GABA), which could act either on astrocytes or smooth muscles GABA-A receptors to cause vasodilation. However, GABA seems to restrain the astrocytic signal propagation. Simultaneously, Neuropeptide Y (NPY) is released most probably to a remote vascular bed (on the left of the illustration) to cause a vasoconstriction and concentrate the blood flow close to the neuronal activation.

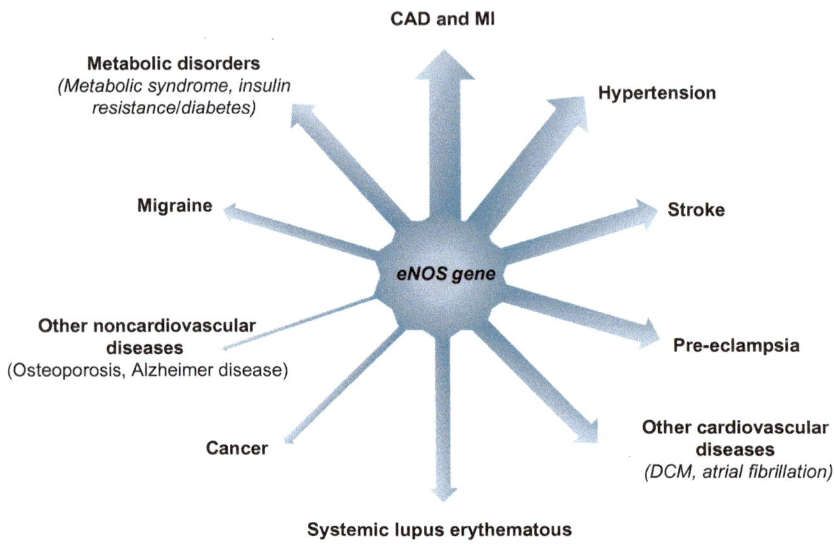

Cecilia Vecoli, Figure 15.1 The gene–disease association network. The width of the arrow represents the strength of association in the network.

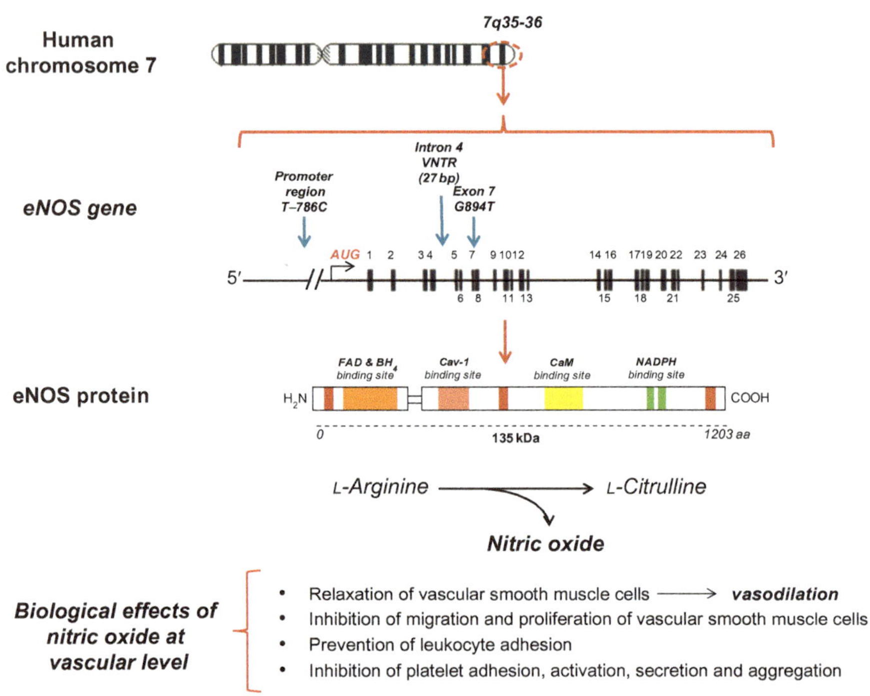

Cecilia Vecoli, Figure 15.2 (A and B) Scheme of the gene encoding endothelial nitric oxide synthase (*eNOS* or *NOS3*). The human *NOS3* gene (located at 7q35–36) contains 26 exons that span 21 kb. Exons are described by number. AUG denotes transcription start site. Three specific polymorphisms in the *NOS3* gene are marked by arrows. (C) Scheme of the eNOS protein. Cofactors: orange, FAD = flavin adenine dinucleotide; BH4 = tetrahydrobiopterin; pink, Cav-1 = caveolin 1; yellow, CaM = calmodulin; green, NADPH = nicotinamide adenine dinucleotide phosphate. (D) Major biological effects of nitric oxide at vascular level.

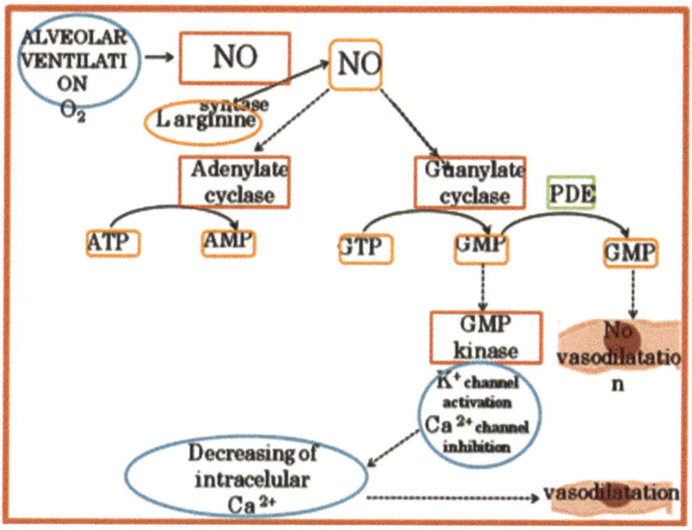

Peter Kruzliak *et al.*, Figure 16.1 Mechanism of NO-mediated vasorelaxation. NO activates soluble guanylate cyclase and stimulates cGMP production. In addition to this major pathway, NO binds directly to protein kinases and phosphatases and changes their enzyme activity. cGMP and cGMP-dependent protein kinase induced by NO promote vasorelaxation to decrease the cytosolic Ca^{2+} concentration in vascular smooth muscle cells by (1) phosphorylation of the Ca^{2+}-sensitive K^+ channel, which causes hyperpolarization and closes the voltage-dependent L-type Ca^{2+} channel, (2) direct phosphorylation of the voltage-dependent sarcolemmal Ca^{2+} channel, resulting in its closure, and (3) inhibition of IP_3-sensitive Ca^{2+} release from the sarcoplasmic reticulum.

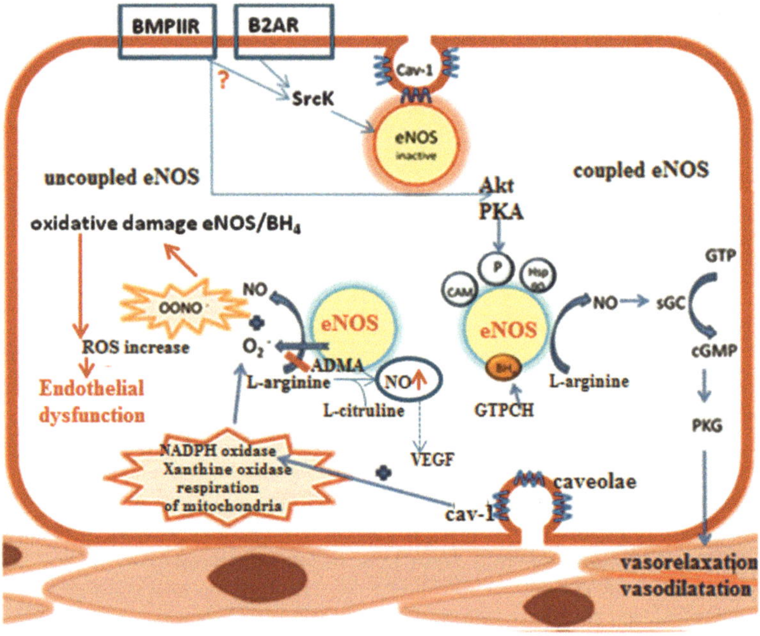

Peter Kruzliak ***et al.*****, Figure 16.2** Endothelial nitric oxide synthase (eNOS) expression and its activity in relationship with the vasorelaxation.

FSC
www.fsc.org
MIX
Papier | Fördert
gute Waldnutzung
FSC® C083411

Kolibri 360 GmbH
Ferdinand-Jühlke-Straße 7
99095 Erfurt, Deutschland
produktsicherheit@kolibri360.de